REMOTE SENSING APPLICATIONS IN AGRICULTURE AND HYDROLOGY

Proceedings of a seminar held at the

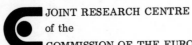 JOINT RESEARCH CENTRE
of the
COMMISSION OF THE EUROPEAN COMMUNITIES

Ispra (Varese) — Italy

in the framework of

21 November — 2 December, 1977

Advanced Seminar on Remote Sensing Applications in Agriculture Hydrology (1977, Ispra, Italy)

PROCEEDINGS OF A SEMINAR HELD AT THE JOINT RESEARCH CENTRE
OF THE COMMISSION OF THE EUROPEAN COMMUNITIES
IN THE FRAMEWORK OF THE ISPRA COURSES / ISPRA (VARESE) / ITALY

Remote Sensing Application in Agriculture and Hydrology

Edited by
GEORGES FRAYSSE
Ispra Establishment, Ispra, Italy

*Published for the Commission of the European Communities,
Directorate General Scientific and Technical Information
and Information Management, Luxembourg*

A.A.BALKEMA / ROTTERDAM / 1980

The texts of the various papers in this volume were set individually
by typists under the supervision of each of the authors concerned.

EUR 6611 EN

ISBN 90 6191 081 1

© ECSC, EEC, EAEC, Brussels-Luxembourg, 1980

Distributed in USA & Canada by MBS, 99 Main Street, Salem, NH 03079

Printed in the Netherlands

Table of contents

PART 2   HYDROLOGY

Preface

Within the framework of Ispra Courses, an Advanced Seminar on Remote Sensing Applications in Agriculture and Hydrology was held in November/December 1977 at the Joint Research Centre of the European Communities, Ispra - Italy. The Seminar was intended for:

- officials responsible for planning and management in public administration (agriculture, forestry, environment, land engineering and equipment, etc.) wishing to acquire the technical knowledge necessary for their work;
- data handling and information specialists desirous of improving their understanding of the applications of their studies;
- specialists in agronomy, forestry, botany and hydrology wishing to complement their knowledge in this new and rapidly evolving field;
- specialists in development of sensors and data acquisition systems;
- students in the earth sciences and agronomy seeking to begin or continue a specialization in the application of earth-remote observation techniques to their specific fields.

Of the twenty-six lecturers asked to contribute to this Seminar, twenty-three are well-known specialists in European universities and institutes, and three are scientists from the U.S.A.; their presentations, accordingly, reflect research and developments internationally. Many of the contributors directed themselves to the difficulties inherent in trying to extrapolate from the wealth of proven remote-sensing techniques as applied in America those appropriate to the quite different structural and ecological conditions existing in Europe.

Of the many seminars and symposia in the field of remote sensing, the Ispra Seminar is one of the very few to date which has limited itself strictly to agricultural and hydrological applications; the success of its formula is proven by the many requests we have received from people who could not attend the 1977 Seminar for copies of the lectures. In response to this flood of requests, and with the support of the Directorate-General of the Scientific and Technical Information of the European Communities at Luxembourg and the organization of the Ispra Courses, we asked the authors to re-write and update their contributions for presentation here. All of them enthusiastically accepted this supplementary task, some even rewriting their contribution completely in order to reflect the rapid progress in this field. We would like to thank all of them for their cooperation.

Our thanks go also to the Directorate-General of the Scientific and Technical Information offices, the Division Ispra Courses and the publisher for their help in the production of this volume. We believe that it reflects the scope and the coverage of the Seminar and trust that it will constitute a useful tool for all those involved in remote sensing of agriculture and hydrology.

Ispra, July 1979

G. FRAYSSE

Part 1   Agriculture

J. GILLOT
Commission of European Communities
Directorate-General for Agriculture, Brussels, Belgium

Potential applications of remote sensing in agriculture

1 INTRODUCTION

One is tending today to regard agriculture as one of the main potential users of remote sensing. It would be interesting to examine the reasons for this trend which incidentally is also beginning to show itself in the United States. At any rate one of the main reasons for the trend would seem to be the need for regular information on the growth cycle of crops, and their sequence in time and space. Agriculture is a dynamic system whose control and management call for rapid and regular acquisition of a great many data on the agronomist's familiar triad : plants, soil and climate. This need is all the more pressing in Europe on account of the diversity of products and production structures which make up a relatively complex pattern, which changes from year to year in response to market demand and the competing claims on land for purposes other than agriculture.

Despite the many constraints, Community agriculture is expanding in real terms, mainly due to technical progress and higher productivity : mechanization, selection of crop varieties, intensive use of fertilizer, improved methods of combating disease and parasites and development of tillage techniques, etc...

Agriculture makes an important contribution to the general economy of the Community and has fully justified the creation of an agricultural policy which, since 1962, has unfortunately constituted the only genuinely common policy of the Member States.

It would seem, therefore, that much can be said for seeking new tools to facilitate the ordering of such a significant aspect of European affairs and thereby meet the ever-growing needs of administrators and decision-makers.

The question arises as to whether in this search such high performance technology as remote sensing should be used. To determine whether remote sensing can solve agricultural problems in Europe, let us first have a look at the context in which it would be used.

2 BIRD'S EYE VIEW OF THE EEC

It extends over 3.000 kilometres from the hills of Scotland in the North to Sicily in the South. The landscape is very varied and the agriculture equally so. For example, there is little in common between a large industrial grain farm in the Paris basin, a farm in the Netherlands or a small holding in Calabria or Sardinia. The Community covers a total of 152.80 million ha with a total UAA of 93.83 million ha, about 60 % of the total area. This area in the Community is broken down as follows :

arable land (% UAA)	50 %
permanent meadow and pasture (% UAA)	45 %
permanent crops (% UAA)	5 %
woodland and forest (% of total area)	21 %

Holdings in different categories are very varied in size. In the United Kingdom there is a predominance of very large holdings (over 50 ha) whereas in Italy holdings of less than 5 ha are in a majority. There is also considerable difference in the breakdown of UAA from one country to another. In Denmark it is mostly arable land whereas in Ireland 75 % is permanent meadow and pasture. Permanent crops are of most importance in Italy and France. France contains 35 % of the Community's UAA, the United Kingdom and Italy each about 20 % and Germany about 14 % (Table 1).

Thus the crop pattern is complex and the size of farms very varied with over 41 % of holdings between 1 and 5 ha and only 6.2 % over 50 ha (Table 2).

Table 1. Principal categories of land use 1974.

COUNTRY	Agricultural used area		Arable land	Permanent grasland
	000 ha	% of total area	% of Agricultural used area	
EUR 9	93.484	61,2	55,7	44,3
Fr. Germany	13.344	53,7	60,5	39,5
France	32.439	59,1	58,2	41,8
Italy	17.503	58,1	70,2	29,8
Netherlands	2.101	56,9	40,3	59,7
Belgium	1.553	50,9	53,0	47,0
Luxembourg	132	51,0	47,1	52,9
United Kingdom	18.637	76,4	38,4	61,6
Ireland	4.846	69,0	26,3	73,7
Denmark	2.928	68,0	90,5	9,5
World	4.551.000	34,0	33,0	67,0

Table 2. Agricultural holdings of 1 ha and over by size groups EUR 9.

Size groups (ha)		Number of holdings (X 000)	X' 000 ha
1	5	2.222 (42,5 %)	5.635 (6,4 %)
5	10	924 (17,7 %)	6.606 (7,5 %)
10	20	914 (17,5 %)	12.963 (14,8 %)
20	50	847 (16,2 %)	25.793 (29,4 %)
	50	321 (6,1 %)	36.781 (41,9)

Despite this, the Community's level of self-supply is satisfactory, though changes will need to be considered to meet the needs of a growing population.

3 THE ROLE OF REMOTE SENSING IN AGRICULTURE WITHIN EEC

Seamingly, the scope for further improvement of crop yields in Europe is not great; the use of fertilizers which is now widespread may be discouraged by higher costs following the energy crisis; the same applies to pesticides.

Future policy will have to take account of these factors and include :
- continued efforts to improve productivity in an ever more dificult and costly energy supply situation,
- optimum use of the soil to produce the crop for which it is best suited,
- improved management of water resources for agriculture,
- improvement and care of the Community's marginal land,
- more effective protection of good farm-land against the demands of urban development, road construction and industry through regional development plans,
- improvement and care of the Community's forest resources.

Under such a policy, we do not dissociate agriculture from its environment : we regard it as one of man's essential activities and not as a separate entity.

Let us see what information is needed to manage agriculture on these lines and to what extent remote sensing can help.

In this context it is difficult to analyse separately the three elements - plants, soil, climate - that form the basis of agricultural production.

But as the cultivated plant is our main concern here it is natural that the discussion should be centred on it.

Most crops grown are annuals if one excludes permanent meadow and pasture and permanent crops. There is thus rapid turnover of sown areas and the crop cover varies appreciably from one year to another.

The crop distribution pattern within arable areas is relatively important having to be linked with rotation. It becomes more important taking into account the large number of small holdings and consequent fragmentation even assuming a reasonably static proportion of arable land as a % of UAA.

This situation necessitates the permanent presence of specialists on the land to revise the statistical data each year.

I won't dwell on this aspect of agricultural statistics which Mr Thiede from the Statistical Office of the Commission will go into later.

When the crops are sown various factors influence their growth and development and consequently the yield : soil, water, climate, diseases and parasites play essential roles that man cannot always anticipate or control.

It is, however, essential to attemt to control and correct at least some of these parameters on a permanent basis. It is known for example, that the amount of water needed by crops is not the same at all stages of growth; the correction of a shortage or maybe an excess at a critical period of growth can have a positive effect on the final yield. Many more examples of this kind could be given to demonstrate the advantage of having a data on yield factors which man can influence available on a continuing basis.

- soil water – a thorough knowledge of the water network of a region is essential for the management of agricultural land, indeed of land in general; it also helps in the siting of pumping stations for the irrigation of areas short of water,
- the water retention capacity of the soil depends on its structure and texture. Information about this capacity over several consecutive years could influence decisions as regards sowing dates, dates for irrigation, even choice of the type of crop best adapted to the water balance of the growing year,
- the wind is an important factor in the evapotranspiration of plants and soil. Hedges used as windbreaks can considerably reduce evapotranspiration and thus completely alter the water balance of the soil and consequently even the type of crop grown on it.

All this is perfectly familiar to agronomists and farmers but if the data are to influence decisions they must be :

1. readily accessible
2. and representative of a sufficiently large region.

I believe I am right in saying that neither of these conditions today is completely fulfilled and one may well wonder whether a special effort should not be made in this area of bioclimatology.

Other biological factors could also be brought under better control if the relevant data could meet the two abovementioned conditions. In particular, if diseases or attacks by parasites could be checked in the early stages when the first symptoms appear on the crop, quick remedial action could be taken. Thorough-going plant health inspection is impracticable with present means so that insecticides and herbicides

Table 3. Yields of some principal crops (X 100 kg/ha).

Product	EUR 9	
	1950	1975
Wheat	19,2	36,4
Barley	22,5	35,9
Maize	15,0	47,4
Total cereals	19,4	37,0
Potatoes	182,0	244,0
Sugar Beet	339,0	401,0

Table 4. Production of cereals - Average 1972-1974.

Country	Total of cereals ('000 t)	Wheat	Barley	Maize
EUR 9	105.676	42.663	34.366	14.724
USSR	185.629	93.230	48.688	11.717
USA	218.994	45.780	8.341	134.488
Canada	34.274	14.656	10.103	2.636
World	1.017.000	361.000	165.000	304.000

Table 5. Population and estimated population growth - Middle of 1975.

Country	Population '000	Projected pop. ' 000		%
		1980	1985	
EUR 9	258.462	261.377	264.752	2,4
USSR	254.382	267.459	280.884	10
USA	213.611	222.769	234.068	9,5
World	3.968.000	4.326.000	4.769.000	20

Table 6. Wooded area and timber production 1974.

Country	Wooded area ('000 ha)	Total prod. of roundwood ('000 cu.m)	Production of sawn softwood ('000 cu.m)	Production of sawn hardwood ('000 cu.m)
EUR 9	32.041	78.219	16.169	7.956
	(0,0077) %			
World	4.126.000	2.511.381	327.056	96.147

are usually sprayed as a matter of routine during the growing cycle, thereby considerably increasing the quantity applied, with undesirable consequences for production prices and the environment.

In view of the role that European agriculture may have to play in a world context, serious attention should be given to safeguarding the heritage of cultivated land which for some years has suffered encroachments from property and highway developments and from industrial activities. It is obvious that the search for more intensive use of lands has lead sometimes to soil degradation either by superficial erosion or organic matter reduction.

Extreme examples of such phenomena are the wastage of some 1/4 mio acres of peatland in the U.K. from intensive arable cropping especially high value cash crops. In Ireland, where the proportion of arable land is limited the depth of peat remaining after mechanical and hand harvesting of approximately one mio acres of peatland will determine the use range of the resulting cut over areas. In an EEC context if sufficient depth of peat is conserved in these areas they could act as an important reservoir of land for horticultural croppings. The evergrowing needs of expanding populations cannot of course continue for ever to be met by an increase in crop yields alone, nor by excessive use of ever more costly fertilizers. Sooner or later Europeans will have to search for new arable land or new types of crops for marginal land, as has already happened in certain over-populated third world countries (Table 3 and Table 4).

It should be borne in mind that average population density varies considerably from one region of the world to another and that although estimated growth rates are low for Europe, they are high for the world population as a whole. Our population in Europe is about 258,462,000 compared with the world population of 3,168 million. In 1980 the figures will be 261,377,000 and 4,326 million respectively, rising to 264,752,000 and 4,769 million in 1985. This represents a growth rate of 2,4 % for Europe and 20 % for the world. Consequently the search for new agricultural land will have to be one of the main concerns of rising generations and the necessary means could therefore be thought of and developed as of now (Table 5).

Agriculture, while increasing its production potential must, as is becoming evident at the same time play a part in counteracting the increasing pollution associated with the proliferation of the human species. It is here that the role of forests in restoring disturbed ecological balances becomes obvious.(Table 6).

The search for land which is suitable for reafforestation and not wanted for agriculture is a long and difficult task and requires prospection that is costly in terms of both men and resources.

The above is no more than an outline sketch of the problems which these responsible for Community agriculture will have to deal with in the future : water and its

management, soil conservation, weather
factors, harvest forecasts, disease control,
search for new land, the environment.

Data on these problems have already been
obtained by remote sensing and put to practical use in the United States and it is
safe to say will be used more in the future
on this continent.

The purpose of this conference should be
to define the scope and limits of a technique as yet little known by its potential
users.

G. THIEDE [1]
Statistical Office, Commission of the European Communities, Luxembourg

2

Agricultural statistics analysis of the main requirements
Conventional and new methodologies

Introduction

1. I am pleased to have the opportunity, at the very beginning of this seminar, to draw your attention to some of the specific requirements of future agricultural and forestry statistics. The discussion should give us an opportunity to define the needs, to examine the many projects already under way in some areas, and to identify the current operational possibilities as well as the areas requiring more research.

2. I have divided my talk into three main sections:

  I. Which areas of study are suitable for remote sensing applications, and which are not
 II. The requirements and problems of an EEC statistician
III. A very brief account of some statistical surveys based on aerial photography which have already been carried out in Europe

I. Areas of study suitable and unsuitable for remote sensing application

3. A large number of projects are currently under way in many different countries throughout the world. I do not know whether there is a central body for systematically recording (and possibly evaluating), (a) all the projects which have been carried out and (b) all ongoing and projected studies. This could be extremely useful for potential users of remote sensing techniques and of the resulting data. It would certainly be desirable from the point of view of the agricultural statistician making plans for the future

1) in collaboration with H.G. Andresen and T.B. Wilson

4. Given the circumstances obtaining in Europe the following areas of study would appear to be most suitable for remote sensing applications. But I would like to stress that these applications may not of course be immediately feasible, and may have to be tested for several years or even decades before practical statistical surveys can be conducted on a reasonably large scale (that is, for a whole country, or the whole Community). This list is therefore by no means exhaustive, and is simply meant to give examples of some of the most important, probable applications.

(A) Agricultural weather data

5. In a paper (Symposium Tel Aviv) which he presented in Tel Aviv last June, G. Fraysse suggested the setting-up of an AGROMET System whereby information on local weather conditions would be collected from a large number of ground stations and transmitted by satellite to a central station for immediate processing. A reporting system of this kind could replace the system already developed by the SOEC (Statistical Office of the European Communities, series "Crop production") for the collection of meteorological data for agricultural purposes in the Nine. Such data on weather conditions can be extremely valuable for short-term crop forecasts, and even have a bearing on agricultural policy especially in extreme situations of the type created by the drought in 1976 in some farming areas of Northern Europe.

(B) Land use statistics

6. One day not too remote it may be possible to compile statistics on land use by means of remote sensing techniques. Obviously, the land use statistics obtained

in this way would only show relatively rough general categories of land use. The categories could be limited in number and probably correspond to the nine level I groups of the Anderson Land Use/Cover classification of the U.S.A. (J.R.Anderson 1977) The emphasis would be on Land Cover rather than Land Use.

At present, surveys of this kind do not have a high priority, are in most European countries only held rarely, and are always very labour-intensive. This sort of general land use survey might be effected by analysis of aerial photographs or other remote sensing of the entire area to be covered (i.e. a complete census). Thereafter the changes occurring over the years could be recorded with the help of suitable sampling methods (with relatively small sampling fractions). These sampling methods could be devised on the basis of information from the basic aerial surveys.

(C) Statistics on areas cultivated

7. Annual surveys of the areas under main crops are essential for harvest and production reports each year. The question arises as to whether remote sensing data such as aerial photographs can be obtained and processed rapidly enough. This information on areas cultivated has to be made available very rapidly, as it forms the basis for calculation of harvest data for the year concerned, and because those responsible for agricultural policy must be provided with approximate harvest figures at an early stage (if possible actually before harvest date). The statistician aims to have as soon as possible data on the last phenological stage which makes a significant contribution to the final biological yield. It would be necessary to include the largest possible number of different crops. Unfortunately the specificity of the remote sensing methods developed so far is limited; and the agricultural and climatic conditions in Europe are not the most suitable for remote sensing but I shall come back to this problem again in the next section.

(D) Statistics on the progress of crops and harvest predictions

8. A further possible application would be the provision of continuous and extremely prompt information on the state of growth of the main crops and the degree of damage caused by pests, disease and other agents which affect the actual harvested yield.

We are all aware that in the past few years so-called "spy satellites" have been successfully used to provide very rapid estimates of the probable grain harvests, including fairly accurate estimates of the harvest in the Soviet Union before the relevant data were published by the official Soviet bodies.

(E) Forestry statistics

9. Some remote sensing techniques are already fairly common in the forestry sector, as most forests cover large areas with more uniform vegetation than is the case in the farming sector. This makes for greater ease of differentiation. I shall come back to this topic later on in my talk.

(F) Fishery statistics

10. Finally, it is claimed that remote sensing surveys could under certain circumstances be used to determine the abundance of stocks in the fishing grounds.

11. However, the types of agricultural statistics mentioned above represent only a small part of our day-to-day work on agricultural statistics. It is therefore worth making some brief mention of the other types of statistics, for which aerial photographs and other remote sensing techniques are out of the question, at least in Europe:

- livestock censuses cannot be conducted by means of aerial photography because a large proportion of the livestock in Europe is kept indoors;

- qualitative features, such as grapes "grown for home consumption "in vineyard surveys, or the varieties of apples "grown for cooking" in orchard surveys, have to be recorded by on-the-spot inspection or questionnaires. Aerial photographs would, at the most, provide rough divisions into categories such as young, medium and old trees from the observation of vigorous or weak vegetative growth;

- surveys of the structure of holdings, which entail the collection of data from individual enterprises on the number, size and structure of holdings, with details of land tenure fragmentation, and many other aspects.

- Figures and utilisation data relating to production factors in agriculture, that is, machines and equipment, fertilizers, plant protection products, etc.

- all collection of <u>demographic data</u> in agriculture, as it is necessary to have details of the age and education of the farmers, the employment position of employees on the holdings, duration of employment and so on.

- <u>Economic data</u> for agricultural holdings, that is, data on expenditure, revenue, sales, prices etc.

12. This incomplete list does not include all the areas covered by agricultural statistics. But I wanted to show you that at the very most only a small fraction of the conventional studies carried out for the purposes of agricultural statistics can be considered as "likely material" for remote sensing surveys. This has implications for the form of the data output from remote sensing which I will discuss in my next section.

II. The requirements and problems of an EEC statistician

13. New methodologies in statistics inevitably create new problems. The statistics derived from these remote sensing observations will have to fulfil certain specific requirements. Here I shall mention the most important in terms of data content and the quality of data.

14. Since changes over time are important for policy makers, the statistician is a conservative. He has to ensure continuity of the data in his series, and where possible, ensure "overlap" to provide a statistical link between different sources of data. The onus of proof is on remote sensing technicians, as the innovators, to ensure comparable results with traditional methods. I say "comparable" and not "identical", because there is also a need for statisticians and technicians to work together to develop appropriate definitions and classifications.

(A) Harmonisation

15. In EEC surveys the survey scope, content and basic concepts are decided by EUROSTAT (this is an abbreviation for the Statistical Office of the European Communities) in collaboration with the Member States and then embodied in EEC Regulations and Directives. But the Member State has in general freedom of choice as regards methods: e.g. postal questionnaire to farmers, interviewers with questionnaire, or survey teams measuring without farmer contact. Statistical information of lower priority for

EEC policy is collected without EEC legislation from long established surveys in Member States who have individual and sometimes differing definitions and classifications. It is the duty of EUROSTAT to ensure that the <u>statistics can be harmonised</u> into an EEC total.

(B) <u>Level of detailed breakdown</u>

16. At present we make distinctions between a total of over 68 agricultural items of land use statistics for the Community. How many items can be identified by remote sensing? 15 or 20 different items?

17. For different policy purposes, the statistician has to provide data at <u>different levels of detail</u> for example: 1. Total land area, 2. Agricultural land area, 3. Permanent cropland, 4. Orchard area by species of fruit eg. apples, citrus, 5. Area of individual fruit varieties such as Golden Delicious, for those products which are marketed by variety. The remote sensing technician must therefore develop and study spectral and temporal signatures to provide an inventory of the land cover/crop types which can be recognised and counted. The same crop can be easy to recognise in certain conditions (eg. vines grown on the "tendone" overhead trellis system in Italy with 100 % canopy) and difficult in other conditions (eg. in the small bush "goblet" form in the Rhone Valley in France with say 10 % ground cover). Mature fruit trees with developed canopies are easier to identify than new plantations, with large bare soil areas.

18. Not all Member States have complete land use statistics of the total national land area. Where they do exist, the national urban classifications are not identical. In a recent study by an Expert for EUROSTAT (L.F. Curtis, R.H. Best 1977) it was found that the various urban land classes in Netherlands (14), UK (8), France (7 major, 30 minor classes), Denmark (9) and Germany (8) could be combined into only 4 standard classes. One of the basic problems of both population and land use statisticians is to find satisfactory definitions of "urban" and "rural". A satellite based land use data system, if operational, could provide a <u>uniform</u> classification at EEC level.

19. It is possible indeed probable that remote sensing will be adopted by some Member States and not by others. The former will tend to develop land "cover" type classifications and the latter to retain

their land "use" classifications. The level of detail must be sufficient in each to permit comparability with the other system and still have sufficient detail for the policy makers. Within the remote sensing systems, there must be common and agreed classifications. Remote sensing data from different altitudes do not have the same degree of resolution, hence of detailed classification.

20. Furthermore, policy decisions are taken at different levels (EEC, national, regional, departmental). The traditional "build-up" of census data from individual units (farms, etc.) satisfies automatically local as well as Community requirements at the same time. In sample surveys of course even traditional methods have greater accuracy at the national level than for individual administrative subdivisions. It is possible therefore that remote sensing methods may satisfy EEC statistical requirements before the requirements of individual Member States.

(C) Data accuracy

21. Town planners, and surveyors have been active in using remote sensing methods for the rapid construction and updating of maps in particular. Their requirements require high spacial resolution to provide accurate demarcation of roads, residential or industrial areas etc. because development control can have costly legal implications. Accuracy is often judged by "ground truth" on individual plots and the percentage of "correct" classification. In the paper given by Mr. Jaakda of Finland in Strasbourg 1977 (Sipi Jaakda 1977) the percentage of correct agicultural classification ranged as follows :

Grass	95 %
Hay harvested	78 %
Barley	77 %
Rye	64 %
Oats	52 %

22. The statistician also must state his accuracy need which is for relatively low spatial resolution at individual sites, provided that the data at the required policy level eg. EEC, national or regional is compatible with traditional sources in aggregate. Compatibility does not necessitate 100 % correspondence but a constant relationship. Errors at the individual sites may be self compensating. Of course bias can exist also and it should be recognized and if possible measured. Bias can exist in traditional systems due for example

to tax benefits or penalties. I hope that we will soon receive reports on the experiments into certain selected Departments in France by comparing remote sensing and traditional methods, in terms of accuracy levels.

(D) European problems

23. Remote sensing techniques have proved satisfactory in the conditions of certain of the major agricultural areas of the world with extensive monoculture and huge fields. But their immediate application without modification and more research may be limited in Europe to certain specific regions such as the grain-growing area of the Paris basin. Elsewhere the European scene consists of small farm units with very mixed production patterns and often small plots.

24. Small fields do not necessarily pose a problem to remote sensing techniques. The average viticultural holding in Germany is less than 1 ha in size and consists of just under 6 plots (nearly 11 plots in Luxembourg) But since they are concentrated predominantly into one monocultural belt along the Mosel and Rhine, much of this crop is suitable for helicopter application of chemical sprays and hence for remote sensing surveys. But there is a problem when small size of fields is combined with polyculture. By contrast large fields and large holdings are very cheap and easy to enumerate by remote sensing but also by traditional methods.

25. Associated crops (permanent, or permanent and temporary) are a feature of agriculture in Italy and in parts of France. They are either continuous (eg. permanent fruit trees) or short term (eg. vegetables in between immature fruit trees, for two or three seasons; or one season if the permanent crop is frost damaged). Traditional surveys have to provide specific solutions (allocation to the principal crop, pro rata, etc.). Remote sensing methods will have to provide and appropriate solution also.

(E) Speed

26. In certain cases, rapidity of result processing is vital. There must therefore be a proper balance between refinement in the methods used for technical data collection, and the amount of interpretation and analysis required. Here I am thinking particularly of work like the surveys of areas cultivated under different crops, which have to be used as a basis for harvest statistics just a few weeks after be ing compiled.

27. The statistician often faces the criticism of producing data too late; he is therefore always receptive to newer methods to give him quicker results even on a partial basis. Under traditional methods, EUROSTAT's finalised EEC data cannot be earlier than that of the latest individual Member State. Remote sensing offers a unique chance to obtain at a stroke global data at the EEC level and to use it quickly providing the correct form of data output is employed.

(F) Form of data output

28. It is necessary to provide links between the data obtained by remote sensing methods and data from other sources. Geocoding is now normal practice in censuses of agriculture, industry and population in order to establish such interrelationships. The usual form of data from remote sensing for use by planners seems to be high resolution printing or plotting on maps or photographs. To satisfy the statisticians' demand for cheap, speedy data which can be linked to other data, it is necessary to produce computer listings or, preferably, input direct into such data banks. "The full potential of remote sensing technology will not be reached until all the information, such as normally contained on thematic maps, can be utilised in a rapid interactive information management system" (Rowley 1977).

(G) Costs

29. The cost of new methods, if they are to be adopted, should be in reasonable proportion to the cost of the conventional survey methods formerly used. Of course the local authorities would no longer have to provide census staff and survey officials if land use statistics (for example) were compiled with remote sensing techniques. On the other hand, one should also take account of the costs of processing at the central survey and analysis stations. What use are the most modern and sophisticated methods if they cannot be applied because the finance ministers cannot or will not provide the necessary funds? It is a fact of life that money is more difficult to obtain for new than for long established purposes.

(H) Specific problems in forestry

30. Forest surveys are generally carried out only at fairly long intervals, e.g. every 10 years. As the rate of change is slow, more frequent surveys are not to be recommended because of cost. The minimum requirement is that Forest surveys cover the following characteristics, with a breakdown by region:

- total wooded area

- composition of the forests, a distinction being made between coniferous and broad-leaved species

- standing volume

31. In addition to this minimum information, however, further data are frequently required to answer a wide variety of questions arising in the field of forestry and timber market policy. Such information, almost as important as the characteristics mentioned above, may include the following:

- (1) category of ownership e.g.
- (2) size and other structural aspects of holdings
- (3) species and age classes
- (4) annual average increment
- (5) socio-economic function for example for protection within the hydrological cycle or for recreation.

All these characteristics may be investigated using conventional surveying procedures by either exhaustive or sample surveys. The reliability of the results for the individual characteristics of course depends not only on the procedure adopted but also on the care with which the survey is carried out. It is not possible to quantify the reliability of forestry survey results in the various Member States. It may, however, be assumed that in most Member States the total wooded area quoted for statistical purposes is accurate to within $\pm$ 3 %.
Of the survey characteristics mentioned, only the total wooded area and the proportions accounted for by coniferous and broad-leaved species can, in the present state of the art, be determined with a sufficient degree of accuracy by aerial surveys. It must therefore be assumed that while aerial surveys are a useful addition to the conventional surveying procedures they are not a complete substitute.

Black and white aerial photography has been successfully used for decades in forest mapping, especially for the purposes of long-term planning.

32. The solution of many forestry problems would be facilitated by the availability of data on the annual change in the wooded area, broken down by region i.e. an annual balance of gains and losses. In this connection, remote sensing could provide much-

needed information. A distinction between broad-leaved and coniferous species would be adequate. As, however, there is only a small change from year to year in the wooded area a high standard of accuracy would be required. If the details obtained were not accurate to within ± 3 % (within 2 o) they would be practically valueless as the annual change would lie within the range of error.

III. Surveys already carried out in Europe by means of aerial photography

33. I shall mention certain more or less large scale surveys which have already been conducted and which were intended to replace, or provide a satisfactory alternative to conventional statistical surveys in Spain, primarily because the infrastructure for agricultural statistics in that country is relatively underdeveloped. Use was made by the Spanish Ministry of Agriculture of experience gained in the USA, especially for the photography of citrus plantations.

34. Citrus plantations were surveyed in 1972 and again last winter with the help of aerial photographs. In both cases the survey was based on a combination of activities. First of all, aerial photographs were taken, and then some characteristic features of the citrus plants concerned were obtained by inspection in the field. In 1972 aerial photographs on a sale of 1 to 8 000 were taken (from a height of 2 400 metres), and were subsequently enlarged to a scale of 1 to 2 000. In the more recent survey, the average flying height was 4 000 metres, which gave original photographs on a scale of 1 to 18 000. These were then enlarged to 4 times the original size, to give working photographs on a scale of 1 to 4 500. The plot boundaries were drawn in on the 1 x 1 metre photographs. Then each picture was divided up into 5 - 7 more manageable "Polygons". Further important characteristics were ascertained in the field by inspection of each plot, and recorded on lists. In all, some 1 million hectares were covered, which includes 230 000 ha of citrus growes and orchards

35. In the light of the positive experience gained in the citrus plantation survey, a survey of fruit has also started. As far as I know, this has not yet been completed, but the procedure will in principle be the same as for the citrus survey.

36. Aerial photographs were also used to produce a register of olive groves. The aerial photographs taken were on the scale of 1 to 16 000. Since olive trees are easier to identify, this scale was quite adequate without enlargement specially as olive cultivation in Spain is usually concentrated in certain main zones. There was no special subdivision of olive plantations into plots, as the differences between them are obviously not very great. Whereas private companies had been used for field inspections for the citrus survey, work in the field for the register of olive groves was carried out by Ministry staff.

37. Hop-growing has also been studied in an aerial survey, although here the areas involved are relatively small. No difficulties were encountered in this survey, because hop fields are easily recognizable from the air.

38. Since 1964 forestry areas have been have been systematically recorded and studied with the help of aerial photographs, proceeding on an area-by-area basis.

39. I am sure that the other speakers will be able to tell us more about the many projects carried out to survey olive groves in Italy. It is clear from Council Regulation 154/75, on the establishment of an olive-growing register, that in the first stages of work information must be collected on the area covered by olive groves and the total numbers of olive trees. We shall also be hearing about the Agreste Project for surveying rice fields and poplar plantations in Northern Italy and the South of France.

CONCLUSIONS

40. In view of the requirements outlined above, it will in many cases not be possible, or at least not for the time being, to find satisfactory applications for remote sensing techniques in agricultural and forestry statistics. Endeavours should therefore be made to reduce the difficulties and overcome the obstacles. New techniques of this kind automatically present one very real advantage : it is much easier to improve and update new techniques than it is to improve older methods.

41. It should be borne in mind that these new survey methods are much more easily accessible for computer analysis of the results. This applies equally to the preparation and interpretation of fresh survey data, subsequent utilization for further analyses, and even linkage with analyses from earlier or later points in time.

42. We at the Statistical Office of the
European Communities are ready and willing
to provide advice and support within the
bounds of our capabilities. At all events,
we are always interested to receive information on the subject and would be grate-
ful if researchers could keep sending us
their reports and research results. Constructing future needs is not possible
without thorough research and experimental
work. During our daily work we should not
forget our task for our, and our children's,
future.

BIBLIOGRAPHY

J.R. Anderson: "Land use classification
  schemes used in selected recent geogra-
  phic applications of remote sensing".
  Photogrammetric Engineering, 1971.
L.F. Curtis and Dr. R.H. Best "A Study of
  Community Land Use Statistics, with par-
  ticular reference to Aerial and Satellite
  Methods". A report for the Statistical
  Office of the European Communities,
  August, 1977
Mr. Rowley of NASA USA "Remote Sensing
  Applied to Regional Planning in the Unit-
  ed States"; Toulouse, France: June 1977
Sipi Jaakda, Technical Research Centre of
  Finland "Monitoring of agricultural and
  forest resources by remote sensing tech-
  niques", 8 June 1977
Statistical Office of the European Commu-
  nities, Series "Crop Production" (Approx.
  10 booklets each year)
Symposium on "The contribution of space
  observations to global food information
  systems" 8-10 June 1977

G.M.LECHI
Istituto per la Geofisica della Litosfera
CNR, Remote Sensing Unit, Milan, Italy

3

Survey of photointerpretation techniques in agricultural inventories
Identification of crops, discrimination of species, biomass evaluation ECC
Typical results

INTRODUCTION

"The image interpretation is defined as the act of examining images for the purpose of identifying objects or surfaces and judging their significance.

Interpreters study remotely sensed data and attempt through logical processes to "detect, identify, measure, and evaluate the significance of environmental and cultural objects, patterns, and spatial relationship" (from Manual of Remote Sensing - American Society of Photogrammetry - Page 869).

Normally in R.S. domain we can dispose of several kinds of images of the same object depending on the wavelength at which the survey has been performed.

In the table 1 the different objects of investigation indicated the most suitable bands are clearly shown.

In the table 2 the possible applications of R.S. techniques for agricultural surveys are shown.

It is possible to split the image interpretation in two, that is classic and automatic interpretation, depending on the importance of the role of the human interpreter on the final result.

In other words in the classical photointerpretation every judgement from the image to the final map is performed by the human interpreter while in the automatic photointerpretation there is a suitable machine or system which provides some judgement on the basis of previous instructions.

1 CLASSIC PHOTOINTERPRETATION

The classic photointerpretation in agriculture is performed normally using the following means:
-B&W aerial photographs
-colour aerial photographs
-I.R. colour or B&W aerial photographs
-multispectral photographs
-satellite images.

In the case of B&W prints the information is related to the variations of gray tones, in value and geometry, or the density variations in value and geometry in the case of the transparencies.

Neverthless it is important to notice that to the B&W paper belongs a dynamic information range of 1:100, while to the transparencies of about 1:1000.

This fact allows a very good interpretation of the transparencies, even if at first sight it seems to be easier to interpret the paper B&W prints.

In the case of colour or false colour images it is important to know the laws of colours composition:

blue
green   fundamentals additive system
red

yellow
magenta   fundamental in substractive
cyan   system

blu+green=cyan=minus red
green+red=yellow=minus blue
blue+red=magenta=minus green
blue+green+red=white.

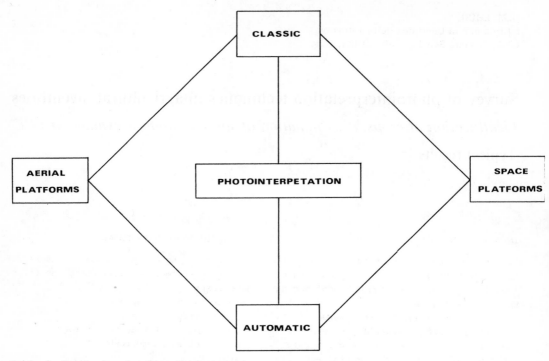

Table 1. Remote Sensing Data Interpretation

Sensor	Without Ground Control Derived Information	With Ground Control	
		Ground Information	Derived Information
Black and White Photography	1. Topographyc variations	1. Elevations and locations	1. (a) Topography (b) Quantitative stratigraphy (c) Engineering geology
	2. Landforms	2. Geomorphic reconnaissance	2. Geomorphology
	3. Lithologic distribution	3. Geologic reconnaissance	3. Mapping of geologic units
	4. Surface structures	4. Geologic reconnaissance	4. Structural geologic maps
Color Photography	1. All of above	1. All of above	1. All of above
	2. Subtle variations in lithology	2. Geologic reconnaissance	2. Detailed mapping of geologic units
	3. Color anomalies	3. Spectral reflectance	3. (a) Alteration (b) Weathering
Color IR Photography	1. All of above	1. All of above	1. All of above
	2. Variations in vegetation	2. (a) Vegetation type (b) Vegetation size (c) Lithology-selective vegetation	2. (a) Vegetation species mapping (b) Vegetation density mapping (c) Differentation of geology as a function of vegetation

Sensor	Without Ground Control Derived Information	With Ground Control	
		Ground Information	Derived Information
		(d) Moisture-selective vegetation	(d) Geologic structures controlling ground water
	3. Vegetation IR reflectance anomalies	3. Phenology	3. Vegetation vigor, stress due to moisture deficiency; salinity; or disease
Thermal Infrared Imagery	1. Radiometric temperature variations	1. (a) Radiometric temperatures	1. (a) Isoradiance maps
		(b) Spectral emissivity	(b) Isothermal maps
		(c) Thermal diffusivity	(c) Gross lithology
		(d) Geologic reconnaissance	(d) Mapping of some geologic units (radiometric units)
	2. Tonal lineaments	2. (a) Moisture sampling	2. (a) Fractures controlling ground water
		(b) Vegetation observations	(b) Fractures controlling phreatophyte distribution
		(c) Topography	(c) Fractures controlling topography
		(d) Surficial geology	(d) Fractures controlling deposits of high materials
	3. Thermal anomalies	3. Geologic reconnaissance	3. (a) Bedrock/alluvium contacts
			(b) Disturbed ground
			(c) Cultural features
			(d) Hot springs, hot ground
			(e) Subsurface openings
SLAR	1. Topographic variations and landforms	1. Geomorphic reconnaissance	1. Geomorphology
	2. Lineaments	2. Geologic reconnaissance	2. Fractures
	3. Textural variations	3. Surface roughness, particle size	3. Gross Lithologies
Microwave Radiometry	1. Relative radiometric temperatures	1. (a) Moisture sampling	1. (a) Relative moisture content
		(b) Density and particle size	(b) Map changes in density and porosity
		(c) Thermometric temperatures, surface and subsurface	(c) Microwave emissivity and penetration depth

Table 2. Possible applications of remote sensing techniques for agricultural surveys

A. Areas of general Applications:

agricultural land-use mapping	rural transportation network	
agricultural land-use change	soil surveys	
agricultural population distribution	water resource surveys	
land-use potential		

B. Areas of specific applications:

1. Applicable to crop surveys:

crop identification	irrigation effectiveness
crop acreage	drough prediction
crop vigor	wedd concentrations
crop density	nematode infestations
crop maturity	insect infestations
growth rates	disease infestations
yeld forecasting	location of disease-resistant species
actual yeld	
planting dates	frost damage
harvesting dates	storm warning
soil fertility	flood warning
areas of fertilizer application	fire surveillance
effects of fertilizers	fire control
soil toxicity	damage assesment
soil moisture	water availability
excessive salinity	location of canals
water quality	detection of heat
soil requirements	in silos ecc.

2. Applicable to range surveys:

forage-species identification	soil fertility
delineation of forest types	soil moisture
condition of range	weed infestation
carrying capacity	insect infestations
forage yield	disease infestations
growth rates	wildlife inventory
times of seasonal change	effects of wildlife
development potential	rodent damage
location of water	fire surveillance
water quality	fire control
drought prediction	trafficability
extent of erosion	conditions of fences
identification of toxic species	

3. Applicable to livestock surveys:

cattle population	distribution of animals
sheep population	animal behaviour
pig population	health of animals
poultry population	disease identification
age-sex distribution	types of farm buildings
(from Estes)	

In the false colour images (yellow filtered during exposure) is fundamental to know the hue of the various layers, and the relative spectral sensitivity:

band	resulting colour
blue	black
green	blue
red	geen
near infrared	red

The use of the false colour film instead of the normal colour is due to its particular spectral sensitivity related to the vegetation: in fact the vegetation cover reflects strongly the near infrared light, so that it is important to utilize a sensor able to detect and describe such a radiation.

In the classic photointerpretation there are some elements to be taken into account; these are:

-Size of the objects. The interpreter can exclude " a priori" some areas too small or too large in comparison to those of possible identification. Obviously the scale plays a fundamental role in the "size problem".

-Shape of the objects. It is one of the most useful elements, after the important and not so easy training of the interpreter to look at the objects from a vertical point of view not horizontal as usual.

-Shadow. The shadow provides a lot of information on the shape and size of the objects, depending on the sun-angle, the high of the objects; it emphasizes the contour of the objects but simultaneously obscure important ground information.

-Colour. Is an additional information ve-

ry important for helping the interpreter
if there are some ambiguities in objects
of the same shape and size.

We already treated the way to interpret
the false-colour images.

-Texture. "Texture in images is created
by tonal repetitions in group of objects
which are often too small to be descerned
as individual objects. Texture, the visual
impression of roughness or smoothness created by some objects is a valuable clue in
interpretation" (from Manual od Remote
Sensign - American Society of Photogrammetry - Page 875).

In agricultural inventories, where not
often it is possible to distinguish every
single plant, the texture analysis becomes
a fundamental tool in reconnaissance.

-Pattern. Is the characteristic which is
related to the regular displacement of the
boundaries of the objects. Pratically every human agricultural culture is regularly
planted, so that it is easy to apply during
the interpretation the pattern recognition
criteria.

-Site.The geographical location of the
objects tell us information regarding the
probability cf existance of such an object
we are looking for.

Some parameters are characterizing the site, that is aspect, topography, geology,
soil, vegetation and shapes of human cul-
tures.

-Association. The association of objects
is normally of a great help identifying man-
made objects. In agricolture inventory it
is not a fundamental element.

-Resolution. The resolution in general
is the element which allows the interpreter
to distinguish an object from its background.

It is possible to split the resolution
roughly in two: spatial or geometrical resolution and spectral resolution.

The spatial resolution is not so important
for the problems involved by agricultural
inventories, while the spectral resolution
is very important, either in classic or automatic interpretation.

The reflection of the sunlight on a canopy depends of many factors the principal
being the nature of the canopy.

The resolution in general is a very complex parameter to be considered, but just

to give an idea in an aerial photograph -
at the scale 1:10.000 - the combination of
spatial and spectral resolution allow us to
distinguish objects 20 cm sized.

On the other band - for instance - the
Landsat 1 or 2 have a nominal geometrical
on ground resolution of 80x60 m, but it
is possible to detect objects even smaller,
if the contrast with the background is
enough.

-Stereo vision. Normally in the photointerpretation the so called stereopairs are
used. It's matter of two aerial photographs
taken with some overlapping (60%) in such
a manner that an area is taken by two different looking angles.

By means of the help of a stereoscope it
is possible to have the impression of the
third-dimension.

The stereoscope is a machine which allow
the eyes to look indipendently at the two
different images, so that the brain can re-
construct a new, complex, image where the
impression of the third dimension appears.

The stereo-vision is a fundamental element
in photointerpretation as it is very easy
to understand.

-Multispectral photographs. In some cases
for agricultural problems it is useful the
multispectral photography, that is a series
of B&W photos, taken simultaneously in different spectral bands.

It is possible then to combine the various
spectral plates with different colours in
an additive colour viewer; this kind of machine allows the realization of some chro-
matic contrasts, in false colour in order
to emphasize some areas in comparison to
others.

2 AUTOMATIC PHOTOINTERPRETATION

The automatic photointerpretation in agriculture from remotely sensed data started
some years ago when the data from multispectral scanners (either from aircraft or
spacecraft) were available.

As it is well known a scanner normally
has a poor geometrical resolution - compared to the photo camera - but has a very
good spectral resolution.

From aircraft "normally" there are scanners employing 11 channels, distributed
in the visible, near I.R. and thermal infra-

red bands of the electromagnetic spectrum: of course there are scanners with a greater or reduced number of bands, depending on the scope for which such a machine has been designed.

Speaking about satellites for agriculture purposes, it is obvious consider the Landsat satellites: two of those are working on four bands of the electromagnetic spectrum (0.5-0.6/0.6-0.7/0.7-0.8/0.8-1.1/micron) with a geometrical resolution of 80x60 meters on ground, from an altitude of about 920 km, the Landsat 2 and Landsat 3.

The data of a survey - normally in digital form - are recorded on magnetic tape and then they can be analized using a digital computer.

In order to perform an automatic or semiautomatic classification starting from colour or false colour aerial transparencies, it is mandatory to scan the image in order to convert the chroma and density description into three electrical signals, corresponding to the blue, green and red.

In any case, either starting from digital data or from a digitized image, the problem of the automatic interpretation is to reproduce the logical processes of human reconnaissance performed by the eye-brain system, by means of a computer, following the criteria as shown in the classic photointerpretation.

-Unsupervised classification. At the moment the unsupervised classification are the most used techniques for automatic recognition of the objects on the ground, particulary, concerning the land-use applications.

The main aim is to detect and to recognize these objects by their spectral features taken by means of sensors which are placed on a satellite or on a plane; it is even possible to assemble very similar objects.

It is very important to remember that with these methods we are not introducing parameters concerning the ground spectral signatures but these are already established "a posteriori" during data elaboration.

For this reason there are some different methods for the classification, which are more or less sophisticated.

An exemple is given by the "maximum likelihood" and by the "clustering".

The decisional value of "maximum likeli-

hood" uses the probability density of the spectral characteristics of the targets, while the clustering, as the name implies, is a technique for grouping pixels together, which is based upon their distribution in the spectral space.

The mistake percentage is generally prestabilished for both classification methods and they can vary from two to twenty percent.

Further, there can be used other classification methods and we will mention the following:

A) the canonical "maximum likelihood"

B) the principles of linear decision

C) the principles of decision that make use of parallelepipede surfaces.

-Supervised classification. By with this term is intended the grouping techniques that are similar to the "unsupervised classifications", but they vary from the first as make use of the spectral features collected directly during the "ground-truth" which after wards are used at the "training" stage.

Besides, we obtain more accurated classification with elaborated data.

In this case we will have results that depend on the exact ground-truth and also will depend on the principle of data sempling "in loco".

The most sophisticated criteria for classifying objects in a supervised mode are based on Bayesian statistical methods.

So, if the probability density function for patterns pertaining to different fields are known, each pixel (picture element= minimum detected area) may be classified according its likelihood of belonging to one of the classes.

During the elaboration stage the classes must be clearly distinguishable specially if they are referring to the test-site, which knowledge is normally used as reference.

This knowledge is generally easily acquired through the traditional ground-truth.

In case of inacessive areas we can apply upon plane flights at low altitude; but these have the disadventage of rising the survey-costs.

For the moment the selection of significative test-areas is the only critic-stage of the supervised classification.

-Final considerations. The efficency of both methodologies depends highly upon the surveys means we makes use of, upon the aim one desires to reach and upon the cost one can afford.

Generally the supervised methods are more reliable, while the unsupervised methods have essentially an heuristic character, being less accurate.

3 BIOMASS EVALUATION

During 1973 a certain number of radiometric measurement were performed on the lysimeters of the JRC Euratom-Ispra (Nasa contract called Agreste Project).

In the lysimeter cells the same variety of rice was treated with different amounts of fertilizers. It was possible to check the strong influence of Nitrogen on the bio-mass production.

In spite of the limited amount of collected data it seemed possible to establish a linear correlation between total overground bio-mass and the ratio between reflectance of Landsat channels 5 and 7 (0,6-0,7 and 0,8-1.1 micron bands).

From this result it was not possible to state that this ratio of reflectances is an indicator of the standing bio-mass because in the lysimeter cells the variations in bio-mass were obtained only by variation of Nitrogen: it is well possible, even probable, that in these conditions the radiometer may see more the "quality" of the bio-mass (for istance the amount of Chrophyl depending on the amount of Nitrogen) than the "quantity" of bio-mass.

This hypothesis is supported also by the fact that the linear correlation mentioned above has been observed lately in the rice-growing cycle, when strong overlapping of leaves could make the overall horizontal green surface quite independent on the actual total bio-mass.

For these considerations it seemed convenient to devote the campaign of ground tests of 1974 mainly to the attempt of separating the effects of actual bio-mass and Nitrogen on radiometric measurements.

At this purpose two different sets of experimental fields have been considered: fields in which the same variety of rice has been cultivated starting from the density of plants, but treated with different amounts of Nitrogen; fields in which the same variety of rice has been cultivated with the same amount of Nitrogen, but starting from different densities of plants.

The results obtained from the analysis of ata collected on the rice fields in Vercelli (fertilization test area in North Italy) can be summarized as follows:

-The differences in amount of Nitrogen are constantly detected by the four Landsat channels. It is known from literature that content of Nitrogen and content of Chrophyl in the bio-mass follow a linear correlation. Then it is possible to state that reflectance in channel 7 and 6 increase, while reflectance in channel 5 and 4 decrease with the content of Chrophyl.

-The growth of the rice plants produces significant variations in the reflectances, with, then, can be considered useful indicators of the vegetation development.

-The ratio enhances all the effects produced by vegetation characteristics on reflectance.

-The final total overground bio-mass is a linear function of Nitrogen given as fertilize.

-The ratio is a function of total overground bio-mass. The function which can link the ratio observed in different periods of the vegetative cycle to the final total overground bio-mass, can be assumed always linear.

But the coefficient of correlation is very close to 1 in a stage close to flowering time, while it continuously decreases going toward harvesting. This coefficient, though still greater than 0,9, assumes a minimum value when reflectance and bio bio-mass amount data refer to the same day of survey (harvesting time). Since we have no significant data about bio-mass in the previous days we cannot be absolutely sure that this linear correlation holds all over the vegetative cycle, but we can state that the best time for a production prediction is located by flowering-early ripening stage.

-The rice yield depends on Nitrogen fertilization. The "efficiency" of production (ratio between grain and total bio-mass) cannot at this stage of the research be determined by radiometric means (from Agre-

ste Project Report - Commission of the
European Communities - Agazzi,Franzetti -
n.15 - July 1975).

REFERENCES

Agazzi & Franzetti 1975, Agreste Project-
   Internal Report 1975.Ispra,Italy,JRC
American Society of Photogrammetry 1975,
   Manual of Remote Sensing.
Brooner,W.G., R.M.Haralik & I.Dinstien 1971,
   Spectral parameters affecting automated
   image interpretation using bayesian pro-
   bability techniques, Proceedings of the
   Seventh Symposium on Remote Sensing of
   Environment.3:1929-1949, Ann Arbor,
   University of Michigan.
Colwell,J.E. 1972, Uses of Remote Sensing
   in the inventory of agricultural crops.
   Willow Run Laboratories, pp.34. Univer-
   sity of Michigan, Ann-Arbor.
Estes,E.J. 1974, Remote Sensing.S.Barbara,
   California, Hamilton Publishing Company
Gates,D.M., H.J.Keegan, J.C.Schleter &
   V.R.Weidner 1965, Spectral properties
   of plants, Applied Optics. 4(1,2):11-20.
Harnapp, V.R. & C.G.Knight, Remote sensing
   of tropical agriculture systems, Procee-
   dings of the Seventh International Sym-
   posim on Remote Sensing of Environment.
   1:409-433, Ann-Arbor, University of
   Michigan.
Kelly, B.W. 1970, Sampling and statistical
   problems, Remote Sensing with special
   reference to agriculture and forestry,
   pp.324-353. Washington,D.C., National
   Academy of Sciences
Krumpe, P.F. 1972, Remote sensing of terre-
   strial vegetation: a comprehensive bibliography. Knoxville, Ecology Department,
   University of Tennessee.
MacDonald, R.B., R.Allen, J.W.Clifton, M.
   E.Buaer, D.Landgrebe & J.D.Erickson 1972,
   Results of the 1971 Corn Bligh Watch
   experiment, Proceedings of the Eight
   International Symposium on Remote Sensing
   of Environemnt. In press, Ann-Arbor,
   University of Michigan.
Steiner, D., K.Baumberger & J.Mauer 1969,
   Computer-processing and classification of
   multi-variate information from remote
   sensing imagery, Proceedings of the Sixth
   International Symposium on Remote Sen-
sing of Environment.2:895-907, Ann-Arbor,
   University of Michigan.
Welch, R.I. 1963, Photointerpretation Keys
   for classification of agricultural crops,
   paper presented at Symposium on Uses of
   Aerial Photography in Agriculture, Cali-
   fornia, Sacramento.

4

G. HILDEBRANDT

Abteilung Luftbildmessung und -interpretation, Universität Freiburg, Germany

Survey of remote sensing techniques in forestry

1 INTRODUCTION

The importance of remote sensing for forestry is only to be estimated by knowing its characteristics and accomplishment of tasks. At present one can observe overestimations as well as understatements of remote sensing for forest aims - both, because of lacking knowledge in forestry tasks or of the possibilities and limits of remote sensing for their corresponding solution.

Forestry is - in contrast to the simple exploitation - orientated to the cultivation and the maintenance of forests for securing its sustained, that is lasting and perpetual use as recreation forest, protection forest, source of income etc. Independently of the level of intensity in forestry, forestry is above all marked by four specific qualities:

- by the multiple functions of forests
- by the long-dated growth and periods of production
- by the vast extension of areas to be cultivated
- by the high risk and the dependence on nature and site

Tasks and characteristics of forestry require, for securing the sustained yield, optimazion of economic management and minimization of risk, periodically recurrent planning in the level of stand, management units and country, as well as a constant observation of the spacious of production. Fundamental principles for planning are besides knowledge of the conformity with growth of forests site conditions and the regional-specific functions of forests, as well as the market conditions, above all actual information about the distribution of forests within the planning area, about the condition, the territorial order and the accessibility of the forest, as well as volume and structure of the timber

and therewith about the present growth potential.

This enumeration permits to recognize that many-sided information from most different sources are required. An essential part of the regularly recurrent need of information is to be attained by aerial photos and other remote sensing data. From this the importance of remote sensing technologies are to be deduced. They must be differently judged, both, individually and locally, in dependence on information need (specification, precision, temporal disposition, repetition rates), on the kinds of forests and forestry (e.g. tropical rain forest in contrast to northern conifer forest and on the disposition of already available information or functioning terrestrial information systems).

Figure 1 should represent a model trying to offer a conception about the contribution of information by aerial photos and other remote sensing data, so as to satisfy the information need for forestry aims: The model takes into consideration the different planning levels (abscissa) and the in the rule given relation between intensity of forestry and size of the inventory- and planning areas as well as the information need, disposition of present information, respectively (case a and b).

2 FIELDS OF APPLICATION AND INFORMATION NEED (MODEL)

From the above-mentioned points it is easily to be understood that in forestry tasks one needs possibilities of a useful application of remote sensing, where inventory and inquiry of condition and development are in question. The useful application of remote sensing is therefore above all

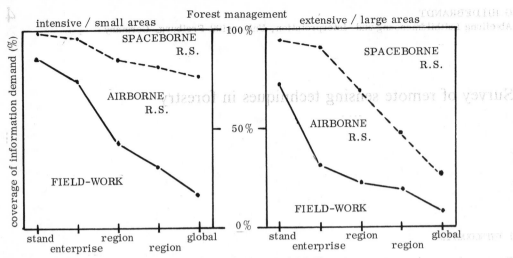

Forest management

intensive / small areas   extensive / large areas

Figure 1: Possible covering of information need (model)

emerging when concerned with:
1. large area inventories serving the purpose of forest- and land-use-policy,
2. inventory and mapping as basis regio-
   nal- or management planning
3. for several actions concerning forest-
   and landscape protection
and besides this, also, when concerned with
other forestry tasks depending on other
condition data, like, for example, landscape planning, wild-life management, etc.

Other applications of the information need
may be orientated towards:
 - the size of areas or parts of different
   categories (percentage of forest,
   forest types or tree species, age-clas-
   ses),
 - the site of certain area categories or
   extensively towards the spatial order
   in the forest area or a region,
 - the kind and accessibility (opening-up)
   of forests
 - the sylvicultural structure, cover type and sylvicultural structure of the
   stands, as well as of the conditions
   of the stands (age, density, disease,
   mortality, etc.),
 - the hight, structure and, in appropriate cases, the local distribution of
   timber volume and its growth
 - site conditions of forest areas (geology, geomorphological factors, soil,
   water supply).
Thereby it can be asked for the status quo
of these factors or, because of comparative
interpretation of former photos for changes or consequences of certain events. Many of the enumerated factors are to be gained directly -- and thereby often reliably -
out of adequate aerial photos or other remote sensing data.Some of them are to be

appreciated only indirectly - and then often with a minor degree of reliability.
Again, some of them remain obscure to remote sensing (e.g. quality of the trunk,
natural regeneration under the canape of
the old stand, specific soil conditions).
In the following the most important applications shall be considered in detail.
Thereby, I shall talk about practical
experiences or about already possible
operational application. Uncertain future
prospects will not be mentioned.

3 REMOTE SENSING FOR MANAGEMENT INVENTORY
  OF FOREST CONDITION AND MAPPING

In sustained managed forests and those to
be led to a sustained management, regularly - between 5 to 20 years -- recurrent
plannings of cutting budgets, or regeneration, among others of management actions
are realized. The application of aerial
photos within this field of activity is
nowadays a self-evident practice. Other
remote sensing data have not yet been applied in this field, neither in Germany nor
anywhere else and have at present no
chance to be applied. In detail, the aerial
photos serve for the purpose of
1. reconsideration of present maps and
   supplementary, in appropriate instances, first mapping;
2. acquisitions of synoptic information
   regarding the forest to be managed
   (topography, distributions of forest
   and other land use categories (transportation system, spatial organization
   in the forest, etc.) and later on for
   the purpose of an exhaustive investigation of the whole area conditions

and its incorporation in the landscape;
3. orientation in the district to be
planned and organizations of the
planning itself;
4. characterization of individual forest
sites;
5. choice of inventory methods of timber
volume, in appropriate cases, for the
purpose of stratification of forest
areas for later terrestrial timber
cruise;
6. description of the conditions including quantitative data (e.g. density
areas covered by various species, tree
heights) or of the timber volume itself;
7. support of recollection and supplementary information resource in case of
lacking information during the domestic interpretation of photo investigation results;
8. document of the stand and district
history

The aerial photo does not replace the terrestrial work. It is rather of great consequence that an adequate combination of
field work and aerial photo interpretation
is realized. It is known from a comparative study (Hildebrandt, 1957) as well as
from practical experiences that, for the
orientation in the district, for the planning of forest arrangement and the stand
descriptions, 30% to 60% of time will be
saved, in case that aerial photos are used
as supplementary information resource. The
time reduction, naturally, also depends on
the district conditions, the quality of
aerial photos and on the forester's working style, as well. The economizing within forestry mapping may amount to 70% and
more, within new measurements to 50%,
compared to the conventional terrestrial
measuring and mapping.

An improvement of the work is in many
cases also possible. In the aerial photo,
relations and numerous details are to be
recognized, which could only be taken in
fragmentary and such an incomplete manner,
in the case that only field work was
applied.

It is surprising that, after 50 years of
otherwise successful forestry aerial photo
interpretation, the recognition of tree
species by means of aerial photos is still
limited. This is, above all, applicable to
the distinction of hardwood species the
reflexion of which, during certain times
in the year, are so similar that a sure
decision is very often not at all possible.
In case that the distinction of hardwood
species is very important for the interpretation, the time of taking photos in

the year is to be found, in which phenological differentiation of the types is maxe-
mal (compare, e.g., Hildebrandt 1957, Dörfel 1977). European and North America experiences show over and above that the recognition of trees on infrared-colour-pho-
tos is much more possible, than in photos
of other film material. MSS-data have the
same object-conditioned limits.

As texture characteristics for the recognition of tree species are at least as important as the spectral signature, MSS data can only offer similar good results as
IRC aerial photos, if the same ground resolution is obtained. Recent data analysis
shows that MSS data are to be taken from
1000 m height, so as to gather the same texture information as IRC aerial photos of
forest stands, which have been photographed from 4000 m height (Masumy and Hildebrandt 1977). Special problems of interpretation regarding the recognition of tree
species are offered by the tropical rain
forest with its, locally by hundreds of
counted tree species (compare for example
Stellingwerf 1971).

Inspite of such problems of interpretation,
the recognition of stand differences and
therewith the delimitation of homogeneous
stand is possible. This delimitation of
standing growth is a manifold task of classification. Characteristics as tree specie,
proportion of mixture, age density, and
site quality are entering as factors of
decision, and there are also economical
and sylvicultural factors to be taken into
consideration. Stand boundaries between
80 and 95% are visible from the air. About
the meaningful delimitation is normally
only to be decided by means of spectral
"classical" interpretation of adequate aerial photos (scale, film/season, photographi-
cal quality). The process of decision is
at the most to be transferred to computer-
aided classification, in case of extensively managed forests having scarcely differentiated stand conditions.

3.1 Maps for the management

Thematic maps, showing the organization,
the stand classification and the opening-
up of forests, are indispensable for the
forestry management. Requirements in regard
to the information of such maps are thereby,
naturally, dependent on the intensity of
forestry.

Aerial photos have already been applied
in the early twenties with the preparation
of forest maps. They have been for a long
time an obvious basis of the forest mapping everywhere in the world. Depending

27

on the aim of applications, black and
white coloured maps, or picture maps
as controlled mosaic, or the orthophotomap, respectively, are prepared. For countries with intensively managed forests
the role of remote sensing may be eluci-
dated by the German survey practice.

Topographic details of forest base maps
1:5000 and the economic maps 1:10.000 are,
as a rule, derived from up-dated large
scale topographic maps. Property boundaries are adopted from cadastrial maps or,
in appropriate cases, are measured terrestrially. These boundaries, which must
be mapped with great precision, cannot be
reliably interpreted by aerial photos, due
to the coverage of tree tops. The stand
boundaries to be mapped, based largely
on aerial photos of flat and hilly regions as well as uplands, are as a rule,
with sufficient precision and economically
incorporated from the aerial photo into
the forest map by simple partial transfer
using a sketch master. In mountainous regions stereophotogrammetrical evaluation
is indispensable. Where the orthophotomap
is part of the forestry mapping systems
these boundaries can be directly traced
from them.

. Increasingly, also photomaps are being
successfully used as economic maps instead of, or in addition to traditional
coloured maps (Voß 1970, Dexheimer 1971).
Thereby it has become clear that the majority of foresters, to whom both maps forms
are available, prefer the photomaps.

A new feature is the development of
methods of digital mapping of stands.
Through aerial photo interpretation on a
stereo plotter, the boundaries to be mapped,
which formerly were marked on aerial photos
after expert interpretation, are now approximated by a discrete series of points
and the coordinates of the points are registered. After transformation of these
coordinates into the coordinate system of
the area survey, the automatic mapping and
calculation of area follow (Kölbl 1976,
Talts 1976).

Computer maps, developed on the basis of
MSS data, have not yet found an opportunity
for use as forest economic maps in the case
of intensively cultivated and small area
forests. This applies to airplane as well
as to Landsat data.

The counterpart to European forest economy
maps in North America and other countries
are forest cover type maps or forest stand
maps. Like in Europe, in these countries
too, both, line maps and photo maps are
also in use. Their information content
differs from that of Central European maps
in consequence of different needs arising
from forms of forest management, practiced

(Peerenboom 1975, fig.2)

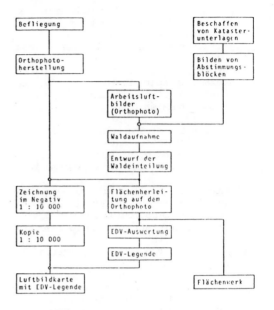

Figure 2
Preparation of an orthophotomap as a forestry economic map including area calculation (Peerenboom 1975), corresponding to the
practice of the forestry administration of
Koblenz

in these countries and the much larger
spatial relations. Classification itself
is based even more extensively on aerial
photo interpretation than in European practice. Interpretation problems are essentially the same as in Europe.

Photo maps, increasingly as orthophoto
maps (Heller 1976a, 1976b) and experimentally as stereophoto maps (Kalensky 1976a,
1976b), have made more headway in North
America than in Europe.

In spite of many sufficient results Computer maps based on MSS data are at present
only experimentally but not yet being used
in the routine practice as forestry maps
for management. In regard to this possibility, there is indeed no lack of experiments using airplane and, above all, Landsat data. In general they lead to the conclusion that thematic computer maps are not
yet useful for the type of map discussed
here until now. Their potential for application lies, as far as forest tasks are
concerned, in other areas.

3.3 Timber volume estimation for forestry
    planning

A special and interesting problem of fore-

28

stry aerial photo interpretation is the assessment or measurement of useful timber volumes. Already in the twenties of this century, procedures to such assessment were proposed in North America and in Germany (e.g. Ziegler 1928). The direct assessment of timber volume from aerial photos is possible for a given forest type by means of correlative relationships between the timber volume and measurable, estimable or countable stock parameters in the aerial photos, such as tree height, density, number of tress or age category. Corresponding aerial photo stand volume formulas and stand volume tables must be developed separately, according to forest type and growing area. In given cases they are successfully applicable only for specific aerial photo material (scale, film type). Acceptable estimates can have a margin of error with ± 10%. In cases, where the volume of the entire forest is aduced from terrestrial sampling it is often stratified. These stratifications can be realized according to information from aerial photos, such as forest types, stand height, density. Sample plots can then be optimally so divided that, in a given area, precision in estimation can be reached or a required degree of precision can be attained with a minimum of samples.

4. REMOTE SENSING FOR FORESTRY INVENTORIES OF LARGE AREAS

Inventories of large areas can be orientated to the first exploration of widely unknown regions of this earth or can - as country-wide, regional- or preinvestment-inventories - contribute in getting information about the potential of forests for decisions in forest policy and policy of development. The periodically recurrent inventory of large areas - practiced in many countries as an important instrument of forest political control - leads to the observation of qualitative and quantitative development of the forest potential, the areas, timber volume and growth.

The results of these inventories of large areas are very often noted only in form of statistics and analysis. The possibility of registering very large areas by satellite photos or small scale radar imagery in a short time and also periodically recurrent, has helped in advancing the development and has opened new prospects of the inventory technique of large areas. Various inventory methods have been developed in the practice of inventory of large areas. They reach from purely terrestrial sampling methods, where air pho-

tos are at the most used as help of orientation within the inventory area to nearly pure remote sensing inventories, where field working is only included for the preperation of interpretation or classification and, in appropriate cases, later for the field checks. However, in the majority of methods, these are combined inventories, which are based on remote sensing information, as well as on terrestrially registered data. Information for area inventories and, in appropriate cases, for the stratification of following terrestrial timber volume inventories are obtained. Thereby the detection and quantification of parts, area sizes and local occurences of different forest types, markable area species, condition- or volume classes and others. The necessary classification of the areas takes place either for the whole inventory area or by sampling of strips, plots or points. The data, needed for timber volume and growth inventory are however, measured on sample plots in the forest.

In other - also successfully - practiced inventories of large areas the timber inventory is accomplished by two steps in form of a double-sampling. Thereby numerous sample plots are chosen from air photos and assessed according to the timber volume/hectare. From the quantity of air photo sample plots a part is chosen as subsample. The timber volume is then measured in these selected plots in the forest.

Such two-stage-inventory-models are leading to a multi-stage inventory, including or not including satellite data. These models are at present repeatedly discussed and also experimentally applied (above all Langley et al. 1969, finally 1976). At most times they are statistically following the model per p.p.s. (p.p.s.= probability proportional to size sampling, Hansen, Hurwitz and Madow 1961) sampling methods. At present they are also examined and adjusted to their utility for middle european forest conditions. Also in inventories of large areas the aerial photo as means of remote sensing has plaied a most important role. Without doubt, this still applies in todays practice. First practical and multiple experimental successes have been reached for purposes of large area inventory by multispectral scanner data and radar imagery. The present stage of technology shows the following realizable possibilities for their practical application in forest- and vegetation inventories.

1. Landsat MSS data
The separation of forest area is quite

possible. In favourable cases, analog as well as digital classification leads to statements about the participations of forest areas in an inventory area with a precision of about 90 to 95%. Grave classification mistakes occur in case that transitional forms such as bush area, very loose formations and wide plantages or large aerial young stands are existent. In such cases the reflexion of parts of the soil or of the herbaceous vegetation can lead to false classification. Similarly, the classification following coniferous and deciduous stands is quite possible. If only the investigation of coniferous area is involved, the classification precision can reach through digital evaluation 95% in a favourable case. Deciduous stands as "class" are properly interpreted between 60 and 85%. The reachable classification precision on one hand depends on object conditioned factors, as for example species and kinds of mixtures, different density aspect, differences in illumination etc. and on the other hand, on the applied analysis technique.

It is also obvious that Landsat data taken at different seasons, produce results differing in their interpretability. Differences in phenological development at a given time - for example in dependence on the height, of locations - lead without fail to false classifications if an adequate stratification of the inventory area is not given by an interactive interveniance of the interpreters. On other occasions it has been shown that a well chosen multi-temporal evaluation of Landsat data in the classification of coniferous and deciduous stands may improve the result by 2 to 5% points (e.g. Kalensky and Wightman 1976).

Besides the above-mentioned rough classification of forest areas certain, locally interesting tree species can, in appropriate cases, be classified, even with already clearly restricted reliability, by computer aids. Some examples: In Northern Italy it was possible to inventory the at least four years old poplar stands, which are important in this region, with a 83% and 95% accuracy (Lapietra and Megier 1976). In North American test areas in 1976, Hoffer reached between 60 and 80% correctly classified pixels. (It is misleading to use the number of pixels classified within the right category as a measure for the validity of the classification results. The user is interested in the inventory results, which for the individual categories follow from the pixels, correctly and incorrectly assigned therein. Also certain forest types in tropical regions, like for example

mangrove forests and auracaria stands (in Brazil), pine - or eucalyptus plantations can be classified and can, for example, be separated from rain forest types. First results in South West Germany suggest that for large area inventories it could be possible with still sufficient precision to classify spruce/fir stands, spruce/beech/fir stands, pure pine forests and two types of deciduous forests. Nevertheless, there are occuring the same difficulties as described by M.P.Meyer (1972) after extensive researches in Minnesota: "... areas where the vegetation consists of complex mixes, fringes, and transitions between type, its (=Landsat data) applicability is doubtful" and he concluded, "The classification accuracy of forest land cover type was inadequate for extensive (or intensive) use by field level resource managers." As far as forest economical inventory works in Maine are concerned, a recent report by Ashley (1977) arrives at a similar conclusion. There is an agreement among experts that Landsat data 1 and 2 are essentially useful for reconnaissance inventories and, only under certain forest conditions, for preinvestment inventories, but then with a special relation to aerial photos with an appropriate scale. From MSS scanner data in narrower areas one hopes to gain improvements in forestry classification.

2. Airborne-MSS-Data

Analysis of results show also in this case object-conditioned interpretation problems and trouble factors. They are caused by different forms of reflexion behaviour, even within the same tree species or type vegetation, as well as by various shadow structures in the stands etc. The considerable extent of ground resolution results especially in the forest in typical textures, which are very helpful in the case of visual interpretation, but which can lead to considerably false classification in the case of computer based classifications, or the development of algorithms for recognizing characteristic textures and their insertion in existing programs will lead to progress in these cases.

Furthermore, we must consider that, as a result of the object specific directional reflexion, in certain cases signature variations of the same object will appear along the scanner line (Reichert 1976). The problematic factors are evident in a LARS study (Coggeshall et al. 1974). MSS data taken from 6000 m above ground were evaluated. Indeed, rather high percentages of correctly classified pixels were obtained for the six categories of the test area (table 1, columns 2). The inventory results, which can be deduced from the figures given

Tab. 1 (according to data given by Coggeshal et al. 1974

class	percentage of correct classification	classification results			
		in absolute values		in percent	
		true val.	clas.res.	true val.	clas.res.
deciduous	83,9	32252	27119	64,7	54,5
conifer	89,8	88	3233	0,2	6,5
water	96,2	339	342	0,7	0,7
forage	64,1	11760	8028	23,6	16,1
corn fields	97,5	2679	5295	5,4	10,6
soya fields	92,5	2676	5777	5,4	11,6
		49794	49794	100,0	100,0

by Coggeshall et al. shows that they, as a whole are still inadequate (tab. 1).

The most favourable results of a detailed classification based on airborne MSS data up till now has, to my knowledge, been presented by Jaakkola (1976). He was able to classify with success six classes, namely, clear cutting areas, pine seeding stands, pine mature timber, spruce mature timber, birch seeding stands and birch mature stands.

3. Radar imagery
In the last seven years, ever since SLAR was made available for civilian purposes, millions of square kilometers of forest surfaces in Central- and South America and West Africa have been photographed for reconnaissance inventories. Radar imagery reached significance in a short time everywhere in humid tropics, where nearly always clouds exclude remote sensing in the visual or the infrared part of the spectrum. The most spectacular and well-known project is the RADAM project, which for the first time made it possible to obtain a surveillance as rough as it may be, and a thematic mapping catalogue of the Amazonas basin. The radar image interpretations are practiced visually, like the classical aerial photo interpretation, and it is also using the same interpretation techniques. Rougher classes are to be identified essentially in larger areas of existent forest types. Differences in grey tones and texture point out various vegetation or forest formations. The interpreter, however, has thereby to take into consideration the differences caused by the mapping system, if objects are viewed along-track and across-track and situated near or far to the antenna.

The geomorphology and the water system are well recognizable in radar imagery. The possibilities of connection of certain forest - or vegetation forms existing in hardly touched areas of nature with physiographical conditions are therefore important for the interpretation success.

Finally, the radar photo interpretation cannot manage in the same w a corresponding field work in order to reach the knowledge for specific project orientated interpretation keys. Likewise, the forest- and vegetation maps of the Amazonas areas, derived from the RADAM project, are the result of a meaningful, well thought combination of radar image interpretation, preceding and accompanying field survey and aerial photo interpretation as well as conclusions from the relation between physiographical conditions and vegetation forms to be expected.

In regard to the Columbian tropical rain forest, it was possible for Sicco-Smit to delimit swamp, grass land, shifting cultivation, mangrove- and low swamp forests, shrub savanna, "gallary" forests, wetland forest types of the floodplain, dryland forest types of the flat and low hill region, as well as the mountains region with an accuracy, which is important for a first survey inventory.

David Francis, one of the most experienced interpreters having shown a variety of interpretation possibilities, concludes as follows, "This is not to say, that RADAR imagery should replace conventional aerial photography in all areas of tropical vegetation. If weather conditions and finances permit, the author's preference would be for normal colour or false colour photography at a scale of around 1:15.000." But then Francis continues, "But over 20 years experience in the humid tropics has shown that this can rarely be achieved over commercially valuable forests in the humid tropics. This a method which allows useable imagery to be obtained irrespective of weather conditions (cloud or haze), and equally effectively by night or day, according to a precise timetable and at an exact cost (because survey aircraft are not standing by at a daily rate waiting for suitable weather) has a place in forest survey in the tropics. Even more so when large forest areas are being evalua-

ted or changes in their boundaries are being monitored." (Francis 1976 p. 84).
  4. Infrared colour aerial photos with
    ultra small scales
Besides the not yet concluded development
of the MSS-, SAR- and the beginning of
CCD-acquisition and the photo procedure
technology, aerial photos from 12.000 to
20.000 m flight height with high resolution IRC-films have now again come into
considerable discussion.

  Aldrich and Greentree (1971), Nielsen and
Wightmann (1971), Lauer and Benson (1973),
Aldrich (1975) and others have done research in the interpretation possibilities
of such photos serving for purposes of
forest inventories of large areas. As a
partial system of multiple stage inventory for future European forest (and land-
use-) inventories in 1973 photos from
10.000 m flight height have been recommended (Hildebrandt 1973). Also Heller (1976a
/b) has recently outlined the increasing
importance of ultra small scanless IRC
aerial photos for large area forest inventory and observation tasks.

  The forestry information content regarding the differentiation between forests
and non-forests as well as between coniferous and deciduous forests is similar
to the satellite. But in cases of transitional forms ultra small scale IRC photography can be better than satellite data.
Higher ground resolution of aerial photographs allows in addition a further interpretation of the forest conditions (density, mixture, disease, etc.).

5. REMOTE SENSING FOR FOREST PROTECTION
5.1 Discovery, delimitation and quantifi-
    cation of forest damage

As far as photo scale and ground resolution
of the photo are sufficient, vegetation or
soil damages, which occur as area damages,
shown in typical photo figures (form, delimitation, size) are to be recognized and
delimited in aerial photos of every kind
as well as in MSS data. Damage areas of
this well recognizable kind could have for
example resulted by storm, erosion, forest
fire, drought, inundation, etc.

  Damages in forest stands causing mortality or disease of individual trees, without
immediately causing changes in the areas,
but changing the reflexion characteristics
of the leaves and changing the sight of
the crown, are, on the contrary, normally
only to be recognized by colour aerial
photos and data with corresponding scales.
The best practical results of damage inventories or relevant specific studies until now, have been produced by stereosco-

pic evaluations of infrared colour aerial
photos.

  In the majority of cases only the damaged
trees as such are seen. The cause of damage can be interpreted in many cases through
the introduction of additional information
or through expert deduction which goes
beyond the visible.

  At this point, a word concerning the
question of pre-visual identification of
plant damage seems to be necessary. Primarily, plant damages, which have already become
visible in nature, can be discovered. In
this process remote sensing has the great
advantage, in so far as it enables us, in
an early stage, to recognize incipient
changes in the top of the crowns, where
many damages first appear. Such changes
are very often, either not visible or only
partly visible from the ground. This, however, is not yet pre-visual identification
of damages. Such pre-visual recognition can
theoretically then be expected when a disease or a stress situation produces morphological changes in the structure or in
the water content of the leaves. In both
cases this can occur simultaneously without foliage discolourations or dropping
of leaves or other visual appearances. In
the first case (morphology), according to
present day knowledge, the spectral reflexion would have to change between = 0,7
and 1,3μm and in the second case (water)
between = 1,3 and 2,5 μm. A few reports
in the literature seem to confirm this hypothesis, but numerous questions are still
to be answered.

  Quite apart from such special, still open
questions, regional and local damage inventories in forests have been conducted in
the last ten years everywhere in the world.
In this process predominant use was made of
large- and middle scale infrared colour
aerial photos and of traditional stereoscopic interpretation methods. In forestry
and gardening damage assessments of this
type were aimed at tree- and stand damage
through:
    - insects (of various types)
    - phyto-pathological causes
    - industrial emissions (e.g.F, SO_2 etc.)
    - salt accumulation in the soil
    - droughts or sinking of ground water
      levels
    - floods etc.
    - effects of forest fires, storm and
      winds and other forces.
Interpretation has been on the whole successful. Mortality can be grasped almost
totally, concerning trees. In middle scale
pictures and, concerning groups, in small
scale photos. Only sheltered trees escaped
detection. Different degrees of vitality
or damages are to be classified under cer-

tain circumstances (refer to Kenneweg 72, Murtha 1976 and others). In many cases great significance is attached to the correct choice of the time of the survey and the scale. In regard to evaluation goals, optimal photo- and data acquisition is decisive for the success of a damage assessment. This is commonly still underestimated, as well as difficulties of interpretation, arising from the multiplicity of the appearance of damages in the photo, as well as from the subjectivity of interpretation. The latter occurs in connection with the limited human capacity to distinguish and to recognize numerous extremely subtle colours values. It is pertinent to try to overcome the above-mentioned difficulties by means of objective colour and light measurements in the photo or through digital evaluation of MSS data. Drastic differences of dipictions of healthy, sick or dead trees, which, indeed, could also be clearly interpreted can, in fact, be classified digitally. With less manifest symptoms however, digital interpretation shows limitations, on the one hand, in the above-mentioned and, on the other hand, in problems conditioned by the system. The association of specific symptoms of damage or degrees of damage with specific colours in the photo or with specific spectral intensity values in an MSS data, cannot, unfortunately, be absolutely defined. Caused by different lighting conditions of individual trees in the stand, by the strong directional dependency on their spectral reflexion behaviour and by the atmospheric conditions prevailing at the acquisition, remote sensing data can vary depending on the point, where the object is seen in the photograph and on the scanner line, respectively. These variations can be such that the differences, caused by disease can be camouflaged by depictional difference, which as it is, appear in forests.

5.2 Thermal imagery in forest fire surveillance

Remote Sensing by means of thermal imagery has gained practical significance in forestry in the areas of surveillance and control of forest fires. In the case of present-day available scanners, the temperature resolution lies at 20° between C 0,05 to 0,1° C. It differs in the different temperature levels. With this information of possible temperature resolution nothing is said about the actual possibility of discovering forest fires or the observation of known fires. The discovery of an only 0,46 qm large, burning, non-smoking trunk section through a ther-

mal imagery taken from 7000 m heights is guaranteed, shows the enormous capacity of the system, but cannot be generalized as an option. The recognition of small fire sites of mouldering ground fires in dense forests is often but not always possible. Fundamental researches by Wilson, Hirsch, Madden and Lsensky (1971) have cast considerable light on this problem.

1. The succes of recognition depends on the type of forest, structure of the stand, crown form and density explain these differences. The most successful recognitions have been achieved in forest types consisting predominantly of pine or aspen whereas the results of a douglas fir is less successful.

2. In all types of forests the success of recognition unmistakably depends on the angle of recording. The successful recognition is maximal, when the fire spot is situated directly beneath the aircraft. With increasing Nadir-deviation the success of identification diminishes. This takes place in various degrees depending on the resp. forest type.

3. The successful recognition of beginning fires depends on the size of the fire spots. The success in the above-mentioned American experiments with relatively small experimental fires was always higher when the five partial fires were situated on the circumference of a larger circle than in those situated on the circumference of a smaller circle.

4. As the foliage increased or as the paths of the rays through crown space became longer as photo angles increased, the signal received became correspondingly weaker until it was no longer recognized as a fire.

In order to assess the possibility of success of discovering small beginning forest fires one must consider, in addition to the above-mentioned stand characteristics and recording conditions, also the morphology of the area under observation and the inclination of slopes in relation to the direction of observation.

IR-rays cannot, indeed, penetrate rain clouds but they can penetrate dry fire damp. Forest fires, which have already affected rather large forest areas or have even evolved as ground- and trunk-fires, can thus be observed even in the presence of strong smoke, providing, no rain clouds lie beneath the altitude of the air plane. The development and the perimeter of the fire front, the path direction and the speed of the expanding fire, the origin of new fires beyond the fire front is well possible in real time.

The US Forest Service has several airplanes in permanent use, which are equipped

	1975			1976		
airplane	number of missions	number of forest fires	number of flight hours	number of missions	number of forest fires	number of hours
Merlin III	26	13	105	53	15	117
King Air	133	30	428	80	25	238
Queen Air	51	16	227	13	7	83

Tab. 2

with a two-channel IR-Scanner (3-6µm, 8-14µm) arranged especially for forest fire observation. For 1975 and 1976 the airborne infrared reports of the US Forest Service show for example the following applications in forest fire-control and combat.

After three years of full operational deployment the national fire report comes to the conclusion: "Although comparatively new, IR detection has proved its usefulness. Accurate and timely intelligence on location of the fire perimeter is a vital component to success in campaign fire management. During critical periods of a fire , perimeter and spot fire locations must not only be accurately known, but the means of obtaining this information must be rapid and timely. With this intelligence, the fire boss and his staff can plan their strategy with respect to tactical employment of fire resources and logistical support. IR scanning is not meant to solve all the fire boss's problems, but is intended to supplement the present method of ground and air-reconnaissance. Fire intelligence requirements can be met with airborne infraredscanners that can rapidly and accurately map fires either day or night and through dense smoke. Fire perimeter and spot fire information from IR scanning should remove one of the fire boss's problems."

A real time thermo vision system (e.g. AGA Optronic THV 750 superviewer) seems appropriate for the observation along the boundary of the burnt area after the forest fire.

After the abatement of an open forest fire such system can be very useful for observing and controlling smouldering fires, which can last for days, particularly ground fires, which are still smouldering underground without giving smoke.

Literature

Aldrich R.C., Norick N.X., Greentree W.J., IN: Evaluation of ERTS 1 Data for Forest and Rangeland Surveys. USDA Forest Serv. Res. Paper PSW-112.67op.

Allen, P.T.E., The use of Side Looking Airborne, RADAR Imagery for Tropical Forest Surveys. FAO Paper. FO:MISC. 75/1o.,1975

Ashley M.D., Morin L., Spray Block Mapping Control for Spruce Budworm using Landsat and High Altide Remote Sensing. Symposium Image Proceeding, Graz 1977

Coggeshall M.E., Hoffer R.M., Berkebile J.S. A Comparison between Digitized Color Infrared Photography and Multispectral Scanner Data Using ADP Techniques. LARS Inform. Note o33174 Purdue Univ. , West-Lafayette 1974.

Dexheimer, W., Erfahrungen bei der Einführung forstlicher Orthophotokarten in Rheinhessen-Pfalz. Allg. Forsteitschrift 1975, S. 14 - 18.

Dörfel, H.J. Phänologie als Einflußgröße für die Fernerkundung verschiedener Vegetationsformen. Symposium Flugzeugmeßprogramm Hannover 1977.

Fagundes, P.M. Pinto, M.N., Natural Resources Inventory ISP Comm. VII WG 4 Final Report. XIII Congress ISP, Helsinki 1976, 44p.

Francis, D.A. Possibilities and Problems of RADAR-Image Interpretation Vegetation and Forest Types with Particular Reference to the Humid Tropic. In Remote Sensing in Forestry, Proc. Symp. IUFRO Subj. Group 6.o5, Oslo 1976, S. 79-86

Hansen, M.H. Hurwitz, W.N. Madow, W.G. Sample Survey Methods and Theory. New York 1963/64

Heller, R.C. Color and False Color Photography, its Growing use in Forestry. In Application of Remote Sensors in Forestry, Freiburg 1971 S. 37-56

Heller, R.C. Remote Sensors for Airborne and Spaceborne Imagery in Remote Sensing in Forestry, Proc. Symp. IUFRO Subj. Gr. 6.o5, Oslo 1978, S. 37-52

Heller, R.C. Natural Resource Surveys, Proc. XIII Congress ISP, Helsinki 1976

Hildebrandt, G., Forsteinrichtungsarbeiten mit Hilfe von Luftbildern, Forst und Jagd 1957, S. 58-64

Hildebrandt G., Zur Frage des Bildmaßstabs und der Filmwahl bei Luftbildaufnahmen für forstliche Zwecke. Arch. f. Forstwesen 1957, S. 285-3o6

Hildebrandt G., Zum Einsatz von Erderkundungssatelliten für suprationale Inventur der Wälder und landwirtschaftlicher Nutzflächen. Raumfahrtforschung 1973 S. 164 - 168.

Hoffer, R.M., Techniques and Application for Computer-aided Analysis of Multispectral Scanner Data. In Remote Sensing in Forestry, Proc. Symp. IUFRO Subj. Gr.6.o5 Oslo 1976, S. 1o3-114.

Jaakola, S., An Automated Approach to Remote Sensing oriented Forest Resources Surveys. In: Remote Sensing in Forestry, Proc. Symp. IUFRO Subj. Gr. 6.o5, Oslo 1976, S. 147 - 156.

Kalensky, Z.and Wightman, Automatic Forest Mapping Using, Remotely Sensed Data. In Remote Sensing in Forestry, Proc. Symp. IUFRO Subj. Gr. 6.o5, Oslo 1976, S. 115-136.

Kalensky, Z. Automation of Thematic Mapping based on Remote Sensing and Computerized Iamge Processing. In Remote Sensing. Int. Train. Sem. Lenggries 1976.

Kenneweg H., Die Verwendung von Farb- und Infrarot-Farbluftbildern für Zwecke der forstlichen Photo-Interpretation unter besonderer Berücksichtigung der Erkennung und Abgrenzung von Kronenschäden in Fichtenbeständen. Diss. Freiburg 1972, 253 p.

Kölbl, O. Digital Stand Mapping. In Remote Sensing in Forestry, Proc. Symp. IUFRO Subj. Gr. 6.o5 Oslo 1976, S. 447-459

Langley, P.G. Aldrich R.C. Heller R.C., Multistage Sampling of Forest Resources by uding space photography. Vol. " Agr. Forest, and Sensor Studies, Proc. 2. nd Ann. Earth Resources Aircraft, Review NASA MSC, Houston, Texas 1969, 19-1 to 19-21.

Langley, P.G. Multistage Variable Probability Sampling, Theory and Use in Estimating Timber Resources from Space and Aerial Photography. PH. D. Diss. 1972, Ann Arbor, Michigan, 1o1 pp.

Langley, P.G. Sampling Methods useful to Forest Inventory, when Using Data from Remote Sensing. In, Remote Sensing in Fc restry, Proc. Symp. IUFRO Subj. Gr. 6.o5 Oslo 1976 p. 313-322.

Lapietra, G. Megier, J., Acreage Estimation of Poplar Planted Areas from Landsat Satellite Data in Northern Italy. In Remote Sensing in Forestry, Proc. Symp. IUFRO

Subj. Gr. 6.o5, Oslo 1976, p. 157-17o.

Lauer, D.T. Benson, A.D. Classification of Forest Lands with Ultrahigh Altitude Small Scale, False Color Infrared Photographs. Proc. Symp. IUFRO Subj. Gr. 6.o5 Freiburg 1973

Masumy S.A., Hildebrandt, G. Analyse der Textur als Erkennungsparamter zur Identifizierung von Waldtypen. Symposium Flugzeugmeßprogramm, Hannover 1977

Meyer, M.P., M. Read, Landsat Digital Data Application to Forest Vegetation and Land Use Classification in Minnesota, IAFHE RSL Research Report 77-6, 1977, Remote Sensing Laboratory, College of Forestry, University of Minnesota, Final Report.

Murtha, P.A. A guide to Airphoto Interpretation of Forest Damage in Canada. Can. For. Serv. Ottawa, Publication No. 1292, 63 p.

Murtha, P.A. Inventory and Monitoring of Forest Diseases and Damages by Remote Sensing - Considerations about Promising ways to do it. In, Remote Sensing in Forestry, Proc. IUFRO Subj. Gr. 6.o5, Oslo 1976, p. 385-396

Nielsen, U. Wightman, J.M. A New Approach to the Description of the forest Regions of Canada Using 1:16o.ooo Color Infrared Aerial Photography. Can For. Serv. FMR-X - 35, 25 p.

Peerenboom, H.G. Erfahrungen bei der Einführung forstlicher Orthophotokarten aus der Sicht der FD Koblenz. Allg. Forstzeitschrift 1975, p. 14 - 18

Reichert, P., Vegetationskundliche Auswertung multispektraler Scanneraufzeichnungen. In, Remote Sensing in Forestry, Proc. Symp. IUFRO Subj. Gr. 6.o5 Oslo 1976, p. 191-2o2

Reichert, P. Auswertung multispektraler Scanneraufzeichnungen für forst- und landwirtschaftlichen Inventuren. Symposium Flugzeugmeßprogramm, Hannover 1977

Smit, Sicco, Experiences with the use of SLAR in forest and land-use classification in the tropics. In Remote Sensing, Int. Train. Sem. Lenggries 1976

Stellingwerf, D.A. Aspects of the Use of Aerial Remote Sensors in Tropical Forestry. In Application of Remote Sensors in Forestry. Freiburg 1971, S. 89-98.

Talts, J., Persönliche Mitteilung Sommer 1976, USDA Forest Service, Airborne Thermal Infrared Report. 1974 and 1975 Fire Season

Voß, F. Zur Herstellung von Forstbetriebskarten mit Hilfe maßstäbiger Luftbildkarten und automatischer Rechen- und Kartieranlagen... Allg. Forst- und Jagdzeitung, 197o, S. 153-16o.

Wilson, R.A. Hirsch, S.N. Madden. F.H., Lasensky, J., Airborne Infrared Forest

Fire, Detection System. Final Report.
USDA For. Serv. Res. Paper Int.-93.1971
Zieger, E. Ermittlung von Bestandesmassen
aus Flugbildern mit Hilfe des Hugershoff
Heyde'schen Autokarthographen. Mitt.
Sächs. Forstl. Versuchsanstalt, Tharandt
1928, S. 97-197

COLETTE M. GIRARD
Institut National Agronomique Paris-Grignon, France

5

Application of photointerpretation technique to the classification of agricultural soils, choice of the sensor, use of the results

What is a soil ? It is the result of the decay, alteration and organization of the upper layers of the earth crust under the effect of life, atmosphere and energy exchanges.

A soil might be studied for various reasons. One might be interested in :
- study of soils birth and evolution
- soils cartography
- arable layer utilization.

The two last points will be considered in this paper through photointerpretation techniques. With these techniques, two types of data will be mentionned : the soils characteristics directly detected, and the undirectly detected soils characteristics.

The first are just <u>seen</u>, the last are <u>deduced</u>.

1 DIRECTLY DETECTED CHARACTERISTICS

These characteristics show through the reflectance curves of bare soils, they may be : water content, chemical composure organic matter content...

They will be illustrated by some examples.

1.1. Texture and water content

Experiments done in LARS show for soils samples with different textures the influence of water content over reflectance (figure 1, a, b).

For Clay soils at different water content values, the reflectance characteristics are the same through visible and near infra-red where the water absorption bands show at 1,400 and 1,900 nm.

It is not the same for Sandy soils. The dry sample does not show through its reflectance curve the water absorption bands in the near infra-red. The difference is due to the soil texture. Clay contains

water in its chemical formula, and even a dry clay soil contains water.

It is not the same for sand.

Reflectance %

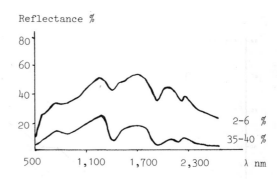

Reflectance %

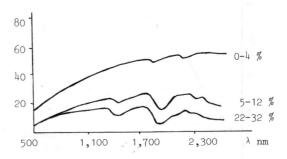

Figure 1. Soils reflectance
  a) Clay soils at two water content levels
  b) Sandy soils at three water content levels.

1.2. Chemical components

Reflectance measurements have been done the same day over three soils with different CaCO3 and Fe contents, but with the same water content.

    a greenish soil   CaCO3 57 % Fe 3.3 %
    a yellow soil   CaCO3 65 % Fe 1.3 %
    a light soil   CaCO3 67 % Fe 0.6 %

In the visible, the light soil is more reflective than the others, the greenish soil being the less reflective. In the near infra-red, the reflectance curves follow the same distribution (fig. 2). It seems that for the same water content, a high content in CaCO3 and a poor one in Fe increases the reflectance, while a less CaCO3 content and a high Fe content decreases the reflectance, both in the visible and the near infra-red.

1.2. Organic matter and surface roughness

Measurements have been done in the field over soils containing more or less organic matter, and with a smooth or a rough surface.

For soils with high organic matter content (4.10 %), the reflectance is always low. The surface roughness influences the reflectance, but the water content seems to be without influence (fig. 3).

Soils with lower organic matter content have higher reflectances. When the surface is smooth there is no difference between the dry and the moist soil, but for two samples of the moist soil, the reflectance is lower for the one with a rough surface (fig. 4)

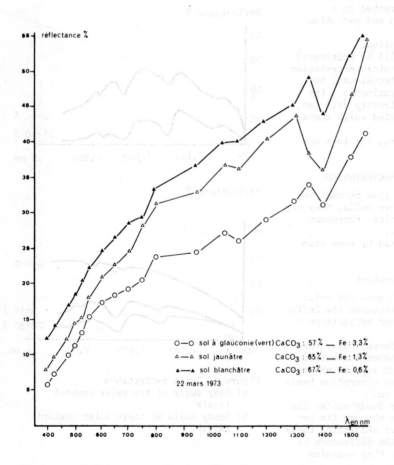

O—O sol à glauconie (vert) CaCO3 : 57 % — Fe : 3,3 %
△—△ sol jaunâtre   CaCO3 : 65% — Fe : 1,3%
▲—▲ sol blanchâtre   CaCO3 : 67% — Fe : 0,6%
22 mars 1973

Figure 2. Reflectance of three soils. Measurements done in the field.

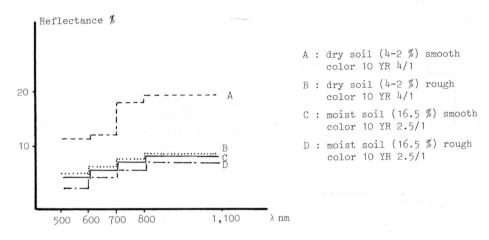

A : dry soil (4-2 %) smooth
    color 10 YR 4/1

B : dry soil (4-2 %) rough
    color 10 YR 4/1

C : moist soil (16.5 %) smooth
    color 10 YR 2.5/1

D : moist soil (16.5 %) rough
    color 10 YR 2.5/1

Figure 3. Reflectance values for a dark soil organic matter 4.10 %

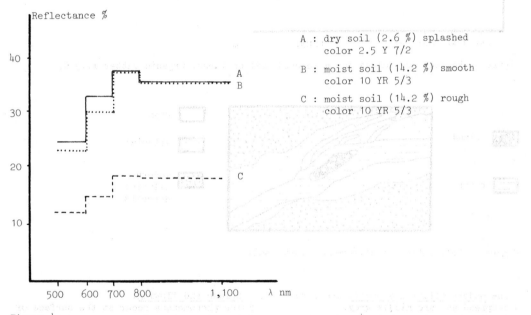

A : dry soil (2.6 %) splashed
    color 2.5 Y 7/2

B : moist soil (14.2 %) smooth
    color 10 YR 5/3

C : moist soil (14.2 %) rough
    color 10 YR 5/3

Figure 4.

When the state of the surface is the same, the reflectance of the dry soil is then higher than for the fresh soil (fig. 5). All the factors, influencing the reflectance curves, are reponsible, in the visible part of the spectrum, of the various hue, value, and chroma of the colour seen by the human eye.

For example chalk lightens the colour, but every light area is not due to chalk content. Organic matter, iron are reponsible of darker colours, but very often the colour is the result of different interactions of these factors.

The soil texture influences also the colour. The following example shows how to recognize three type of alluvial deposits on a panchromatic photograph (fig. 6). The flood deposits are light grey, nearly white, they have an elongated shape linked with the level contours.

The alluvial deposits along the river are light grey and spread more or less from the river banks.

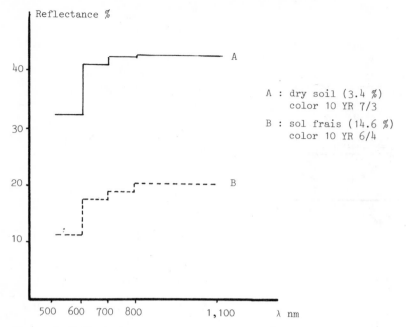

Figure 5. Reflectance values for a leached soil on loess. Organic matter I.73 %.

In the figure:

Reflectance %

A : dry soil (3.4 %)
    color 10 YR 7/3

B : sol frais (14.6 %)
    color 10 YR 6/4

x-axis: 500 600 700 800 ... 1,100   λ nm

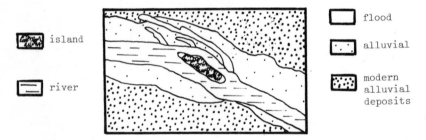

island

river

flood

alluvial

modern
alluvial
deposits

Figure 6. Delineation of different alluvial soïls.

The modern alluvial deposits are more widespread and are middle grey.

Samples taken in each type and analyzed show that the flood deposits contain mostly sand (83 %), 32 % are calcareous sands. They are poor in organic matter and clay.

The modern alluvial deposits contain 66 % of clay and loam. Coarse elements are not numerous (31 %). Limestone is not abundant (7 %), but the content in organic matter is rather high (2.7 %).

Alluvial deposits along the banks contain as much tiny elements as the modern deposits (55 %). It is the same for organic matter (2.2 %). The content of sand and of limestone is the same than for the flood deposits. This deposit is in transition between the others.

Splash and flowing.
These phenomenons occur at the surface of soils with a high loam content. The surface becomes as smooth as a mirror as all tiny particles are moved by the rain and oriented in the same way.

The soil reflectance increases, and it shows on photographs as very bright areas.

2 UNDIRECTLY DETECTED CHARACTERISTICS

These characteristics are not directly seen, they are deduced through visible characters. Some elements "read" the soils, they are : vegetation, morphology, human influence.

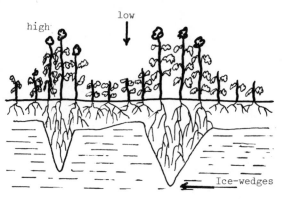

Figure 7. Plants reading an ice-wedges net.

2.1. Vegetation

1. Soil depth

As plants grow in the soil, the roots might explore a certain depth, and give an information on what is going under the surface. But all plants do not "read" the soil in the same way. For example a cereal is a best indicator than a betroot because its roots explore a greater depth.

2. Ice - wedges net

Such discontinuities as ice-wedges inherited from old quaternary may be shown by differences in the stage of growth.

Colza growing over a wedge will blossom earlier. On a color photograph the net will appear in yellow against a green background (fig. 7).

3. Substratum differences

Once again growing of vegetation and its ripening may be altered by soil differences.

Vegetation "gaps" are due to plants missing. If these gaps occured just after germination, they may be detected long after, and they reveal a discontinuity in the substratum (fig. 8).

The figure show what happens with lens of "sables de Lozère". They are made of coarse sand and gravel stuck together by clay. Usually they are difficult to locate as they are scattered and lay at a certain depth between loam and limestone.

They can be detected by the means of vegetation on aerial photographs.

2.2. Morphology

1. Soil depth

On a false color photograph taken in March over an area in "Bassin Parisien" different white and grey tones may be interpreted. First there is a thin layer of snow which appears like white patches and remains on the higher or windy areas.

A gaz pipe, buried, crosses a valley, the mark of the trench is visible. Its depth is constant, so according to the morphology, it crosses different soils and substratum (fig. 9).

.On the plateau the mark is dark, snow has melted, while it is remaining everywhere else. Loam is thick (A).

.On the slopes the mark is white when the limestone is near, it is dark when the colluvium is thick (C, D).

.On top of the slope the mark is light, surrounded by dark tones when the trench goes through a thin loam moved by colluvium. (B).

2. Evaluation of limestone depth

Erosion, on steep slopes, show deep soil layers. In Northern Africa, for example, limestone, white, appears under the red

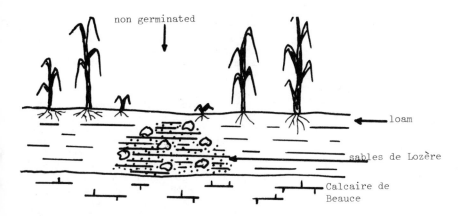

Figure 8. Plants reading substratum differneces.

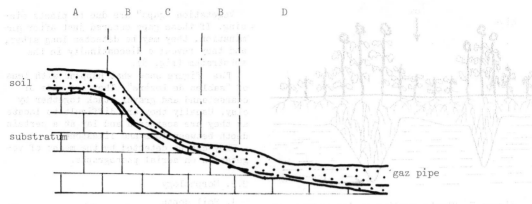

Figure 9. Reading soils through morphology and a trench.

clay. On aerial photographs it shows as white lines on the slopes, topped by grey tones.

3. Textural variations

Textural variations may appear by differences in soil water retention. In South Eastern France (Plateau de Valensole), small streams cut a plateau. Thalwegs are visible downstream. Upstream, flowing areas are visible, because of their different texture which retains water better. They are darker on the photographs.

2.3. Human influence

1. Existing, or not, organic horizons

Clearing done on a wide area enables to see the soil, when it is accompanied by a levelling, deep layers of the soil may be visible.

Photographs of Champagne show after a clearing, dark, sinuous narrow marks, they look like an hydrographical net. When there was still a forest, small organic elements gathered in the small thalwegs. After clearing and a small levelling the old thalweg shows in grey against the surrounding white chalk.

2. Porosity

If aerial photographs are taken two or three days after an heavy rain, differences in the soil drying are visible and give drainage information. The soils where infiltration is quick are a lighter grey than the others.

Interpretation is sometimes difficult because of the fields limits which introduce often false contours.

3. Saturation

On aerial photographs, natural and man-made drainage looks like fish bones shapes inside fields. Its limits are always the field limits.

These tracks in loamy soils indicate a pseudogley horizon at middle depth.

3. CHOICE OF THE SENSOR

3.1. Emulsions

The four commonly used emulsions may be used in soil study and give different results.

1. Panchromatic

As it is employed from many years, it is the best known emulsion and people are used to get a lot information from it.

The different grey tones are interpreted, there may be some misinterpretation, but the interpreter is used to imaginate the soils and the colors in terms of grey tones.

Soil scientist is used to that emulsion and do much work with it.

2. Black and white infra-red

With that emulsion, wet soils appear darker than dry ones. It is useful to interpret differences in soil moisture.

For example a dry yellow sand may be separated from a wet white sand. The former will appear light grey, the latter dark grey.

3. Colour

Colour rendition on a photograph is much relative. It depends on the film age, the way it has been processed... but a colour photograph is useful. A light colour may indicate limestone, a change in the colour (inside one photograph, not between two photographs) may show soil differences.

Human eye catch better hue than value and with that emulsion, the interpreter may be bothered by lines such as field limits which will hide real soils limits.

4. False-colour

This emulsion is interesting because it presents advantages of both black and white infra-red and colour films.

Bare soil appears blue. Very wet areas are dark blue, drier areas are pale blue.

Vegetation appears very clearly, even if the vegetation cover is sparse or light. The difference between reddish or bluish colours is great for the eye.

Because of its spectral characteristics, this emulsion enables to distinguish details inside water, shoals for example. Like on black and white infra-red one can differenciate easily water from land and marsh.

Like colour film, colour rendition varies a lot, and colour comparisons between different photographs are not recommanded.

The choice between emulsion is sometimes difficult. It seems that under temperate climate the best choice may be panchromatic and false-colour. If one must use only one emulsion, false-colour, is often the one to choose.

3.2. Date

A photograph gives an instantaneous view of a landscape. According to the date, the view of the same place may be quite different. For France the best meteorological conditions to study soils, are two or three days after a rain of medium intensity, or after the melting of snow.

When the soils are covered by a crop, the begining of growth and the ripening are interesting periods.

3.3. Season

It is an important factor which has an effect upon interpretation. Examples are given for France.

1. Winter

During that period, soil is usually bare. If there is a crop, its covering is light.

Soil surface is directly visible and any difference in its aspect may be interpreted.

2. Spring

At the begining of that season, melting of snow and some agricultural works magnify contrasts.

The end of the season is not a good period as the crops cover mostly the soil surface ; when the soil is still bare, works like rolling, hinder all surface differences.

3. Summer

After the harvest and the removal of plants remainings, it is possible to interpret soil characteristics. Through stubbles some aspects of the soil are visible.

A light tilling gives good indications of the soil to the interpreter.

4. Autumn

Agricultural soils have different aspects at that season. They might be bare, still covered by crops (corn, betroot, alfalfa...), have been ploughed and already sown.

The best period to take photographs varies according to the crops cultivated in the area and to the agricultural works calendar. A new tilling or a too deep one are not interesting as they show a very rough surface, where the soil differences are hidden.

The interesting periods to take photographs useful for the soil scientist are short, and scattered all over the year. When planning a flight, one must keep in mind the agricultural calendar, and be aware of the meteorological conditions.

3.4. Scale

It is important to consider different points.

1. Perception of all elements

The scale must be great enough to enable to see all elements necessary to do the interpretation.

For example to do a drainage study it is necessary to detect natural, man-made drains, flowing areas... It is also necessary to see the links existing between these elements, which is possible at a smaller scale than previously.

If only one scale, has to be choosen, because of costs, it must not be too small nor too great.

2. Limits drawing

Interpretation limits are drawn directly from the photograph. Very often, limits drawn from a photograph need to be enlarged to be readable. The choice of the photographs scale is not good. The scale must have been greater than the scale of the final map.

3. Scale of the photograph, scale of the work

The scale of the work imposes the scale of the map. The photographs have been choosen at a compatible scale to do interpretation and drawing.

Some details seen on a photograph may correspond to a different scale than the scale of the work. They must not be drawn despite the fact that they are quite visible.

In conclusion it appears that it is impossible to indicate a good scale to interpret soils.

For each work it is necessary to determine what will be the scale of the aerial photographs. If it is possible, the interpretation work is better when done using both photographs at a large and at a small scale

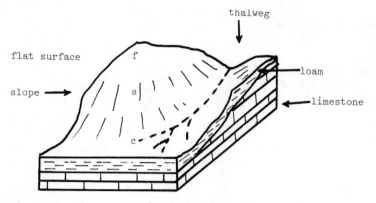

Figure 10. Diagramm for interpretation of an area.

The following chapters will give examples of utilisation of photointerpretation.

4. USE OF PHOTOINTERPRETATION IN SOIL CARTOGRAPHY

A map is structuration of different elements, collected by various means. When mapping one has :
    -to search the elements and their definition,
    -to structure the elements.

4.1. First step

Before begining to map an area, it is necessary to collect all information linked to that area. It consists of :

    -topographical, geological, botanical... maps.
    -all books describing aspects of the area.
    -aerial photographs.
    Between these sources of information, aerial photographs are the easier to obtain.

4.2. Second step

Aerial photographs present raw material which needs interpretation. It is necessary to recognize elements and then to replace them inside a coherent structure.

4.3. Third step

From this structure, homogeneous areas are determined. They are marked and the contours delineated, from the photograph. The limits are copied, if necessary, on a topographical map.
    For all homogeneous area on the photograph is given a description using photointerpretation data.
    It does not mean that all area drawn from the photograph is really existing in the field. The use of such a work is to prepare the work in the field. Each area must be studied by a profile, at least. So the field trip may be scheduled from the photograph.

4.4. Fourth step.

It is the most important one, the field work. If the photograph is recent, and at a rather large scale, 1/25,000, 1/15,000, it shows details not visible on the topographical map. It may be used then to take one's bearing, and also as a drawing item.
    Field work with the photograph enables also to verify the interpretation hypothesis. If some limits are wrong, they are then modified, if some limits are missing, they are added.
    It is then, that the description of the area, with photointerpretation data, is compared with the pedological description of the profiles. This step establishes links between field and photograph.

4.5. fifth step

The links drawn between soil and photointerpretation data are studied and analyzed. They give birth to a new structure from which the pedological map will be drawn.

5. PHOTOINTERPRETATION EXAMPLES

5.1. The whole processus : from the photograph to the profile

1. Identification

As it is shown on figure 10, the area is in an agricultural field with a small slope (s) ending in a flat surface. A small thalweg (t) runs across the area and ends in (c). These morphological aspects are clearly visible on the aerial photographs.
    Several flights have been done over this area, at different seasons, heights, and using different emulsions. The aspect of each element : s, t, c, flat surface (f), field (F), is given in the following table.

Season	Emulsions	ELEMENTS				
		f	s	t	c	F
Winter	P	white	whitish grey	light grey	middle grey	middle grey
	C	white	white-ochre	light ochre	ochre	ochre
Spring	F-C	white	bluish white	blue	blue	blue
	C	white-green	light green	white & green	dark green	green
Summer	F-C	white	bluish white	white & red	dark red	pink
	IR	middle grey	middle grey	middle grey	dark grey	middle grey

2. Interpretation

The geological map indicates for the field: limestone or sand in horizontal stata. As the limit of the flat surface is parallel to the level contours, it indicates that the strata are horizontal.

If the white patch (f) was due to particularly filtering sands, it would appear in light grey on black and white infra-red, because of dryness, but it is not the case.

On false-color, the same patch appears in white, instead of blue around. That means a light coloured cropping out rock.

The interpreter may conclude that the patch is due to limestone. In the center of the patch, the very white color induces an out crop of limestone. Its content in the soil is so high that it prevents a normal vegetation growth. In full growing period, the patch remains white. The limit is cut, the patch is due to an out crop, not, to a colluvial deposit.

The darkest area is at the end (c) of the small thalweg. It contains less limestone than the other parts, in its surface horizon. But deep horizons may contain more limestone as they are nearer to the out crop. Coarse elements are brought by flowing along the thalweg.

Crops are vigorous at the end of the thalweg, that means that thinner elements concentrate on that point. The organic matter content is higher on top of the profile than at the bottom, and higher in the dark than in the light area.

3. Analytical results

From data obtained by photointerpretation, a classification of pedological character (limestone, organic matter, loam, sand...)

can be made and a toposequence be drawn (fig. 11).

4. Profile sketch

Since interpretation is done, it is interesting to sketch the profile which may be found in the interpreted area. It enables to understand better interpretation once in the field. Verifications, rectifications are easier to make, as only a part of the interpretation may be wrong. For the soil scientist the interpretation understanding is better when directly comparable to a profile.

5.2. Soil interpretation from different factors.

1. Ploughing

Very often ploughing enables interpratation of lower horizons. If the organic upper horizons are thin, the lower horizons, taken to the surface are easily detected as their colour is ligter or deeper. Clayey horizons B, horizons A2 of podzolic soils, or leached soils are clearly visible that way.

On the other hand bed-ploughing may indicate a damp area.

Tilling give information of the soil horizons, but differences in the soil aspect may be due only to some characters of tilling.

.Depth : on a photograph, different tillage operations done the same day are in different colours. A medium winter plou-ghing (25 cm deep) has a darker aspect than a surface tillage by a canadian cultivator (10 cm deep).

The deeper a soil working is, the more visible it will be.

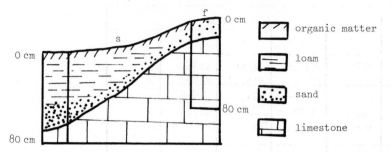

Figure 11. Interpretation sketch of a toposequence.

It is sometimes impossible to differenciate on a photograph a field covered by stubble from a field worked by a rotary hoe (Semavator type). This machine modifies only the first cm, and the soil surface is not too much disturbed. Furthermore, organic matter is scattered, and it hides the soil surface.

.Date : some observations have been done on an aerial photograph taken on March, 5, 1971.

Some fields had received a medium ploughing on November, 20, 1970, others had received the same ploughing on February, 25, 1971. Nearby were fields not tilled since the previous summer.

The fields with no autumn tillage were of a light colour. The fields with a winter ploughing were of a medium colour, and the fields with a spring ploughing were of a very dark colour.

The date of a tillage affects much its aspect on a photograph. The more recent it will be, the darker it will appear.

As on the same photograph standed fields with the same ploughing but at different dates, and fields with different soil workings at the same date, it was possible to draw the following table.

From this table it appears that the date of the soil workings influences more the aspect on the photograph, than the ploughing depth.

Tillage operations are visible on photographs because they modify the upper horizons (A or B) and because they increase porosity.

A more porous soil heatens quicker during day, and cools quicker at nightfall.

All these factors affect the aspects taken by the same soil. One must be very carefull in interpreting different soils colours, or tone on a aerial photograph.

2. Drainage

The settlement of a man-made drainage is always visible on photographs. When it is completely buried, its detection is difficult.

But on some flights drainage nets appear, they look like fish bones. It may be possible then to draw the draining scheme and to count the drains space, and determine main drain direction.

Drainage is a costly operation. It is done in precise conditions. Presence of drainage indicate in some regions : wet, loamy-clayey non porous, hydromorphic soil...

3. Irrigation

Like drainage, it is a costly operation undertaken only in precise conditions. It is linked to climate conditions, but also to soil water supply.

An irrigated area may indicate a dry, sandy, pervious, highly porous, thin soil..

4. Cultivated areas

Constraints due to soil characteristics are sometimes so imperative that crop distribution may indicate elements for a soil cartography.

For example meadows may be found on alluvial, immature, humid soils. "Savarts", are found in Bassin Parisien, on dry, colluvial soils on slopes. Vineyards may be found on colluvial or stony soils.

But very often crop distribution is the result of History, and woods may be found on acid, infertile, soils on slopes, as well as

Image value	Stereoscopic effet	
	Low	High
Great	stubbles (semavator)	Canadian (Nov. 20)
Medium		past ploughing (Nov. 20)
Low		recent ploughing (Feb. 20)

46

on fertile soils. Their presence is then due to cynegetic reasons, or simply because there were enough cultivated soils.

5. Agricultural lots

Lots, their shape, location, illustrate adaptation of man to environment. Old lots, even abandonned from many years are visible on aerial photographs. They indicate that the soils were fertile, and so had a certain depth, or were on a gentle slope.

The whole lot distribution is often linked with soil units. On alluvial deposits lots have an elongated shape, while polyhedric shapes and check patterns indicate a somewhat rich soil.

Large, rather shapeless lots are established on poor soils : thin or hydromorphic soils.

Lots boundaries correspond sometimes to soil boundaries.

All these remarks may be wrong if there has been a land consolidation.

6. Crops

As it has been written previously crops give indications on soils characteristics, but soils limits do not always fall in with fields boundaries. Moreover some fields works may lead to soil misinterpretation.

For example alfalfa is cut many times a year, like other forage. There may be differences between fields sown with the same forage, not because of soils differences, but because of mowing date differences.

6. INTERPRETATION OF MULTISPECTRAL IMAGERY

It is possible to interpret multispectral

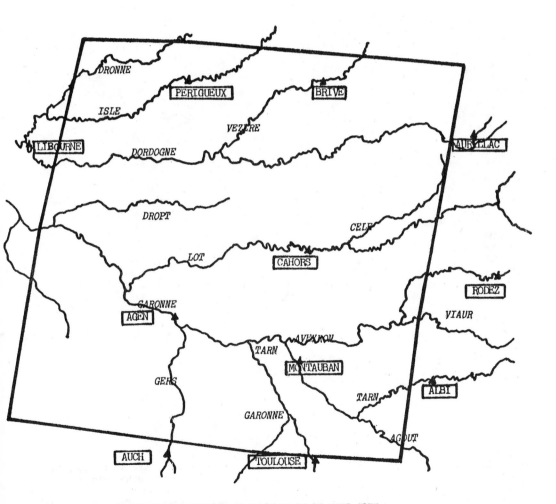

LOCALISATION DE L'IMAGE LANDSAT DU 23 MARS 1973.

47

images but the interpretation is different from photointerpretation.

6.1. Interpretation factors

On multispectral images, the interpretation factors are different from those used for photointerpretation. The factors consist of :

1. Density values

They show as different grey tones on the visualisations of the different channels. They correspond to the mean radiance of the radiances of the different objects situated inside the pixel. If the objects are identical, the mean radiance gives a good indication, but if the objects are very different the mean radiance cannot be compared with the radiance of known objects. Nevertheless, the risk of error being known, it is possible to interpret wide classes of elements through the density values, that is to say, radiance values.

For example, it is possible to separate objects as : water, bare soil (more or less wet), cereals, forage, different sorts of trees... from the density values on visualisation.

The more homogeneous and the more widespread the objects are, the more accurate the distinction will be (fig. 13).

2. Structure

The structure seen on images has not the same signification than on photographs. Because of the accuracy of the document, the structure of a cultivated field cannot be seen while it is clearly visible on photographs at the same scale (under and near 1/15 000).

The structure seen on multispectral images is the result of spatial distribution of the different density values. It has usually no relation with structures of natural or anthropic features. The interpretation of structure on multispectral images is a delicate one.

6.2. An example of interpretation

To illustrate what have been written previously, and to show what can be done, an interpretation of Landsat images will be given.

1. The images

The images selected correspond to three channels of the scene of March 23rd 1973 taken over the area : BRIVE à TOULOUSE and part of LES LANDES till LE CAUSSE de LIMOGNE (fig. 12).

A color composite of channels 4,5 and 7 had been made by l'Institut Français du Pétrole. It consists in coding in yellow the more reflective areas on channel 4, in green the less reflective areas on channel 5, and in red the less reflective areas on channel 7.

2. Signification of the composite

On the color composite one can see different colors, according to the code, these colors were compared with the colors that would be affected to known objects.

The following table (fig. 13) give some indications which are only true for THIS composite and THIS precise date.

It is also impossible to correlate the color with a radiance. The color is the result of three colors combination and it cannot be quantified.

Notwithstanding these remarks it is easier to interpret one color composite instead of comparing three different black and white images.

That is to say that it is possible to get valuable information through a very esily processed imagery and without sophisticated data treatments.

3. Interpretation of the composite

The composite is at a scale of 1/500 000. Wide areas may be characterized from it by their color and their structure.

According to the scale and the pixel size color traduces great types of soil utilization (non cultivated, bare soils, meadows...), the structure is due to morphology and rivers patterns.

These areas have been called "rural landscape", and they have been defined from the composite using the following data :
  Morphology
  Rivers patterns
  Structure of the image
  Soil utilization.

These data have been put together on a description file to ensure an uniform definition for every rural landscape (fig. 14).

A map of these rural landscapes has been drawn from the color composite (fig. 14). Each landscape has been given a number in such a way that alikness between landscape was enhanced.

It works in the following manner, when the number begins by :
  1 different crops, small fields, a few forests
  2 an elements mosaïc, many deciduous forests
  3 alluvial plain, bare soils, meadows
  4 many meadows, few other crops
  5 meadows, deciduous forests
  6 evergreen forests, a few wet bare soils

| CHANNELS | VEGETATION | | | | SOILS | | WATER | CITIES |
| | FORESTS | | Moors | Meadows | Dry | Wet | | |
	deciduous	Evergreen						
4 positive	–	–	–	–	+	ε	–	–
5 negative	ε	+	+	+	–	–	+	–
7 negative	ε	+	ε	–	–	+	+	+
Resulting color	Pink	Brownish green very dark	Dark green	Green	Yellow	Orange red	Very dark red	Red

Figure 13. Color given to different objects through the color composition LANDSAT imagery, March, 23rd 1973.

+ means a strong answer
ε means a weak answer
– means no answer

49

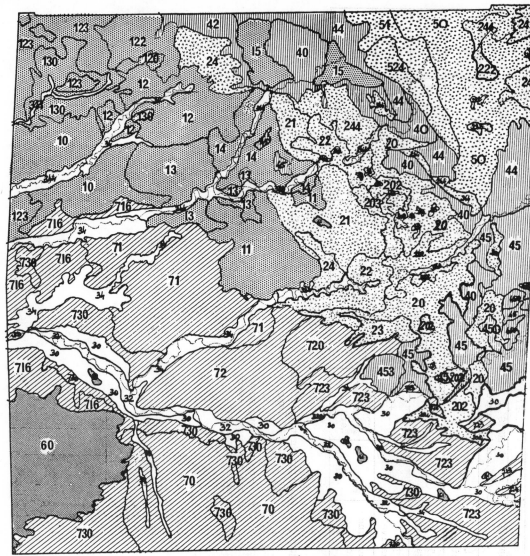

INTERPRETATION D'UNE COMPOSITION COLORÉE   LANDSAT 1 : 23 MARS 1973
SUD-OUEST DE LA FRANCE

7 different crops, small fields, deciduous forests.

The data on soil utilization are important in the description because they are
clearly and directly visible on the image.

On the contrary variables such as climate, geology, pedology....are not directly
visible on the composite and they must be
deduced by interpretation.

For example, on this image the soil utilization is linked with some geological
levels, so a geological interpretation may
be done from the image.

For soil science the problem is nearly
the same : soils are not often directly
visible, and they are then interpreted
through their utilization.

Satellite images are profitably studied
for soil management and conservation.

4. Comparison with "petites régions
agricoles"

The "petites régions agricoles" have been
defined by the Institut National de la
Statistique et des Etudes Economiques in
1954. These "petites régions agricoles"

have boundaries much dependent of the administrative boundaries (parish, district...), they group parishes presenting common characters : cropping system, relief, type of substratum... Their rôle was to be test zones, representative of an area where the same trend of agricultural production could be proposed. Beyond, they were considered as basic units for great regional managements and economical surveys.

Unfortunately, the boundaries of "petites régions agricoles" were more dependent on administratives limits than on natural constraints. Their contents based on inventories and inquiries became quickly obsolete.

So there is a great need for a synthetic document that can bring knowledge, and traduce agricultural environment.

Rural landscapes are defined with data linked with environment and soil utilization. They are described at a comparable perception level than the "petites régions agricoles". The boundaries are easily obtained by interpretation of color composites.

The content is somewhat imprecise with only one composite. It could be improved with the use of two or more images taken at different dates. Of course, all information concerning areas smaller than the pixel are to be collected through other means, such as great scale image or photographs.

Through these images it is possible to appreciate for a wide area the resultants of environmental characteristics and of social and economical constraints.

Inside a rural lanscape, differences in soil utilization and hydrographical, morphological characteristics are weaker than between two rural landscapes, even if the soil types cannot be named. The limits drawn from the image correspond to the stronger contrasts.

This type of map is terribly needed in agricultural management. It can be quickly obtained, its cost is low, and the actualization is easy. They must bring to a greater use of satellite images, even with a poor resolution, in European countries.

7. CONCLUSION

Aerial photographs are very useful to interpret and map soils, but the interpretation must be done very carefully because of risks of error as it has been described.

In anyway, the use of photographs do not prevent a study in the field, but shortens it as the points to verify or to describe are best choosen.

The photographs must be considered as an essential tool for the soil scientist.

Remote-sensing techniques give also information about soils, but they are not so well known as photographs, and if they are used for research work, they are not so much used for operational work.

Among the remote-sensing techniques, the more often used are certainly multispectral images taken from satellites.

8. REFERENCES

Actes 1er Colloque "Pédologie et Télédétection". 31 Août - 9 Sept. 1977. Ed. GIRARD. Publ. ISSS.

GIRARD C-M - GIRARD M-C. Applications de la Télédétection à l'étude de la Biosphère. Masson. Paris. 1975.

Ch. C. GOILLOT
Institut National Agronomique Paris-Grignon
Laboratoire de Télédétection de l'INRA, Versailles, France

6

Significance of spectral reflectance for natural surfaces

1 A FEW BASIC PRINCIPLES OF PHYSICS TO REMIND.

1.1 The electromagnetic radiation (EMR)

It is well known that the cutting of the electromagnetic spectrum into various "types" of radiations :
- γ
- X
- ultraviolet;
- visible;
- infrared (near, medium, thermal, far);
- hertzian (microwave, high frequency, UHF, VHF,...)
- ... a.s.o.
is a rather arbitrary segmentation of the same kind of radiation : the electromagnetic radiation, which later on will be quoted EMR.

1.2 Interaction of the EMR with the matter

When the EMR strikes the matter, some physic phenomenons take place :
- a part of the incoming energy is brought back in the space containing the source: there is REFLECTION ρ.
- a second part enters the matter and disappears therein ; it is ABSORPTION α.
- a third part, at last, having passed through the matter, is transmitted beyong it : there is TRANSMISSION τ.

The budget of this INTERACTION between the EMR and the matter may be written if 1 is the value of the incident radiation:

$$\rho + \alpha + \tau = 1$$

This equation simply conveys the conservative property of the energy.

The actual lecture deals only with the first of these phenomenons : the REFLECTION.

1.3 The REFLECTION

A thorough analysis of this phenomenon shows up that it is more complex than it appears at first sight and especially when the question is about natural surfaces.
Basically two main types of reflection are to be considered (see fig. 1) :
1. The specular reflection which occurs with perfectly smooth (polished) surfaces, such as still water. This type of reflection is described through the laws of Descartes.
2. The diffuse reflection, occuring with perfectly rough surfaces (known as diffusing surfaces), such as freshly fallen snow. The perfect case is described by the Lambert's law (lambertian surface).

In common situation, the actual reflection is always in between these two sheer cases of reflection and some combination of them : the resultant of the all direction reflection from the actual surface is given by the so-called "indicatrix" (see fig. 1) which gives the value of the energy measured in each direction θ as reflected by a surface when lighted by an incident flux I_o.

Rem. : the emettrice of a Lambertian surface is a sphere.

N.B. It is customary to consider the reflection as a "surface property" of the matter; in fact, a thorough analysis of the phenomenon brings out that the radiation always PENETRATES the matter and that the EMR interaction with the material medium takes place in a layer the thickness of which depends mainly on :
- the wavelength of the EMR;
- the electrical properties of the stuff.

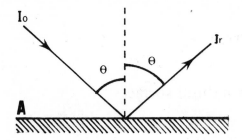

Perfect specular reflection
(Descartes's law)

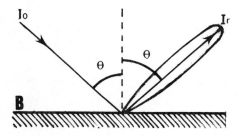

Pseudo specular reflection

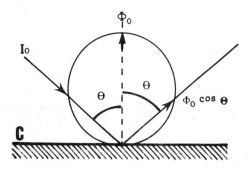

Perfect diffuse reflection
$\Phi = \Phi_0 \cos \Phi$

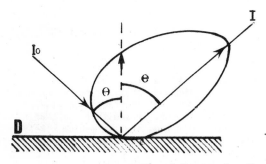

Diffuse reflection
Non lambertian surface

—— Indicatrix

Fig. 1 Types of reflection

3. We shall not consider here
- the diffraction, which occurs when the surface is affected by a periodical structure (such as a grating). The mutual contribution of the ordered elements of this structure gives rise to an enhancement of the EMR in given directions, away from the specular direction.
- the spontaneous afterglow of the matter such as the FLUORESCENCE, which occurs afterward the excitation by the incident radiation has been removed.

More information on this two last phenomenons and their laws will be found in Physics handbooks.

We shall bear in mind that the reflection phenomenons are due to the interaction of the EMR with the matter and that this interaction finds expression in alterations of the incident radiation in :
.direction;
.intensity lowering (see absorption or extinction coefficients);
.spectral composition (colour or spectral "signature" of the matter);
.polarization (increasing or decreasing);
.Phase.

1.4 The variables to describe the reflection

Two classes of variables are used :
- relative variables, also called "coefficients" which are by far the most employed (ALBEDO, REFLECTANCE);
- absolute variables, defined with energy units (RADIANCE). These last ones are less employed in Remote Sensing activity.

1. Relative variables. They come from the ratio of two variables of the same nature thus resulting in a dimension-less coefficient which is often turned into a percentage.

. The ALBEDO or "whiteness coefficient" A which for a given surface is the ratio :

$$\frac{\text{all the light reflected by the surface}}{\text{all the light incident on the surface}}$$

Due to the angular dependance of the reflection phenomenons, which has been reminded above for the two types of reflection, it is noteworthy how often this variable is improperly used.

The evaluation of such a coefficient implies that the sensor is able to measure, first, the whole spectrum of the incident radiation with even sensitivity and, second, the whole radiation reflected in all direction by the irradiated surface. The last condition indeed is met, neither in airborne nor in spaceborne remote sensing.

. The REFLECTANCE COEFFICIENT $\rho(\theta)$ is the ratio of the reflected radiation in a specified direction θ to the incident radiation. As the reflection is a function of the wavelength, the spectral reflectance coefficient $\rho(\theta,\lambda)$ is the value of this ratio at a given wavelength λ. The value of this ratio is often given for a small interval $\Delta\lambda$, with $\Delta\lambda = 1$ μm.

Rem. : Without any special specification it will be understood that the coefficient is "global", i.e. includes both specular and diffuse components in the measured reflected energy. Otherwise will be mentionned, respectively :
. specular reflectance coefficient
. diffuse reflectance coefficient
When the value of each of these terms is known separately ;
. hemispheric (spectral) reflectance coefficient

$$\int_{2\pi ster} \rho(\lambda,\theta)d\theta$$

when the reflected radiation is considered in all directions θ of the source space (2π steradian). It is obvious that integrating now for λ (spectrum of the source) leads to the albedo value.

2. Absolute variables. The measured values must now be compared to a standard. Only will be mentioned here the most commonly used in Remote Sensing : the SPECTRAL RADIANCE.
. SPECTRAL RADIANCE $N(\lambda)$ (See fig 2 for definition).

It is worth noting that when absolute measurements take place in Remote Sensing, the reflective surface is considered as a source of radiation, even in the visible. Whereas absolute values are scarcely reffered to in the studies on reflectance
- apart from special works on the instrument technology -, on the other hand, they are fundamental for studies involving the emittance of the surface (thermal infrared f.i.)

1.5 Measuring the reflectance

1. Photography. As first approximation, a photography may be considered as an image of the scene reflectance.
. The colours in each point are due to the spectral alteration of the incident solar radiation by each surface element of the targets in the scene;
. The optical density of the film is related to the intensity of the reflected radiation.

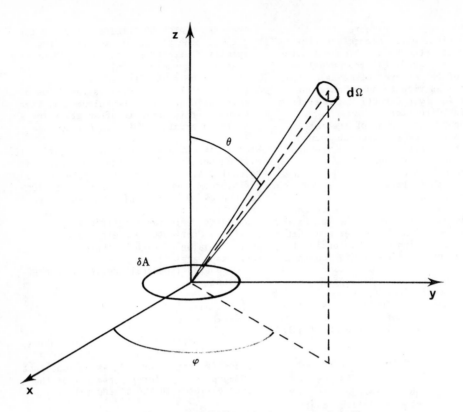

Fig. 2 Definition of the spectral radiance

$$N(\lambda) = \frac{\delta[\delta\phi(\lambda)]}{\delta A.\ \cos\Theta.\delta\Omega\delta\lambda} \quad (w.Sr^{-1}.m^{-2}.\mu m^{-1}) \quad \text{(per unit of spectral band } \delta\lambda\text{)}$$

where :

δΦ : total power (or radiant flux w) radiated by the surface element δA in the elementary solid cone angle δΩ of direction Θ and φ

Rem.   For a Lambertian surface N is Θ and φ independant, thus the distribution of radiant flux from the surface is found to be spherical (See fig. 1)

The fig. 3 shows what kind of informa-tion can be derived from the simultaneous use of two types of black and white emul-sions : a panchromatic one, and an infra-red one, the spectral sensitivities of which being limited at 0,7μm and 0,9μm res-pectively.

However, this "sensor" is unstandardisa-ble in actual practice and thus unsuitable for quantitative radiance measurements. Nevertheless, it provides easily a lot of useful informations, the implementation of which is unexpensive as long as no machine processing is involved (i.e. if confining to photointerpretation technics).

2. Photoelectric instruments. Two cate-gories of instruments are in use for mea-suring the reflectance :

Reflectance spectrophotometers, such as f.i the well known and thoroughly used DK2 Beckmann. They are based on the re-flective integrating sphere.

These instruments measure the hemisphe-ric spectral reflectance of a sample stuck on a window of the sphere and lighted by an artificial source of radiation. The set up of these instruments makes them unappropriate for field measurements, they are mainly suited for laboratory studies. They have been broadly used to determine the specific spectral reflectan-ce of a wide variety of natural surfaces and to study the variation of this reflec-tance in relation to the state of the sample. They supply a leading and invalua-ble indication to prospect for a remote sensing mission :

Panchromatic film

Infrared film

I

Cesar Wheat

Beets

II

Fig. 3 Site of GRIGNON - Yvelines (France)

Experimental parcels of the S.E.I.

The infrared photography shows the differential reflectance in the near infrared (700 < λμm < 900) of Wheat and beets in I.

Notice : the reflectance inversion in II (see 2.3 in text).

- the sensitive spectral bands;

- the amplitude of reflectance variations
which can be expected (see further at §
2. Plant reflectance).

. The spectroradiometers. They are of two
kinds :
- with fixed spectral bands (as f.i. the
Exotech). The scanners, either airborne
or spaceborne, to date, belong to this
category·To n channels correspond n fixed spectral bands.
- with shifting spectral bandwidth (as f.i.
Isco or SpectralMaster types).

Rem. When increasing the number of channels
and narrowing the bandwidth of each spectral band of a fixed spectral band radiometer, the operating mode of a shifting
spectral radiometer is approximated.

The fig. 4 shows the reflectance of a scene as recorded by a 9 channels scanner with
spectral bandwidths of 50 and 90 nanometers.
The vegetation surfaces are well identified
yet, due to their genuine spectral reflectance as given further (see fig. 5).

1.6 Energy sources in reflectance remote
sensing

They are of two kinds :

1. Natural ones i.e. in order of importance :
. The sun : punctual and localised source
(by clear sky) the spectral emission of
which stretches from 300 to 3500 nanome-
ter as for the part useful in remote sensing with actual technology (atmospheric
absorption has also to be taken into account).
. The atmosphere (eventual clouds included)
It is a diffused, stretched source.
. The surrounding of the considered point
in the scene. The reflective and emissive
properties of surrounding structures such
as trees and hedges, contribute in fact **to**
the irradiance of the elements in the scene. Thus the albedo of a bocage land is
less than the one of the open land (Plessey Radar 1972).
2. Artificial : (in so called "active")
remote sensing) :
. Radar (using microwaves)
. Laser-Lidar (visible in the broad sense).

TO CONCLUDE :
These few reminders and considerations
of physics are essential to understand how
remote sensing makes use of the reflectan-
ce phenomenons (see fig. 6) and explain
why discrepancies may arise when various
measuring methods are implemented, considering :
- the surfaces in study;

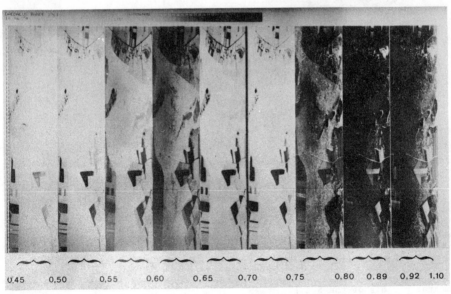

0,45   0,50   0,55   0,60   0,65   0,70   0,75   0,80 0,89 0,92 1,10

Fig. 4 Multispectral scanner DAEDALUS (9 channels)

Simultaneous images of a scene in 9 spectral bands (7 of 50nm and 2 of 90nm bandwidth).
Each element of the scene is to be considered in relation to the reflectance caracteristics of natural surfaces -specially of vegetation- as given farther (See Fig. 6 and 7).
The bare soils are easily recognized due to their evenly increasing reflectance from
450 to 1 100 nanometers.

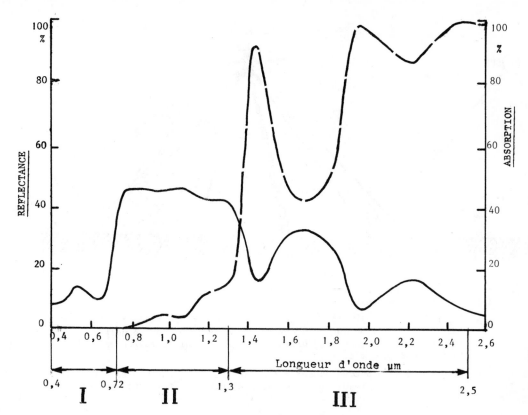

Fig. 5 Typical reflectance of a green and spectral absorptance of water comparatively.

    ——————— Reflectance (leave)   — — — — — Water absorption

(After L.A.R.S. Purdue Univ[y]., 3, Res. Bull. n°884, sept. 1968)

- the instruments employed;
- how these instruments are used.
(See fig.8.)
  As a general rule, the reflectance value of a surface at a given wavelength depends on :
. The geometry of the set: source, surface and sensor;
. The spectral composition and the incident flux polarisation (IRRADIANCE) on the target surface.

2 VEGETATION REFLECTANCE

  The study of the vegetation reflectance involves four different and complementary points of wiew which have led to distinct works :
- On one hand :
. The reflectance of plant parts (single organs - pigments);
. The reflectance of sets (canopies).
- On the other hand :
. The nature and the state of the plants;

. The structure and texture of the set.
  The synthesis of these four aspects, mainly attempted by means of models is not fully achieved to date.

2.1 Reflectance of plant organs :

  1.Nature of the plant. Numerous measurements have been performed on this topic by many authors most of them working with a spectrophotometer.
A compiling of such data carried out at the University of Michigan (1971) has gathered the spectral reflectances of some 2600 kinds of plants.
The fig. 7 shows the spectral reflectance of some species belonging to various groups.
These studies lead to three general conclusions :
- For a plant in its normal state -i.e. typical and healty, the spectral reflectance is specific of the group, the species and even of the variety at a given stage in its phenological evolution.

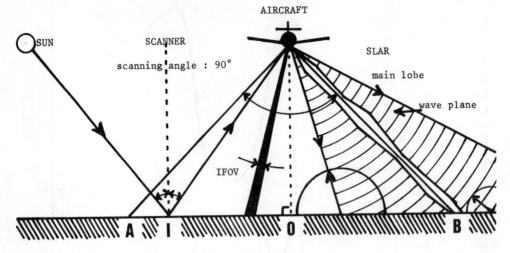

Fig. 6

IFOV : instantaneous field of view
SLAR : side looking airborne radar

a) depending upon the relative position of the sun and the aircraft, each of the points of the scene is "seen" by the scanner in various conditions of reflectance (specular in I, forward diffusion in A, normal diffusion in O, backward diffusion in B. See fig. 1).

b) with the SLAR - active sensor - the radiation the antenna receives back from the scene always results from a backscattering of the irradiated surface.

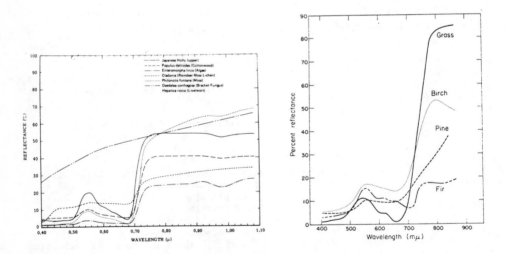

Fig. 7

Global spectral reflectance of 11 species belonging to various groups (leaves or other organs) as measured by a spectrophotometer)
(After Gates 1955)

60

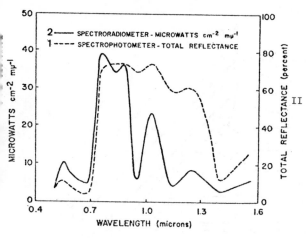

Fig. 8   Spectral reflectance a measured
          with :

1 ----- a reflectance spectrophotometer
        using an integrating sphere, in
        the laboratory on cotton leaves.

2 ——— a terrain spectroradiometer poin-
        ted verticaly from 300 meter above
        cotton plants.

The observed difference is due to the type
of reflectance measured in the two cases
(See 1.5, 1.2). The second one is closer
to the conditions met in remote sensing
measurements (After Myers V.I. 1975).

- The chlorophyl plants feature a typical
  spectral reflectance the general aspect
  of which, for a healthy plant, in the range from 400 to 2600 nanometers, is given
  in fig. 5 :
  It shows 5 striking features concerning
the absorptance :
. high in the ultraviolet and the blue;
. reduced in the green;
. high in the red;
. very low in the near infrared (700 < λnm
  < 1500) along with high reflectance and
  transmittance.
The very abrupt increase in reflectance
near 700nm and the fairly abrupt decrease
near 1500nm are present for all mature, h
healthy green leaves.
. Very high, further in the infrared for
  λ > 3000nm.
  Each of the regions which have been quo--
ted I, II and III of this curve has been
correlated with morphological caracteristics
of the leaves (Gates, 1971).
1   : pigment absorption zone (chlorophylls,
      Xanthophyll , carotenoids) : the ab-

sorbed energy is strong especially
in ultraviolet-blue and red bands,the
reflectance and transmittance are weak
    In this part of the spectrum, takes
place the photosynthetic activity for
which the absorbed energy is used.

II   : Multidioptric reflectance zone : the
       reflectance is high while the absorp-
       tance remains weak; all the unabsor-
       bed energy (30% to 70% according to
       the kind of plant (see fig. 7) is
       transmitted. The reflectance is essen-
       tially due to the internal structure
       of the leaf which the radiation pene-
       tratesand is of physical more than
       chemical nature (Fresnel's random
       reflection). Apart from a contribution
       of the waxy cuticle, the magnitude of
       the reflectance depends primarly upon
       the amount of spongy mesophyll(LARS,
       1968).

III : "Hydric zone". The amount of water
      inside the leaf brings on the pattern
      of spectral reflectance with water
      specific absorption bands at :
      . 2660 nanometers (fundamental vibra-
        tion of water molecules);
      . 145 and 1950nm (the more intense
        bands);
      . 1900, 1400, 1100 and 900 nm (combi-
        nation bands, progressively weaker)

      Liquid water in a leaf is largely the
      cause of the strong absorption through-
      out much of the infrared.
IV   : Beyong 2500nm, the reflectance becomes
       less than 5%. Gates and Tantraporn
       (1955)have studied this spectral region
       which, due to atmospheric absorption,
       is allowed to remote sensing data ac-
       quisition from 3μm upwards only, whe-
       re the plant becomes a quasi blackbody.

*More detailed information will be found in
books on vegetation physiology (see f.i.
French C.S. 1960) relating to the spectral
properties of the various pigments and tissues along with the fluorescent properties
of these pigments and the contribution of
this phenomenon to the overall radiance.*

Rem. The experienced absorption bands are
broader than the accountable absorbents
would have let anticipate. This spectral
broadening is due to the internal diffusion
of the light inside the tissues the structural dimensions of which are close to the
wav, length magnitude thus giving rise mainly to the Mie diffusion and secp,darily to
Rayleigh diffusion.

61

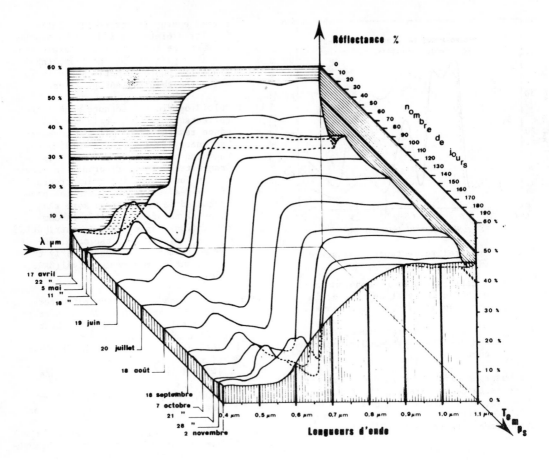

Fig. 9

Spectral reflectance vs time of Oak leaves (Quercus alba) from april to november
(After Gates 1971)

Moreover, depending upon the surface state of the organ : smooth, dull, glabrous
or pubescent (Gausman H.W., Cardenas R.
1968), the reflectance is more or less
Lambertian (Breece H.T. 1969).
- There are numerous factors, either internal to the plant or external, coming from
the environmental conditions, which have
an influence on the specific spectral reflectance.
Internal factors : Besides the specific
morphological factors of the considered
species, the spectral reflectance varies
according to :
- the age of the organ, in relation to
the time evolution of the pigments and
structures (See fig.9). A detailed discussion about the pigmentary and structural causes of this kind of spectral
reflectance variation will be found in

Gausman and Al (1969) mainly concerning
the cotton leaves.
- the location of the organ in the plant
- at the top or the base - especially
for the trees;
- the state of turgescence (See fig. 10)
which itself is in relation with external
factors as mentioned further in the next
paragraph.

. External factors : the influence of these factors on the spectral reflectance
is due to the alteration they bring the
plant proper characteristics about water
content and turgescence, mesophyll structure, evapotranspiration, pigmentation,
metabolism). These external factors are
connected :

- at ground level

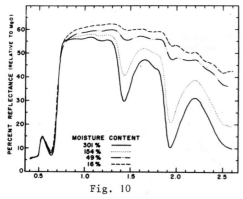

Fig. 10

Spectral reflectance of sycomore leaves in various state of moisture content. (After D.S. LOWE)

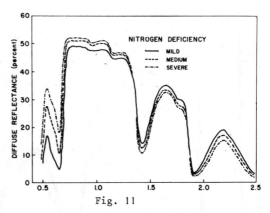

Fig. 11

Impact of a nitrogen deficiency on the spectral reflectance of peartree leaves.

to the water availability for the plant (lack of which induces a moisture stress);

to trophic mineral ions availability (otherwise specific deficiency diseases occur) with specific evidence for nitrogen (See fig.11), iron (chlorosis), potassium, phosphorus, calcium or magnesium

toxic mineral salts - effect of water salinity (Ward 1969 and Thomas 1966).

- from atmosphere :

climatic factors (winds, air moisture content, temperature, sunshine conditions) on which depend the CO_2 acceptance and the evapotranspiration;

seasonal variations (See above : plant aging);

toxic pollutants (especially fluor, SO_2)

mineral deposits (dusts).

- by biological pathogenic agents :

parasites (See fig. 12);

predators;

- to irradiance incident angle (solar h height) from which a diurnal variation of the reflectance results (Salmonson V. V. 1971). This last aspect will be mentioned further on the subject of canopy reflectance.

2.2 Reflectance of vegetation canopies

1. New variables : Despite the fact that the spectral reflectance as measured on separate organs by means of a spectrophotometer built up a fundamental and very useful information, they do not allow a straightforward interpretation of the information obtained by remote sensing on a scene.

Reflectance measurements taken on single plant leaves or branches do not permit reliable prediction of spectral reflectance or photographic tones.
New parameters are needed to undertake the spectral reflectance of a canopy taking into account those of separate organs, then of the plant itself.
These new parameters now describe the characteristics properties of the whole of which only elements are known, they also appear to be sensible to new phenomenons and the resulting reflectance actually observed varies according to :
- how the vegetation covers the soil, thus giving rise to a composit reflectance in between these specific of the soil and the vegetation (See further § 3). The ratio describing the actual rate of cover may be related to the "leaf area index (LAI)",works by Allen 1965 have shown that for some plants, the maize f.i., the LAI is in correlation with the height of the plant.
- the bearing of the plant : the orientations taken by the leaves are not at random but specific of the plant;
- the structure of the canopy :
. The incoming radiation penetrates inside the vegetation cover by multiple

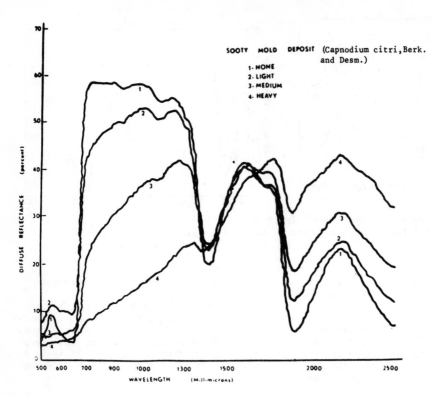

SOOTY MOLD DEPOSIT (Capnodium citri,Berk.
and Desm.)
1. NONE
2. LIGHT
3. MEDIUM
4. HEAVY

Fig. 12

Fungus attack (capnodium citri) on leaves of citrus
(After Hart and Myers 1968)

reflections and transmissions, acting
more specifically on the infrared part
of the spectrum, as stated before (See
above § 1-6).
The resulting canopy reflectance is
consequently less than the one featured
by single leaves and rarely exceeds 30%.
The rate of diffusion increases with
the leaves irregularities and the sun
elevation; Bonner (1962) found an exponential drop of the radiation vs the
thickness of the cover (Lambert's law
concerning the attenuation of a beam).
the various attacks the vegetation undergo do not evidence the same way :
some, such as the moisture stress or
the corn blight touch primarily the
foot so that the observed reflectance
alteration arises later than for others
which first touch the top (f.i. cochi-
neal attack on pines (Pinus maritima).
- the irradiance declination - therefore
the time and the season -. Generally,
the more vertical to the canopy the
more intensively the radiation penetrates the cover. A strong angular depen-

dance features the reflectance as a function of wavelength (Allen W.A., Richardson A.J. 1968) and according to the structure of the canopy, thus to its phenological state. In order to set up the canopy reflectance from the spectral properties of single organs, reflectance models
have to be built up.

2. Reflectance models for vegetation cover
(See A.S.P. especially p. 99-107 and PLES-
SEY RADAR 1972 vol. 2).

To date, 6 main models can be listed which
have been developped to describe the interaction of the EMR with a typical canopy
taking into account the properties of its
constitutive elements (mainly leaves),as
well as its structure (stochastic arrangement of the elements or stacking them
up...)
They all have the Kubelka-Munk model
(1931) as a common basis.

- Kubelka-Munk model (K.M. model) : this
model derives the diffuse reflectance

of stacked leaves or of a vegetation cover from the reflectance and transmittance of a single leave. It is a two parameters, one dimensional representation of the diffuse radiation interaction with the canopy. The basic hypothesis involved are :
. The canopy have infinite lateral extent to avoid edge effects;
. The reflection quickly becomes diffuse within the canopy;
. The leaves are uniformly distributed;
. The dimensions of the leaves are much smaller than the height of the canopy. The derivation of the equations will be found in PLESSEY RADAR (1972) vol. 2, appendix 3.2, p. 245-247.

Although these conditions never apply rigorously it is noteworthy than :

The KM relations apply accurately to the case of typical leaves stacked on the window of a spectrophotometer.

The KM theory has a good likelyhood of representing the irradiance within a plant canopy illuminated by an overcast cloudy sky. However discrepancies occur when the specular reflectance increases inthe cover as in a corn canopy (Allen and Brown 1965).

According to various hypothesis, approximations and edge conditions, some authors have worked out special solutions for this model (See A.S.P., 1975) :
- Solution with two parameters - namely the absorption and the scattering coefficients known as the Allen-Richardson Model (Allen Richardson, 1977);
- Solution with three parameters by Silberstein (1937);
- Solution with four parameters of Ryde (1931);
- Others ought to be mentionned here such as these of Duntley (1942) which introduces eight parameters and includes the specular reflection; Duncan and Al (1967); Anderson and Dennead (1969); Allen and Al.(1970) which generalizes the Duntley's equations.

When experienced, these last models lead to results satisfactory often within a few % of standard deviation.

The main significance of these models lies in their ability to provide informations concerning the vigour and the maturity of the vegetation as derived from the actual reflectance of the canopy.

3. Other models.

Others models have been developped which introduce the very structure of the vegeta-ted cover, in order to correlate the reflectance to new parameters considered as outputs of the model :
- Evapotranspiration models, including the advection (Perrier 1976).
- Growth models which as input call on reflectance transitions due to specific phenological change. These models lead to the evaluation of the biomass, even of the crop production of the vegetated cover Kanemasu (1977), Malet (1977).

3 COMPOSITE REFLECTANCE

It is worthy of note that in natural conditions, the elementary area on ground as resolved by the IFOV of a scanner f.i. (the pixel in the image) is most of the time, composed of mixed elements (vegetation + underlying soil). This leads to spectral reflectance curves without the clear maximum and minimum values otherwise typical for vegetation. This mixing occurs when the density of the plant or of the cover as a whole is too low, or at the transition between two crops...a.s.o.) The resulting reflectance of the heterogeneous area appears as being in between the typical reflectances of the constitutive elements, it shows a "composite reflectance" $\rho(\lambda)$ which can be derived from the proper reflectances $\rho_\alpha(\lambda)$ of each of the elements α. Thus : the reflectance of a soil area A which is partially covered with vegetation may be expressed as a composite reflectance ρ

$$\rho = \frac{A_v \rho_v + A_s \rho_s}{A} \; ; \; \text{with } A = A_v + A_s$$

ρ_v and ρ_s being the specific reflectances of the vegetation and the soil respectively.
. In the visible usually $\rho_s > \rho_v$;

. In the near infrared conversely : $\rho_s < \rho_v$

This explains the inversions of reflectance spoted II in fig. 3.
This explains also why the reflectance of a crop changes rapidly (we spoke above of "reflectance transitions"), when flowering, although the flower elements themselves are not resolved individually.
A composite reflectance is all the more difficult to interpret since the spatial variability and the area resolved are both greater.

4 THE REFLECTANCE AS STATE VARIABLE

We have seen how the spectral reflectance

$\rho(\lambda)$ depends - at a given instant - on a fair number of variables or parameters x

$$\rho(\lambda) = \rho_{\lambda_1}(x_i) \qquad i = 1, 2, \ldots, n$$

$$+ \rho_{\lambda_2}(x_i)$$

$$+ \ldots\ldots$$

$$+ \rho_{\lambda n}(x_i)$$

$\lambda_1, \lambda_2, \ldots \lambda_n$ being the spectral bands of the n channels of a multispectral scanner f.i.

In a spectral space, $\rho(\lambda)$ may be considered as the resulting vector of the various components $\rho_{\lambda i}$. The interaction, at the same time, of the various variables involved in a phenomenon gives rise to a reflectance resultant $\rho(\lambda)$ which may be considered as a STATE FUNCTION if there is a one to one correspondence between x and ρ for all the n values of i (i = 1, 2, ...,). Then, the ρ values may be used as state parameters.

Mathematics provides a DEFINITION for the differential of such a function ρ

$$d\rho = \left(\frac{\partial \rho}{\partial x_i}\right)_{x_j} dx_i$$

$$= K_i \qquad dx_i$$

where the K_i are function of the x_i and the subscript x_i on the partial derivatives ( ) indicates that all x_j are held constant except the one in the derivative considered.

It follows from $d\rho = \Sigma(K_i dx_i)$ being an exact differential expression that, for a finite variation between two states A and B, the values of the integral

$$\int_A^B d\rho = \rho_B - \rho_A = \rho(^B x_i) - \rho(^A x_i)$$

is independant of how the reflection has changed between the two states A and B, $^B x_i$ and $^A x_i$ being the values of the ith parameter in both states A and B. If ρ is a state function we know that, for i : 1, 2, 3 f.i. :

$$\frac{\partial K_1}{\partial x_2} = \frac{\partial K_2}{\partial x_1} \; ; \quad \frac{\partial K_2}{\partial x_3} = \frac{\partial K_3}{\partial x_2} \; ; \; \ldots a.s.o.$$

vice versa if this condition is fulfilled, then ρ is a state function.

Thus $\oint d\rho = 0$

Later on, the reflectance has changed

$$\rho'(\lambda) = \rho'_{\lambda_1}(x'_i)$$

$$+ \rho'_{\lambda_2}(x'_i)$$

$$+ \text{----------}$$

$$+ \rho'_{\lambda n}(x'_i)$$

Some of the x parameters may have changed whereas others remain constant. This would imply for exactness that as many independant measurements are needed as there are x different parameters, to have the reflectance equation solved, thus allowing an "unsupervised" analysis of the datas to be performed. In fact, all these variables are unequally significant and when their respective weight is known a first approximation can be made acceptable, by only retaining the most significant ones.

It follows that :

- The UNSUPERVISED analysis requires that "STATE VARIABLES" be found which can be combination of single physical characteristics ;

- The number of independant equations has to be increased in relation to the number of variables involved in a phenomenon
  . by increasing the number of spectral bands (channels), hence the multispectral analysis; it must be born in mind howere that the "information content of the signal itself being limited (Kondratyev K.Ya. and Al. 1973) the number of significant spectral bands is also limited,
  . by comparing datas taken at two or more dates, because many types of features exhibit unique changes with the passage of time, hence the temporal or multidate analysis.

Obviously, practical considerations always restrict the theoretical demand so that Remote Sensing, most of the time, seeks for compromise solutions, one of them, to date very common, consists in resorting to groundtruth (SUPERVISED ANALYSIS).

TO CONCLUDE

It can be infered from the various phenomenons outlined how a remote sensing mission has to be planed for a definite finality :

- Choise of spectral bands (pitch value in the spectrum) ;
- Choice of spectral bandwidth, according to its information content - in the sense of the SHANNON'S information theory - in view of the target characteristics in each of the spectral bands ;
- Optimal irradiance conditions (day of the year and time);
- Sensitive periods in the vegetative cycle of the plants.

Otherwise, when the whole system itself is right away given - as it is the case with present satellites which, in fact, are a real compromise - it must be born in mind how the data have been acquired and consequently what precautions are to be taken in processing them, in order to bring out the significant and useful information.

BIBLIOGRAPHY

AGAZZI A., 1977, Utilisation de la variabilité spatiale de la réflectance pour déterminer les stades fiénologiques du riz, Tel Aviv, Cospar, 20th plenary meeting

ALLEN W.A. , RICHARSDON A.J. , 1968, Interaction of light with a plant canopy, Pr. 5th Int. Symp. Remote Sensing of Environment, Ann Arbor U.S.A. p. 219-232

A.S.P., 1975, Manual of remote sensing, Falls Church Va, 2 vol 2144/XXVIp. *This book collects more than 4800 bibliographical references which constitute an almost complete list, in 1974, concerning all the aspects of Remote Sensing.*

BREECE H.T., 1969, Bidirectional scattering characteristics of health green soybean and cirn leaves in vivo, La Fayette, Purdue Univy. U.S.A.

BUZNIKOV A.A. , KONDRATYEV K.Y.A , SMOTKY O.I., 1975, Reflectivity of the earth's surface near the interface of two uniform areas, Univ. Tennessee, Pr. Remote Sensing earth Res. IV, p. 161-175

CHIAPALE J.P. , 1975, Theory and eseperiment in microclimate modification by regional seal and land roughness changes, San Francisco, Summer Computer Simulation

FRENCH C.S. , 1960, Encyclopedia of plant physiology, Ruhland Ed. Springer p. 252-297 *particularly*

GATES D.M. , TAMTRAPORN W. , 1955, The reflectivity of deciduous trees and herbageous plant in the infrared to 25µm, science 115, p. 613-616

GATES D.M., 1971, Physical and physiological properties of palnts, Univ. of Michigan, in : Compilation of Univ. of Michigan

GAUSMAN H.W. , CARDENAS R? , 1968, Effect of pubescence on reflectance of light, in : pr. 5th Int. Symp. Remote sensing of Environment, Ann Arbor U.S.A. , p. 291-297

GAUSMAN H.W. , ALLEN W.A. , CARDENAS R., 1969, Relations of light reflectance to cotton leaf maturity, in : Pr. 6th Int. Symp. Remote Sensing of Environment, Ann Arbor, Michigan. U.S.A., p. 1123-1141

GIRARD C.M., 1976, Etudes multispectrales à faible altitude en Agronomie, G.D.T.A. Toulouse, 1, p. 59-65

KANEMASU E.T. , RASMUSSEN V. P. , 1977, Using Landsat data to estimate evapotranspiration and yield of winter wheat, Tel Aviv, Cospar, 20th Pl. Meet.

KONDRATYEV K.YA. , VASILYEV O.B. , IVANYAN G.A., 1973, On the optimum choise of spectral intervals for remote sensing of environment from space, In : Pr. Remote sensing of environment from space, In : Pr. Remote sensing Earth Resources, 2, Univ. Tennessee, U.S.A., p.417-434

L.A.R.S. , 1968, Report, 3, Res. Bull. n° 844

MALET Ph., 1977, Amélioration possible des modèles statiques et dynamiques de prévision de récolte par les données de Télédétection, Cospar 20th Pl. Meet.

N.A.S., 1970, Remote sensing with special reference to Agriculture and Forestry, _ashington DC U.S.A., 424 + XIIIp.

NASA, 1971, NASA Earth resources spectral information system, compilation, Univ. of Michigan U.S.A.

PACE W.H. , DETCHEMENDY D.M. ,1973, A fast algorithm for the decomposing of multispectral data into mixtures, Univ. Tennessee U.S.A. , in : Pr. Remote sensing earth Resources IIp. 831-848

PLESSY RADAR, 1972, Multispectral scanning systems and their potential application to earth resource surveys, ESRO Contract n° 1673/72 EL. 7 vol. *particulary vol 2*

SALMONSON V.V. , 1971, Airborne measurements
of reflected solar radiation, Remote sensing of Environment, 2, n° 1, p. 1-8

SUITS G.M. , 1976, The cause of azimuthal
variations in directional reflectance of
vegetative canopies, Remote sensing of
Environment, 2, p. 175-182.

C.DE CAROLIS & P.AMODEO
Istituto di Patologia Vegetale, Università degli studi, Milano, Italy

Basic problems in the reflectance and emittance properties of vegetation

1 The reflection of sun radiation by leaves is - as it is known - relatively low in the visible portion of the electromagnetic spectrum (o.4 - 0.7 micron).This fact is usually ascribed to a high absorptance of radiation in this region by leaf pigments, particularly chlorophylls.

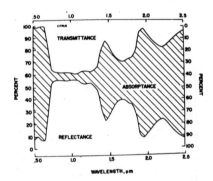

Fig. 1

In the infrared, reflectance increases very quickly to about 0.8 micron, then it remains high to about I.3 micron. This high reflectance seems to be due to the fact that in this region radiation absorptance by the pigments is very low. Moreover, the internal structure of the leaf blade seems to acquire great importance for reflectance.

In the spectral region from I.3 to 2.3µ (near thermal infrared), the internal leaf structure appears to be less important than water content in the tissues. In fact the rather linear behaviour of the curve in this region is interrupted

by peaks of lower reflectance at about 1.45, 1.95, and 2.6µ , corresponding to the main water absorpyance bands.

In the far thermal infrared (from 2.6 to 25µ) leaves show a high absorptance percentage of radiation again and about 75% of the total absorpted energy is emitted, and about 25% of it is dissipated by convection and transpiration. In this range, reflectance is low, ranging (according to the wavelength, the plant species, etc.) from 0 to about I5%.

2. The considerations exposed so far are true only for a normal, mature and healthy leaf, while sensible variations in the spectral response of leaves can occur if we change the previously indicated conditions. Moreover, it must be remembered that there are deep differences between the spectral response of a single plant and that of canopies, differences due to several factors which we will not consider here.

The factors which affect the spectral response of the leaves are leaf structure (in its broadest meaning), maturity, pigmentation, sun exposition, phyllotaxis, pubescence, turgidity (water content), nutritional status, diseases, etc. On the base of the considerations precedingly done, among these factors the most important seem to be pigmentation, nutritional status, internal structure of leaves and water content.

2.1 For what concerns the influence of pigments on the spectral response of vegetation, it has been seen that leaf pigments, including chlorophylls, carotenoids (carotens and xanthophylls) and antocyanins, can markedly affect light absorptance and hence reflectance of plant leaves.

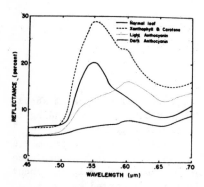

Fig. 2

KLESHNIN and SHUL'GIN (1959) found that the chlorophyll of green leaves usually absorps 70 to 90 % of the light in the wavelength region around 0.450ɲ (blue) and around 0.675ɲ (red). In the region around 0.55ɲ (green) a reflection peak of about 20% occurs. Hence, absorption is smallest.

Reflectance depends also upon the pigment content: low content results in higher reflectance and vice-versa. Moreover, different leaf pigments show different spectral response, according to each pigmentation condition as seen for Coleus leaves (HOFFER and JOHANNSEN, 1969),(see fig. 3).

The same experience was carried out by the same authors in 1969, on leaves of tuliptree, a normal deep green one and a bright yellow one, due to the normal autumn breakdown of the chlorophylls, thus letting the presence of the carotens and xanthophylls become evident. In fact, the yellow leaf showed sharp increase in reflectance starting at 0.50ɲ.

In the case of variegated leaves of Philodendron, the part lacking pigments reflects the visible light much more because of absorptance decrease (WOLLEY,

1971). On the contrary, a decrease of reflectance is observed in the infrared because of the lower thickness of the white part of the leaf compared to the green one, (see fig.4).

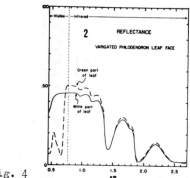

Fig. 3

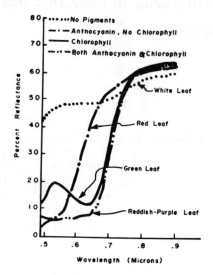

Fig. 4

MOSS and LOOMIS (1952) investigate on the relationship between leaf color and reflection spectra over the 0.4 to 0.7ɲ wavelength interval. It is interesting that yellow and orange leaves show greater reflection in the green than does the green leaf. However, the high reflection throughout the yellow and red accounts for their characteristic colours,(see fig. 5).

2.2 Another factor which greatly affects

70

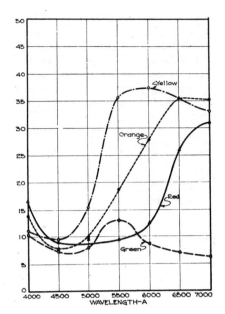

Fig. 5

the spectral response of vegetation is the nutritional status of plants.

THOMAS and OERTHER (1972) study the relation between nitrogen content in leaves of sweet pepper (Capsicum annuum L.) and reflectance, in the wavelength interval from 0.5 to 2.5µ.

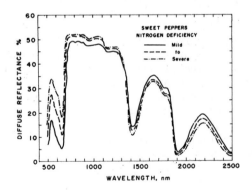

Fig.6

As nitrogen deficiency increases, reflectance increases from 0.5 to 0.7µ, and does the same from 0.7 to 1.3µ, but decreases from 1.3 to 2.5µ.

The increase of reflectance from 0.5

to 0.7µ is due to the fact that light absorptance in this spectral region is greatly affected by pigments concentration, which depends on the nitrogen concentration. By lowering the nitrogen content, a decrease in the chlorophylls and consequently a reduced absorptance of radiations (hence an increase of reflectance) are expected.

Reflectance increase from 0.7 to 1.3µ might probably be related to an increase in the intercellular spaces (GAUSMAN and others, 1969). Reflectance decrease from 1.3 to 2.5µ might be directly related to a greater water content of leaves (THOMAS and others, 1966).

In order to study the effect of nutrient deficiencies on the spectral response, AL-ABBAS and others (1974) measured reflectance, transmittance and absorptance spectra of 'normal' and six types of nutrient-deficient maize leaves, at 30 selected wavelengths from 0.5 to 2.6µ .

Chlorophyll concentration of leaves in all nutrient-deficiency treatments was lower than that of leaves in the control, and consequently absorptance was lower in the range from 0.53 to 0.75µ .

N-deficient maize had the least amount of chlorophyll and was followed in order of increasing chlorophyll content by S, Mg, K, Ca, P and normal maize plants.

In the infrared region from 0.75 to 1.3µ, reflectance and transmittance of the leaf is generally associated with leaf structure and morphology. Among all treatments, K-deficient leaves had the highest reflectance and the lowest leaf thickness and leaf moisture content.

The spectral reflectance and transmittance in the wavelength interval between 1.3 and 2.6µ is related mainly to leaf water content. This experience showed a positive correlation between water content and percent absorptance at the 1.45 and 1.95µ. In fact, S, Mg, and N-deficient, which have a higher percent moisture, showed an increase of absorptance at these wavelengths.

The amount and spectral distribution of energy that a leaf radiates depends upon its temperature. Therefore, a thermoscope was used to obtain thermograms of normal,

N-deficient and S-deficient leaves. Both
S and N-deficient leaves were warmer by
0.4 and 0.9°C respectively than normal
leaves.

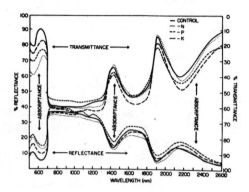

Fig. 7

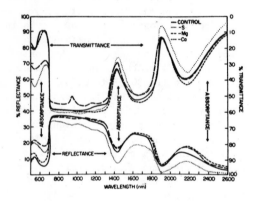

Fig. 8

Differences in the light reflection,
transmission and absorption rates of
leaves were observed among rice strains
in relation to their chlorophyll and nitrogen content during an experience carried
out by TAKANO and TSUNODA (1970). In the
visible region, both reflection and transmission rates were negatively related to
the chlorophyll content. Also nitrogen
content affected reflection negatively
but the close relation between spectral
response and nitrogen content seemed to
be of an indirect nature as there was a
close association of the nitrogen content
with chlorophyll content.

2.3 The problem of the influence of leaf
structure on light reflectance is very
important, because it involves many factors (for example, the type of plant
cell walls; intercellular spaces; epider-
mal, palisade and mesophyll cells etc.).
Excellent reviews on this argument are
available.

In this field, an example is the fact
that the vacuum infiltration (water) of
citrus leaves markedly reduces reflectance
over the 1. 35 to 2.5μ wavelength interval, by eliminating the role og hydrated
cell wall-air interfaces.

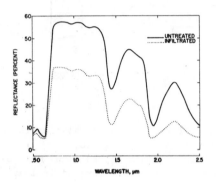

Fig. 9

2.4 As for the leaf water content, the
problem will be considered in the section
concerning thermal infrared.

3 We now give a short description of the
various types of diseases, trying to point
out the possibilities of spectral response
variation for each group of diseases. It
is only an attempt, as literature, as to
this subject, is very poor.

3.1 -Trophic diseases- (Grape powdery
moldew, wheat rusts, grape downy mildew,
cereal smuts). The chlorophyllian or
assimilation function is primarly affected.
The organs attacked by the pathogens
show a mould layer formed by the pathogen's
vegetative reproduction structures.
Besides changes in the chlorophyllian
function, resulting in a decrease of
chlorophyll content, the pathological
event also leads to other changes concerning the whole plant, hence not limited
to the areas where the pathogen is present.

72

In particular, an increase in the respiration and transpiration rates is observed, with evident consequences on the water content. Moreover, a greater nutrient demand resulting in an increase of dry weight, hence in the modification of the leaf structure is shown.

From what mentioned above, all parameters previously indicated as basic in the determination of the vegetation spectral response, that is pigment content, leaf structure and water content, show to be directly affected by this group of diseases. Thus, variations in the spectral response in the three wavelength intervals respectively interested are to be expected, and, in particular, a greater reflectance from 0.4 to 0.7 μ, due to a decrease of pigment content.

The spectral response in the far IR is expected to change too, due to the higher metabolism in the affected areas, which influences leaf temperature.

Variations in the leaf inner structure should not be so important in the case of powdery mildew, as the pathogen lives on the leaf surface, affecting the sole epidermis. In this case, a further modification may occur in the visible region, due to the presence of the mould formed by the pathogen on the leaf.

3.2 -Auxonic diseases- The growth capability of the plant, or of a part of it, is mainly affected. The agents which can cause this type of diseases are several and various: nutritional stress, chemical agents (for instance herbicides), bacteria, fungi and - of great interest- viruses. Examples are: peach leaf curl, rice yellowing, grape fan-leaf, corn smut, crown gall.

Like the pathogenous agents, also the symptomatology of this kind of diseases presents eterogeneous characteristics. Hence it does not seem possible to give general indications on the expected spectral response.

However, common characteristic is a decrease in chlorophyll content, thus a variation of the response in the visible region is expected, due also to the change in the morphology of the plants.

3.3 - Necrotic diseases - (rice blast, tobacco wild fire, southern corn leaf blight, pear scab).

Pathogens attack all the living tissues of the plant, causing the death of their cells and feeding themselves on them, after altering them.

The symptomatology is characterized by necrotic spots on leaves and fruits and by cankers on the lignificated organs.

The diagnosis is sometimes difficult as also alterations caused by chemical compounds and particularly by gaseous pollutants and by urban and industrial waste may cause similar effects and hence are to be placed within this pathological section.

The pathogen's presence results in an alteration of pigment content, leaf structure and water balance.

On this ground, variations in the leaf spectral response in the whole wavelength interval here considered, that is from 0.4 to 2.6 μ, can be expected.

An experience has been carried out by Safir and others (1972) on the reflectance of healthy and deseased corn leaves. Leaves inoculated with Helmintosporium maidis,(causal agent of southern corn leaf blight),were observed in the wavelength interval from 0.4 to 2.6 μ.

Reflectance of infected areas shows to be always higher than that of healthy areas and differences are more evident in the absorptance bands of chlorophylls (0.5 - 0.7 μ) and water (1.45 and 1.95 μ).

These differences have been related to a decrease of chlorophyll and water in the diseased tissue, (see fig. 10).

3.4 - Vascular diseases - These diseases are characterized by the location of the pathogens along the vascular elements, thus interfering with the plant water supplying.

Examples are tomato wilt, "mal secco" of citrus and elm dutch disease.

The symptoms consist of various levels of chlorosis and expecially partial or total wilt.

Agents of this group of diseases are mainly fungi and bacteria.

Obviously, as the importance of water stress becomes predominant, in vascular

73

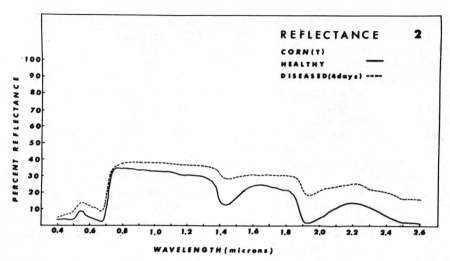

REFLECTANCE 2

CORN(T)

HEALTHY ———

DISEASED(4days) ----

Fig. 10

diseases the change in the spectral response interests mainly the regions of
the near and particularly thermal infrared
that is from 1.3 to 2.5 μ.

The variation will be particularly evident at the three peaks of highest water
absorptance, that is at 1.45, 1.95, 2.6 μ.

However, as a water stress always leads
to a modification of the leaf inner structure, the spectral response will probably
change also in the wavelength interval
from 0.7 to 1.3 μ, that is mainly affected
by this parameter..

Such symptoms as chlorosis, moreover,
result in a lower pigment content. Hence,
the spectral response in the wavelength
interval from 0.5 to 0.7 μ is expected
to change too.

3.5 – Lytic diseases – The tissue structure which keeps cells together and the
lignificated parts are here interested,
(for instance, wheat foot-rot, grape grey
mould).

The main characteristic of this group
of diseases, from the symptomatological
point of view, is the formation of "rots",
as an effect of the tissue disgregation.

The water balance of the plant and the
leaf inner structure are mainly interested
and, consequently, also the pigment content of the affected areas is modified.

Hence a variation of the spectral response in the whole wavelength interval
considered, that is from 0.5 to 2.5 μ,

should occur. In particular, a higher
reflectance in the visible portion of
the spectrum is expected, due to a lower
pigment content.

3.6 – Epiphytic diseases – Here, the alterations regard functional processes and
particularly photosynthesis. Besides
typical diseases, caused by the so called
epyphytic plants which utilize the host
as a support, such as Ivy, by fungi
(sooty moulds) or by lichens, also alterations caused by dusts coming from industrial plants are to be considered in
this group.

The plant suffers from an indirect
damage caused by ivi's vegetation and the
mould's thick mycelium which form a
shady screen on the assimilation organs,
that interferes with the photosynthetic
activity and gas exchange.

As to the spectral response of a diseased plant, variations should occur
in the wavelength interval from 0.5 to
0.7 μ, that is in the visible portion of
the electromagnetic spectrum, due to the
presence of extraneous substances or vegetation on the leaf's surface, which
alter the leaf's normal reflection.

In particualr, by means of the IR
tecnique, it is possible, for example,
to detect the presence of ivy vegetation
in a coniferous wood as the spectral
response of deciduous trees differs very
much from that of coniferous trees.

74

4 One of the problems that recently have been of greatest interest to the researchers, is that of verifying if the identification of diseased plants, before the symptoms become evident, is possible.

Considering that reduced leaf transpiration is usually accompanied by increased leaf temperature, great attention has been addressed to the measurements carried out in the thermal infrared.

HEINNEKEN and ALGER (1977, reported by OLSON, 1977) used an AGA Thermovision 680 system to determine apparent temperature of tree crowns of severely stressed loblolly pines. Stressed trees were as much as 2°C warmer than control trees, even though the crowns of the stressed trees showed no visible symptoms of stress.

Remote sensing temperature measurements of healthy and young tree decline-affected citrus trees were made with a Barnes PRT 10 infrared thermometer to test feasibility of making previsual diagnosis of affected trees (EDWARDS and DUCHARME, 1974). The highest temperature difference was 1.3°C, but the temperature of trees in early decline was virtually identical to that of healthy ones.

DE CAROLIS and others (1974) study the radiation behaviour of some areas of rice fields affected by a virus disease. The approach to the research consisted in using both an on-ground thermocamera and an air-borne IR-dual channel scanner facility. Either in the on-ground and from aircraft surveys the diseased rice areas behave like thermally anomalous areas compared to the healthy ones. The thermal contrast corresponds to some °C of blackbody temperature.

Further on, in the attempt of finding out if the diseased rice plants showed an anomalous temperature behaviour before the yellowing of the leaves, DE CAROLIS and LECHI (1976) study experimentally virus-infected rice plants in lysimeters. The first results seem to point out that an anomalous high temperature can be measured in the infected areas few days before any visible symptoms of the disease appears.

5 Another problem of grat interest is the assessment of crop diseases by means of infrared false color aerial films.

Plants may exibit visible symptoms under advanced stress and therefore these symptoms may be recorded by conventional color photography.

However, some conditions exist in crops which are not visible by normal means but may be detected by using wavelengths beyond the visible spectrum, that is in the "photographic infrared".

The value of Photographic infrared (false color photography) for aerial sampling of crop conditions, forests and characteristics of some soils, lies in its ability to reproduce as a "colour" those wavelengths which are not visible to the eye.

Infrared-sensitive colour film is a multi-emulsion sensor which records green and red from the visible portion of the spectrum, in addition to near-infrared.

The three layers that compose the film are the same three colours required for conventional photography: red, green and blue.

A blue image is produced from those portions of the scene which reflect green light; a green image results from red reflection; a red image is indictive of an infrared reflection.

The reason green plants are not reproduced as a blue colour is related to the fact that the level of sensitivity in the infrared response layer is about five times that of the green response layer and, as a result, the green record is overwhelmed.

Soil may be seen as bluish-green, depending on its surface moisture, sun angle and angle of view.

Water, on the other hand, is usually shown as black because it is an efficient absorber of infrared radiation.

It is the variation in tone and colour in the red or magenta image which is assessed to determine plant vigor.

Once it was thought that the variation of infrared reflection from plants was due to chlorophyll brakedown in the mesophyll tissue prior to the appearance of visible symptoms of stress or infection; now, however, studies indicate that the behaviour of infrared reflectance or absorptance is dependent on the optical characteristics of the air-water interface within the cells.

The fact that infrared reflection from plants is reproduced as red is not really important as the process is only a method of photographic image enhancement.

It has been demonstrated that density levels in photographs can be related to known infection levels of disease in experimental plots.

These relationships can be utilized in the development of crop loss assessment methodology for determining disease levels necessary for yield prediction.

As well, work is being done to separate optically the colour response layers contained in false colour films in order to determine which layer provides the most information recorded for a particular disease or crop condition.

Enhancement techniques then allow analysis by computer.

For instance, JACKSON and others (1971) study Potato late blight intensity levels as determined by microdensitometer studies on false colour aerial photographs.

Comparisons of the two sets of data show that there is a linear relationship between microdensitometer readings and field-disease ratings.

JACKSON and WALLEN (1975) study field beans infected with bacterial blight, by means of microdensitometer measurements of sequential aerial photographs.

The objective of this work was to ascertain if optical density measurements of sequential aerial photographs of plots of diseased and healthy field beans made throughout the growing season could be utilized for remote assessment of disease severity or complement field observation methods. Assessments are made by plant pathologists who examine diseased plants and estimate disease severity using growth-stage and disease assessment keys.

Results show that direct measurements of density levels obtained from sequential photographs is not realistic or feasible with the techniques used in early stages of disease onset, but in later stages of the disease, notable optical density differences are measurable.

ALI and AGGARWAL (1977) describe a system designed to analyse aerial colour infrared photographs of citrus orchards. The input of the system is a 35mm transparency of the infrared aerial photograph mounted in a three colour film digitizer which is computer controlled.

During an experience on Cabbage (Brassica oleracea), THOMAS and GERBERMANN (1977) found that nitrogen and water stressed plants showed a significant correlation with dry matter yields, and that it was detectable by means of infrared color photographs as, in the visible spectral region N and water stress decrease the absortive coefficient. The film's optical density was significantly affected by the variation in the amount of biomass produced, thus allowing yield prediction.

6 Several works exist concerning plant water stress, either in the visible and expecially in the near and thermal infrared region. Of these, we report here only a few.

Spectral response of vegetation is greatly affected by water content. THOMAS and others (1971) investigated the effects of plant water stress on reflectance of light from leaves of several important agricultural crops. An increase in reflectance was observed as leaves changed from a fully turgid to a wilted condition. The increase in reflectance in the visible region of the spectrum $(0.4 - 0.7 \mu)$ does not always occur, as, often, the influence of water is masked by the leaf pigment content. In the 0.75 to 1.35 μ near infrared region, the spectral properties of a leaf are associated mainly with its morphology, which changes with the loss of water, resulting in a slight increase of reflectance. The greatest change occurs in the spectral region beyond 1.35 μ, where water is the main factor affecting reflectance; reflectance increases as the water content decreases.

The greatest change in reflectance occurred at 1.45 μ (water absorption band), thus reflectance measurements at 1.45 μ are a better predictor of leaf water status.

Moreover, in cotton the greatest change in reflectance was measured when the relative turgidity was below 70%. Within

76

the relative turgidity range from 70 to 80% reflectance changes are small and may not be predictable because of variations among leaves of field-grown cotton due to age differences.

Another experience concerning the effect of water stress on leaf spectral response was carried out by MILLARD and others (1978). They used an airborne thermal scanner to measure the temperature of a wheat crop canopy. Significant within field canopy temperature variability, due to water stress, was observed. But, what's of greatest interest is the fact that water stress conditions undetected in color IR photography, were clearly detected in thermal imagery.

6.1 As we just mentioned, a problem of great interest is the relationship between leaf water content and its maturity which greatly affects its thickness. Theoretically, a leaf can be regarded as a pile of N compact layers separated by infinitesimal air spaces.

GAUSMAN and others (1970) measured the number and size of intercellular air spaces in cotton leaves differing in age and found that maturity increased reflectance beyond 0.75μ. In fact, reflectance of light for the 0.75 - 1.35μ wavelength interval is positively correlated with leaf maturation until an age is reached when leaf thickness and number of intercellular spaces cease to increase and leaf thickness may actually decrease.

Here, a decrease in reflectance occurs and it may be primatly caused by the presence in older leaves of larger cells with increased water content. In the 0.75 - 1.35μ region, leaf spectral response is more affected by internal leaf structure than by its water content. In the wavelength interval from 1.35 to 2.5μ leaf spectral response is greatly affected by the amount of water in the tissues (water high absorption bands occur at 1.45 and 1.95μ), and to some degree by leaf structure.

Both parameters are affected by leaf maturity. In fact, GAUSMAN and others (1971) showed that young cotton leaves contain less water than mature leaves as immature cells in young leaves are

primarly protoplasmic with little vacuolate water storage. During cell growth, cell water-filled vacuoles develop up to a point when, as maturity goes on, they may coalesce to a central sap cavity and protoplasm covers only the cell wall in a thin layer. Thus a decrease in reflectance, as maturity goes on, due to the increase in water content, is expected, until an age is reached when an increase in reflection should be measured, due to a decrease in water content.

Intensive studies are presently in progress to further relate leaf age to spectrophotometrical results.

An airborne spectroradiometer was used by COLLINS (1978) in order to determine the relationship between maturity and crop spectral response over 10 mm wide spectral bands, centered at 0.745 and 0.785μ. A red spectral shift in the chlorophyll absorption edge of heading wheat and grain sorghum was detected, thus allowing discrimination of crop type, maturity and canopy density during the heading stage. Though, no measurements were made in the thermal infrared portion of the spectrum.

7. Another problem that seems of remarkable interest is the measurement of surface temperatures in healthy and diseased plants, not considered within the canopy any more, but as single plants.

Also the measurement of radiance variations between healthy and infected areas of the same leaf should contribute to the evaluation of their energy balance, and therefore the use of "proximal sensing" techniques at laboratory level could represent a new approach to the study of host-pathogen relationship.

DE CAROLIS and others (1975) study the thermal behaviour of leaves of tobacco plants experimentally inoculated with TMV (Tobacco Mosaic Virus), and of bean plants affected by Uromyces Appendiculatus.

The measurements were made with an AGA Thermovision camera in the 2 - 5.6μ band. In the case of the inoculated tobacco plants, an increase of radiance, paralleling the maturation of the necrotic symptoms, was measured. The leaves with the highest number of lesions appeared more

77

radiant than those inoculated with lower concentration of inocula. However, the completely necrotic single lesions appeared less radiant than the sorrounding green tissue.

In the case of the inoculated bean plants, a strong decrease of radiance occurs as the uredia maturation goes on and the pustule eruption takes place.

This effect is due to the crack of the upper leaf epidermis, which induces a transpiration rise, thus making possible modification of the thermal behaviour caused by other host-pathogen relatioships.

8 REFERENCES

AL-ABBAS A.H., HALL J.D., CRANE F.L., BAUMGARDNER M.F. 1974, Spectra of normal and nutrient-deficient maize leaves, Agron. J., 66:16-20.

ALI M., AGGARWAL J.K. 1977, Automatic interpretation of infrared aerial color photographs of Citrus orchards having infestations of insect pests and diseases, IEEE Trans. Geosci. Electronics, GE-15:170-179.

COLLINS W. 1978, Remote sensing of crop type and maturity, Photogramm. Eng. 44:43-55.

DE CAROLIS C., LECHI G.M. 1976, A new way to detect stress condition of the plants by means of thermal infrared. XVI Conv. sullo spazio, Roma, 18-20 marzo, p.135-138.

DE CAROLIS C., CONTI G.G., LECHI G.M. 1975, Investigations on thermal behaviour of plants affected by virus and fungus diseases (Tobacco Mosaic Virus in Nicotiana Tabacum L.cv. Xanthi nc., and Uromyces Appendiculatus (Pers.) Link in Phaseolus vulagaris L.),Proc. Xth Int. Symp. on Remote Sensing of Environment, Ann Arbor, Michigan, 6-10 Oct., p.1219-1229.

DE CAROLIS C., BALDIG., GALLI DE PARATESI S., LECHI G.M. 1974, Thermal behaviour of some rice fields affected by a yellows-type disease, Proc. IXth Int. Symp. on Remote Sensing of Environment, Ann Arbor, Michigan, 15-19 Apr.,p.1161-1170.

EDWARDS G.J., DUCHARME E.P. 1974, Attempt at previsual diagnosis of Citrus young tree decline by use of a remote sensing infrared thermometer, Plant Dis. Reptr., 58:793-796.

GAUSMAN H.W., ALLEN W.A., MYERS V.I., CARDENAS R. 1969, Reflectance and internal structure of Cotton leaves, Gossypium hirsutum L., Agron. J., 61:374-376.

GAUSMAN H.W., ALLEN W.A., CARDENAS R., RICHARDSON A.J. 1970, Relation of light reflectance to histological and physical evaluation of Cotton leaf maturity, Appl. Opt., 9:545-552.

GAUSMAN H.W., ALLEN W.A., ESCOBAR D.E., RODRIGUEZ R.R., CARDENAS R. 1971, Age effects of Cotton leaves on light reflectance, transmittance and absorptance and on water content and thickness, Agron. J., 63:465-469.

HEIKKENEN H.J., ALGER L.A. 1977, Previsual detection of stressed loblolly pine, Pinus Taeda L., Manuscript submitted to the Journal of forestry.

HOFFER R.M., JOHANNSEN C.J. 1969, Ecological potential in spectral signature analysis, In:Remote sensing in ecology, Univ. of Georgia Press, Athens, Georgia, p.1-16.

JACKSON H.R., WALLEN V.R. 1975, Microdensitometer measurements of sequential aerial photographs of field beans infected with bacterial blight, Phitopathology, 65:961-968.

JACKSON H.R., HODGSON W.A., WALLEN V.R., PHILPOTTS L.E., HUNTER J. 1971, Potato late blight intensity levels as determined by microdensitometer studies of false-color aerial photographs, J. Biol. Phot. Ass., 39:101-106.

KLESHNIN A.F., SHUL'GIN I.A. 1959, The optical properties of plant leaves, Dokl. Akademcii Nauk SSSR, 125:1158.

MILLARD J.P., JACKSON R.D., GOETTELMAN R.C., REGINATO R.J.,IDSO S.B. 1977, Crop water stress assessment using an airborne thermal scanner, Photogramm. Eng., 44:77-85.

MOSS R.A., LOOMIS W.E. 1952, Absorption spectra of leaves. I. The visible spectrum, Plant Physiol., 27:370-377.

OLSON C.E. Jr. 1977, Previsual detection of stress in Pine forests, Proc. XIth

Int. Symp. on remote sensing of environment, Ann Arbor, Michigan, 20-26 Apr.
SAFIR G.R., SVITS G.H., ELLINGBOE A.H.
1972, Spectral reflectance of Corn leaves infected with **Helminthosporium** may-
dis, Phytopathology, 62:1210-1213.
TAKANO Y., TSUNODA S. 1970, Light reflection, transmission and absorption rates
of rice leaves in relation to their
chlorophyll and nitrogen contents, To-
hoku J. Agr. Res., 21:111-117.
THOMAS J.R., OERTHER G.F. 1972, Estimating
nitrogen content of sweet pepper leaves
by reflectance measurements, Agron. J.,
64:11-13.
THOMAS J.R.,MYERS V.I., HEILMAN M.D.,
WIEGAND C.L. 1966, Factors affecting
light reflectance of Cotton, Proc. IVth
Int. Symp. on Remote sensing of environment, Ann Arbor, Michigan, p.305-312.
THOMAS J.R., NAMKEN L.N., OERTHER G.F.,
BROWN R.G. 1971, Estimating leaf water
content by reflectance measurements,
Agron. J., 63:845-847.
THOMAS J.R., GERBERMANN A.H. 1977, Yield/
Reflectance relations in Cabbage, Photo-
gramm. Eng.,43:1257-1266.
WOOLEY J.T. 1971, Reflectance and trans-
mittance of light by leaves, Plant
Physiol., 47:656-662.

ARCHIBALD B. PARK
General Electric Company, Beltsville, Maryland, USA

RONALD E. FRIES & ANDREW A. AARONSON
General Electric Company, Beltsville, Maryland, USA

8

Agricultural information systems for Europe

ABSTRACT

Opportunities for increasing and sustain-
ing agricultural productivity and facil-
itating product flow are identified by
the availability of accurate, comprehensive, and timely information on crop
productivity and on the current and potential use of land. The ability of a
country (or region) to accurately forecast the harvest production of its major
crops in a timely manner, coupled with
the ability to analyze its position with
respect to the current world market enables its agricultural planners to make
more rational, timely, and economically
rewarding export-import decisions. Indeed, the lack of current and accurate
agricultural information can be a major
obstacle to the economic development of
any particular country or region.

This paper describes the characteristics
and capabilities of a satellite data based
agricultural information system applicable
to any or all European countries. It
makes use of Landsat and meteorological
satellite data along with various types
of collateral data. System outputs are
both visual and statistical--presenting
acreage, yield, and production information
by crop and by geographic area.

1 SUMMARY

Earth resources data cost benefit studies
have found that more timely and accurate
crop production statistics can be worth
billions of dollars in benefits (ECON Inc.
and Earth Satellite Corp. with Booz-Allen
Applied Research Corp. 1974). Such agricultural application benefits, coupled
with those afforded by the same data being
interpreted and results applied in such

varied disciplines as land use, water
resources, geology, environment, oceanography, and marine resources, suggest that
the potential benefits of timely receipt
and timely analysis of Earth resources
data obtainable from satellites are indeed large.

Europe, a region in which agriculture
is very important, would be ideally
suited for the use of satellite gathered
Earth resources data for crop production
prediction. Since wheat is a major crop
in the countries which make up the region,
this paper concentrates on the application
of Landsat data to wheat production prediction.

Landsat has been shown to be useful for
crop information gathering, including
acreage, yield, and production statistics.
Landsat data, coupled with various types
of collateral data, have been used very
successfully for several years for acreage
determination and more recently have been
applied to crop condition and yield
assessment. Production is calculated by
the merging of acreage and yield data for
a particular geographic area. Thus, there
is every reason to believe that wheat
production could be predicted for/by the
countries of Europe using Landsat data as
the primary crop production prediction
system input.

An operational crop production prediction system would need to be able to
accurately determine both acreage and
yield because past experience has been
that neither of these production equation
(acreage x yield = production) parameters
has remained constant or been reliably
predictable based on historic data.
Approximately 163 Landsat scenes would be

required to cover the major wheat growing areas in Europe.

A series of three crop production prediction systems has been defined based on the extent of Landsat and other data sources utilized:
1. Landsat Crop Production Prediction System ("L System")
2. Landsat + Collateral Crop Production Prediction System ("L+C System")
3. Landsat + Collateral + Meteorological Crop Production Prediction System ("L+C+M System").

The "L System" is essentially a Landsat data alone system, specifically defined as being limited to inputs of Landsat and non-current crop and crop acreage data only. The system approach involves "wall to wall" classification of wheat vs. non-wheat. Derived wheat acreage is coupled with available yield data to calculate production.

The "L+C System" uses both Landsat and collateral data inputs; collateral data being defined as including field level ground truth (field location, crop ID, acreage, yield, phenology, etc.) and all types of available supplemental data (soils, productivity potential, trend statistics, etc.) with the exception of current growing season meteorological data. The L+C System is sample-based (total land area analysis not required) and involves within system prediction of both area and yield at the local level. These local, homogeneous area acreage and yield values are then aggregated mathematically to determine production.

The "L+C+M System" makes use of three major types of data: Landsat, collateral, and meteorological. Functionally it is very similar to the L+C System, the only major difference being the additional input of meteorological (historic and current growing season) data to the yield prediction activity. As with the L+C System, local acreage and yield statistics are generated and aggregated to determine production.

These three systems are evolutionary and modular in nature; that is, the L System is the baseline system from which to grow into the L+C and L+C+M Systems. As additional sources of input data are added, (that is, evolving from the L to the L+C+M System) crop production prediction accuracies will increase. The L+C+M System represents the present state-of-the-art remote sensing data based crop production predicting system.

2 INTRODUCTION

The entrance of the USSR into the world wheat market in 1975 as an unexpected buyer rather than a seller caused world wheat prices to soar and a general disruption of normal market conditions. Each and every day world demand for food increases by 250,000 mouths; at this rate world demand for food will increase by as much as three times its current level in the next 50 years. These and other factors underscore the need for agricultural information systems that can provide accurate and timely crop production information such that domestic and world agricultural policy planners can better manage the world food market.

At present a central world agricultural information system does not exist, therefore making it difficult for commodities to move within world food markets in a timely and efficient manner. At best, timely and accurate crop production information must be relayed on a country by country or region by region basis in order that world food markets operate at an acceptable level of efficiency. The ability of a country to accurately predict domestic crop production in a timely manner, coupled with the ability to analyze its position with respect to the current world market could enable domestic agricultural planners to direct commodities to the most opportunistic markets available. Accurately inventorying domestic production allows agricultural policy planners to make more rational and timely export-import decisions. In general, improvements in the timeliness and accuracies of crop production data can be translated into substantial economic benefits.

In addition to the application of Landsat data to agriculture, other proven applications of Landsat data are listed in Table 2.1. These discipline applications are: land use and mapping, geology, water resources, oceanography and marine resources, and the environment. The potential benefits from these various applications of Landsat data further support the economic justification for a Landsat data processing facility.

3 AGRICULTURE IN EUROPE

The distribution of the major land use categories at the national level is

Table 2.1 Summary of applications of Landsat data in the various Earth resources disciplines. (Short, Lowman, Freden, Finch 1976)

Agriculture, forestry, and range resources	Land use and mapping	Geology	Water resources	Oceanography and marine resources	Environment
(1) Discrimination of vegetative types: Crop types Timber types Range vegetation	(1) Classification of land uses	(1) Recognition of rock types	(1) Determination of water boundaries and surface water area and volume	(1) Detection of living marine organisms	(1) Monitoring surface mining and reclamation
(2) Measurement of crop acreage by species	(2) Cartographic mapping and map updating	(2) Mapping of major geologic units	(2) Mapping of floods and flood plains	(2) Determination of turbidity patterns and circulation	(2) Mapping and monitoring of water pollution
(3) Measurement of timber acreage and volume by species	(3) Categorization of land capability	(3) Revising geologic maps	(3) Determination of areal extent of snow and snow boundaries	(3) Mapping shore-line changes	(3) Detection of air pollution and its effects
(4) Determination of range readiness and biomass	(4) Separation of urban and rural categories	(4) Delineation of unconsolidated rock and soils	(4) Measurement of glacial features	(4) Mapping of shoals and shallow areas	(4) Determination of effects of natural disasters
(5) Determination of vegetation vigor	(5) Regional planning	(5) Mapping igneous intrusions	(5) Measurement of sediment and turbidity patterns	(5) Mapping of ice for shipping	(5) Monitoring environmental effects of man's activities (lake eutrophication, defoliation, etc.)
(6) Determination of vegetation stress	(6) Mapping of transportation networks	(6) Mapping recent volcanic surface deposits	(6) Determination of water depth	(6) Study of eddies and waves	
(7) Determination of soil conditions	(7) Mapping of land-water boundaries	(7) Mapping land-forms	(7) Delineation of irrigated fields		
(8) Determination of soil associations	(8) Mapping of wetlands	(8) Search for surface guides to mineralization	(8) Inventory of lakes		
(9) Assessment of grass and forest fire damage		(9) Determination of regional structures			
		(10) Mapping linears (fractures)			

presented as Table 3.1. In the design of a system which uses remote sensing from satellites it is necessary to define a strategy for the acquisition of the imagery. One of the more important tasks is to create a map showing the distribution of Landsat scene centers over the agricultural areas of interest. There is an interesting facet to the problem in Europe. In looking at the distribution of cereals in the producing countries it becomes obvious that it is necessary to cover the entire land area of the countries. This is because the farms are quite uniformly distributed over the country in terms of Landsat scenes which cover more than 34,000 km^2. It is true that within the scene there are areas of concentration of cereal production but that information becomes valuable in the next task, that of sampling the Landsat frames for interpretive processing. It is not the purpose of this paper to discuss the design of a sampling system for Europe, rather to present a systems level overview of the elements of an approach to the problem. Tables 3.2, 3.3 and 3.4 are, respectively a summary of production, area and yield of cereals in the producing countries in Europe over the ten year period 1967-1976. These data are all taken from FAO Yearbook tables. If one disregards the accuracy of the data for any specific year or any specific country and merely looks at relationships and trends, some very important conclusions can be drawn. One can answer questions concerning the stability of the area planted to cereals as a function of time. This can be inferred by examining the minimum, maximum and standard deviation of the area term for each of the countries. Similarly one can use the yield data in an examination of climatological history to determine whether or not one can account for the dispersion about the mean solely in terms of climate or whether one must look at technological trends to complete the picture. In most countries both factors are at work and certainly for Europe this is the case. The requirement for collateral data sources to provide insight into the contribution that technology has made to the yield term is typical of the need for an extensive library of agricultural research and cultivation practices. It is perhaps axiomatic to state that no single source of data is able to provide all the answers to the demands of an agricultural information system. On the other hand many people have expectations of the contribution of remote sensing that are very much larger than the proportionate contribution that remote sensing can provide.

Finally one can see the relative importance of the two terms, area and yield, on the trends in production. This type of analysis is done in detail to determine how to employ the satellite. Landsat can provide two kinds of input: (1) identification and measurement, and (2) monitoring to assess vigor. If the analysis shows that the area term is relatively stable and that the largest variable is yield then the monitoring function is stressed. If on the other hand there have been major changes in the area planted to crops it is essential that an acquisition strategy permit the identification and measurement of the cereals. In actual practice both functions are carried out. However, it is very cost effective to be able to avoid processing the data. Whenever this kind of saving can be made it should be made.

4 CROP PRODUCTION ASSESSMENT USING LANDSAT

In the past, the U. S. has used periodic aircraft survey data as one of the inputs into models designed to predict harvest area and yield, and forecast production of major agronomic crops. In general, however, aircraft coverage has been sporadic, insufficiently current, and photographic in nature as well as very costly for gathering repeated data over large areas. As a result, automation and operational application of remotely sensed data to the task of inventorying crops over large geographic areas was limited.

With the advent of a multispectral, repetitive coverage satellite that can image the entire world's land surface numerous times each year, the "dream" of an automated crop inventory capability for any individual country and indeed the entire world is now approaching reality. Information from satellites such as Landsat can be integrated into a sophisticated crop assessment system such as the one outlined generally in Figure 4.1. A detailed diagram of the overview portion of the model is contained in Figure 4.2.

Other satellites, including already operational meteorological satellites and some planned for the future, add critical data on weather conditions such as temperature and moisture to the system. The total number of observations and volume of information produced from such a satellite

Table 3.1 Land use categories

Country	Land Area	Arable Land	Perm. Crops	Perm. Pasture	Forest & Wool	Other Land
Europe	472816	127707	14988	87138	152787	90195
Albania	2740	555	93	590	1195	307
Austria	8271	1511	98	2181	3250	1231
Belgium	3282	846	31	791	702	911
Bulgaria	11055	3957	382	1612	3797	1307
Czechoslovakia	12558	5126	130	1748	4506	1048
Denmark	4237	2656	14	277	4844	806
Finland	30545	2641	---	156	22630	5118
France	54592	17200	1610	13450	14610	7722
German DR	10603	4699	237	1359	2952	1356
German Fed.	24403	7538	521	5244	7162	3938
Greece	13080	2940	950	5255	2615	1320
Hungary	9238	5128	367	1275	1545	923
Ireland	6889	1044	4	3820	216	809
Italy	29405	9330	2983	5204	6306	5582
Netherlands	3381	804	37	1241	308	991
Norway	30810	779	13	106	8330	21582
Poland	30459	14781	303	4125	8608	2642
Portugal	9164	3030	590	530	3641	1373
Romania	23034	9741	759	4446	6316	1772
Spain	49963	15821	5012	11088	14944	3098
Sweden	41148	2970	45	700	26424	11009
Switzerland	3977	367	19	1633	1052	906
United Kingdom	24177	6905	76	11642	2020	3534
Yugoslavia	25540	7321	713	6354	9040	2112

Table 3.2 Production (thousands of metric tons)

Country	Min.	Yr.	Max.	Yr.	$\bar{P}$	σP
Europe	183602	70	235391	74	210237	16718.2
Albania	459	70	764	76	550	88.8
Austria	2934	67	4281	76	3511	418.6
Belgium	1565	70	2225	74	1877	200.7
Bulgaria	5279	68	8148	72	7020	916.7
Czechoslovakia	6627	67	10388	74	8558	1143.1
Denmark	5888	76	7261	74	6609	432.3
Finland	2351	67	4008	76	2958	458.1
France	31477	70	42858	73	36034	4051.1
German DR	6456	70	9703	74	7995	912.5
German Fed.	17291	70	22654	74	19873	1576.1
Greece	2596	68	4054	76	3391	429.3
Hungary	7618	70	12555	75	10204	1703.4
Ireland	1196	73	1580	71	1378	102.6
Italy	15038	68	17152	75	16108	669.3
Netherlands	1073	75	1841	67	1424	220.5
Norway	630	67	1128	74	878	212.5
Poland	16371	70	22977	74	19475	2032.9
Portugal	1458	73	1860	71	1621	141.9
Romania	10632	70	19786	76	14337	2399.5
Spain	10237	70	14204	75	12104	1211.6
Sweden	3982	69	6599	74	5135	643.6
Switzerland	625	70	857	74	742	81.6
United Kingdom	13252	70	16395	74	14416	1054.8
Yugoslavia	11621	70	16224	76	13937	1368.0

Table 3.3 Area (thousands of hectares)

Country	Min.	Yr.	Max.	Yr.	$\bar{A}$	σA
Europe	70695	75	73056	68	71833	812.0
Albania	314	70	366	76	334	15.5
Austria	903	67	1026	76	966	34.6
Belgium	404	75	486	67	457	23.5
Bulgaria	2055	74	2259	75	2172	59.7
Czechoslovakia	2575	67	2792	73	2684	61.8
Denmark	1637	67	1777	72	1727	40.7
Finland	1158	69	1360	76	1239	62.6
France	9303	67	9822	74	9544	189.8
German DR	2282	73	2546	76	2353	93.4
German Fed.	4971	67	5304	72	5209	106.1
Greece	1543	76	1705	68	1604	55.5
Hungary	2995	70	3285	74	3178	84.3
Ireland	336	75	386	71	3573	16.2
Italy	5125	75	5955	68	5508	323.2
Netherlands	240	76	436	67	334	69.3
Norway	228	68	300	75	267	25.3
Poland	7766	76	8737	69	8116	692.2
Portugal	1313	76	1694	68	1490	123.3
Romania	5900	70	6656	68	6264	261.7
Spain	7192	75	7519	74	7350	110.6
Sweden	1395	67	1623	76	1517	59.2
Switzerland	163	76	196	73	179	9.0
United Kingdom	3654	75	3821	67	3748	56.9
Yugoslavia	4720	75	5242	68	4957	191.6

Table 3.4 Yield (kilograms per hectare)

Country	Min.	Yr.	Max.	Yr.	$\bar{Y}$	σY
Europe	2568	70	3302	74	2929	250.9
Albania	1434	67	2087	76	1654	217.0
Austria	3196	70	4172	76	3626	320.3
Belgium	3351	70	5078	74	4120	462.4
Bulgaria	2442	68	3863	76	3234	411.8
Czechoslovakia	2574	67	3822	74	3182	369.5
Denmark	3319	76	4187	74	3827	245.4
Finland	1992	67	2947	76	2378	269.9
France	3352	70	4370	73	3769	355.0
German DR	2823	70	3970	74	3367	342.1
German Fed.	3338	70	4275	74	3812	253.3
Greece	1523	68	2628	76	2125	324.0
Hungary	2497	67	3934	75	3208	515.0
Ireland	3463	75	4172	74	3857	242.0
Italy	2525	68	3347	75	2940	269.4
Netherlands	3798	70	5041	74	4318	389.8
Norway	2526	75	5329	70	3301	788.0
Poland	1931	67	2839	74	2349	295.8
Portugal	923	69	1801	71	1092	93.6
Romania	1802	70	3128	76	2290	381.8
Spain	1383	70	1975	75	1647	165.8
Sweden	2680	69	4261	74	3381	369.1
Switzerland	3531	70	4766	74	4144	360.8
United Kingdom	3575	70	4369	74	3845	263.7
Yugoslavia	2311	68	3435	76	2823	360.6

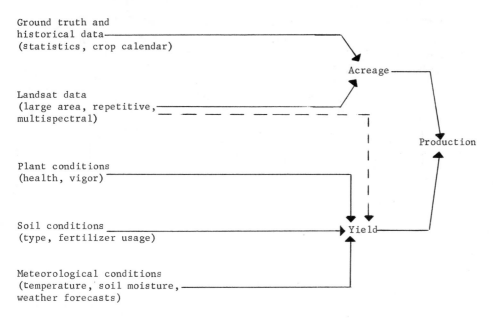

Ground truth and
historical data
(statistics, crop calendar)

Landsat data
(large area, repetitive,
multispectral)

Plant conditions
(health, vigor)

Soil conditions
(type, fertilizer usage)

Meteorological conditions
(temperature, soil moisture,
weather forecasts)

Acreage

Production

Yield

Figure 4.1 General crop production assessment procedure using Landsat.

system and from supplemental data sources are tremendous indeed; however, the digital format in which the satellite data are usually recorded allows computers to process these data directly and thereby reduce the vast quantity of everflowing information to the precise outputs required.

The principles of remote sensing by which crops are recognized and yields assessed from Landsat observations are relatively simple. First, land given to agriculture is generally easy to identify in a Landsat image because it customarily has uniform or regular boundaries giving rise to well-defined shapes such as rectangles, polygons, and circles. Present Landsat satellites can recognize many major crop types with accuracies of 90 percent or higher in fields of 80,000 m^2 (20 acres) or larger and can achieve lesser accuracies in fields as small as 20,000 m (5 acres). Landsat-D, scheduled for launch in 1980, will contain several additional spectral bands and increased spatial resolution, thus providing real potential for significant increase in crop classification accuracy, especially in areas where small fields predominate.

Fallow or unplanted fields are easily distinguished in false color composites by their characteristic blues, tans, or browns--the specific color depends on the type of soil. Growing crops appear in

various shades of pink to red--young plants tend to appear pinkish while vigorously growing full canopy crops are bright red.

However, crops are best identified from computer-processed digital data that represent quantitative measures of radiance (percentage or intensity of reflected light). In general, all leafy vegetation by itself has a similar reflectance spectrum regardless of plant or crop species. The differences among crops, by which they are separated and identified, result from degree of maturity, percent of canopy cover and differences in soil type and soil moisture. In practice, the crops are differentiated in the satellite image or the computer data by using training sets; i.e., the spectral signatures or reflectance responses of known specific crop types are determined for several individual fields in the image and closely similar responses are looked for elsewhere in the image. When data from all four channels of the MSS are used, there are usually enough subtle differences in reflectance from one crop to the next to distinguish them providing that the problem is bounded by a confined farming region where similar crops are grown and similar agricultural practices are followed and where soils are similar. Even within this region the crops may not be separable at one time of the year. They are, however, normally separ-

87

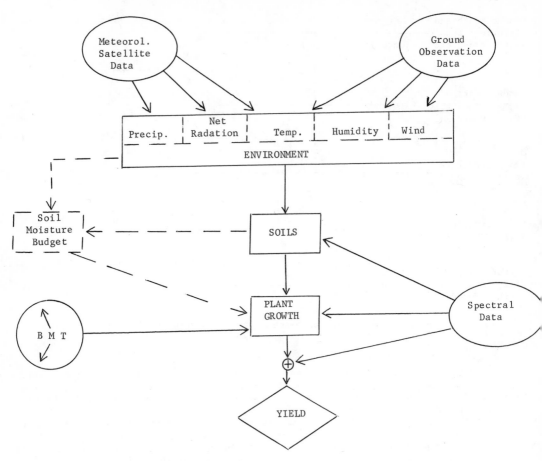

Figure 4.2   Flow diagram for hybrid crop yield model.

able at another time due to the natural differences in planting, maturing, and harvesting dates.

The degree of maturity and the yield for a given crop also influence the reflectance at any stage of growth. This maturity and yield can be assessed as the history of any crop is traced in terms of its changing reflectances. When a crop is diseased or seriously damaged (as by hailstorm), its reflectance decreases, particularly in the infrared, allowing the presence of crop stress to be recognized. Lack of available moisture also stresses a crop, which effect again shows up as a reduction of reflected light intensity in the infrared and usually as a drop in reflectance in the green MSS band (4) and a rise in the red MSS band (5).

For Thematic Mapper (TM) bands on Landsat-D Table 4.1 indicates the expected biological inference by band.

4.1   Acreage

Dr. Morain (1974) used Landsat photographic data to estimate wheat acreage for a ten-county area in southwest Kansas. Comparison with estimates prepared by the Statistical Reporting Service (SRS) of the U. S. Department of Agriculture using conventional, field-level intensive data gathering techniques indicated no significant difference between Landsat and SRS prepared estimates (Table 4.2). Furthermore, the Landsat acreage estimates were prepared in March whereas the data available from SRS represent May and August prepared estimates of harvest acreage.

88

In a study by Dietrich et al. (Fries, Egbert 1975), multiclass identification and acreage determination was performed using Landsat digital data. Two study areas, Williams County, North Dakota and Melfort, Saskatchewan, were examined. Wheat classification results were approximately 99% as tested against complete field by field ground truth data for the 3 x 13 km study areas (Tables 4.3 and 4.4).

In an investigation involving a larger study area, Dr. Bauer (1977) used Landsat data for wheat acreage assessment in the State of Kansas. Several levels of aggregation were examined; namely: county, crop reporting district (an aggregation of several counties), and state.

Landsat data acquired during March to June for the counties in seven crop districts in Kansas were classified; estimates of the area of wheat in each of the 80 counties were made and compared to the corresponding estimates made by the USDA/SRS. The correlation of the USDA/SRS county estimates of wheat area to the Landsat estimates was 0.80. The wheat proportion estimates of 49% of the Landsat county estimates were within ± 5% of the SRS estimates and 81% were within ± 10%. At the crop reporting district level there was a significant difference in the Landsat and SRS estimates in only one of the seven districts. In that district the differences, although small, were all in one direction. For the state, the SRS estimate was 4,555,000 hectares compared to the Landsat estimate of 4,613,000 hectares, a relative difference of only 1.27%.

The coefficient of variation, a measure of the precision of sampling error, of the Landsat estimates was 0.06% compared to 4% for the SRS estimates at the state level. The median coefficient of variation of the Landsat county estimates was 0.6%. At all levels, state, district, and county, the Landsat estimates were extremely precise compared to the corresponding USDA/SRS estimates.

The overall conclusions of the investigation were:
1. Landsat MSS data were adequate to accurately identify wheat in Kansas.
2. Computer-aided analysis techniques can be effectively used to extract crop identification information from Landsat data.
3. Systematic sampling of entire counties made possible by computer classification methods resulted in very precise area estimates at county, district, and state levels.

4. Training statistics can only be successfully extended from one county to other counties having similar crops and soils if the training areas sampled the total variation of the area to be classified.

4.2 Yield

A model based on departure from average weather conditions was used by Morain (op cit) to estimate average yield for a ten-county area in southwest Kansas (Table 4.5). U. S. Weather Bureau published monthly climatological survey data were used as model inputs. These data are published for each of one or more weather stations in each county. The monthly mean of all stations in the ten-county study area was used in solving the yield model equation.

Calculated results using the Table 4.5 Morain meteorological model were 34 bu/A as compared to an average 33.2 bu/A figure obtained by the U. S. Department of Agriculture traditional, time-tested techniques. Furthermore, Morain's model outputs were calculated in June, two months before the U. S. Department of Agriculture statistic was available.

Nalepka et al. (Colwell, Rice 1976) have examined some of the direct relationships between Landsat data and wheat yield for a test site in Finney County, Kansas. In one of their tests, the correlation between a Landsat transformation, the square root of Landsat band 7/band 5 was r=.916. Of the dates tested, Landsat data from April 15 showed the highest correlation with final yield. With regard to the question of whether Landsat data could improve traditional estimates of yield, it was concluded that Landsat estimates of winter wheat yield can be as good as those made by traditional means, even when Landsat estimates are made as much as two months before the estimates using alternative methods.

Nalepka and co-workers also examined the relative utility of Landsat and meteorological data for the preparation of wheat yield estimates for Finney County, Kansas and found that there is important local wheat yield related information contained in Landsat data that is not provided by meteorological data. Differences in important meteorological conditions, temperature and precipitation, over the 30 square mile site were small (May and June coefficient of variation

Table 4.1 Thematic mapper spectral and radiometric characteristics.

Band	Wavelength (μm)	NE	Basic primary rationale for vegetation
TM 1	0.45-0.52	0.008	Sensitivity to chlorophyll and carotinoid concentrations
TM 2	0.52-0.60	0.005	Slight sensitivity to chlorophyll plus green region characteristics
TM 3	0.63-0.69	0.005	Sensitivity to chlorophyll
TM 4	0.76-0.90	0.005	Sensitivity to vegetational density or biomass
TM 5	1.55-1.75	0.01	Sensitivity to water in plant leaves
TM 6	2.08-2.35	0.024	Sensitivity to water in plant leaves
TM 7	10.4-12.5	0.5K	Thermal properties

Table 4.2 Comparative estimates of 1973 wheat acreage for ten counties in SW Kansas as compiled by USDA, SRS and by analysis of Landsat imagery.

County	SRS Acreage Estimate May 1973	SRS Acreage Estimate Aug. 1973	Landsat Acreage Estimate March 1973
Finney	205,000	198,000	239,000
Grant	81,000	87,000	74,000
Gray	157,000	162,000	174,000
Haskell	104,000	109,000	110,000
Kearney	117,000	119,000	115,000
Meade	141,000	132,000	151,000
Merton	91,000	97,000	72,000
Seward	83,000	80,000	78,000
Stanton	135,000	132,000	108,000
Stevens	85,000	87,000	86,000
Totals	1,199,000	1,202,000	1,207,000

Table 4.3 Williams County, North Dakota 3x13km study area results.

Crop Category	Computer Classified Area in Hectares (Acres)	Ground Truth Area in Hectares (Acres)	Percent Correctly Classified
Wheat	1567 (3871)	1578 (3899)	99.3
Fallow	1406 (3475)	1432 (3538)	98.2
Sod	975 (2409)	1064 (2628)	91.7

Table 4.4 Melfort, Saskatchewan 3x13km study area results.

Crop Category	Computer Classified Area in Hectares (Acres)	Ground Truth Area in Hectares Acres)	Percent Correctly Classified
Wheat	1656 (4094)	1675 (4140)	98.9
Fallow	1632 (4033)	1696 (4190)	96.0
Rape	644 (1592)	652 (1610)	98.9

Table 4.5 Yield/acre equation and applicable weather data for the 1972-73 growing season in SW Kansas

$$y=12.215 + .066 + .498X_1 + .631X_2 + .223X_3 - .095X_4 + .191X_5 - .223X_6 - .327X_7 - .101X_8$$

where:

Value in
1972-73

29.00	X_1 = Technology factor measured in years since 1943
10.02	X_2 = Departure from total precipitation from August to March
.14	X_3 = Departure from average April precipitation
-5.30	X_4 = Departure from average April temperature
-1.87	X_5 = Departure from average May precipitation
-2.90	X_6 = Departure from average May temperature
-1.75	X_7 = Departure from average June precipitation
- .50	X_8 = Departure from average June temperature

~0.10). However, wheat yield on the site varied substantially, 21 to 74 bu/A. These differences in yield were likely related to such factors as differences in topography, soils, planting density, fertilization, cropping practices in a field, and irrigation factors which are not accounted for by most meteorological models. The resulting differences in crop condition and eventual yield were however, to a substantial degree, manifested in the Landsat data.

Landsat data was the major source of input to the "Alarm Monitoring Center" (Figure 4.3) established during the Landsat Agricultural Monitoring Program (LAMP) study performed for NASA by the General Electric Company in 1977. The LAMP study focused upon the monitoring of corn in Iowa during the 1976 crop season. Landsat images, meteorological data, and field reports were received and evaluated in near real-time. Manual and computer-aided techniques were used to screen the data for critical crop conditions and changes between each satellite pass. The screening identified alarms which were

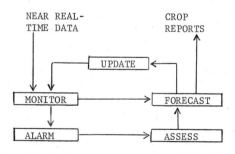

Figure 4.3 Operation of LAMP Alarm Monitoring Center.

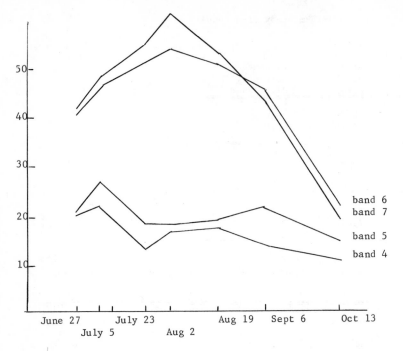

Figure 4.4 Temporal spectral profile for Iowa corn, 1976.

responded to by assessing their source and impact. Repetitive Landsat coverage was used to prepare update reports on alarm conditions and monitor crop recovery.

Landsat imagery over two crop seasons and data on corn physiology, Iowa's climate and terrain, and corn production trends were collected to create a Corn Data Base. The Corn Data Base was used to interpret the Landsat imagery, screen for alarms, and to assess the source and impact of alarms.

During the crop season, several "alarms" were generated by monitoring Landsat imagery and meteorological reports for Iowa. Alarms included heavy rainfalls, hail damage, tornadoes, and drought. Forty percent of the alarms could be identified in the Landsat photographic imagery; in 30 percent of the alarms, cloud cover prohibited immediate Landsat detection. Two of the alarms, tornado damage and drought, were assessed in depth.

Temporal profiles of the spectral signature of corn were developed from computer compatible tapes of seven Landsat scenes (Figure 4.4). This "temporal spectral profile" showed a close correspondence

with corn development. For example, near-infrared radiance peaks in early August, just after silking, and then decreases rapidly after the silking stage as the ears mature and the stalk turns brown.

The LAMP study also examined some of the relationships between Landsat data and yield. Individual band correlations with yield on four different dates are shown in Table 4.6. All MSS bands were significantly correlated with yield on at least two dates. Of the four Landsat bands, band 7 was most highly correlated with yield on all dates, followed by band 6, 5, and 4. The infrared bands were positively correlated with yield, while bands 4 and 5 were negatively correlated with yield. Highest correlation in all bands was on July 23, except for band 5 which was most correlated with yield on August 19. Lowest correlations were in the September 6 image for bands 4 and 5, and in the June 27 image for bands 6 and 7.

Correlation analysis was also used to determine the relationship between grain yield and six Landsat band ratios (Table 4.7). All correlations were positive except for the MSS band 6 over MSS band 7 ratio which was negative in all but the September 6 image. Strongest correlation was found for 6/7 on July 23 (r-.913). In general, the ratios examined were more highly correlated with yield than single Landsat bands, but no single band ratio was consistently correlated higher on all dates.

A least-squares regression was calculated for grain yield using band 7 reflectance from the July 23 Landsat image
$$y=14.397x - 290.226$$
where y is estimated yield in bushels per acre and x is the mean band 7 reflectance of a field.

4.3 Production

Morain (op cit) combined his Landsat data derived acreage (Table 4.1) and meteorological model calculated yield, 34 bu/A, to determine winter wheat production for his ten-county study site in southwest Kansas (Table 4.8). Agreement between Morain's production figures and those of the USDA/SRS is quite good. In addition, the potential for earlier production assessment using Landsat data inputs is a noteworthy finding of this study.

In 1974 a joint National Aeronautics and Space Administration (NASA), U. S. Department of Agriculture (USDA), and National Oceanic and Atmospheric Administration (NOAA) "proof of concept" experiment known as LAICE, Large Area Crop Inventory Experiment, was initiated. Its goal was to develop an experimental crop production forecasting system using satellite remote sensing and meteorological data, test it for a single crop--wheat, and establish the technical and cost feasibility of global agricultural monitoring systems. Achievement of statistical success was to be based upon a 90/90 criterion--90% accuracy, 90% of the time.

LACIE results to date have been a qualified success as presented at the LACIE symposium held on October 23-26,1978 at the Johnson Space Center in Houston, Texas. Soviet spring and winter wheat and U. S. winter wheat production forecasts for 1977 met the 90/90 criterion; however, U. S. and Canadian spring wheat did not. The spring wheat problem was largely due to the abundance of crops which are readily confused with spring wheat by the remote sensor. Principal reasons given for the high accuracy levels achieved in the USSR were the Soviet's large field size and relative absence of confusion crops.

5 WHEAT PRODUCTION ASSESSMENT IN EUROPE USING LANDSAT

Crop production information systems must

Table 4.6 Correlation Coefficient (r) for mean field reflectance and maize yield.

MSS Band	June 27	July 23	August 19	September 6
4	-.153	-.706*	-.659*	-.087
5	-.486**	-.705*	-.748*	-.099
6	.522*	.857*	.722*	.858*
7	.682*	.906*	.804*	.875*

* Significant at the .01 level
**Significant at the .05 level

Table 4.7 Correlation coefficient (r) for mean Landsat band ratio and corn grain yield.

MSS Band Ratio	June 27	July 23	August 19	September 6
4/5	.591*	.606*	.254	.287
6/4	.564*	.900*	.808*	.729*
7/4	.621*	.922*	.841*	.642*
6/5	.558*	.869*	.805*	.137*
7/5	.753*	.907*	.790*	.635*
6/7	-.465*	-.913*	-.819*	-.758*

* Significant at the .01 level
** Significant at the .05 level

provide accurate and timely production data for the efficient operation and stabilization of domestic and international food markets. Successful integration of Landsat data into current agricultural information systems can improve their capabilities and thereby provide more timely and accurate crop production predictions. The level of sophistication or development of Landsat crop models defines the accuracies of Landsat derived crop production statistics. The addition of collateral data sources such as crop phenology/physiology data and soils information and of meteorological data to a Landsat data only crop production information system can be expected to significantly improve crop production prediction model accuracy. Three crop production information systems can be defined depending on the extent of Landsat data and other data sources utilized:

    1. A Landsat crop production prediction system,

    2. A Landsat + collateral crop production prediction system, and

    3. A Landsat + collateral + meteorological crop production prediction system.

5.1 Requirements

An operational wheat production prediction system must be able to accurately determine both acreage and yield in order that accurate crop production predictions be achieved. A number of economic and climatic related factors cause the year-to-year variation in total area devoted to crop production. Farmers decide on the acreage planted to a specific crop after reviewing present and future supply-demand conditions in the domestic and world food markets. If potential profits from wheat production are low compared to profits expected from alternative crops, then farmers will reduce the area planted to wheat in favor of other more attractive crops, providing of course that they have both the information and the option. In most winter wheat areas they at least have the option.

Crop yields also experience extreme fluctuations due primarily to variable climatic conditions. Those genetic research advances that have improved yields of major crops have helped to reduce extreme year-to-year fluctuation.

The extreme year-to-year variability of historical crop production statistics require that a crop production prediction

Table 4.8 Matrix of total estimated production comparing SRS and Landsat data.

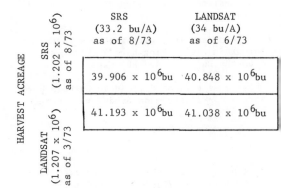

AVERAGE YIELD/ACRE

	SRS (33.2 bu/A) as of 8/73	LANDSAT (34 bu/A) as of 6/73
HARVEST ACREAGE — SRS (1.202 x 10^6) as of 8/73	39.906×10^6 bu	40.848×10^6 bu
LANDSAT (1.207 x 10^6) as of 3/73	41.193×10^6 bu	41.038×10^6 bu

system accurately predict acreage and yield in most countries. Other more specific requirements are: the areal coverage requirement (the number of Landsat scenes required to cover agricultural regions), the Landsat data acquisition requirement (growing season/crop calender), and the stratification requirement (partitioning the agricultural regions into homogeneous strata, such that within stratum variance is less than the variances between strata).

Partitioning the agricultural regions into homogeneous strata, such that within stratum variance is minimized is required to reduce the errors of crop misclassification. The heterogeneous composition of soils, climatic conditions, cultural practices, etc. throughout the agricultural region make it difficult to determine unique crop signatures that can be accurately extended over the entire agricultural region. Development and extension of crop signatures must be limited to relatively homogeneous areas, thereby increasing crop classification accuracies. The methodology of stratifying an agricultural region consists of closely examining historical meteorological/climatological, and local level crop data as well as soils, land cover, topography, geomorphology and crop phenology/physiology data and then partitioning areas such that the land within the stratum bear similar characteristics. Landsat data can be used to refine the stratum boundaries defined by the collateral data, but only if it is available over a variety of agricultural environments; wet vs. dry, several times during the growing season, etc.

5.2 Methodology

As pointed out in section 5 above, a series of at least three crop production information systems can be readily defined based on the extent of Landsat and other data source utilization: 1) A Landsat data alone crop production predicting system, 2) a Landsat plus collateral data crop production predicting system, and 3) a Landsat plus collateral plus meteorological data crop production predicting system.

As is evident from these brief descriptive phases, each of these would be designed so as to have the capability to predict crop production for the crops of interest. However, the production accuracies achieved by operational application of the systems would vary according to the level of inputs utilized; that is, as additional sources of data are added, crop production prediction accuracies should significantly increase. Thus, this series of crop production predicting systems can be thought of as being evolutionary, the Landsat data alone utilizing system serving as the baseline and systems 2 and 3 being more advanced. System 3 would represent the present state-of-the-art remote sensing data based crop production information gathering system.

For simplicity and ease of further discussion of these systems, they will be designated as follows:
1. Landsat crop production prediction system ("L System").
2. Landsat + collateral crop production prediction system ("L+C System").
3. Landsat + collateral + meteorological crop production prediction system ("L+C+M System").

5.2.1 Landsat crop production prediction system (L System)

Definition:

The Landsat Crop Production Prediction System, the "L System," is fundamentally a Landsat data only system; that is, Landsat data will be the only current data analyzed during system operation. The L System will be defined as being limited to inputs of Landsat and non-current crop and crop acreage data. Derived crop acreage will be coupled with the most current yield data available via crop yield gathering methods presently operating to calculate production.

Procedure:

The L System involves "wall-to-wall" (no sampling) classification of wheat in the area of interest. The total land area analysis approach is required because the L System has no field level ground truth data inputs to use for crop signature identification and testing.

An L System functional flow is presented as Figure 5.1. Each of the System's major functions is discussed briefly below:

Cloud Cover Assessment:

Initially, the cloud cover assessment function effort will involve scene cloud cover assessment and identification of 100% cloud cover scenes. Only 90% or less cloud cover scenes will be retained for analysis.

Strata boundaries will be overlaid onto the 90% or less cloud cover scenes and within stratum cloud cover assessments made. Crop acreage data will only be extracted from strata that are not 100% cloud covered.

Crop Signature Extraction/Refinement:

One or more working scenes will be selected from cloud free portions of strata and crop signatures extracted. Histograms of working scene data will be constructed and published crop spectra data used to assess the identity of histogram clusters. Spectral band upper and lower bounds for the wheat clusters will be determined by interactive processing in which initial crop signature bounds are selected and then modified as needed to generate "reasonable" crop area values.

It is expected that in Europe, as has been the case in Canada and the U. S. that wheat will be indistinguishable from other cereal grains. Insofar as classification is concerned cereals vs. all other crops may be the limit of the Landsat contribution. In such a case one must depend on two available historical sets of data. First the historical production trend of wheat vs. other cereals and second, the historical price trend of wheat vs. other cereals. The data for the current year must then be fit to the data and combined to produce the value for current wheat.

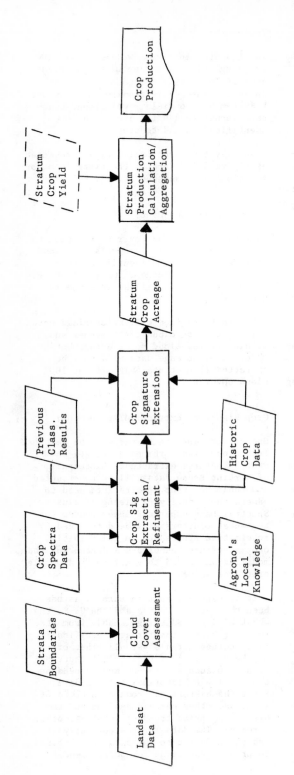

Figure 5.1 Landsat crop production prediction system (L System).

96

The key to the crop signature refinement process is the agronomist and his knowledge (inherent/acquired) of the area being classified. In addition to his agronomic and local experience background, he will make use of previous classification results and crop acreage and trend statistics to assess whether the crop signature being evaluated produces area classification values that are realistic.

Crop Signature Extension:

Within stratum working scene determined crop signatures are used to classify the entire stratum and obtain wheat area for the stratum. This would be accomplished via bulk processing of the stratum polygon.

It needs to be pointed out that the wall-to-wall classification approach being employed by the L System necessitates a 0% cloud cover for maximum accuracy. In cases where a particular stratum does not have 0% cloud cover on a particular date, the crop area of the cloud covered portion of the stratum will need to be estimated. Previous classification results and/or historic local crop acreage data will be the major inputs into this assessment.

Stratum Production Calculation/Aggregation:

Stratum production calculation and aggregation is a purely mathematical activity involving the calculation of production (acreage x yield) for each stratum and then summing these stratum production values to obtain country level production figures. Inputs include stratum level crop acreage and local (stratum representative) crop yield.

Since the L System assumes no current collateral or meteorological data inputs, within system determination of yield will not be possible. Thus, in general, stratum yield will be an outside of the L System determined value which is obtained via presently existing yield determining means. In cases where no current yield data of any kind is available, the L System uses local multiyear average and/or trend data modified by any available local crop condition as well as Landsat crop condition information for generation of a stratum yield statistic.

Outputs:

The outputs of the L System will be several levels of acreage, yield, and production statistics for wheat. Acreage values will be within system generated via Landsat data analysis and could be expected to be usefully accurate by wheat (cereals) heading and maize tasseling time, the times when these crops are most readily (accurately) identifiable using Landsat data. The most important variables impacting acreage accuracy will be the cloud cover conditions at the time of Landsat overpass and the agronomist's local crop and cropping practice knowledge. Yield values are outside of the system inputs and would be expected to be no better than presently gathered and reported yield values. Production values would be expected to be better than non-L System generated production figures, more accurate acreage data being the major contributing factor. Acreage, yield, and production statistics publication levels would include stratum and country as well as other desired aggregations such as county and/or province.

A second type of system output might be a crop acreage, yield, and/or production distribution visual. This visual could be produced for any or all levels indicated above: county, stratum, province, or country.

5.5.2 Landsat + Collateral Crop Production Prediction System (L+C System).

Definition:

The Landsat + Collateral Crop Production Prediction System, the "L+C System" will be described as using both Landsat and collateral data inputs. As used here, collateral data will be defined as including field level ground truth (crop/ field ID, acreage, yield, phenology, etc.) and all types of available supplemental data (soils, productivity potential, trend statistics, etc.) with the exception of current growing season meteorology data.

Procedure:

In contrast to the L System which involves wall-to-wall classification, the L+C System is a sample-based system. Thus, only a minimum size sample need be classified in order to obtain suitably accurate stratum level information, minimum size being determined by within stratum variance. This can be a major advantage when working in a cloud environment because it is much more likely that a portion of a stratum will be cloud free then it is that the entire stratum will be cloud free.

The L+C System procedure will be discussed in terms of two phases, a developmental phase and an operational phase. It is during the developmental phase that the stratum area and yield models used for operational area and yield prediction will be developed and tested. A two phase approach is selected because significantly more and somewhat different types of collateral data are needed for model development than are needed for operational application of developed models. Thus, overall system implementation is more efficient and less costly when a two phase approach is employed.

L+C Developmental Phase:

The developmental phase of the L+C System is directed toward stratum area and yield model development. A flow diagram of the overall effort is presented as Figure 5.2. Each of the major functions outlined is discussed briefly below.

Cloud Cover Assessment:

This function is the same as that described during discussion of the L System. It involves identification of 100% cloud cover scenes, stratum boundary outlining on 90% or less cloud cover scenes, and identification of strata to be further analyzed (90% or less cloud cover, stratum level).

Sample Fields Location:

For stratum model development, each stratum's field level data sample size, "n" needs to be sufficiently large so as to be sufficiently large so as to statistically account (to some level of confidence; for . example, 95%) for its within stratum variance. In fact, additional samples of size "n" will need to be collected from each stratum for verification testing. Furthermore, very comprehensive field level data will need to be gathered: field location, crop type, field area and yield, crop phenology, etc.

It should be noted that recent work on "ground truth" in phenology in terms of

yield has led to the conclusion that observations of the timing and sequence of important phenologic events (emergence, flowering, fruiting, etc.) should not necessarily be coincident with the overpass of a satellite or aircraft. It is much more important to know when the event really occurred and then to interpret the remotely sensed data than to be uncertain of when in terms of the overpass date the biological event occurred. For example, with current Landsat systems it is possible to get 2 measurements within 9 days of the "heading" event. Notice that the word heading is an active verb and can correctly describe the event over a considerable time frame ($>$ 9 days). What is essential for yield models is "the onset of heading," "heading," "the onset of soft dough," etc.

Stratum sample fields from which ground truth data has been or is to be gathered must be located on the Landsat data to be analyzed. Each working scene area used for stratum representative crop signature extraction must have sample fields that total at least size "n" located within its bounds.

Crop Signature Extraction:

Crop signatures will be extracted from the working scene located sample fields. Raw data band values, band ratio data, and transformation index data will be secured for each of the wheat fields comprising the strata samples.

Stratum Crop Temporal Spectral Profile Development:

Sample wheat individual band, band ratio, and transformed index statistics (means and variances) will be computed for each Landsat overpass for which suitable imagery is available and then plotted as temporal spectral profiles for each stratum. Accurate crop phenology data will then be overlaid onto the temporal spectral profiles, and the amplitude and directionality ($\pm \Delta$ differences) will be interpreted.

Stratum Temporal Spectral Profile Analysis:

Statistical analyses will be performed to: 1) compare within stratum temporal spectral profiles constructed from different samples of size "n," 2) compare between stratum temporal spectral profiles, 3) compare year to year variation in temporal spectral

98

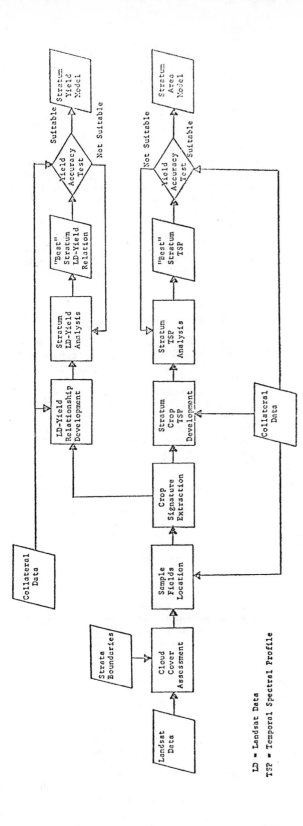

LD = Landsat Data

TSP = Temporal Spectral Profile

Figure 5.2   Landsat + Collateral Crop Production Prediction System (L+C System), Developmental Phase.

99

profiles, and 4) compare individual band, band ratio, and transformed index temporal spectral profiles. The output of these analyses will be an initial cut at a "best" stratum temporal spectral profile as well as providing a guide to the optimum location of stratum boundaries.

Area Accuracy Test:

Collateral data will be used to test the suitability for crop area prediction of the stratum temporal spectral profiles selected. Ground truth data gathered crop acreage values will be compared with those obtained from Landsat data analysis at each of several times during the growing season. A stratum temporal spectral profile capable of accurately predicting crop acreage sufficiently early in the growing season will be selected as the stratum crop area model.

Landsat Data--Yield Relationship Development:

Field level wheat yield data will be obtained for several samples of size "n" for each stratum. Single band, band ratio, and transformation index values extracted from the sampled fields will also be used as Landsat data--yield reltationship development inputs.

Correlation and linear regression calculations will be made to determine Landsat data--yield relationship for each stratum. The following are examples of relationships that will be examined: 1) single band, single date; 2) multiple band, single date; 3) single band, multiple date; 4) multiple band, multiple date; 5) band ratio, single date; 6) band ratio, multiple date.

Stratum Landsat Data--Yield Relationship Analysis:

Landsat data--crop yield relationships developed for each stratum will be compared and similarities and differences from stratum to stratum, from year to year, and among single band, band ratio, and transformation index models made. Crop phenology will also be considered during this model analysis and interpretation activity.

Additional stratum samples of size "n" will be used to examine within stratum repeatability of relationship character-

istics. The output of all these analyses will be an initial selection of a "best" stratum Landsat data--yield relationship model.

Yield Accuracy Test:

Collateral data will be used to test the suitability for crop yield prediction of the Landsat data--yield relationship model selected. Ground truth data gathered crop yield values will be compared with those obtained from Landsat data analysis at each of several times during the growing season. A stratum Landsat data--yield relationship model capable of accurately predicting crop yield sufficiently early in the growing season will be selected as the stratum crop yield model.

L+C Operational Phase:

The operational phase of the L+C System is designed to determine stratum acreage and yield in near real-time and aggregate these in such a manner as to obtain crop production for a country and/or other parcel sizes for which crop production, acreage, and/or yield information is desired. The use of the stratum models developed during the developmental phase will significantly reduce the amount of collateral data required during the operational phase.

A flow diagram of the operational phase of the L+C System is presented as Figure 5.3. Major functions outlined are discussed briefly below.

Cloud Cover Assessment:

Again, this function will be the same as described previously; that is, 100% cloud covered scenes will be identified, stratum boundaries outlined on 90% or less cloud cover scenes, and strata to be further analyzed determined.

Stratum Area Model Operation:

One or more working scenes will be cut from each stratum having 90% or less cloud cover. The area model temporal spectral means and variances determined during the developmental phase using samples of size "n" will be used as operational phase crop signature (gray level range) inputs. The lower and upper limit of a 95% confidence limit will serve as the crop signature.

Crop signatures will be used to classify working scenes and identify areas planted to wheat.

Test Field Evaluation:

A very quick and simple crop signature and area accuracy evaluation test will be performed in which a few known test fields will be located on the crop thematic map output of the area model and examined as to whether they have been properly classified or not. If they have not, an adjustment in the crop signature will be necessary and the thematic map generated from the new crop signature retested. Once acceptable classification results are obtained, the crop signature will be recorded and the thematic map retained.

Overlap Theme Generation:

For the earliest date, the thematic map obtained will be used for initial crop acreage estimation. Since this date would be expected to have the highest variance and thus the grossest (widest gray level range) crop signature, it would be expected that this initial acreage figure would be high. In fact, as the reliability with which a crop can be identified (and thus area determined) increases from planting to approximately the heading of the wheat crop, the sample variance (and likely the classified area) decreases with each succeeding Landsat pass crop classification.

For dates following the earliest one, a comparison will be made between the area classified on the date presently being analyzed vs. the area classified on the previous date(s), pixel by pixel. The acreage figure used for the date being analyzed will be the area of the overlap pixels only.

Area Delta Test:

The guiding principle behind the overlap technique being employed for area estimation is that with each successive time an individual pixel is classified as being a particular crop it is more likely that the pixel really contains that crop and its area should be counted. Inherent in the procedure, however, is the fact that by definition the acreage estimate obtained will be highest for the earliest date and decrease, or remain the same, with each succeeding date. Consequently, final

acreage accuracy could be expected to be the value obtained on the earliest date for which there is no longer any significant acreage delta when compared with the succeeding pass.

Stratum Yield Model Operation:

Stratum yield model Landsat data inputs will be secured from the overlap theme data obtained during the crop area estimation activity. Collateral data inputs will be those specifically required for a particular stratum's yield model operation. One major type expected to be generally required would be general stratum--level crop phenology data.

Yield Accuracy Evaluation:

An evaluation of the reasonableness of stratum yield model outputs, yield values, is made using collateral (mainly historic crop statistics) data. A limited amount of ground truth and/or other current crop condition information will also be gathered and considered during the evaluation activity.

Stratum Production Calculation/Aggregation:

Just as for the L System, L+C System stratum production calculation and aggregation is purely mathematical activity involving the calculation of production (acreage x yield) for each stratum and then summing these stratum production values to obtain country level production figures. Inputs include stratum level crop acreage and stratum representative crop yield. Working scene determined area(s) are converted into stratum areas using verified crop signatures and bulk classification (if stratum has no cloud cover) or mathematically by relating working scene classified crop area to total stratum crop area using the relationship between working scene land area and stratum land area.

Outputs:

The outputs of the L+C System will be the same as that of the L System, several levels of aggregation of acreage, yield, and production statistics for wheat. Acreage values will be within system generated and could be expected to be quite accurate by, or perhaps even significantly before, wheat heading. Yield values are also within

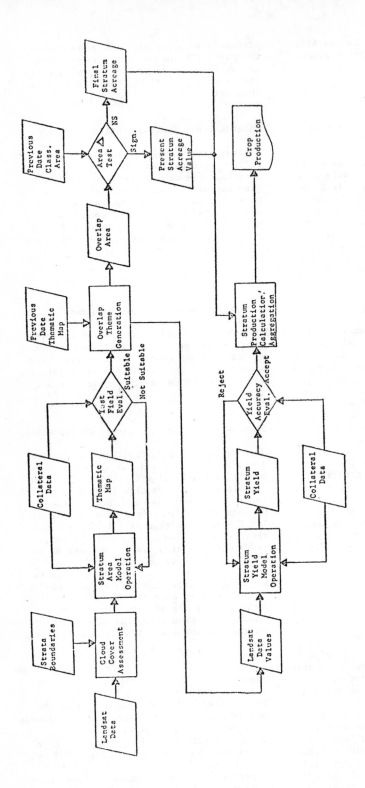

Figure 5.3 Landsat + Collateral Crop Production Prediction System (L+C System), Operational Phase.

102

system determinations and will be as good as the Landsat data--yield relationships that exist for each stratum. Although the limited work done to date using Landsat for yield assessment makes probable accuracy difficult to estimate, it is felt that L+C System stratum yield values will be significantly better than L System values. Consequently, production values could be expected to be notably better than L System generated production values.

A second type L+C System output might be crop acreage, yield, and/or production visuals at various scales.

5.2.3 Landsat + Collateral + Meteorological Crop Production Prediction System (L+C+M)

Definition:

The Landsat + Collateral + Meteorological Crop Production Prediction System, the "L+C+M System," will be described as using meteorological data in addition to Landsat and collateral. Thus, all types of significant crop production impacting data will be input into the L+C+M System.

Procedure:

The L+C+M System will be an exact duplicate of the L+C System except for the additional input of meteorological data and the slightly modified analysis procedures which occur as a result. The L+C+M System will also be a two phased, developmental and operational system.

L+C+M Developmental Phase:

As with the L+C System, the L+C+M System developmental phase will be directed toward stratum area and yield model development. A flow diagram of the overall effort is presented as Figure 5.4.

The only difference between Figures 5.4 and 5.2, the developmental phase of the L+C System, is the input of meteorological data to the stratum yield model development activity. These data include raw data and derived parameters such as temperature, precipitation, evapotranspiration, growing degree units, etc. The objective of the developmental phase will be to optimize the stratum yield models by identifying the most important types of meteorological data inputs and Landsat-meteorological relationships for accurately determining crop yields.

L+C+M Operational Phase:

Like the L+C System, the operational phase of the L+C+M System is designed to determine stratum acreage and yield in near real-time and aggregate these in such a manner as to obtain crop production for a country and/or other parcel sizes for which crop production, acreage, and/or yield information is desired.

A flow diagram of the operational phase of the L+C+M System is presented as Figure 5.5. The only difference between the operational phase of the L+C+M System, (Figure 5.5) and the operational phase of the L+C System (Figure 5.3), is the input of meteorological data to stratum yield determination. Near real-time meteorological data of the type(s) identified as most valuable for accurate stratum yield prediction will be input into the stratum yield model.

Outputs:

As for the L and L+C Systems, the outputs of the L+C+M System will be several levels of aggregation of acreage, yield, and production statistics for wheat. Both acreage and yield will be within system generated and could be expected to be quite accurate, yield accuracy improvement over the L+C System resulting from the input of meteorological data. Consequently, production statistics could be expected to achieve state-of-the-art accuracies.

A second type of L+C+M System output might be crop acreage, yield, and/or production visuals at various scales.

6 EVALUATION OF CROP PRODUCTION PREDICTION SYSTEMS

Each of the three Landsat driven crop production prediction systems described in section 5 has the capability to predict wheat production in/throughout Europe. The production accuracies achieved with each of the three systems vary according to the level of data inputs utilized; that is, as additional sources of data such as collateral and meteorological data are added to the base system (L System) production prediction accuracies should significantly increase. The capabilities of each of the three systems to estimate crop acreage, predict yield and thus determine total crop production will be discussed in more detail in section 6.1,

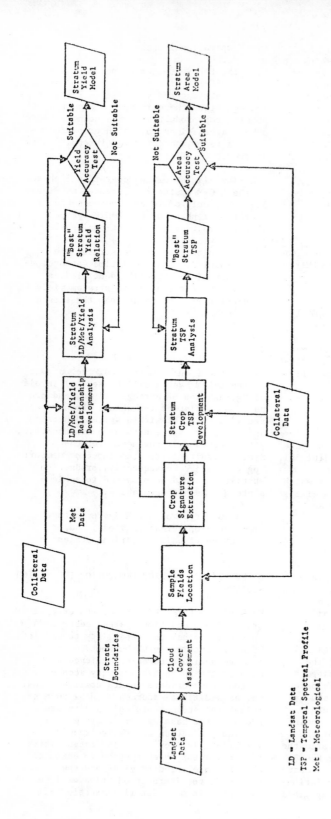

LD = Landsat Data
TSP = Temporal Spectral Profile
Met = Meteorological

Figure 5.4 Landsat + Collateral + Meteorological Crop Production Prediction System (L+C+M System), Developmental Phase

L System Evaluation; 6.2, L+C System
Evaluation; and 6.3 L+C+M System Evaluation.

6.1 L System Evaluation

Acreage:

By definition the L System is limited to
analysis of current and historical Landsat
data and non-current crop and crop acreage
data. Without the application of extensive
ground truth information such as field ID,
crop phenology and other related data, the
development of extendable crop signatures
are truly only as "good" (accurate) as the
operator's local agricultural knowledge
and computer handling expertise. Although
stratifying or partitioning an agricultural
region into smaller more homogeneous regions
will improve crop classification accuracies,
acreage estimation is still based on judgement rather than on statistical principles.
Without current collateral data the acreage
classification for wheat cannot be verified.
The L System does not contain a function for
testing the accuracy of stratum crop classification.

Individual stratum acreages are determined by wall-to-wall stratum crop classification. The presence of clouds over a portion
of a stratum during a Landsat overpass
prevents crop classification of the cloud
covered area and thus complete wall-to-wall
classification of the entire stratum. The
operator then has two choices in a cloud
cover situation:
1. If the pass is not critically important to the L System's crop production
prediction process, then wait until the
next Landsat overpass to generate the
required statistics.
2. If the pass is critically important
to the L System's crop production prediction process then historical crop acreage
statistics derived from previous classification of Landsat data would be analyzed in
an effort to estimate crop acreages under
the cloud covered area.

The L System will provide significantly
more accurate acreage data than is currently
available for the countries of Europe. Over
time, experience will improve crop classification accuracies and therefore improve
crop acreage estimates.

Yield:

Since the L System assumes no current
collateral or meteorological data inputs,
determination of yield will not be possible
within this system. Stratum yield determination will be an external system calculation which is obtained via presently
existing methods. In cases where current
crop yield data is not available, the
L System will gather and use local multiyear average or trend data modified by
readily available local crop condition
information for generation of a stratum
yield statistic.

Production:

Total country by country wheat production
is calculated through the following
relationship (equation 6.1):

$$\sum_{i=1}^{N} \sum_{j=1}^{N} TP_{ij} = \sum_{i=1}^{N} \sum_{j=1}^{N} (A_{ij} \times Y_{ij})$$

where TP = Total Crop Production
      A = Acreage
      Y = Yield
      i = Stratum (1,2, ...N)
      j = Crop (Wheat, etc.)

The A (acreage) component of the production formula is determined via the
L System, while "Y" is estimated by
present methods used by the various
European countries. Wheat production is
determined for each stratum and then
aggregated to the entire country. Crop
production estimates would be expected
to be better than current non-L System
generated production figures, more
accurate acreage data being the major
contributing factor. Acreage, yield,
and production statistics can be aggregations such as country and/or province.

Additional Applications/Outputs:

An additional agricultural application
of the L System is to assess the areal
extent of episodal weather related and
pest induced phenomena. Events such as
insect infestation change the spectral
characteristics of vegetative material
and thus allow Landsat analysts to
determine event areal extent. The temporal coverage by Landsat also allows
analysts to monitor crop recovery, that
is, they can continually monitor the
spectral changes of the crops over time
and sense when crops return to a more
normal condition.

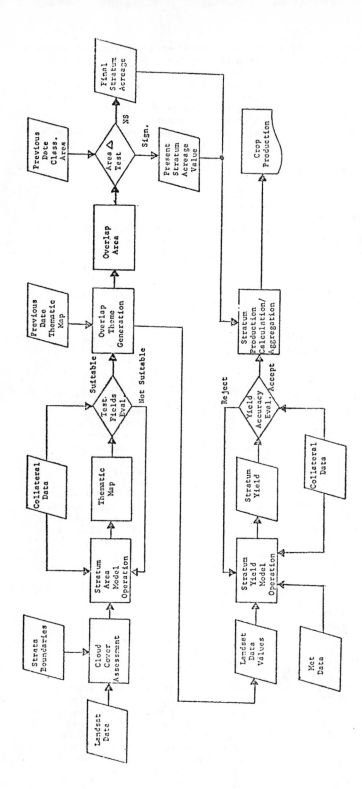

Figure 5.5 Landsat + Collateral + Meteorological Crop Production Prediction System (L+C+M System), Operational Phase.

A second type of system output might be a crop acreage, yield, and/or production distribution visual. This visual could be produced for the county, province, stratum or country levels.

6.2 L+C System Evaluation

Acreage:

The Landsat + Collateral Crop Production Prediction Systems, the L+C System, uses both Landsat and collateral data inputs. As used here, collateral data will be defined as including field level ground truth (crop/field ID, acreage, yield, phenology, etc.) and all types of available supplemental data (soils, productivity potential, trend statistics, etc.) with the exception of current growing season meteorology data.

Unlike the L System which uses a wall-to-wall crop classification technique to determine crop acreage, the L+C System uses a sample-based procedure to estimate crop acreage. Only a minimum size sample need be classified in order to obtain suitably accurate stratum level information. This can be a major advantage when working in a cloud environment, because it is much more likely that a portion of a stratum will be cloud free than it is that the entire stratum will be cloud free.

In the L+C System, crop signatures are developed during the developmental phase and are based on the spectral characteristics of sample fields within strata. The variances of sample field spectral characteristics define the lower and upper bounds of the crop signature. Unlike the L System which bases development of crop signatures on judgement, the L+C System bases the development of crop signatures on sound statistical principles that can be tested and verified. Test fields will be used to verify the accuracy of the crop signatures developed in the L+C System. After crop classification, operators can train on selected test fields and determine if these fields are included or excluded from the crop classification. If excluded, minor signature adjustments are implemented to include these test fields as a part of the crop classification.

The acreage estimates determined during the operational phase of the L+C System should be more accurate then outputed by the L System. The acreage models developed during the developmental phase of the L+C

System should prove to be both reliable and accurate. During the operational phase stratum acreage models are constantly adjusted and improved with current Landsat crop data, thus over time the accuracies of these models should significantly increase.

Yield:

During the developmental stage of the L+C System stratum yield models are developed and tested. Once analysts achieve accepted levels of accuracy these stratum yield models are input into the operational phase of the L+C System. Crop yield predictions obtained via these Landsat crop yield models will be significantly more accurate and more timely than achievable through the present yield prediction system. Although the limited work done to date using Landsat for crop yield assessment makes probable accuracy difficult to estimate, it is felt that L+C System stratum yield values will be better then L System values.

Production:

Production estimates could be expected to be more accurate and timely than L System generated production values. The higher accuracies achieved with stratum acreage and yield estimates combine to significantly increase stratum production accuracies. Landsat crop yield models are formulated to provide timely crop yield predictions which then directly improve the timeliness of production estimates. Both components of the production formula (equation 6.1) are determined from within the L+C System.

Additional Applications/Outputs:

Like the L System, the L+C System has the capability to assess the areal extent of episodal weather related and pest induced phenomena. The L+C System also has the capability to assess the full impact an event has on crop production. Because the L+C System is a total system, that is, it can estimate both crop acreage and yield, the impact an event has on total production can be accurately assessed. Timely crop condition visuals at the county, province, stratum or country level can be used to locate areas that suffer from drought, insects, etc. Necessary precautions or strategies can be recommended and provided to the afflicted areas.

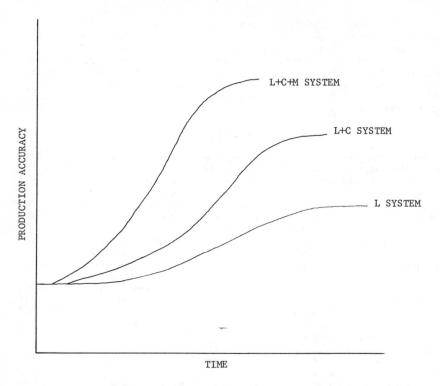

Figure 6.1 Crop production prediction accuracy comparison.

6.3 L+C+M SYSTEM

Acreage:

The Landsat + Collateral + Meteorological
Crop Production Prediction System, the
L+C+M System, uses Landsat, collateral and
meteorological data inputs. Thus, all
types of significant crop production impacting data will be input into the L+C+M
System.

The L+C+M System uses the same acreage
estimation technique as the L+C System
and therefore would be evaluated similarly
(refer to section 6.2).

Yield:

The addition of current meteorological
data to the yield modeling activity should
significantly increase the accuracy and
timeliness of crop yield predictions.
During the developmental phase of the
L+C+M System significant collateral and
meteorological based variables will be
identified and included in stratum yield
models. An increase in the accuracy of a

stratum model would be attributable to the
addition of significantly important
meteorological variables. Important yield
related variables include soil moisture,
rainfall, temperature, growing degree
units, etc. The stratum yield models
developed and later used during the operational phase will achieve state-of-the-art
accuracies.

Production:

Production estimates achieved with the
L+C+M System could be expected to achieve
state-of-the-art accuracies. The greater
accuracies achieved via the stratum yield
models make the L+C+M System the optimum
system for accurate and timely crop production predictions. The L+C+M System is
a total crop production prediction system,
because all components of the production
formula (acreage x yield) are computed
from within the System, and the accuracies
achieved are virtually unmatched. Figure
6.1 graphically shows the production
accuracies achievable with each of the
three crop production prediction systems.
The graph serves only to represent relative

accuracy/time differences among systems
and is not meant to show absolute differences. The L+C+M System is an optimal
system and therefore requires less time
to achieve given accuracies. Also the
L+C+M System can expect to achieve the
highest accuracies possible.

Additional Applications/Outputs:

The L+C+M System, like the L+C System has
the capability to assess the areal extent
and production impact of episodal weather
related and pest induced phenomena.
Timely crop condition visuals at the
county, province, stratum and/or country
level can be used to locate areas that
suffer from drought, insects, etc. Expected
levels of rainfall and temperatures for a
stratum can be overlaid on these crop
condition visuals to better assess the
effect weather will have on crop production.
These visuals (maps) provide government
officals with information on the extent
and severity of agricultural impacting
events. Necessary precautions or strategies can be recommended and instituted
to reduce potential production losses.

H.HANUS
Lehrstuhl Allgemeiner Pflanzenbau, Universität Kiel, Germany

9

Regression agromet yield forecasting models

1 INTRODUCTION

The variation of yields from one year to the other is a main problem in agricultural production. In extreme cases with poor yields whole nations may be suffered from hunger as happened few years ago in the Sahel-zone. Therefore it will be of high importance to forecast yields in advance so early as possible, so that preventive measures may be started.

With respect on consequences three facts in yield forecasting are important:

1. The mean accuracy of forecasts
2. The reliabiliy of forecasts, especially in years with extreme weather conditions
3. The earliest date, when sure forecasts can be made.

For the Federal Republic of Germany a method for yield forecasting was developed in 1967. After some years of examination this method was taken over from the Central Statistical Office in Wiesbaden for official forecastings. Up to now forecastings of yields of the main agricultural crops (cereals, potatoes and sugar beets) were sufficient.

Since 1973 it was examined, whether this method could be adapted for other countries with divergent climatical conditions. In cooperation with the Central Statistical Office of EEC in Luxembourg and with FAO studies were made for the countries of the EEC and for Turkey and Argentina.

2. THEORETICAL MODEL

The theoretical principles of the developed method may be explained in the easiest way using an example. Yields of experimental plots, which are managed in all years in the same way (constant crop rotation, cultivation, fertilization, always the same variety etc.), show a great variation according to different weather conditions in the single years.

In this case yields are only a function of the different weather conditions. In long term experiments there may be sometimes a slope of yields with time because soil becomes more poor or fertile. Therefore the possible trend of yields also must be considered.

The yield function reads in this case as following:

$$\hat{Y} = f\,(T,\ x)\ [1]$$

$\hat{Y}$ = estimated yield

T = time variable (number of year within the period of reference)

x = weather

But weather in one year is not a simple variable. Weather conditions are normally described by several individual weather elements, such as temperatures, sunshine duration, precipitation etc. Equation [1] therefore has to be enlarged to

$$\hat{Y} = f\,(T,\ x_1,\ x_2\ \dots\ x_k)\ [2]\ ,$$

wherein the index 1.... k describes the different weather elements.

A further question is the duration of the influence of weather on yields. The end of this influence is defined through harvest date, but the beginning is not true, because influence of weather on yields many start before sowing time (warming and wetting of soil, mineralization of nutrients etc.). On the other hand the influence of weather changes with time. Therefore growing season has to be subdivided into different parts, and the question is, how to define those periods. With respect to processes of yield formation phenological periods would be the best classifikation. In single cases such a subdivision of the growing season may be possible, but for forecasting mean yields of a country normally phenological periods are not useful because the mean yield consists of parts from areas in which phenological periods are very different. Moreover phenological data are mostly not available. Therefore normally calendar units are used, for example monthly means for decades or pentades.

The yield function [2] has to be enlarged again to

$$\hat{Y} = f \ (T, \ x_{11} \ ...,x_{22} \ ...x_{kl}) \ [3]$$

wherein the index 1 ...l represents the different periods of time independent of that, which type of units is used.

Formula [3] describes the yield function of a small plot or of an uniform area with equal weather conditions. If mean yields of larger areas or of a whole country shall be forecast regional differences in weather conditions must be considered. That means, that weather data of several observation stations have to be used. In this case single weather factors in formula [3] get a third dimension, so that the final equation reads as following:

$$\hat{Y} = f \ (T, \ x_{111} \ ...,x_{222} \ ...x_{klm}) \ [4]$$

wherein:

$\hat{Y}$ = estimated yield
T = Time variable (number of the year within the period of reference)
x = weather factors
1 ... k = weather elements
1 ... l = periods of time within the growing season
1 ... m = number of the station

Theoretically it may be expected, that yields can be forecast if the following requirements are fulfilled:

1. The selected weather elements must enclose all factors which may influence yield or at least the most important of them.

2. The growing season has to be subdivided into such parts, which are closely correlated to yield formation.

3. If yields of larger areas have to be forecast regional differences in weather conditions must be considered, using data of several observation stations.

4. The mathematical procedure has to quantify relations between weather and yield very exactly, whereby not only main effects but also all important interactions of weather factors have to be considered.

It's easy to see that the yield of a single year depends on a great number of individual weather factors.

The structure of the equation is similar to a multiple regression function, and therefore the best mathematical procedure for estimation of yields from weather data will be multiple regression. With respect to the great number of independent variables (single weather factors) which may influence yields, one needs a large sample of historical data. However, such large samples are mostly not available. In these cases the full regression model cannot be used. The following chapter therefore deals with possible mathematical procedures.

3. MATHEMATICAL PROCEDURES

The problems of yield forecastings from weather data occur from the point, that the number of independent variables (individual weather factors) normally exceeds sample size (number of years in which yields and weather are measured). In these cases a possible procedure would be the stepwise regression with foreward selection. But special investigations show, that a formal selection according to statistical parameters does not lead to sufficient results. There may be two points for explanation. First of all, a reduction of variables, which influence yields, means a loss of information. Forecasting yields with help of multiple regression models means to define a single point within a multidimensional distribution. Each loss of information by an exclusion of weather data deminishes the accuracy of forecast values generally. Especially in extreme years such weather factors may become significant which normally are not of high importance. On the other hand just in years with extreme weather conditions early and axact forecasts of yields are more important than in normal years. Therefore mathematical procedure has to make sure that no losses of information occur, respectively the loss of information should be minimized.

Each yield of a larger country depends on a great number of individual weather factors in a three dimensional matrix (weather elements x periods of time x observation stations), as described in equation 4 . Because sample size is normally much smaller than the number of weather factors the full regression model must be divided into smaller parts.

With respect to the necessary number of degrees of freedom each regression equation should not include more independent variables than degrees of freedom remain. That means, that normally only regression models can be used which include only one dimension of the weather factors.

Three different models are possible:

number of
possible
equations:

1. $\hat{Y} = f (T, x_{11m} \ldots x_{k1m})$ 1 x m

2. $\hat{Y} = f (T, x_{k1m} \ldots x_{klm})$ k x m

3. $\hat{Y} = f (T, x_{kl1} \ldots x_{klm})$ k x l

Explanations see equation [4] , page 2.

Each group of equations considers the same main effects of weather factors on yield but in each case other interactions. Group 1 for example considers the influence of rain on yield if temperatures at the same time are high or low, group 2 the effect of rain in one month if rain in other months was high or low, and group 3 finally considers the effect of rain in

one month of one region on the mean yield of a country dependent on rain in other regions at the same time. Former investigations showed, that regional differences in weather conditions and their interactions are less important than interactions in regression models of group 1 or 2, but interactions mostly dominate main effects. Therefore different types of regression models should be used which consider different types of interactions.

Because the full regression model cannot be used to forecast the mean yield of a country one gets a more or less number of individual estimates of the yield according to the used equations, and it is necessary to calculate a mean value for the final forecast. This is possible by calculation of the arithmetical mean or by weighing the individual values. The first procedure is easy, but if a weighted mean shall be calculated it is an important question which weights shall be used.

For weighing the individual forecasts the variance around the regression will be a possible factor. In this case the weight of a single value can be calculated from the relations between the minimum variance around the regression which can be reached from any regression equation and the variance of the equation which is used for the actual estimation of yield.

$$\text{weight} = \frac{S^2(\text{min. of all regr.equat.})}{S^2(\text{of equat. used for calc.})}$$

Compared with arithmetical means weighted means according to the procedure described above will lead to a higher accuracy of forecast of about 10 percent. Summarizing individual forecasts to a mean generally leads to a lower error compared with the error of the best single equation, because over- and underestimations of the yield occur for the same year. After decreasing to a minimum which amounts about 80 percent of the error of the best individual equation the mean error of forecasts increases again and exeeds finally the error of the best individual estimation. The number of individual forecasts therefore has to be limited to reach a point near the minimum error.

How many equations finally should be considered depends on differences of the variances around the regression of equations which shall be used. These equations should be ranged at first according to variances around the regression. Summarizing the deviations between calculated and measured yields within the sample of historical data including successively results of the ranged equations decreases the mean error as described above. Previous investigations showed, that the minimum error of simulated forecasts and within the sample will be reached with the same number of individual results. With respect to the results within the sample it therefore can be deduced, how many equations are necessary to minimize the error of forecasts.

A second point is of interest. Calculating successive means of estimates using equations with increasing variances around the regression over- and underestimations of yields in the same years compensate each other so that the mean error of forecasts decreases below the error when using only one equation but the best of all. This effect of compensation between results of different regression equations can be calculated in advance by correlations between the residues within the sample. If deviations between measured and calculated yields are low or negative correlated with residues of another equation the compensation effect will be much greater than in cases of high positive correlations.

For selection of equations both criteria, the expected accuracy of forecasts according to the variance around the regression, and the possible compensation effect according to the correlation of the residues between measured and calculated yields, should be considered.

With respect to accuracy of forecasts the selection procedure normally leads to a number of equations which is smaller than the necessary number of equations with respect to reliability. To be sure that also in extreme years no extreme errors of forecasts occur the final estimation should be based on as many equations as possible. The mean error of forecasts after reaching the minimum point increases on-

ly very slowly and the loss of accuracy is not very high but extreme deviations between measured and calculated yields in single years will be reduced.

There are other possibilities too to improve accuracy and reliability of forecasts but investigations showed that the weakest point of the developed method are not lacks in mathematical or statistical procedures but lacks of the sample. Because sample size often is only small and exeeds seldom 15 - 20 years one cannot be sure that the sample includes all possible weather cycles also extreme ones. Because regression equations are only true within the variation of weather conditions which is recorded by the sample the representance of the sample is the most important factor for forecasting yields from weather data. Enlarging sample size by summarizing yield and weather data from a few years but from different parts of a country into a common sample does not increase the representance of the sample because possible variation of weather between different years cannot be replaced by the variation of weather conditions between different regions of a country in the same years, at best only in very large countries with different climatical zones.

4. MATERIAL AND METHODS

The developed method of forecasting yields from weather data is an empirical method. With help of historical data of yields and weather conditions relations between weather and yields can be quantified by multiple regression. The first question therefore is, which weather data are recorded or easily available in years in which also yields are measured. Because the theoretical model is very flexible various weather data can be used and tested whether they are sufficient or not. The situation is normally very different from country to country. In Table 1 data are shown which are used to forecast yields in various countries. As can be seen basic data were very different, also sample size, level and variation of yields. Seven different weather elements were used in Germany but only a few in Italy and Turkey. The mean yield varies from

12 dt/ha in Turkey to 44 dt/ha in the Netherlands. Also the variance differed in a wide range from 1,6 to 7,8 dt/ha or relatively from 10 to 28 percent. Only precipitation was used in all countries, while days with rain, sunshine or wind speed were available only in a few.

The method was developed with yield data from plots of 100 m^2 of a long term experiment at the University of Bonn. The used weather elements were the same as for Germany.

Calculating mean yields of a country in all cases regression equations only included weather data of a single station and these data were correlated with the mean yield of the country including a time variable in each equation.

Two different models were used:

1. Regressions with all weather elements of a single month and separate equations developed for each month from January to July.

2. Regressions with a single weather element but including all monthly values from January to July.

In all cases monthly values were used, means or sums respectively. Generally linear regressions were calculated. Although it cannot be excluded that nonlinear models will lead to results of more accuracy special investigations led to greater errors due to the reduction of degrees of freedom. Only when sample size is large enough and the variance of weather conditions is very high nonlinear regressions will give better results than linear ones.

After calculating all possible regressions according to the both possible models those equations were selected which let expect the most accurate forecasts (equations with the lowest variance around the regression (sum of squares of deviations between measured and calculated yields, divided by degrees of freedom)). Using the data of these equations new regressions were calculated by excluding one year of the sample after the other. In this way forecasts for all years of the sample could be simulated. Results of these simulated forecasts with the different equations were summarized to successive means to get a final forecast for each year.

An example for this procedure is given in Table 2 using data of Italy. With the first equation (data of Catania, April/May) a mean error of 1,2 dt/ha did occur. The second equation (Turin, minimum temperature) led to an error of 1,8 dt/ha. Summarizing results of both equations the mean error decreases to 0,98 dt/ha because of compensation of errors in some years (1954, 1955 etc.).

Calculating a mean error for a single year considering results of several equations the sign of deviation has to be taken into account because over- and underestimations can compensate each other. Calculating a mean error of forecasts for the whole sample it is only allowed to use the absolute values of deviation because an overestimation of the yield in one year cannot be compensated by an underestimation in another year. Therefore mean errors of several years are always without signs.

5. RESULTS OF FORECASTINGS OF YIELDS

A general problem of forecastings of yields is the fact, that a single point within a multidimensional system has to be defined.

On the other hand the available sample size is seldom full representative for all possible weather conditions. Statistical parameters therefore are often not true and an exclusion of weather elements, although they are not significant, leads often to a great deviation of forecast values. An example is given in Table 3. Regressions with different weather data led to equations with nearly the same multiple B (r^2) but forecastings of yields showed a large variation according to the used equation and also when weather factors are excluded if they are not significant.

A backward selection of weather data according to statistical parameters does not result the best forecasts as can be seen from figures in Tab. 3. Also a stepwise regression with foreward selection of weather data from a large matrix but on base of a relatively small sample size results greater errors of forecasts than regression models which include only a small part of the whole matrix of weather data.

The efficiency of different regression models can be seen from results in Table 4. Using the whole matrix of weather factors (weather stations x weather factors x month) or two dimensional parts of it led to higher errors than models which used only data of a single weather station and a single weather element or a single month (model 5 and 6 in comparision with models 1-4). A comparision of model 4 with models 2 and 3 showes moreover that it is more important to consider interactions between different weather conditions within the whole growing season than regional differences in the weather cycle.

Forecasting yields one has to consider that deviations between measured and calculated yields within the sample are much less than outside of it. This general effect is shown in Table 5. With decreasing multiple B (r^2) errors increase both within the sample and simulating forecasts. Errors of forecasts are generally higher and increase more rapidly than errors within the sample. Errors of forecasts can be calculated from errors within the sample by the following formula

$$Y = -0,45+1,93 \cdot x \quad [5]$$

Y = error of forecasts

x = error within the sample (mean differences between measured and calculated yields)
Errors of forecasts nearly twice as high as within the sample. These relations have to be considered when selecting equations for forecasts.

The method of forecasting yields from weather data was developed on base of a long term experiment with yields of plots of 100 m². Results in Fig. 1 show that yields of these plots could be forecast very exactly, although yield differed in a wide range from year to year.

In a second step the mean yields of the Federal Republic of Germany were calculated using data given in Tab. 1. Final forecasts of wheat at the end of the growing season are shown in Fig. 2. In years before 1968 these forecasts were simulated, in the later years data are forecasts in advance which were made by order of the government. As can be seen from Fig. 2 the deviations between measured and forecast yields are very similar in both cases simu-

lating forecasts within the sample or forecasting yields in advance. The deviations are somewhat greater than in the experiment but always sufficient also in extreme years. Both the mean accuracy and reliability are acceptable.

Another point is of interest: At which moment certain forecasts can be made? In Fig. 3 the mean relative errors of forecasts are shown which could be reached at different points of time forecasting yields of the experimental plots and mean yields of Germany. Although in both cases at harvest time the same relative error can be reached the decrease of error with time is much higher at experimental plots. But in both cases at the end of April the mean error of forecasts is already as low as at the end of growing season. That means that yields can be forecast relatively early with a high accuracy. These results could be attested through forecasts of yields of Germany in the last years since 1968. Only in very extreme years forecast values did change in the later season. Because weather conditions after the moment of forecasting are mostly similar to the mean conditions of the historical sample the early forecasts of yields are mostly correct.

After using the developed method in Germany with good success it was tried to applicate this method also in countries with other climatical conditions. In cooperation with the Central Statistical Office of the EEC and with FAO in Rome studies were made for all countries of EEC, Turkey and Argentina.

In Fig. 4 results of forecastings are shown for some countries with very different levels and variation of yields of wheat. The recorded errors were very similar to those in Germany. Only in a few years with extreme situations larger errors did occur. But generally results were sufficient.

Results for other crops than wheat are given in Tab. 6. Although the absolute values of errors are also important a better comparison between countries and crops can be made on base of relative errors because the influence of different levels of yield on the recorded errors is eliminated. Figures given in Tab. 6 are mostly errors deduced from calculations of yields within the sample. Former investigations showed that normally mean errors of forecasts amount about 80 percent of the standard error of the equation with the lowest variance around the regression. simulated forecastings with wheat show, that these values can be reached really. As can be seen from Tab. 6 the mean errors of forecasts are very similar in all countries and for all crops. From these results it is possible to deduce that the developed method does work under very different conditions.

Only in one case the mean error of forecasts was not acceptable (Italy, spring wheat). Although the absolute value was not too extreme it could not be accepted with respect to the mean level of yields. Detailed investigations showed that the great mean error of forecasts was mostly influenced by extreme deviation in two years (1961 and 1967) in which yields were extremly high (s. Tab. 7). But in both years the acreage of spring wheat was about twice as high as normal. Therefore it has to be assumed that the extremly high yields in these both years were not a result of very favourable weather conditions but mostly influenced by better soil conditions in the enlarged area.

This example may illustrate that the developed method cannot work if variation of yields does not mainly depend on differences of weather conditions but also on influences of other factors which cannot be taken into account.

6. SUMMARY

A method is described to forecast yields from weather data using different multiple regression models. This method was developed on base of yield of wheat and weather data of an experimental plot from a long term experiment and then applied to forecast the mean yields of Germany. Later yields of other crops and other countries were forecast in the same way.

If no other factors than weather did influence yields a mean error of forecasts of about 5 percent with a range from 2 - 7 percent according to different crops and countries could be reached. Forecasting yields of cereals using only weather data till the end of April led to errors which were only a little higher than at harvest time.

The principles of the developed method which is used since 1968 for official forecasts of yields by the government of Germany may be summarized as following:

1. All weather elements which are important for yield formation must be considered

2. The growing season has to be subdivided into smaller parts. Although phenological periods would be the best interval also calendar units can be used. The investigations were based on monthly values.

3. Regional differences in weather conditions must be considered using weather data of various observation stations.

4. To quantify the relation between weather and yield a large sample of historical data is necessary. Only if the full variation of yield and weather conditions are recorded by the sample exact forecasts can be made. The greater the lacks in the representance of the sample the higher the errors of forecasts.

5. With respect to the theoretical model of the relations between weather and yield multiple regression will be the best mathematical procedure to forecast yields. For the present investigations only linear regression models were used.

6. The number included variables into the regression equation has to be limited. With respect to the necessary generalisation no more variables should be included than degrees of freedom remain.

7. A formal selction of weather factors according to statistical parameters with help of stepwise regressions does not lead to sufficient results if the matrix of weather data for selection exeeds the sample size.

8. The best results were obtained when the whole matrix of weather factors was divided into smaller parts and regressions were calculated using only weather data of single weather stations and these results were summarized afterward.

9. The errors of forecasts can be deduced from the variance around

the regression of equations which shall be used. The mean error amounts of about 80 percent of the standard deviation of the equation with the lowest variance.

10. The decrease of the mean error below the best individual result when summarizing several forecasts is due to compensation effects between over- and underestimations in the same year.

11. Also from results within the sample it can be deduced which and how many equations should be used. Summarizing deviations between measured and calculated yields the mean error decreases to a minimum and increases again if more results of equations with increasing variances around the regression are included. With respect to the accuracy of forecasts the number of equations should be reduced near to the minimum error. If the reliability of forecasts especially in years with extreme weather conditions is of higher importance the number of used equations should be enlarged so far as the mean error increases only slowly.

12. The developed method is very flexible against weather data required for forecasts. Whether the available data are sufficient or not can be tested very easily and depends on errors which still can be accepted.

Tab. 1. Used weather data in different countries to forecast yields.

country	sample	yields of wheat $\bar{x}$	s	$s_{\bar{x}}(\%)$	sta- tions	weather data (monthly values)[*] 1	2	3	4	5	6	7	8
Germany	53-71	34,6	5,70	16,5	13	x	x	x	x	x	x	x	-
France	51-72	28,2	7,85	27,8	18	x	x	x	-	x	x	x	-
Great Britain	51-70	34,6	7,20	20,8	16	x	x	x	x	x	-	-	x
Italy	53-69	19,8	2,58	13,0	13	x	x	-	-	x	-	-	-
Netherlands	53-70	44,5	4,64	10,4	6	-	x	x	x	x	x	x	-
Turkey	60-73	11,8	1,56	13,2	11	-	-	x	-	x	-	x	x

[*] used data = x
1 = max. temperatures
2 = min. temperatures
3 = mean temperatures
4 = sunshine

5 = rain
6 = wind
7 = rel. humidity
8 = days with rain

Tab. 2. Differences between forecast and measured yields in hundred kilogramms per hectare (Italy winter wheat)

	Catania Apr./May	Turin Min. Temp.		Brescia Feb./March		Bari Max./Temp.		∅
1953	1,6	2,4	4,0	1,2	5,2	0,2	5,4	1,4
1954	0,6	- 1,7	- 1,1	- 0,2	- 1,3	0,5	- 0,8	- 0,2
1955	- 0,1	0,4	0,3	1,9	2,2	1,4	3,6	0,9
1956	- 0,9	0,9	0,0	- 1,8	- 1,8	- 1,2	- 3,0	- 0,8
1957	- 3,0	1,3	- 1,7	0,0	- 1,7	0,9	- 0,8	- 0,2
1958	1,2	- 2,0	- 0,8	0,2	- 0,6	0,2	- 0,4	- 0,1
1959	1,6	- 5,6	- 4,0	1,0	- 3,0	1,5	- 1,5	- 0,4
1960	- 3,6	- 3,2	- 6,8	- 2,0	- 8,8	- 3,6	-12,4	- 3,1
1961	- 0,7	- 1,4	- 2,1	- 2,5	- 4,6	3,2	- 1,4	- 0,4
1962	0,6	- 1,4	- 0,8	- 1,9	- 2,7	- 1,1	- 3,8	- 1,0
1963	1,8	- 1,2	0,6	- 0,7	- 0,1	- 2,1	- 2,2	- 0,6
1964	1,0	2,2	3,2	1,7	4,9	1,5	6,4	1,6
1965	0,8	2,6	3,4	0,7	4,1	0,4	4,5	1,1
1966	- 0,7	2,1	1,4	- 4,1	- 2,7	2,6	- 0,1	0,0
1967	- 0,4	- 0,2	- 0,6	2,5	1,9	1,1	3,0	0,8
1968	0,3	- 0,8	- 0,5	1,2	0,7	- 1,9	- 1,2	- 0,3
1969	- 0,7	- 1,5	- 2,2	- 1,4	- 3,6	1,0	- 2,6	- 0,7
mean* succ.	1,2	1,8		1,5		1,5		
mean*			0,98		0,98		0,78	0,8

* of absolute values

Tab. 3. Calculated yields of wheat using different weather data before and after elimination of weather factors which are not significant

		used weather factors			
		x(31)	x(31)	x(31)*	x(11)
		x(32)	x(32)*	x(61)*	x(21)*
		x(33)	x(33)*	x(22)*	x(62)
		x(24)	x(24)	x(24)	x(43)
		x(44)	x(44)	x(44)	x(14)
		x(54)	x(54)*	x(15)	x(24)*
		x(65)	x(15)	x(65)*	x(34)*
		x(26)	x(65)	x(16)	x(44)
		x(66)	x(26)	x(26)*	x(15)
		x(17)*	x(66)	x(66)*	x(45)*
		x(27)	x(27)*	x(47)*	x(16)
			x(67)	x(67)	x(67)
mult. B (r²)	before elimination	0,91	0,91	0,92	0,92
	after elimination	0,91	0,90	0,85	0,91
calculated yield	before elimination	72,4	60,6	32,5	29,1
	after elimination	74,9	69,0	43,4	28,5

eliminated, for not significant *

Tab. 4. Comparison of different models for forecasting of yields
(Bavaria, wheat, 1953-1971, dt/ha).

1. Whole matrix (weather stations x weather factors x months)

	equation No.					
	1	2	3	4	5	Ø 14 equations
deviation	1,9	2,6	2,8	4,2	3,8	
success.mean	1,90	1,87	1,98	2,20	2,09	2,26

2. Matrix (weather stations x weather factors per month)

	Jan.	Feb.	March	Apr.	May	June	July
deviation	3,2	4,2	4,3	1,7	2,8	4,1	3,2
success.mean	3,20	3,55	3,51	2,88	2,78	2,87	2,82

	Apr.	May	Jan.	July	June	Feb.	March
deviation	1,7	2,8	3,2	3,2	4,1	4,2	4,3
success.mean	1,70	1,99	2,18	2,21	2,42	2,63	2,82

3. Matrix (weather stations x month per weather factor)

	humid.	rain	sunsh.	max.T.	mean T.	min.T.
deviation	1,8	2,2	2,8	3,4	3,9	4,1
success.mean	1,80	1,89	1,97	2,25	2,58	2,78

4. Matrix (weather factors x months per weather station)

	Regensb.	Würzbg.	Ansbach	Augsbg.	Weihenst.	Bayreuth
deviation	1,8	2,1	2,2	2,4	2,7	2,7
success.mean	1,80	1,85	1,92	2,00	2,10	2,13

5. Matrix (weather factors per weather station and month)

	Weihenst. April	Ansbach April	Augsbg. April	Ansbach May	Würzbg. April	Ø 13 equations
deviation	1,4	2,1	2,4	2,6	2,1	
success.mean	1,40	1,44	1,56	1,49	1,53	1,68

6. Matrix (months per weather station and weather factor)

	Augsbg. humid.	Regensbg. rain	Weihenst. humid	Ansbach rain	Augsbg. rain	Ø 13 equations
deviation	2,1	2,3	2,1	1,8	2,7	
success.mean	2,10	1,98	1,83	1,56	1,68	1,62

Tab. 5. Relations between parameters of the sample and results of simulated forecastings of yields (France, wheat, 1951-1972)

Station No.	weather factor	mult. B.	diff. measured-calculated sample	forecastings
1	wind	0,96	1,32	1,98
2	rain	0,95	1,33	1,87
3	wind	0,95	1,36	2,28
4	wind	0,95	1,46	2,29
5	rain	0,95	1,52	2,75
1	humidity	0,94	1,41	2,38
6	sunshine	0,94	1,51	2,59
7	humidity	0,94	1,52	2,30
8	wind	0,93	1,68	2,57
9	humidity	0,92	1,61	2,65
3	humidity	0,92	1,65	2,84
10	sunshine	0,92	1,65	2,99
7	sunshine	0,92	1,69	2,73
11	rain	0,92	1,85	2,91
12	rain	0,92	1,97	3,70
1	rain	0,92	2,02	3,69
2	wind	0,90	1,85	2,95
10	rain	0,90	1,96	3,04

Tab. 6. Possible accuracy of forecastings of various crops in countries
of the EEC

| | mean errors of forecasts | | | | | | | | | |
crops	absolute (dt or t/ha)					relative (% $\bar{x}$)				
	F	NL	I	GB	D	F	NL	I	GB	D
w-wheat	1,4	1,4	0,8		1,6	5 (5)	3 (3)	4 (4)		(4)
s-wheat			2,3					19		
wheat total				1,4					4 (5)	
w-rye		1,1			0,9		4			(3)
rye total	0,6		0,4	0,8		4		2	3	
w-barley	1,4	1,0			1,5	6	3			(3)
s-barley	1,2	2,5			1,8	4	7			(5)
barley total			0,4	1,4				3	4	
oats	0,9	1,4	0,6	0,7	1,8	4	4	4	3	(5)
cereals	0,8	1,0				3	3			
maize	2,4		1,0			7		3		
rice			1,8					4		
potatoes (t/ha)	0,8	1,0	0,2	0,8	0,9	4	3	2	4	(3)
sugar beets (t/ha)	2,0	2,1	0,8	1,8	2,4	6	5	2	6	(5)

( ) reached with simulated or real forecasts

Tab. 7. Acreage and yields of spring wheat in Italy

	Acreage (1000 ha)	yield (dt/ha)
1955	67	9,6
1956	66	8,2
1957	61	10,0
1958	77	9,8
1959	97	10,1
1960	89	8,6
1961	214	20,0
1962	75	10,2
1963	74	11,6
1964	68	10,5
1965	63	12,6
1966	54	12,7
1967	154	25,9
1968	56	14,1
1969	59	14,6
1970	56	14,0

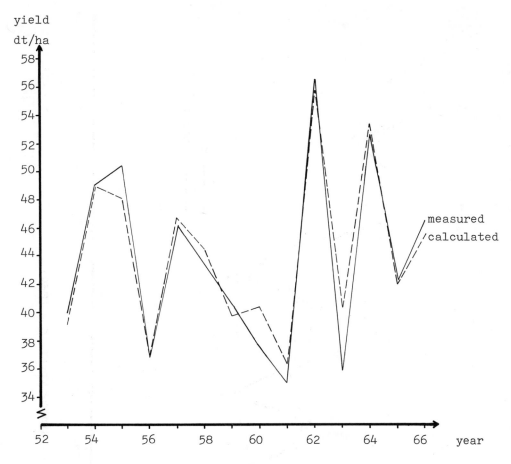

Fig. 1. Measured and forecast yields of winter wheat in the long term
experiment at Dikopshof (1953-1966, mean January-July)

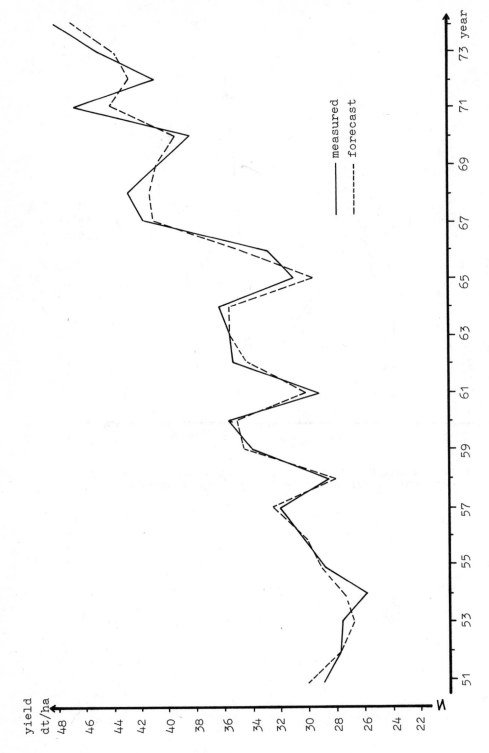

Fig. 2. Measured and forecast yields of winter wheat in Germany

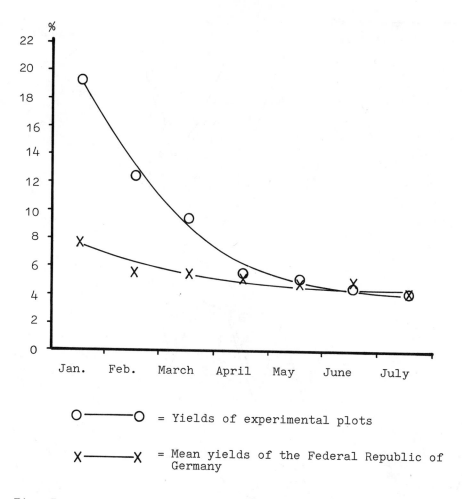

O———O = Yields of experimental plots

X———X = Mean yields of the Federal Republic of
Germany

Fig. 3. Relative errors of forecast wheat yields during
growing season

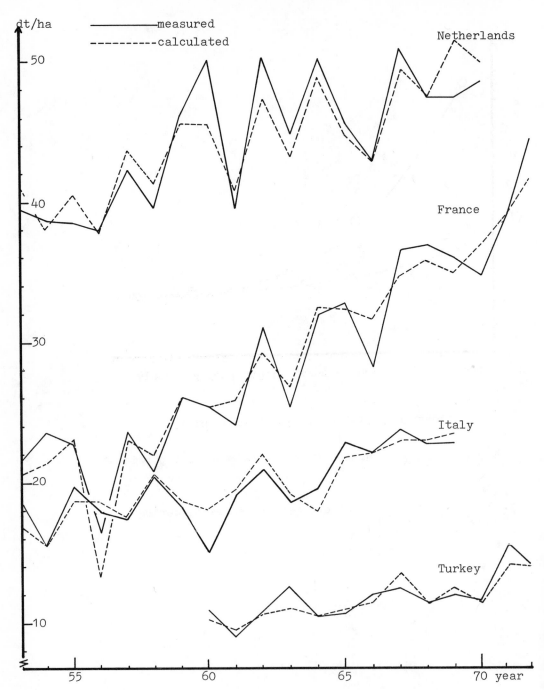

Fig. 4. Results of forecastings of yields for different countries
        (wheat)

126

Ph.MALET

INRA, Station de Bioclimatologie d'Avignon, Montfavet, France

10

A short review of biological agromet models

INTRODUCTION

Biological Agromet models may be classified in various ways.

Wallach (1975) and Baier (1977), for instance, propose both to classify according to the following computation methods:
1. Crop growth simulation methods
2. Statistically-based crop weather analysis models
3. Multiple regression on yield models.

Mc Quigg (1976) prefers a classification according to aims :
1. Biological and physical knowledge synthesis
2. Yield forecasting.

In this lecture, I shall mainly insist on the different logical approaches of the models.

All of them try to describe a phenomenon by the mean of comparisons, images, and analogies. A model never represents reality, but only the image you have of it ; and this image depends much upon the way you want to use the model.

So, it is always better to begin by examining how the models simplify the reality.

1. ANALYTICAL MODELS

A model may be restricted to represent a partial phenomenon (for example the mineral alimentation of a crop) and exclude all the others phenomenons which contribute to the yield. This type of model is only an analytical tool.

Analytical approach is generally used when modelled crop grows either under controlled conditions or within a nearly homogeneous environment. That approach supposes that various submodels have been previously elaborated in order to simulate energy and mass ($H_2O - CO_2$) exchanges inside vegetation : for example, the soil-plant-atmosphere model (S.P.A.M.) of Shawcroft and al. (1974).

1.1 Growth analysis

These models are starting from growth output in order to elucidate how climate, energy partition process, physiological process interact and fit together.

These models generally use simulation technics, i.e. a set of differential equations based on available knowledges of growth processes such as photosynthesis, transpiration, respiration (Chen and al 1969, de Witt and al 1971) and nutrients uptake (Mc Kinion and al 1975). The typical logic of such simulation models is described on the graph relative to"the growth logic", but generally without both genotypic control and development blocks : these are substituted by only flow recorders, set behind the assimilation block valves, which control their opening ; for these models assume that either vegetation is in a steady stage of development or the development does not interfere in growth, or both (Fig. 1).

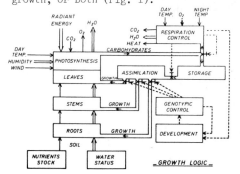

GROWTH LOGIC

That is why growth models alone cannot yet forecast the yields.

1.2 Development analysis

Generally, development models only aim at forecasting some important phenological dates such as emergence, heading, flowering, maturity. So, Cross and Zuber (1972) count twenty two methods only to predict the maize flowering dates in U.S.A. These methods are based on the very simple idea of "heat units".

Robertson (1968) for wheat proposed a more sophisticated development model but which is still based on statistical analysis. Paltridge and Denholm (1974) for grass and Yokoi (1976) for trees attempted to explain the switch from vegetative to reproductive growth. But Vincent et al (1976) showed that it cannot be considered as a pure switch.

So, Vincent et Malet (1977) elaborated a new wheat development logic based on knowledge of biochemical processes which control the organogesis (initiation of leaves, roots, tillers and floral organs). According to this model, organs development depends on control centers which start to work one after the others (Fig. 2)

its working period, a control center sends also an inhibitors flow which modulates the rate of cells division in primordia. The inhibitors stock inside the working control center is renewed by a flow which comes from the leaves and depends upon the climate (temperature and photoperiod). But the stock finally runs dry and the above waiting control center takes the relay.

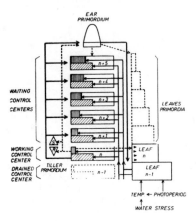

WHEAT VEGETATIVE DEVELOPMENT
LOGIC

The climate interferes also with the time which is necessary to fill up each center with its activators stock. When it reaches a certain threshold, the order does not go any more to the apex bottom but into the ear. The development of floral organs begins and neither leaves nor new control center develop (Fig. 4) ; tillering starts also.

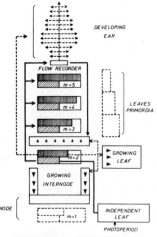

WHEAT FLORAL DEVELOPMENT
LOGIC

WHEAT DEVELOPMENT MODEL _(DETAIL)_

During the vegetative development(Fig.3), each center orders, by mean of a quick emission of activators at the beginning of its working, to form a new leaf primordium and a future control center at the bottom of the apex. So, during all the vegetative period, the number of resting control centers just above the working one remains constant (five). During

128

Even if this scheme is more complex than the other models, it still is more simple than reality. And it shows that there is no mathematical model without assessing previously the relational diagram of behaviour.

1.3 Microclimatical and agronomical analysis

There are also analytical studies on the distribution (i) of some climatical factors (temperature, light) inside the crop canopies, and (ii) of the water and mineral alimentation. These sub-models aim at computing the factors which really act on the plants.

1.4 First conclusion

Analytical models are often elaborated in order to verify the usefulness of concepts. In that case, it is necessary to keep a check on the relative sensibility of the model to each of its parameters (Reed and al. 1976), but this check is very seldom undertaken.

The analytical scheme shows very well the set of models which are necessary to explain a yield : (i) a development model which governs,(ii) microclimatical and agronomical models, and (iii) the growth models corresponding to each phase of development.

2 MODELS WITH A SIMPLIFYING AND FORECASTING LOGIC

2.1 Forecasting problems

2.1.1 Error

$$F = Y + D \qquad \text{(Fig. 5)}$$

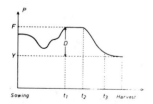

=POTENTIAL YIELD (P) EVOLUTION

So :
$$E_F = \sqrt{(E_{YS})^2 + (E_{YM})^2 + (E_{DE})^2}$$

E_F = forecasting error

E_{YS} = yield or growth sampling error

E_{YM} = yield or growth measurement error

E_{DE} = deviation estimate error

2.1.2 Time-trend and/or technological models

The time-trend model proposed by Thompson (1969) is widely used in agromet models. Analytical technics models are used too (Vinczefry 1975, Pitter 1977), but scarcely. Still, technological models are very useful in order to define the normal climatological years, the bad ones and the good ones with regard to each concerned crop.

2.2 Crop weather models based on analytical submodels

Crops are sensitive to soil water stock and not directly to rain. So, from analytical model, it is hepful to keep soil moisture submodels in simplified models. Many authors (Baier and Robertson 1967, Baier 1973, Minhas and al. 1974, Bridge 1976, Pitter 1977) take care of this problem in their simplified agromet yield models.

This kind of model cannot easily use remote sensing data.

2.3 Bioclimatical models

From growth analytical models and according to some imbreeding works (Hsu and Walton 1971, Grignac 1973), yields depend on growth process. So, some authors (Ueno 1971 cited by Murata 1975, Vinczefty 1975) mix growth and climatological variables in their models.

Haun (1974) proposes a good rough outline of this kind of model ; growth indices, together with pre-season precipitation are incorporated in a multiple regression, in order to forecast wheat yields.

This approach makes it possible to use remote sensing data : there is indeed a relation between the remote sensing data (reflectance and microwave responses) and the growth states. It is unfortunately not a direct relationship : it is a sequence of two kinds of relations : (i) between the remote sensing data and the structures of canopies, (ii) between these structures of canopies and the growth states of plants.

Therefore, it is necessary to study both relations, in order to interpret the relation between the growth and the remote sensing data.

At last, relationships between growth and development processes lead to an interesting evolution of the growth statistic distribution (Fig. 6) in optimal conditions (Malet 1975). This model is hepful to detect any irregular growth evolution. This approach is perhaps applicable to remote sensing too.

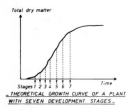

THEORETICAL GROWTH CURVE OF A PLANT WITH SEVEN DEVELOPMENT STAGES.

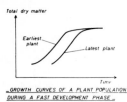

GROWTH CURVES OF A PLANT POPULATION DURING A FAST DEVELOPMENT PHASE.

3 CONCLUSION

This overview of biological agromet models shows that there is still a gap between analytical aims and yield forecasting aims. Crop yields models took perhaps no sufficient care of the analytical models results. On the other hand, analytical approaches scarcely studied the changes which are necessary for use over large areas with climatological and biological data available from most parts of the globe.

4 REFERENCES

Baier, W. 1973, Crop weather analysis model : review and model development, J. Applied Meteor. 12:937-947

Baier, W. 1977, Note on terminology of crop-weather models, W.M.O. Expert meeting on crop-weather models, Ottawa, 11-15 October 1977:12p.

Baier, W. and G.W. Robertson 1967, Estimating yield components of wheat from calculated soil moisture, Can. J. Plant Sci. 47:617-630

Bridge, D.W. 1976, A simulation model approach for relating effective climate to winter wheat yields on the great plains, Agric. Meteor. 17:185-194

Chen, L.H., B.K. Huang & W.E. Splinter 1969, Developing a physical-chemical model for a plant growth system, Trans. A.S.A.A. 12:698-702

Cross, H.Z. & M.S. Zuber 1972, Prediction of flowering dates in maïze based on different methods on estimating thermal units, Agron. J. 64:351-355

Grignac, P. 1973, Relation between yield, components of yields of durum wheat and certain morphological characters, Proc. Sect. Cereals Eucarpia and Plant, 14-18 May 1973:275-284

Haun, J.R. 1974, Prediction of spring wheat yields from temperature and precipitation data, Agron. J. 66:405-409

Hsu, P. & D.D. Walton 1971, Relationships between yield and its components and structures above the flag leaf node in spring wheat, Crop Sci. 11:190-193

Malet, Ph. 1975, Relations entre la vitesse de croissance et les paramètres statistiques dans les populations de mesures de croissance, Note interne INRA Bioclimatologie Montfavet : 16 p.

Mc Kinion, J.M., J.W. Jones & J.W. Kesketh 1975, A system of growth equations for the continuous simulation of plant growth, Trans. A.S.A.E. 18:975-979, 984

Mc Quigg, J.D. 1976, Modelling the impact of climatic variability for the purpose of estimating grain yields. Modelling Climate-Plants-Soils, Proc. Symp. Univ. Gueph 20-21 April 1976:4-18

Minhas, B.S., K.S. Parikh & T.N. Srinivasan 1974, Towards the structure of a production function for wheat yields with dated inputs of irrigation water, Water Resources Res. 10:383-393

Murata, Y. 1975, Estimation and simulation of rice yield from climatic factor, Agric. Meteor. 18:117-131

Pitter, R.L. 1977, The effect of weather and technology on wheat yields in Oregon, Agric. Meteor. 18:115-131

Paltridge, G.W. & J.V. Denholm 1974, Plant yield and the switch from vegetative to reproductive growth, J. Theor. Biol. 44:23-24

Reed, K.L., E.R. Hamerly, B.E. Dinger & P.G. Jarvis 1976, An analytical model for field measurement of photosynthesis, J. of Applied Ecology 13:925-942

Robertson, G.W. 1968, A biometeorological time scale for cereal crop involving day

and night temperatures and photoperiod,
Int. J. Biometeor. 12:191-223

Shawcroft, R.W., E.R. Lemon, L.H. Allen,
D.W. Steward & S.E. Jensen 1974, The
soil-plant-atmosphere model and some of
its predictions, Agric. Meteor.
14:287-307

Thompson, L.M. 1969, Weather and techno-
logy in the production of wheat in the
United States, J. of Soil and Water
Conservation 24:219-224

Vinczefty, L. 1975, Crop forecasting of
sward, Int. Symp. on Crop forecasting
Kompolt-Hungary 16-18 June 1975 : 14 p.

Vincent, A., J. Koller, E. Farcy and
L. Suty 1976, Fonctionnement intégré
du blé, Note de travail, Stat. Amélior.
Pl. Dijon : 30 p.

Vincent, A. & Ph. Malet 1977, Principes
de fonctionnement intégré du blé,
Bull. liaison Département de Génétique
et d'Amélioration des Plantes, n° 5:
282-288

de Witt, C.T., R. Brouwer & F.W.T. de
Vries 1971, A dynamic model of plant
and crop growth. In Potential Crop
production, A case study. Mareing
and J.R. Cooper (Ed.) Heinemann
Eductional Books, London :117-142

Yokoi, Y. 1976, Growth and reproduction
in higher plants. I. Theoretical
analysis by mathematical models,
Bot. mag. Tokyo, 89:1-14

ROGER M. HOFFER
Purdue University, West Lafayette, Indiana, USA

11

Computer-aided analysis techniques for mapping earth surface features

1 INTRODUCTION

The need has long existed for a capability
to obtain reliable information over large
geographic areas in a timely manner. Such
a need is present in many discipline
areas. Because of the very rapid changes
in the condition of agricultural crops and
the influence of crop yield predictions on
the world market, the need for accurate,
timely information is particularly acute
in agricultural information systems. For
these reasons, as remote sensing technology has developed over the past few years,
the potential for using this new technology has been receiving increased and widespread attention.

It is apparent that high flying aircraft
or satellites are capable of collecting
enormous quantities of data over vast geographic areas in a relatively short period
of time. Such masses of data can be collected using a variety of sensor systems,
each of which has its own particular advantages, as well as disadvantages. However, there is a major step between the
collection of the data and the reduction
of this data into useful information. A
key factor in developing remote sensing
technology, therefore, involves the data
analysis techniques that can most effectively reduce the masses of data collected
into the type of information which is required by the user.

One data analysis technique that has
been developed over the past decade involves use of digital computers and the
application of pattern recognition theory
to multispectral scanner (MSS) data, obtained by either aircraft or satellites.
This particular approach was conceived and
developed by the Laboratory for Applications of Remote Sensing (LARS) at Purdue
University (LARS Staff, 1968). This
technique is directed at more effective
utilization of the capabilities of computer systems in mapping and tabulating
earth surface features over large geographic areas in a timely manner. This
approach is not an attempt to develop a
totally automatic data processing system
but rather is directed at developing procedures for computer-aided analysis of
remotely sensed data. Experience has
shown that input by knowledgeable interpreters is a key and essential ingredient
to the effective analysis of multispectral
scanner data. One could therefore refer
to these techniques as "automated", but
it would be incorrect to call them "automatic", because the man-machine interaction is a definite requirement in using
this type of analysis system.

In the development of computer-aided
analysis techniques, multispectral scanners have been utilized as the primary
data source because data from these sensor
systems can be easily quantified and subsequently processed by a digital computer,
and also because the multispectral data
format is ideally suited for pattern recognition analysis. Many of the analysis
techniques developed were based on the use
of aircraft scanner data.

The launch of Landsat-1 (formerly ERTS-1)
opened a new dimension in our capability
to obtain data at any time of the year
(depending on cloud cover). Since the
primary data collection system on Landsat-
1 is a 4-band multispectral scanner, there
has recently been a considerably increased
interest in computer-aided analysis techniques for processing this data. Several
new capabilities for handling this type of
data have been developed, offering con-

siderable potential for meeting agricultural inventory and land use mapping needs. It is the purpose of this paper to describe some of the basic as well as the newly developed techniques for working with multispectral scanner data obtained by aircraft or by satellite.

2 PROCESSING AND ANALYSIS OF MSS DATA BY DIGITAL COMPUTER

Current procedures for digital processing and analysis of data from multispectral scanner systems involve five primary areas of activity:

- Data Reformatting and Preprocessing,
- Definition of Training Statistics,
- Computer Classification of Data,
- Information Display and Tabulation, and
- Evaluation of Results

2.1 Data Reformatting and Preprocessing

Procedures for reformatting and preprocessing of the data do not involve any data analysis per se, but simply involve changing the characteristics of the raw data so that it is in a better format for the analysis sequence. With aircraft data, collected by a multispectral scanner system such as that shown in Figures 1 and 2, the major reformatting activity that is frequently encountered involves converting the data from an analog to a digital format. This is accomplished by a specially designed A/D (Analog-to-Digital) conversion system. The frequency of digitization, level of detail to which the data is digitized, and other factors involved in the A/D conversion can significantly influence the characterisitics of the digital data.

In preprocessing the data, a number of things can be done to change (and hopefully improve) the data characteristics and quality. For aircraft data, these might include such changes as scan-line averaging, geometric corrections, sun-angle corrections, and data calibration. The rotational velocity of the scanner and the ground-speed of the aircraft must be correctly synchronized during collection of MSS data to avoid over-lap or under-lap of the individual scan lines. Very often when the aircraft is flying at low altitudes, the individual scan lines are over-lapped to a large degree. When this occurs, only the nth scan line of data is utilized in some cases, so that the over-lap does not cause serious geometric distortion in the data. In other instances, the scan lines can be averaged (using one of several

possible weighting formulas) in order to improve the signal-to-noise (S/N) characteristics of the data.

Geometric corrections of the data are often applied to aircraft-obtained MSS data because of the geometric distortions caused by the differences in look-angle of the scanner system as the Instantaneous Field-of-View changes from one side of the scan line to straight down to the other side of the scan line. The resultant data on both sides of the scan-lines are compressed in relation to the data obtained directly below the aircraft (See Figure 1). A common result of this geometric distortion caused by the rotational motion of the scanner mirror system and the data compression on the edges of the data is the so-called S-curve seen in roads or other linear features that cross the flight path of the aircraft at an angle. Figure 3 shows a good example of this type of distortion. Such distortions in the data can be corrected in the data reformatting phase of the data processing sequence.

Due to the relatively low altitudes of aircraft data collection missions, severe sun-angle effects have often been found in the data. These are similar to the back-scatter and hot spot (or forward-scatter) effects often seen on aerial photos. However, in collecting scanner data, if the aircraft is flying in a direction such that the individual scan lines are pointed toward or away from the sun, severe sun-angle effects can result. Averaging of the columns of scanner data along the entire flight line, and then adjusting the radiometric values of each data point in each scan line can largely correct these effects (Landgrebe, 1972).

Data calibration can be utilized to adjust the amplitude of the data from one channel (or wavelength band) to the next through the use of calibration pulses in the scanner data. One of the most useful data calibration procedures involves thermal infrared scanner data. If the scanner system being utilized is equipped with hot and cold calibration plates, the thermal infrared data can be adjusted in relation to the energy being emitted by these plates, and accurate values of the radiant energy being emitted by the ground can be obtained. This procedure is particularly useful for remotely obtaining temperatures of water bodies (Bartolucci, et. al., 1973).

Thus far in this discussion of data

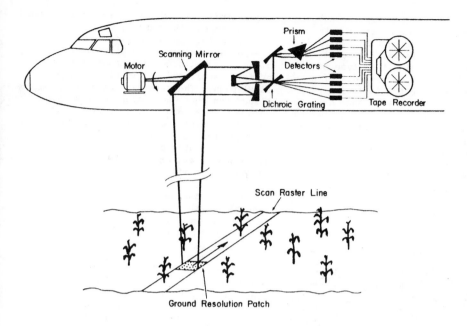

Figure 1. Schematic of a Multispectral Optical Mechanical Scanner.

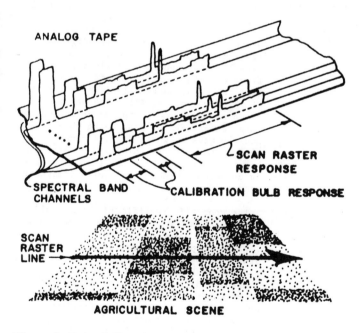

Figure 2. Data Collection Format, Using a Multispectral Scanner and Multiple-Track Recorder.

135

reformatting and preprocessing, we have focused our attention on aircraft data. However, since the launch of Landsat-1 in July 1972, several capabilities for reformatting and preprocessing this type of scanner data have been developed. Of particular importance are the following procedures: reformatting the scanner data to allow a full Landsat frame to be contained on a single data tape, geometrically correcting and scaling the data to a common map base, and overlaying multiple sets of digital data (including sets of Landsat data obtained on different dates, and overlaying Landsat data to other data bases).

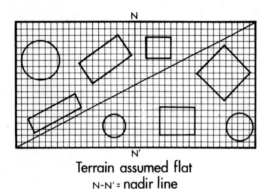

Terrain assumed flat

N-N' = nadir line

Resultant geometry distortion

Figure 3. Geometric Distortion Due to Scale Compression on the Multispectral Imagery (from Tanguay, 1969).

The reformatting of the Landsat data is, in most cases, a fairly straightforward procedure that allows the subsequent data analysis activities to be carried out in a much more effective manner. Originally, the data is received on four, two, or one data tape(s) which contain, in total, a single frame of Landsat data. In the case of four tapes, the data is in 7-track, 800 b.p.i. format, and each tape covers an area of approximately 46 kilometers (km) along a scan line (cross-track from the orbital path), and 185 km along the orbital path. The reformatting procedure allows the four data tapes to be merged onto a single 1600 b.p.i., 9-track data tape containing the entire 185 x 185 km area (approximately 115 by 115 statue miles, or an area of approximately 8,475,485 acres, or 3,429,896 hectares, or 13,243 square miles).

It could also be noted that each resolution element or "pixel" (as it is frequently called) of Landsat data represents an area on the ground of slightly over 0.46 hectares (1.1 acres) (approximately 56 by 79 meters, or 184 by 256 feet). An entire frame of Landsat data contains 2,340 scan lines of data and each scan line contains 3,560 resolution elements, resulting in a total of 7,553,520 individual data resolution elements. Since the data is collected in each of four wavelength bands, there are over 30,000,000 individual reflectance measurements contained on the data tape representing a single frame of Landsat data. The amplitude of each reflectance measurement ranges from 0 to 64, due to the method in which the signal is initially digitized (2^6 digitization accuracy). Since the satellite is capable of collecting a new frame of data every 25 seconds, one sees that the total data collection capability of this type of satellite scanner system is enormous! Growing appreciation for the large quantities of data being obtained and the ways in which digital computer systems can process this type of data has been a major factor in the recent interest in computer-aided analysis of multispectral scanner data.

The geometric rectification and scaling of Landsat MSS data has been a particularly significant data processing procedure. At LARS, this procedure was developed in 1973, and involves a five step sequence in which a frame of Landsat data is rotated, deskewed, and rescaled to the specifications of the user (Anuta, 1973). The development of this procedure was necessitated by the geometric distortions found in the basic "system corrected" or "bulk" Landsat-1 data tapes. The geometric distortions were caused in part by the orbital path followed by the satellite and by the earth's rotation during the time in which the scan line of data is collected. Although NASA applies some 14 different operations to the geometric rectification of Landsat imagery, the data tapes are not corrected for geometric distortions (Anuta, 1977). With such distortions present, the data analyst often could not be sure of his location in the data in relation to a particular location on the ground. The only input required for the basic geometric correction program is the reformatted data tape, and the latitude and longitude of the center point for the frame of data involved. The use of the "system corrected" or "bulk" data tapes allows these geometric correction procedures to be carried out without any loss in the radiometric quality of data. The usual output of the program is a geometrically corrected data tape which, if every resolution element is displayed on a standard computer line printer, results in a 1:24,000 logogrammatic printout, oriented with north at the top. Use of this scale allows the analyst to overlay the printout directly on 7½ minute U.S. Geological Survey topographic maps, or other 1:24,000 scale maps or images. This has proven to be extremely beneficial in helping the analyst locate various features of interest in the data (Hoffer, et.al., 1973).

Since the development of the initial geometric rectification program, a "precision" geometric correction procedure has been developed in which ground control points are located in the Landsat data, and a registration procedure is applied to the data to correct for the geometric distortions and provide an even more accurate output product from a cartographic standpoint. The mathematical details of these procedures have been fully described by Anuta (1973 and 1977).

A procedure for overlaying multiple sets of Landsat digital data has also been developed and offers considerable promise for many types of applications. In this procedure, the geometrically corrected data tape from one date is overlaid with a geometrically corrected data tape from a second, third, or more dates. A data set can thus be developed that allows the differences in spectral response between two different dates to be effectively utilized in identifying various cover types on the surface of the earth, or in delineating areas where changes have taken place in the earth surface cover. For example, in one study in the Rocky Mountains of Colorado, data from six different dates throughout the year were overlaid, and the increase and decrease of the area of the snowpack was determined as a function of date. Thus, in that particular

study, the overlay procedure resulted in a single data tape containing all 24 channels (4 wavelength bands x 6 dates), so that the analyst could easily determine the change in snow cover for any particular location on the ground from one date to another. This capability for overlaying multiple sets of satellite data has many potential applications in land use monitoring, forestry, agriculture and other disciplines where there is a need to determine changes in the characteristics of earth surface cover. It should be pointed out, however, that the ability to effectively utilize the overlay products has not yet been completely defined. Basically, the reason for this is that data on the tape is simply a set of reflectance measurements. Thus, just because there is a change in spectral response from one date to the next does not necessarily mean that a change in spectral response could be due to normal seasonal changes, such as trees leafing out, crops maturing, or even such things as water standing in fields after a heavy rain. It is therefore very important for the human interpreter to evaluate the changes in spectral reflectance recorded by the scanner system in order to differentiate those that are due to normal seasonal changes or weather conditions just prior to the time the data were obtained from those actual land use changes of importance.

Another major type of data overlay procedure that has been developed involves the use of Landsat data in conjunction with other ancillary data sources such as topographic data (elevation, slope, aspect), land ownership, political boundaries, watershed boundaries, soil type boundaries, etc. In overlays of these types, the result is a data base file which can be easily and effectively manipulated by the resource manager to combine various portions of the data base with the Landsat classification results, as needed. This particular capability for data manipulation has tremendous potential for the future use of digital computer systems in a wide variety of applications.

2.2 Definition of Training Statistics

The essence of computer-aided analysis of multispectral scanner data involves "training" the computer to recognize a particular combination of numbers representing the reflectance in each of several wavelength bands from a particular material or cover type of interest. After a good set of training statistics has been developed, the computer is programmed to classi-

fy the reflectance values for every resolution element for which reflectance measurements were obtained by the multispectral scanner system.

The real key to an accurate computer
classification of the data lies in the
definition of a set of training statistics
which are truly representative of the
spectral characteristics of the various
earth surface features present in the
multispectral scanner data. This leads to
the need for the analyst to develop a
thorough understanding of the spectral
characteristics of earth surface features.
Without an in-depth understanding of the
spectral characteristics of the various
materials with which he is involved, the
analyst will not be able to be truly
effective in developing an optimal set of
training statistics to utilize in classifying the MSS data.

As indicated in Figure 4, different
earth surface features have different
amounts of reflectance in the various
wavelength bands. One finds, however,
that the spectral reflectance characteristics of a particular material are not
always unique (Hoffer and Johannsen, 1969,
and Sinclair, et.al., 1971). In many
cases, different species of green vegetation have very similar reflectance
patterns, and one also finds a certain
degree of variation within any particular
species. Therefore, it is not a straightforward procedure to define a particular
spectral pattern that can be used to train
the computer. Thus, one of the first

questions encountered in computer-aided
analysis of multispectral data involves the
identification of the type of categories
or classes of material that the computer
should be trained to recognize. Basically,
there are two conditions which must be met
by each class involved in an analysis of
remote sensor data using pattern recognition techniques:

• The class must be spectrally separable
from all other classes
• The class must be of interest to the
user or have informational value.

In working with multispectral scanner data,
one soon finds that often the classes of
informational value cannot be spectrally
separated at certain times of the year.
One reason for this is that various species of green vegetation have very similar
spectral characteristics, even though their
morphological characteristics may be quite
different. The need for a class to be
both separable and have informational
value therefore leads to two quite different approaches in training of the computer
system.

The first approach is referred to as the
"supervised technique" and involves use of
a system of X-Y coordinates to designate
to the computer system the locations in
the data of known earth surface features
that have informational value. For example, at a certain X-Y location in the
data is a field of corn, at another is a
field of soybeans, another contains wheat,
pasture, etc. (See Figure 5). This super-

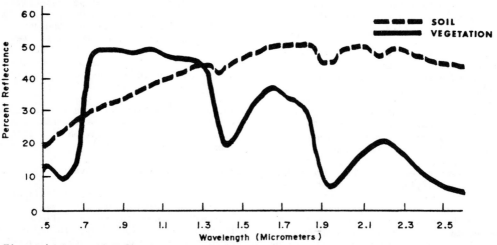

Figure 4. Spectral Reflectance Curves for Turgid Green Vegetation and Air Dry Soils.
These curves represent averages of 240 spectra from vegetation and 154 spectra from air
dry soils. This type of data is frequently referred to as representing the spectral
signature for the materials of interest (from Hoffer, 1971).

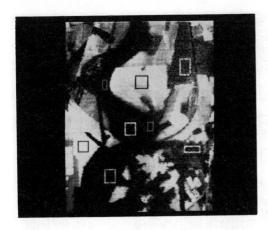

Figure 5. Illustration of Digital Scanner Data Showing Locations of Training Areas.

vised technique or "training sample" approach has been used quite effectively for agricultural mapping (LARS Staff, 1968, Bauer and Cipra, 1973). One must be constantly alert, however, to recognize situations in which two classes may be of great informational value and interest, but which cannot be separated spectrally. In such situations these classes can be combined if there is a sound, logical reason for doing so (e.g., wheat and oats could be combined into a class defined as small grains, but wheat and soybeans could not be logically grouped).

A second approach to training the computer system involves the "clustering" technique (sometimes referred to as the "non-supervised" technique). In this approach the analyst simply designates the number of spectrally distinct classes into which the data to be classified should be divided. The computer is programmed to classify the data into the designated number of spectral classes and then print out a map indicating which resolution elements in the data belong to which spectral class (see Figure 6). The analyst then simply relates this classification output map to known surface observation data, and determines which materials actually are represented by each of the different spectral classes (e.g., spectral Class 1 is wheat, Class 2 is corn, Class 3 is bare soil, etc.). The difficulty with this approach lies in the inability of the analyst to knowledgeably define the number of spectral classes present. Also one finds that the classes of most interest often have subtle

spectral differences, whereas many of the other classes present in the data may be easily separated spectrally but are of little informational value or interest to the user. Experience at LARS has indicated that a combination of the two systems seems to be the most satisfactory and most effective procedure to follow. This is particularly true in wild land or other spectrally complex geographic areas.

A so-called "modified clustering approach" has recently been developed and has proven to be extremely effective in working with satellite multispectral scanner data, both from the Landsat and Skylab scanner systems (Fleming, et.al., 1975). In this method, several small blocks of data are defined, each of which contains several cover types. (See Figure 7). Each area or data block is first clustered separately, and the spectral classes for all cluster areas are subsequently combined. In essence, the modified cluster approach entails discovering the natural groupings present in the scanner data, and then correlating the resultant spectral classes with the desired informational classes (crop species, cover types, vegetative conditions, etc.). This technique is particularly useful in wildland areas, or where the fields are small, or where the cover types and spectral classes are complex. In most cases, less than one percent of the data involved in the analysis is used for the training phase.

Whichever method is utilized to develop the training statistics, one must always keep in mind that these data simply represent the spectral characteristics of the various cover types or earth surface features on the ground. Many of the data characteristics that a photo interpreter would utilize to identify a particular cover type of interest (such as shape, size, texture, shadow, association, etc.) are not used in the computer classification of the data.

2.3 Computer Classification of the Data

After the training statistics are defined, using the supervised, (non-supervised), modified-clustering or some other method, the data must be classified, using one of several classification algorithms. The essence of the theory behind most of the classification algorithms is illustrated in Figure 8. As this figure indicates, different earth surface cover types often have distinctly different spectral response

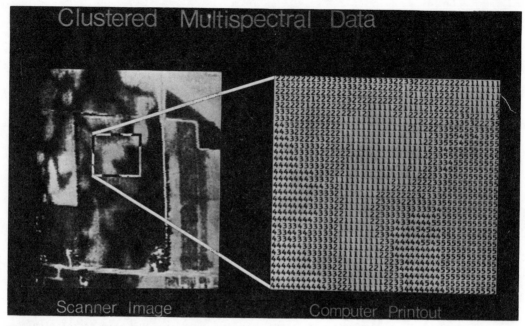

Figure 6. An Agricultural Area in Which the Clustering Algorithm Has been Used to Define Five Spectral Classes.

Figure 7. Digital Display Example of Landsat Data Illustrating the "Modified Cluster" Approach to Developing Training Statistics.

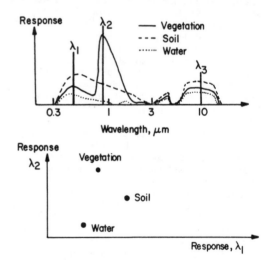

Figure 8. Spectral Data of Basic Cover Types Represented in Two-Dimensional Feature Space.

patterns. Multispectral scanners are designed to measure the relative reflectance or emittance in designated wavelength bands, as indicated by λ_1, λ_2, and λ_3. By plotting the relative reflectance (response values) of vegetation, soil, and water for λ_1 versus λ_2, one sees that these cover

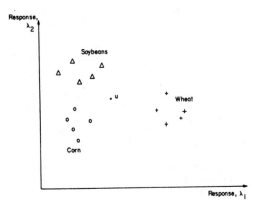

Figure 9. Agricultural Crop Species Defined as Training Samples in Two-Dimensional Feature Space.

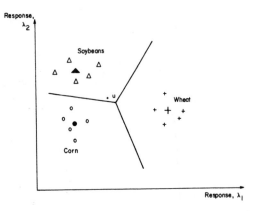

Figure 10. Minimum Distance to Means
Classification.

types occupy very different locations in
two-dimensional space. Of course, in the
real world, agricultural materials will
have some degree of variability, and will
therefore tend to occupy an area in such
a two-dimensional space diagram, rather
than a single point. A theoretical example of such a situation is illustrated in
Figure 9. In this case, response values
for data known to have come from soybeans,
corn, and wheat have been plotted in two-
dimensional space. These data would be
considered the training set. An unknown
data point has been plotted at Point "u"
near the center. The problem now is to
define a classification algorithm that
will divide this two-dimensional feature
space into regions that can be used to
"classify" any unknown data points. Any
unknown data point would therefore be
classified into one of the three categories
for which the computer has been trained.
One of the simplest classification algorithms is the "minimum distance to the
means" in which the mean of each group of
data is calculated, and then boundaries
are defined which define the minimum distance between the means (as illustrated in
Figure 10). In this case, the unknown
Point "u" would be classified by the computer as soybeans, since it fell into that
portion of the two-dimensional feature
space.

Of course, in the actual analysis of
multispectral scanner data, the computer
is not limited to working in two-dimensional feature space, but works with
numerical data vectors that can represent
many wavelength bands of data as more and

more wavelength bands of data are involved, and as the classification algorithm becomes more complicated, the
classification time (and therefore cost)
can increase significantly. A good example of the relationship between number of
wavelength bands and classification time
is shown in Figure 11. This figure also
shows that the classification accuracy
does not increase linearly, or even increase at all, as the number of wavelength bands utilized is increased.

In the classification of multispectral
scanner data, many different algorithms
are available for use. However, one algorithm in particular--the maximum likelihood algorithm--has been used with considerable success by LARS researchers and
others throughout the world. This algorithm has generally been found to be
rather universally applicable to any cover
type and produces classification results
that usually have a relatively high degree of accuracy (as compared to other
classification algorithms). Therefore,
the maximum likelihood algorithm is often
utilized as the "standard" against which
the classification results from other
algorithms are compared.

In the classification sequence involving the maximum likelihood algorithm,
the spectral response data associated with
each resolution element sensed by the
scanner system is examined by the computer
and assigned to one of the spectral
classes defined during the development of
the training statistics. These classification results are then stored on magnetic

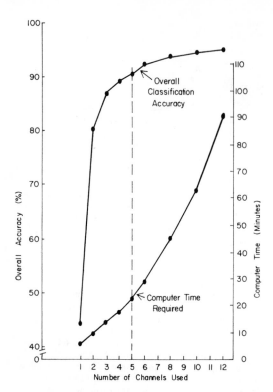

Figure 11. Overall Classification Accuracy and Computer Time Required as a Function of the Number of Channels Used. (From Coggeshall and Hoffer, 1973.)

similar spectral characteristics, and then the area within the boundary is classified as a single spectral class. This technique is somewhat similar to the so-called "Per-Field" classifier, except that with the Per-Field classifier, the analyst has to define the field boundaries, whereas with the ECHO classifier, the computer is programmed to define the boundaries. Other algorithms having somewhat similar characteristics have also been developed by other groups throughout the world.

The primary point that should be stressed in discussing computer classification of MSS data is that no matter which classification algorithm is utilized, the training statistics must be representative of the spectral characteristics of the various cover types present in that data set. If the training statistics are not representative, the classification results will not be satisfactory, no matter which classification algorithm is utilized (i.e., garbage in = garbage out)!

2.4 Information Display and Tabulation

Digital computer-aided analysis of remote sensor data allows a variety of output products to be obtained. These can basically be divided into the two following categories:

  • Display of classification results in map format, and
  • Display of classification results in tabular format.

Maplike displays of the computer classification results can be obtained in one of several different formats. One easily obtained type of output is the so-called "logogrammatic" display obtained from the computer line printer. For this output, the analyst selects various symbols to represent each of the different classes of interest such as C for corn, S for soybeans, D for deciduous forest, F for coniferous forest, W for water, etc. Such an output would be similar to that shown in Figure 12. When the symbols represent the various cover types that have been classified, such a map output would be referred to as a "thematic" map, since each symbol represents a different "theme" or cover type. One could also obtain a map showing the location of the various spectral classes defined in the training procedure, even if the cover type or characteristics of the various cover types were not yet known. In this case, the map would be referred to as a "spectral" map.

tape, and the analyst can subsequently display these results in a variety of output formats. A key point in this classification procedure is that each data point or resolution element in the data is classified and displayed independently. There are many other algorithms that can be utilized (such as the parallel-piped approach used in the G.E. Image 100 system, the "layered" classification techniques, the "levels classifier", and others) to classify the data on a point-by-point or individual resolution element basis.

There are also several classification schemes which have been or are being developed which allow groups of spectrally similar resolution elements to be classified as a single unit. One such technique (developed at LARS) is referred to as the "ECHO" (Extraction and Classification of Homogeneous Objects) classifier. In this technique, the computer is programmed to define the boundary around an area having

142

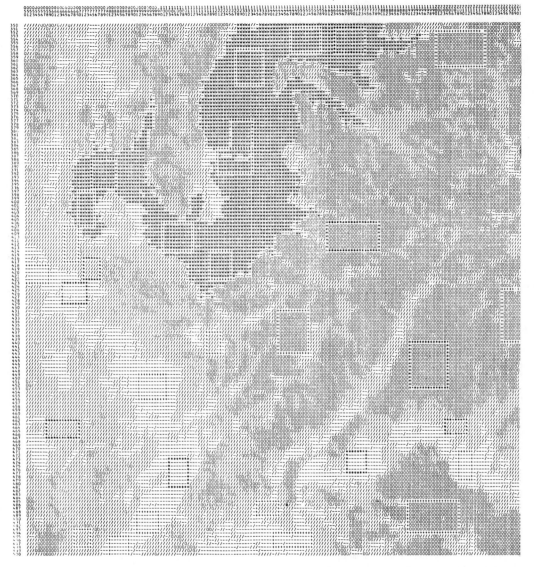

Figure 12. Line-Printer "Thematic" Map Showing Classification Results Based on a "Super-vised" Approach Using Landsat Data. W = Water, O = Forest, - = Bare Soil, and / = Pasture. Training areas are outlined by * and Test areas are outlined by + symbols.

143

Figure 13. LARS Digital Display Unit with Light Pen for Interfacing with the Data.

Similar types of outputs, but having a very different appearance can be obtained through the use of a digital display device (as shown in Figure 13) in which each of the different classes of interest are displayed as a different color or possibly as a different tone of grey, depending on the type of display device available. This output provides a much more photographic-like output of the classification results and also has the advantage of displaying much larger geographic areas on a relatively small image. Many, many other types and formats of output devices are also available, including such things as the ink-jet plotter, digital film writers, Calcomp plotters, "varian" dot grid plotters, and others.

Tabular outputs of the classification results are particularly useful for acreage determinations for any particular area of interest. In this case, the analyst designates to the computer the X-Y coordinates representing the boundary of a test area (such as a quadrangle, county, watershed, etc.). The computer then summarizes the number of data points classified into each of the various cover type categories. Since each data point or resolution element of satellite data represents a certain area on the ground (approximately 0.46 hectares per resolution element for Landsat), this conversion factor can be applied to determine the number of hectares of each of the various cover types of interest. The percentage of the entire area covered by each of the cover types of interest can also be rapidly and easily calculated.

2.5 Evaluation of Results

Classification of large geographic areas can be accomplished very rapidly using computer analysis techniques. However, one must be able to verify the accuracy of such computer classification results. Are the resultant classification maps and tables reasonably accurate, and do they have a reasonable degree of reliability? Several different techniques have been developed and utilized to evaluate such computer classification results. Our experience at LARS has been that a combination of three different techniques provides the best overall indication of the classification accuracy. A qualitative evaluation of the classification results can be obtained by visually comparing the classification to an existing cover type map or to aerial photos of the region. Although the method is subjective, it does provide a quick, rough estimate of the accuracy of the classification. However, quantitative evaluation techniques allow more definitive evaluations of the computer classification results to be obtained.

One quantitative evaluation technique involves a sample of individual areas of known cover types which are designated as "test areas." Test areas are similar to training areas, in that they consist of blocks of data, designated by X-Y coordinates, for which the actual cover type has been determined using some form of reference data (e.g., visual observation on the ground, interpretation of large scale aerial photos, etc.). However, test areas are not utilized in developing the statistics on which the computer is trained. To avoid any possible bias on the part of the analyst, the test areas should be located prior to the classification and, if possible, should be located by means of a statistical sampling design. In many cases, however, the location of test areas tends to be governed by the availability of adequate reference data and also by an evaluation of the importance of statistically determining the classification performance in relation to the cost of obtaining this information. No matter which method is used in defining the location of test areas, the procedure followed is for the cover type classification obtained by the computer for the various test areas to be tabulated and compared to the actual cover type present in the test areas. This tabulation can involve the individual test areas or can be summarized for the entire set of test areas. An example of such a classification

Table 1. Computer Classification Results Showing Test Field Performance for Aircraft Data Using Five Wavelength Bands (from Coggeshall and Hoffer, 1973).

Serial Number ------ 831219706   Classified - Sept. 6, 1972

Channels Used

Channel 4   Spectral Band 0.52 - 0.57 μm
Channel 6   Spectral Band 0.58 - 0.65 μm
Channel 9   Spectral Band 1.00 - 1.40 μm
Channel 10   Spectral Band 1.50 - 1.80 μm
Channel 12   Spectral Band 9.30 - 11.70 μm

Classes

Class		Class	
1	Deciduous	4	Forage
2	Conifer	5	Corn
3	Water	6	Soybean

Test Class Performance

Group	No. of Samples	Percent Correct	Number of Samples Classified Into:					
			Deciduous	Conifer	Water	Forage	Corn	Soybean
Deciduous	32,252	92.2	29,745	1,001	0	378	245	883
Conifer	88	96.6	3	85	0	0	0	0
Water	339	98.2	1	2	333	3	0	0
Forage	11,760	85.5	22	7	2	10,052	413	1,264
Corn	2,679	90.7	1	4	0	191	2,431	52
Soybean	2,676	95.7	10	0	0	91	15	2,560
Total	49,794		29,782	1,099	335	10,715	3,104	4,759

Overall performance: 45206/49794 = 90.8%

Average performance by class: 558.9/6 = 93.2%

---

performance evaluation tabulation is shown in Table 1.

A second quantitative method of evaluating the computer classification results is to compare acreage estimates obtained from the computer classification of satellite data to those obtained by some conventional method, such as manual interpretation of aerial photos. If an adequate number of relatively large areas are summarized, a statistical correlation can be obtained, thereby enabling a quantitative comparison between the computer-derived acreage estimates and the acreage estimates derived from conventional sources. Figure 14 is an illustration of such a comparison, in this case using a classification involving Skylab data.

There are many variations and refinements that can be incorporated into the various analysis and evaluation techniques. However, the general approaches described above have been found to be most effective for computer-aided mapping of general agricultural and forest cover types, utilizing aircraft, Landsat, or Skylab multispectral scanner data.

3 COMPUTER-AIDED ENHANCEMENT OF THE DATA

Computer-aided enhancement of multispectral scanner data can assume many forms, and is sometimes utilized in lieu of computer classification of the data. Therefore, this aspect of data processing could be considered as an alternative to the definition of training statistics, classification, and display of results, as discussed above, although the reformatting and preprocessing, and the evaluation of the results obtained might still be involved. It is important to note that usually the enhancement of MSS data produces an image that must then be manually interpreted. In this regard, computer-aided enhancement is quite different from computer-aided classification.

One of the more common procedures for computer-aided enhancement of multispectral scanner data is to display individual wavelength bands of the original aircraft or satellite data through an appropriate color filter and obtain a "false color composite" of the imagery. With such procedures, one is combining three wavelength bands of scanner data onto a single image. Because the Landsat scanner system obtains two wavelength bands of data in the visible

145

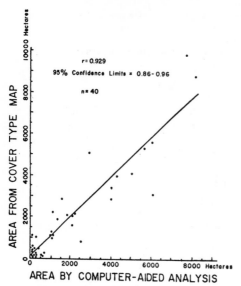

Figure 14. Acreage Estimates of Cover Types Obtained by Computer-Aided Analysis of Skylab MSS Data, Compared to Aerial Photo Interpretation Estimates.

portion of the spectrum (green and red), as well as two in the reflective infrared wavelengths, one can obtain a combination of wavelength bands similar to that involved in normal color infrared film. Therefore, "false" color infrared composites of Landsat scanner data have been widely utilized. The reflected infrared portion of the spectrum allows many earth surface features to be effectively enhanced. Such "false" color infrared composites can be made with computer analysis and enhancement procedures through the use of a digital display device such as that shown previously in Figure 13. Of course such false color composites can also be made through the use of other optical enhancement devices, of which many types are on the market and are used very effectively.

Another type of enhancement that can be achieved with digital computer analysis techniques involves computing the ratios between individual wavelength bands, which are then displayed in various color combination formats. In such instances, one can use a variety of ratio combinations. For example, you could obtain the ratio of a red band compared to an infrared band, or vice-versa; the ratio of two thermal infrared bands could be obtained; or one could add two visible bands together and two infrared bands together, then compute and display the ratio results.

In addition, there are many other more

complex mathematical functions that can be applied to the data to obtain various types of enhancement effects. However, unless there is a theoretical basis for such functions, their value is questionable in terms of developing an operational procedure for processing multispectral scanner data in a reliable, predictable manner.

Along with the simple display and enhancement of the multispectral scanner data, one can also use digital computers very effectively to obtain an enlargement capability of small scale MSS satellite data. This enlargement of digital data has proven to be of particular importance in obtaining detailed images of the scanner data that have not been degraded by optical enlargement techniques. An example of such an enlargement sequence is shown in Figure 15. In essence, each frame of Landsat scanner data contains 3,560 sample points or resolution elements per line of data and there are 2,340 lines of data per frame. However, the digital display unit shown in Figure 13 has a capability of displaying only 800 resolution elements, or so-called pixels (picture elements) per line, and 500 lines of data. Therefore, to fit an entire Landsat frame of data on the screen requires that a subsample procedure be utilized in which only every fifth line and fifth column of data is initially displayed. Such an image is considerably degraded from the original but it does allow one to get a general look at the entire frame of scanner data. This has often proven useful when only the digital data tape was available. One could also display a portion of the entire frame using every fourth line and fourth column, or third line and third column or second line and second column, etc. If one displays every line and column over a small portion of the frame, a very high quality image can be obtained for that small portion of the entire Landsat frame as shown in Figure 15. A procedure has also been developed to display each resolution element measured by the satellite scanner system as four picture elements on the display screen (as shown in Fig. 15), or even 16 picture elements. Thus, even though an individual element might be too small to discern with the eye on the display screen because that resolution element is displayed as a single pixel, by displaying it as a cluster of four or sixteen pixels, a large enough area of the display screen is occupied by that resolution element so that it can been seen. In this way, very high quality images can be obtained for small geographic areas

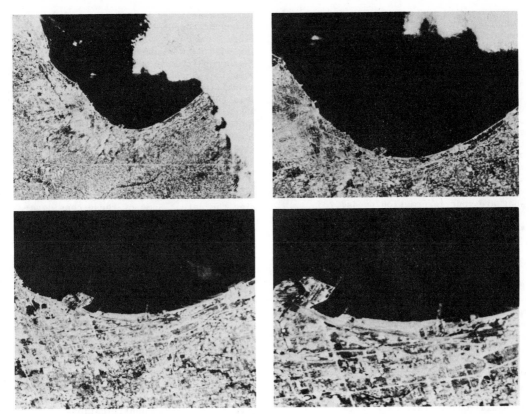

Figure 15. Examples of Computer Enhanced Landsat Data, Showing the Potential for Significantly Enlarging Small Sections of the Landsat Data. (Area shown is the southern tip of Lake Michigan and City of Chicago.)

imaged by the Landsat scanner system. Details not easily seen on the original Landsat imagery (1:1,000,000 scale) are brought out effectively through this enlargement procedure. Many times there is also a tremendous amount of detail actually sensed by the Landsat scanner system that is not apparent until the data was put through such an enlargement procedure.

In addition to these general types of enhancement procedures, there are many other more specific types of enhancement routines to which the data can be subjected, depending on the particular application and type of data involved. The key point involved in the use of enhancement procedures as compared to classification procedures, however, is that the classification requires that the computer be trained, whereas most enhancement procedures are fairly straightforward and can be applied with much less man/machine interaction being required.

4 ADDITIONAL CONSIDERATIONS IN COMPUTER-AIDED ANALYSIS OF MULTISPECTRAL SCANNER DATA

Thus far, we have considered the procedures for handling multispectral scanner data, but very often questions are raised concerning the applicability of one analysis technique versus another, or questions come up concerning the best time of the year to obtain remote sensor data for mapping and tabulating a particular cover type. Therefore, we need to examine a few of these questions, and compare the alternatives.

4.1 What Sensor System or Analysis Technique is "Best"?

In examining the possibilities for using remote sensing to help obtain accurate and timely inforamtion involving agricultural resources, one must review the advantages and limitations of the various

147

remote sensing instrumentation and analysis techniques, so that the user needs can be met with the most efficient system possible.

As an initial step in defining the type of remote sensor systems and analysis techniques which might be most effectively utilized, one needs to define his user requirements, or information needs. In doing so, it is often helpful to consider a series of questions related to the characteristics of remote sensor systems. These can frequently be grouped into spectral, spatial, and temporal considerations, as shown below.

What are your information needs???

● Spectral Considerations

- What are the earth surface features or cover types of interest?
- Can they be spectrally separated from the other associated cover types?
- Which wavelength regions or bands are most useful for spectrally differentiating and identifying the cover types of interest?
- Are the spectral characteristics of these earth surface features unique or more distinctive during some particular time of the year?

● Spatial Considerations

- What size area is involved?
- Do you need complete coverage of the area, or will a sample of the entire data set be adequate?
- What format is required for the results -- maps and/or tables?
- If maps are the final product, what scale and degree of accuracy is required?
- What are the spatial characteristics of the cover types of interest, as compared to the characteristics of the instrumentation involved?

● Temporal Considerations

- How frequently is the information required?
- What time of year is best (or required) for obtaining the information?
- Are there special diurnal considerations involved in the data collection?

To be most effective, computer-aided analysis techniques require a quantitative data acquisition system. In most instances, multispectral scanner systems are the most suitable for providing quantitative data inputs to computer-aided analysis systems. As previously indicated, MSS data can be classified by computer-aided techniques or computer techniques can be utilized to enhance the data and then the enhanced data is manually interpreted. Photographic data rather than MSS data can sometimes be obtained, in which case manual interpretation is usually the analysis technique utilized. Questions are often raised concerning the relative advantages or disadvantages of manual interpretation as compared to computer classification of the data. To develop the most useful operational system for analyzing multispectral scanner data, one must therefore consider the advantages and disadvantages of the two approaches to analyzing the data. Some of these considerations can be summarized as follows:

● A computer classification (using multispectral scanner data, probably from spacecraft altitudes) would potentially be suitable if:

- Geographic area of interest is very large (state, country)
- Informational categories of interest are spectrally separable
- Spectral characteristics of data are relatively simple (spectral classes are reasonably homogeneous and relatively few in number)
- Spatial relationships are not required to achieve identification

● Manual interpretation would probably be more suitable if:

- Geographic area is relatively small (town, county)
- Requirement for detailed spatial information exceeds capabilities of multispectral scanner systems. (In this case, photographic data would have to be utilized).
- "Convergence of evidence" principal is required to identify features of interest
- Spectral characteristics of data are complex and difficult to characterize

From these types of considerations, it is clear that the "best" data collection system and analysis technique is largely a function of the particular problem with which the user of the information is involved. What may be most suitable in one situation may be quite unsuitable in another situation. Therefore, each situation must be evaluated independently, based on its own information requirements.

From this discussion, it becomes apparent that there are three basic components that are involved in the use of any remote sensor system:

User
Requirements

		Data
Data		Processing
Acquisition		and Analysis

These three components are very closely interrelated, and the user requirements as well as the data acquisition system must be taken into account when deciding which data processing and analysis procedure to utilize. In every case, however, it is the User Requirements that are (or at least should be) the driving force behind the entire operation!

4.2 Temporal and Spatial Aspects of Spectral Response Patterns

Agricultural crops are particularly noteworthy in terms of very rapid changes in spectral characteristics at various times in the growing season (Sinclair, et.al., 1971). In late May, winter wheat presents a fairly solid canopy of lush green vegetation to the remote sensor, but by late June in central Indiana the same winter wheat will be golden brown and nearing maturity. Two weeks later, it has probably been harvested and one will see only the highly reflective yellow straw. In many cases, another two weeks' delay will allow many weeds and green understory vegetation to mix with the straw and an appearance very much like grazed pasture or perhaps hay is observed. Such rapid changes bring out the importance of understanding the seasonal changes of the crops or other earth surface features of interest.

Many times, people have discussed the usefulness of "crop calendars" which can be developed for any particular geographic area and will describe the general characterisitics of the various cover types of interest as a function of the time of year. However, one must remember that such crop calendars can vary from one year to the next, depending on weather conditions of that particular year. This is particularly true early in the spring when excessive rains can cause significant delays in planting dates in any particular year. It should also be recognized that such crop calendars can be developed more effectively in areas of the world where seasonal changes are distinct. For example, in the United States there are relatively

narrow periods of time during which corn or soybeans are planted, grown, and subsequently harvested. However, in tropical regions, seasonal restrictions are not as apparent. For example, in Southeast Asia, rice is planted and harvested over a rather wide time span. Such a broad time frame can be even more complicated by second plantings of rice and by distinctly different agricultural practices from one region to the next.

Along with gross seasonal changes in spectral characteristics of agricultural crops and other earth surface features, one can also encounter more short-term temporal variations, such as differences in spectral response at different times of the day or night, and caused by a host of variables. Differences in sun angle certainly needs to be considered, as do variations in atmospheric attenuation. In some cases, vegetation that is not under moisture stress early in the morning will show severe symptoms of moisture stress later in the day. A good example of this is the way in which the top leaves on soybean plants will tend to turn bottom-side-up in the early part of the afternoon on many warm midsummer days. This apparently is a protective mechanism which occurs when the plant cannot obtain sufficient soil moisture. Since the lower surfaces of the soybean leaves are a much lighter green than the top of the leaves, the entire soybean field becomes a lighter green, more highly reflective vegetative surface. Thus, by reducing the amount of solar energy absorbed by the plants, the moisture stress is minimized to a certain extent.

In considering seasonal changes of agricultural crops, one also encounters the problem of definition of a particular cover type of interest. Suppose you are interested in utilizing remote sensing to map acreage of soybeans throughout Indiana. At what stage of development are you going to define a particular agricultural field as being "soybeans"? Do you call Field X a field of soybeans after the beans have been planted? Or after emergence? Or when the beans are four inches high? Ten inches? Or is it not until the soybeans are covering 25% of the ground surface? Or 50%? Etc., etc.??? Work with multispectral scanner data from aircraft has shown, for example, that the difference in reflective characteristics between corn and soybeans is very small. This has also been substantiated with laboratory spectral measurements (Sinclair, et.al., 1971; Hoffer and Johannsen, 1969). To put it another way, green vegetation tends to look very much like other green vegetation. We found that the ability to spectrally differentiate between corn and soybeans during the time period near the end of June was highly correlated with the percentage of ground cover rather than any large distinctive difference in the spectral characteristics of the soybean and corn vegetation itself. Early planted corn fields could be reliably distinguished from soybean fields because of higher percentage of ground cover and smaller percentage of exposed soil involved in each resolution element measurement by the scanner system. However, corn planted later in the season was found to occupy about the same percentage of the ground surface as the soybeans did, and therefore, if one was facing the situation where 50% of the ground surface was bare soil and the other 50% was green vegetation, it did not seem to matter whether the green vegetation consisted of soybeans or corn. The very small spectral differences that do exist between soybean and corn canopies could not be distinguished because of the more powerful influence of the soil/vegetation mixture that was being measured by the scanner system! This type of problem has haunted many people involved in the analysis of multispectral scanner data of agricultural crops, particularly if the data was obtained early in the growing season. Careful selection of data sets to avoid such confusing problems whenever possible seems to be the most logical solution to these types of problems at the present time.

Another aspect of spatial variability of spectral signatures involves geographic variability of various categories or crop species of interest. One finds that the same crop species does not have the same spectral response pattern in all geographic locations on any one date. For example, wheat may be harvested in southern Indiana where it has reached maturity but has not yet been harvested in central Indiana and perhaps is still immature and green in northern Indiana and southern Michigan. Thus in discussing the spectral "signature" concept, it is impossible to define a single spectral response pattern that will be applicable for the same crop species in all geographic areas at any one time.

Another associated aspect of geographic variability of agricultural crops involves the idea that not all crop species are found in all geographic locations. Therefore, knowledge of location from which remote sensor data was obtained can prove quite useful in attempting to identify a particular crop species, even though the spectral response pattern of that crop may not be well known at that time of the year because of lack of "ground truth" data. For example, if one analyzed a Landsat frame obtained on August 15th, and found that 40% of the total area of this Landsat frame consisted of small, rectangular fields of green vegetation, one would conclude that these were agricultural crops of the same species. If the interpreter then knew that the data came from north central Indiana and Illinois, and he had a reasonable knowledge of the agricultural characteristics of this area, he would immediately conclude that the crop species described by that particular spectral response pattern was corn rather than cotton or rice or some other crop species which does not grow in this geographic area to that extent. This points out the need for knowledge of the geographic location over which the remote sensor data was obtained and for knowledge of the agricultural characteristics of that particular geographic region, or in other words, there is a distinct need for a man-machine interaction in the process of analyzing remote sensor data.

4.3 Data Banks or the Extrapolation Approach?

Another subject that often arises in computer-aided analysis of remote sensor data involves two rather different concepts used in training the computer to recognize a given set of spectral response patterns (assuming for now that a supervised training mode is being utilized). These two concepts could be termed the "data bank" and the "extrapolation mode" for obtaining data to be used in the training procedure. In the "data bank" approach, the concept is to have many sets of spectral signatures available in computer storage. Such a "data bank" would include signatures for all different cover types of interest. When a new set of remote sensor data is obtained, one simply selects the signatures for the cover types of interest from the existing data bank, and uses these as training samples to classify the entire data set. The advantage is a great reduction in the amount of time required to train the classification processor as to the characteristics of the spectral signatures involved in each new set of data. Such a system would be highly desirable and, in theory, could work under certain selected sets of conditions. However, in the real world situation, such a concept does not appear to be practical in most cases, particularly for those cover types

150

or earth surface features that have distinct temporal variations in spectral response. The reaons for this involve the temporal effects upon the spectral characteristics of the earth surface features which are measured and which one is attempting to classify. Many studies have shown that the spectral characteristics of different types of vegetation change drastically as a function of time, both for a given growing season and from one year to the next. Plant maturity causes a nearly continuous change in spectral characteristics of agricultural crops. For instance, what is often measured early in the growing season is the amount of bare soil in proportion to the amount of green vegetation present in the instantaneous field of view of the scanner, rather than a unique spectral signature of the different crop species. Since the farmer must depend upon weather conditions to govern the date of planting for the various crop species, the height of the plants (and therefore the canopy coverage and the vegetation-soil relationship will be quite different from one year to the next as of the same date on the calendar. The weather conditions governing the germination and rate of growth of the vegetation after planting are also important, and will cause considerable variation in spectral response throughout the growing season and from one growing season to the next. There are also difficulties in determining differences in atmospheric attenuation from one flight mission to another, and in being able to calibrate sensor data to a high degree of radiometric precision.

In view of such problems of variation from one flight mission to the next, it would seem that an extrapolation mode is much more logical for obtaining spectral data with which to train the computer for the classification task. In this case one simply uses the data collected under the existing atmospheric conditions and conditions of plant maturity, stress, growth, etc., abstracts the training samples from this same data set and then proceeds with the classification and analysis task. Of course, this approach assumes that a reasonable amount of ground truth or reference data is avilable for that particular flight mission.

5 SUMMARY AND CONCLUSIONS

When one considers the current situation of our agricultural and other natural resources, along with the predictions of continued population growth, a person cannot help but be concerned about the manner in which we will be utilizing these resources during the next few decades and beyond. One of the major factors which will influence many decisions on use of these resources is the amount and detail of information which is available about the extent and condition of the resource base.

It becomes apparent that computer-aided analysis techniques are required if we are to take full advantage of our ability to collect data at frequent intervals over vast geographic areas. Further developments in the handling of spatial and temporal data, in addition to spectral data will bring significant improvements in computer-aided data analysis. It also seems clear that both satellites and aircraft will continue to be used in the collection of remote sensing data, and that ground observations will continue to be a necessary and integral requirement for accurate data analysis. However, even though major improvements in computer-aided analysis techniques will be achieved, man's contributions will always be required in the analysis of remote sensor data. A good understanding for instance, of the energy-matter inter-relationships and the spectral reflectance and emittance characteristics of the various cover types of interest will become increasingly important. Also, there will continue to be a major requirement for accurate and detailed manual photo interpretation in many application areas. All of these will be crucial in the maintenance of an effective man-machine interaction in the data analysis process.

Remote sensing has already proved to be a very useful tool in several areas of application in the management of our natural resources, and the potential applications seem almost unlimited. It is our goal to develop a technology that will aid in obtaining the best possible information concerning the location, extent, and condition of our resource base, and to make this information available in a timely manner to the people involved in the management of our agricultural, forest, water, and other resources.

6 REFERENCES

Anuta, P. 1973, Geometric Correction of ERTS-1 Digital Multispectral Scanner Data, LARS Information Note 103073.

Anuta, P.E. 1977, Computer-Assisten Analysis Techniques for Remote Sensing Data Interpretation, Geophysics 42(3): 468-481.

Bartolucci, L.A., R.M. Hoffer, and T.R. West 1973, Computer-Aided Processing of Remotely Sensed Data for Temperature Mapping of Surface Water from Aircraft Altitudes, LARS Information Note 042373.

Bauer, M.E. and J.E. Cipra 1973, Identification of Agricultural Crops by Computer Processing of ERTS MSS Data, Proceedings of the Symposium on Significant Results Obtained from ERTS-1, Goddard Space Flight Center, p. 205-212. Greenbelt, Maryland.

Coggeshall, M.E. and R.M. Hoffer 1973, Basic Forest Cover Mapping Using Digitized Remote Sensor Data and ADP Techniques, LARS Information Note 030573.

Fleming, M.D., J. Berkebile, and R. Hoffer 1975, Computer-Aided Analysis of LANDSAT-1 MSS Data: A Comparison of Three Approaches Including the Modified Clustering Approach, Proceedings of the Symposium on Machine Processing of Remotely Sensed Data, p. 1B-54 to 1B-61. Purdue University, West Lafayette, Indiana. Also available as LARS Information Note 072475.

Hoffer, R.M. and C.J. Johannsen 1969, Ecological Potentials in Spectral Signature Analysis, P.L. Johnson (ed.), Chapter 1 in Remote Sensing in Ecology, p. 1-16. University of Georgia Press, Athens, Georgia. Also available as LARS Information Note 011069.

Hoffer, R.M. 1971, Remote Sensing Potentials for Resource Management, Proceedings of the Third International Seminar for Hydrology Professors, p. 211-227. Purdue University, West Lafayette, Indiana.

Hoffer, R.M. 1972, ADP of Multispectral Scanner Data for Land Use Mapping, Invited paper presented at the 2nd UNESCO/IGU Symposium on Geographical Information Systems. Ottawa, Canada. Also LARS Information Note 080372.

Hoffer, R.M. and LARS staff 1973, Techniques for Computer-Aided Analysis of ERTS-1 Data, Useful in Geologic, Forest and Water Resource Surveys, Proceedings of the Third ERTS-1 Symposium, 1(A): 1687-1708. Goddard Space Flight Center, Washington, D.C. Also LARS Information Note 121073.

Landgrebe, D.A. and Staff 1972, Data Processing II: Advancements in Large Scale Data Processing Systems for Remote Sensing, Proceedings of the 4th Annual Earth Resources Program Review, II: 51-1 to 51-31, NASA MSC Publication No. 05937.

LARS Staff 1968, Remote Multispectral Sensing in Agriculture, Laboratory for Applications of Remote Sensing, Vol. 3. Research Bulletin No. 844, Agricultural Experiment Station and Engineering Experiment Station, Purdue University, West Lafayette, Indiana.

Sinclair, T.R., R.M. Hoffer and M.M. Schreiber 1971, Reflectance and Internal Structure of Leaves from Several Crops During a Growing Season, Agronomy Journal 63(6): 864-868. Also available as LARS Information Note 122571.

Tanguay, M.G. 1969, Aerial Photography and Multispectral Remote Sensing for Engineering Soils Mapping, Joint Highway Research Project, Report No. 13, Purdue University, West Lafayette, Indiana.

F. BECKER
Groupe de Recherches en Télédétection Radiométrique
Université Louis Pasteur, Strasbourg, France

12

Thermal infra-red remote sensing principles and applications

1 INTRODUCTION

Until recently, remote sensing in agriculture and forestry has not received as much attention in thermal infra-red bands as in the visible or near-infra-red ones. Several reasons can be put forward, for instance :

- the relative difficulty of detection which imposes, in order to get significant thermal resolutions, pixels having very large dimensions compared with the characteristic dimension of the European fields. For satellite platforms these pixels are of the order of a kilometer (0,5 km for HCMM).

- the relative complexity of interpretation of the data because there is a competition between several processes which are not easy to separate.

In spite of these difficulties, many works have already been done in this field (Reeves 1975 - Cihlar 1976). They show the usefulness of thermal infra-red bands not only because they give informations which are complementary to the visible and microwave ones, but also because they allow the measurement of specific properties such as the surface temperature.

For a long time, thermal infra-red data have been recorded and used as a particular image of the ground and analysed or interpreted as such. This qualitative approach gave already interesting insights on the behaviour of the surface of the earth for many applications in which a qualitative description of the phenomena is sufficient (current flow mixings, windtrace on the ground, more or less irrigation or evaporation, wet zones, etc...). Nevertheless, such an approach is not sufficient for quantitative analysis because it takes into account only a part of the information

namely the relative contrast, and it is very dependent on the recording procedure (calibration of the instrument, atmospheric perturbations, etc...) so that comparisons between different sites, or between different periods on a given site could not be made on a quantitative basis.

Fortunately, great progress have been made both on the instrumentation techniques which allow now the recording of the data on a digitalized basis, with correct calibration and on the theoretical models for interpreting the observed features. A new area of applications of infra-red remote sensing is now being opened and will be discussed in these lectures.
The aim of these lectures is also to give ideas on the different problems any remote sensor is faced with and to give some indications on the different manners to overcome these difficulties or at least to take care of them. In fact, it is impossible to give a complete and detailed analysis of all these points in a few pages. They will be discussed therefore rather synthetically and the details are to be taken from the literature of which a part is quoted at the end.

This lecture which is the point of view of a physicist, will be divided into four parts : in the first part (section 2), the general principles of thermal infra-red remote sensing and the general problems arising in the realization of experiments and in their interpretation will be presented. At this end, will be introduced the different parameters, their interactions and the physical phenomena which lead to the observed and recorded signal. In order to lighten the text a more detailed description of what is going on with the corresponding equations is given in an appendix.

To see whether it is worthwhile to solve the problems and the difficulties inherent to remote sensing, we shall present the properties of the main parameters occuring in thermal infra-red remote sensing namely the emissivity, the albedo and the surface temperature. In section 3 , some utilization of the spectral emissivity as a tool for analysing soil, bare soils or vegetated soils will be discussed while section 4 will be particularly devoted to all the questions concerning the surface temperature. This temperature may be useful to three respects : analysis of the surface of soils (bare or vegetated), analysis of underground structure composition and texture and finally analysis of heat and mass transfer between the ground and the atmosphere.

The different applications previously suggested are really useful if the necessary informations can be extracted from the data recorded by remote sensing. Therefore, in section 5, we shall analyse different conditions required to get as good data as possible in order to feed correctly the theoretical models to deduce from them the interesting parameters.

It will not be possible to discuss here active infra-red remote sensing, although it is a very interesting domain, because it is not yet very much used. Nevertheless from place to place some connections with active measurements will be made.

2 GENERAL PROBLEMS INVOLVED IN THERMAL INFRA-RED REMOTE SENSING

2.1 General principles of remote sensing and perturbations

The earth can be though to be a physical system which is periodically excited by the sun and which responds in two ways to this external excitation as it is seen on figure 2-1.

i) First it reemits by reflexion a part of the energy which it receives from the sun, mainly in the visible and near infra-red part of the electromagnetic spectrum. The main parameter driving this reflexion is the albedo.

ii) Secondly it emits directly an electromagnetic radiation mainly in the thermal infra-red part of the spectrum (but also in the micro-wave bands !). This emission depends on the emissivity and the temperature of the surface which

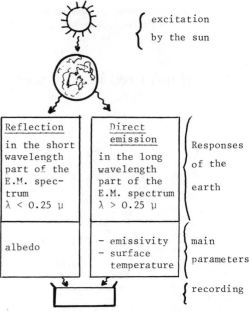

Figure 2 -1. Physical principles of remote sensing.

depends on the balance between the various fluxes of heat and mass exchanged at the boundary between the upper layer of the ground and the lower layer of the atmosphere.

In infra-red remote sensing, one is therefore concerned with two types of physical phenomena which are strongly connected to each other : electromagnetic (E.M.) radiation and heat and mass transfers. The E.M. radiation takes place essentially in the atmosphere and at the boundary between ground and atmosphere. It does not penetrate inside the ground for thermal I.R. bands (a few micrometers !). Heat and mass transfer processes take place mainly inside the ground and at the boundary between the ground and the atmosphere. The different fluxes of mass and energy exchanged between the ground and the atmosphere are sketched on figure 2 -2.

Strictly speaking, the remote sensing technique is based on the measurement of the up-ward E.M. radiation which we shall discuss now. The heat and mass transfer fluxes are not directly measured by remote sensing techniques and methodology, but indirectly since they influence the surface temperature through the thermal properties of the ground and its environment. They will therefore be discussed in

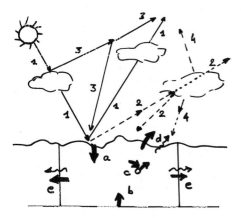

Figure 2 -2. Sketch of the fluxes occuring in thermal infra-red remote sensing. The figures and the thin arrows indicate the E.M. fluxes, the letters the heat (large arrows) and mass transfer (wiggling arrows) fluxes.

(1) Flux directly emitted by the sun and re-emitted by the ground by reflexion. (short wave length $\lambda < 2.5 \mu$). A part of the solar flux is absorbed or scattered by the atmosphere (which will be heated) before it reaches the ground (a part of this flux is absorbed in the ground which will therefore be heated. A part of this re-emitted flux is again absorbed and scattered by the atmosphere. This is the short wave length direct flux.

(2) Flux directly emitted by the ground in the I.R. bands. A part of this flux is absorbed and scattered (back) by the atmosphere (long wave length $\lambda > 2.5 \mu$).

(3) Flux coming from the sun and scattered by the atmosphere and partly re-emitted by reflexion by the ground. This last flux is also absorbed or re-scattered by the atmosphere. This is the short wave length diffused flux.

(4) Flux directly emitted by the atmosphere (long wave length flux).

All these fluxes are discussed in detail in Appendix A.

a) Net coming heat flux due to radiative process.

b) Net geothermal heat flux.

c) - Latent heat flux in the ground (Vaporization of the ground water)

   - Different fluxes exchange inside the ground - sources or sink of heat.

d) Heat fluxes exchanged with the atmosphere comprising

   - sensible heat (convection, ...)

   - latent heat (evaporation, evapotranspiration, ...)

mass fluxes : exchange of water or water vapor between the ground and the atmosphere (evaporation, condensation, evapotranspiration, rain, ...)

e) Heat and mass transfer inside the ground (the method to calculate these fluxes are discussed in Appendix B).

The coupling between the radiative and heat fluxes appears in (a) and (d) which are both strongly dependent on the micro-climatic conditions at the surface.

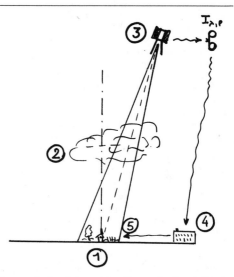

Figure 2-3. Schematic flow of the information in remote sensing.

section 4 and we shall see how they can be looked at by remote sensing. The general set up of a remote sensing experiment is shown on figure 2-3.

In the thermal infra-red bands one measures the properties of the observed scene (a pixel) using the E.M. radiation as the vehicle of the information. It is emitted by the observed pixel ① modified by the atmosphere ② , partly recorded on the platform ③ which transforms it into an output signal $(I_{\lambda,p})$ which is processed either in flight or in the laboratory. It is then interpreted ④ with the help of theoretical models and skill of the scientist, the results are checked during the pre-operational period by ground-truth measurements ⑤ .

The aim of remote sensing research is to avoid these ground truth measurements during the operational period and to produce nevertheless the correct informations

155

for the needs of the different users. Generally speaking, the electromagnetic radiation is mainly specified by five quantities which depend on time :
- its intensity which gives the amount of power carried by the E.M. radiation,
- its phase which says somehow how it is carried,
- its wavelength or more generally its spectrum, i.e. the relative intensity of the radiation in each unit of width of wavelength at the different wavelength,
- its polarization which indicates how the E.M. radiation vibrates,
- its direction of propagation.

Of these five characteristics only three are actually used on a large scale in infra-red remote sensing because the radiometers now in operation, measure only the intensity, the direction of propagation and a very small part of the spectrum of E.M. radiation at a given time. This produces a large waste of information. Therefore tentatives are made to measure the polarization (Deschamps and Tulpin 1976, Deschamps 1977, Gjessing 1978), or to improve the spectral resolution of the radiometer using for instance interferometers (Schaper 1976) or increasing the numbers of I.R. channels (Schaper 1976, Monge and Sirou 1976).

The phase has only useful applications in active measurements with lidars (Gjessing 1978, Collis and Russell 1976). Active infra-red measurements may provide very interesting informations because powerful lidars can be built in this domain ... but this is another story and it is not possible to discuss a so large subject in the scope of these lectures.

In these lectures we shall therefore restrict ourselves to passive infra-red remote sensing based upon energy, spectrum and direction of propagation measurements. The main impact of the applications will be on ground and not on atmosphere, although this one has to be taken into account.

As it has been discussed in the appendix and illustrated on fig. 2-2, the energy emitted by the pixel, comes from three different parts :
- the energy coming from the incident direct solar flux and reflected by the pixel,
- the energy coming from the downward atmospheric flux and reflected by the pixel,
- the energy directly emitted by the pixel.

These different contributions are mixed up and one of the problems of remote sensing is to separate these contributions.

The energy arriving at the window of the radiometer is not the emitted energy. It comprises the energy emitted by the ground and attenuated by the atmosphere plus the energy directly emitted by the atmosphere. Here again these contributions are mixed-up and have to be separated. Generally the atmosphere produces also geometrical distorsions due to refraction and turbulence. They can be neglected to first order in passive infra-red radiometry because the actual dimensions of a pixel are large enough not to be very affected by these distorsions. To take care of atmospheric perturbations specific ground truths and atmospheric sounding will have to be performed.

The output signal produced by the radiometer is again disturbed by the spectral response of the instrument which must be calibrated and by the scanning procedure itself. Furthermore, the image obtained is distorted by the rapid variations of the platform attitudes due to local atmospheric turbulence (this is less important for space platforms). To take care of all these perturbations, radiometric and geometrical corrections have to be performed.

A special attention should be paid to the in-situ ground truth measurements to make them comparable with remote measurements otherwise it will be very difficult to make corrections between the remotely recorded and the ground recorded data. Let us discuss these different points more carefully in the following section.

2.2 The different factors involved in infra-red remote sensing

From the brief review reported on Appendix A, it is possible to gather the different factors which have to be considered in remote sensing in order to obtain usable data. These factors can be divided into two great classes, namely :
- the factors which are not at the disposal of the experimentalist,
- the factors which can be chosen by the experimentalist.
The most part of these factors are shown on Table 2-1.

i) Factors which are not at the disposal of the experimentalist

Ground parameters (useful)		Perturbative parameters (not useful)		at disposal of the observer	
Thematic parameters	Physical parameters influenced by the thematic parameters and leading to the measured radiance	Geomorphological parameters of the scene	Meteorological parameters Microclimatic parameters Atmospheric parameters	Logistical parameters Platform	Logistical parameters Radiometer
Plant - nature - type - dimension of the fields - age - phenological state - moisture stress - evapotranspiration - temperature - soil stresses - distribution of the leaves - LAI - physiological disorder - insect damage - nutrient deficiency etc... Soil - moisture - surface temperature - salinity - type and nature - mineral composition - structure - texture - surface roughness etc ... Underground - temperature - nature - faults - geothermal flux - moisture etc...	✱ Optical constitutive parameters - complex index of refraction - dielectric constants - electrical conductibility } emissivity reflectivity albedo ✱ Thermal parameters - thermal conductibility - heat capacity - volumic mass - composition } thermal inertia, thermal diffusivity → temperature distribution, surface temperature ✱ Mechanical and hydrological parameters - soil moisture conductivities, - liquid water capacity ✱ Geometrical parameters - shape of boundaries - roughness - dimensions of the characteristic element of medium	- altitude - relief - x slopes - x direction of the slopes - dimension of a characteristic structure	Atmospheric - aerosols distribution - molecular composition } air emissivity transmittivity - water content profile - air temperature profile Meteorological - sun position - cloud coverage - cloud temperature - rain - turbulence → variations of attitudes of platforms Micrometeorological (in the limit layer) - wind speed - air temperature - relative humidity	- season - time (hours in day and night) - frequence of measurement - velocity v - altitude h - attitude	- spectral bands (λ, $\Delta\lambda$) - spectral response $f(\lambda)$ IFOV $d\omega$ - integration time τ - total field of view - $\frac{v}{h}$ ratio (for a scanner) - type of recording

Measured quantity $I_\lambda(\ell, \theta, \varphi, t)$

Table 2-1 : Different types of parameters influencing directly or indirectly the recorded out-put signal of the radiometer

We divide these factors according to their direct usefulness for a given problem :

a) The "useful" factors, i.e. those which are related to the description of the phenomena of interest. They are the :

- thematic parameters, i.e. those which describe directly the problems of interest (wetness, roughness, type of plant, production, evapotranspiration, health of plants, temperature, etc...)

- physical parameters, i.e. the parameters which describe the physical processes leading to or describing the thematic properties of the ground (spectral reflectivity, albedo, spectral emissivity, dielectric constant, coefficient of turbulent transfer, etc...).

These parameters describe the emission of E.M. radiation and the heat and mass transfer. They are divided into :

- optical parameters (or E.M. parameters)
- thermal and hydrological parameters
- geometrical parameters.

These parameters describe the properties of the pixel itself and of its boundary conditions (a pixel is not isolated in nature !).

b) The "perturbative" factors, i.e. those which modify the measurement (this distinction is somehow arbitrary because the strong inter-connection between the parameters make some of them "useful" for one application and "perturbative" for another one). They are the :

- atmospheric parameters (including all the meteorological parameters)
- geomorphological parameters.

ii) Factors which are at the disposal of the experimentalist (logistical parameters)

Again we can distinguish two categories :

a) The parameters related to the platform.

b) The parameters related to the instrumentation. These parameters are represented in fig. 2 -4 which represents also the flow chart of the information for experimentation and interpretation. Different examples of these parameters are presented on Table 2-1.

The conclusion of this brief analysis is that there are a few parameters recorded (namely the various radiances averaged over the band width of the radiometer, cf. appendix) and many, many parameters leading to this output signal. The main problems which arise from the remote sensing method and which have to be solved in order to extract significant informa-

tions from these radiances are briefly presented in the following section.

2.3 The different problems inherent to remote sensing data inversion

One can divide these problems into three classes, each having theoretical and experimental aspects :

i) Methodological problem
For each pixel the recorded quantities $I_{\lambda_p}(h,\theta,\phi,t)$ are not very numerous (generally one or two at a given time). Unfortunately they depend on many parameters, the so-called useful factors as well as the so-called perturbative parameters. How to extract from these quantities the useful factor and how to separate their specific contribution to the measured data ? Furthermore, these functions are known with errors. How to correct ? Is the solution unique ?

ii) Problems of reliability (or what is the physical meaning of the measured quantity ?)
Secondly the physical quantities which may be extracted from the output signal are averaged quantities over a given surface of the ground. How these remotely measured parameters are related to the ground parameters ?

iii) Theoretical model (or have the measured quantity a practical interest ?)
Thirdly, once the inversion procedure is assumed to be correctly performed one has to relate the remotely sensed data with the thematic parameters which are finally the interesting parameters for the users. Are these relations unambiguous ? What to do to precise them ? In this part there is also a very complicated inversion problem. These problems are schematically shown on Table 2-2 and some of them are now discussed in more details.

2.3.1 Methodological problems

These problems are of two types :
i) Obtain good spectral average radiances at the ground level.
ii) Deduce from these radiances the average emissivity and the surface equivalent temperature.

To solve the first type of problems it is necessary to look at Table A 1 of Appendix A which shows the complicated relation between the output signal of the radiometer (which is the input in all interpretation procedures) and the interesting physical quantities. We shall only

Table 2 -2 : Theoretical and experimental problems to solve
        a tentative chart for interpretating the data

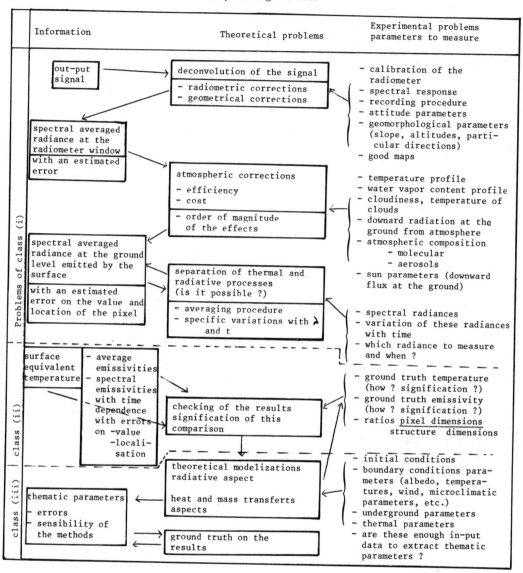

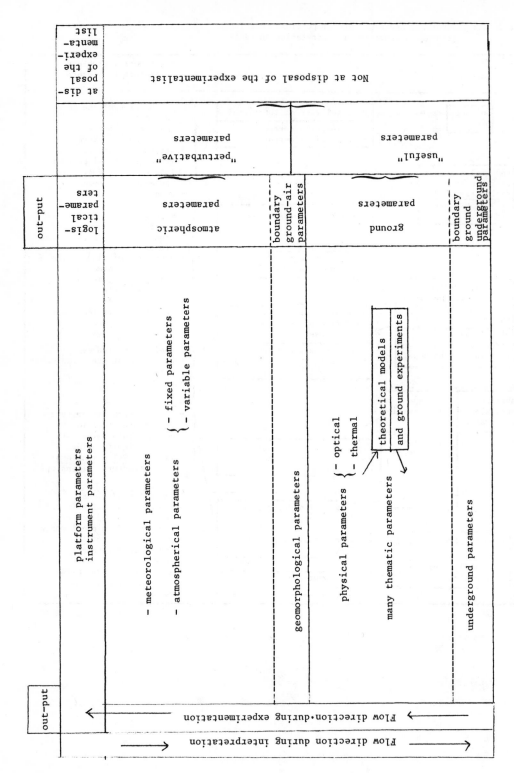

Figure 2-4 : Place of the different factors in the flow shart of information

quote here some aspects of the atmospheric corrections and the separation of emissivity and temperature mixing effects. All the problems connected with the calibration of the radiometer as well as with the instrumental deconvolution will not be discussed here (see a short discussion on § A31 of appendix A).

a) Atmospheric perturbations
First of all, the difference between the ground radiance and the radiance at the radiometer window may be very large (several degrees in black-body equivalent temperature difference). Furthermore it depends :
- on the position of the pixel (for instance the correction is not the same at the center and the edges of the image) (see fig. 2-5)
- on the wavelength (see fig. 2-6)
- on the altitude of the platform (see fig. 2-6)
- on the temperature and emissivity of the pixel
- on the temperature of the atmosphere.

These points are easily shown by formula (A-29) of appendix A (Becker, Imbault and Pontier 1977, Becker a 1978) which gives the ground radiometric temperature at the altitude z, $T(z)$. It is written

$$T(z) - T_S = A(\lambda) \frac{W(z)}{\cos\theta} |\theta(z) - T_S| \quad (2-1)$$

where $\theta(z)$ is an effective atmospheric temperature given in the appendix while $W(z)$ is the precipitable water. Since this formula reproduces correctly, to first order, the data as it is seen on fig. 2-5 and 2-7, it may be used to understand what is going on :
- if $\theta(z) > T_S$ the atmosphere emits more than it absorbs. The observed temperature will be larger than the true one (this is observed on fig. 2-6)
   if $\theta(z) \equiv T_S$ no correction
   if $\theta(z) < T_S$ the atmosphere absorbs more than it emits.

If it is possible to assume that $\theta(z)$ is constant over the image, this is not possible for the ground. Therefore, the contrast itself is modified and attenuated. In this model, the difference of the radiance emitted by two points is multiplied by a constant factor (if $\cos\theta \simeq \cos\theta_e \simeq \cos\theta$)

$$T_1 - T_2 = (1 - \frac{AW}{\cos\theta})(T_{S_1} - T_{S_2})$$

In this model aerosol effects have been neglected as well as clouds effects. In not clear atmospheres, these have to be taken into account.(Practical uses of this model are given in section 5).

- The amount of absorption due to aerosols is shown in fig. A5 of the appendix (Farrow 1975). It depends on the visibility. If this visibility is too low it becomes impossible to realize any remote sensing experiment.

- The effect of dispersed clouds perturbations is not simple to calculate ; it may amounts to (0.15°C) effect. (See for instance Hodora and Wells 1977).
These clouds modify also the local heating of the ground and will appear as shadowed regions.

b) Separation of emissivity and temperature
Again the equations discussed in the appendix show that the emitted radiance depends on the emissivity and the temperature of the surface. Therefore a contrast observed on an image may come from an emissivity difference, as well as from a temperature difference or both. How to separate these effects ?
The main idea is to take into account the differences of the variation with the time and the wavelength of the surface temperature and the emissivity. This question will be partly answered in section 5. Let us give some figures to show the importance of these effects.

It is easy to see that in the band 10,5 - 12,5 μ, i.e. at maximum emission, for 23°C, the relative variation in emissivity $\frac{\Delta\varepsilon}{\varepsilon}$ which compensates the relative variation of surface temperature $\frac{\Delta T}{T}$ is given by

$$\frac{\Delta\varepsilon}{\varepsilon} \simeq 5 \frac{\Delta T}{T} \quad (2-2)$$

Using Stephan formula, i.e. emission for the whole spectrum, one has $R = \varepsilon \sigma T^4$ which gives

$$\frac{\Delta\varepsilon}{\varepsilon} \simeq 4 \frac{\Delta T}{T}$$

The emissivity ε in Stephan's formula is not the same as the emissivity in the band width 10,5 - 12,5 μ. This point will be discussed in section 3.

From (2-2) one sees that a variation of 1% on ε (i.e. for instance from 0.98 to 0.99) is equivalent to a variation of 0.6°C in temperature for $T \simeq 300°K$ i.e. 23°C. Therefore, it is not necessary to spend a lot of time in correcting atmospheric perturbations or to measure the temperature with great accuracy if ε is

161

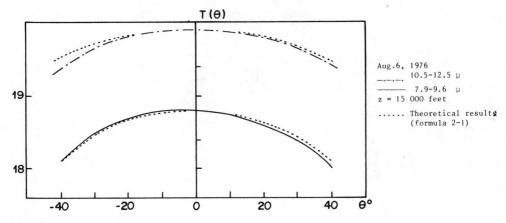

Aug.6, 1976
10.5-12.5 μ
— · — · —.
——— 7.9-9.6 μ
z = 15 000 feet

...... Theoretical results
(formula 2-1)

Figure 2-5. Angular dependence of the atmospheric attenuation. The data have been
taken above the Mediterranean Sea with
the radiometer ARIES. The theoretical
calculations are made with formula 2-1.
(From Becker and Pontier 1978)

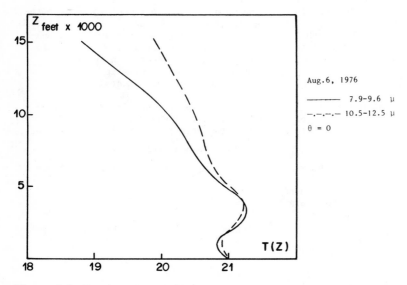

Aug.6, 1976

——— 7.9-9.6 μ
—·—·—·— 10.5-12.5 μ
θ = 0

Figure 2-6. Wavelength dependence of the
atmosphere attenuation. Same data as
above. (From Becker, Imbault and Pontier
1977)

162

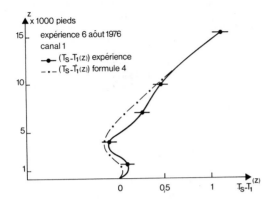

Figure 2-7. Theoretical prediction of the atmospheric perturbation from formula (2-1) and the experimental data. (From Becker, Imbault and Pontier 1977)

not known with an equivalent accuracy. This point is sometimes confusing. As it will be discussed in section 4, a relative variation of emissivity of 1% does not change very much the true surface temperature. What is changed here is the accuracy of the measurement : a true temperature of 300°C for a ground with an emissivity of 0.98 would appear to be at 299.4 if the emissivity is thought to be 0.99 and the difficulty is that ε is not constant over grounds. It depends on the moisture for instance. Inversely one sees that a small contrast in ε is detectable since actual radiometers can measure differences of 0.1°C or even less.

2.3.2 Problems concerning the physical meaning of the measured quantities

We shall restrict ourselves to the parameters occuring in thermal infra-red remote sensing, the emissivity, the transfer coefficients and the surface temperature.

i) Signification of the surface temperature and the surface emissivity

Let us consider some examples :

a) For a really homogeneous flat ground at thermal equilibrium, the surface temperature is well defined (although it is not easy to measure it and to know exactly what physical quantity is actually measured). This case is close to what happens if the dimensions of the pixels are small with respect to the characteristic dimensions of the structure of the ground.

b) For a non-homogeneous bare not flat soil, the question becomes more complicated. The point is that the temperature observed from the spacial platform is an average over a surface corresponding to several homogeneous parts of the ground.

For instance, in the case of fig. 2-8 it is easy to show from the definitions of appendix A that the observed radiance R_λ^{obs} is

$$R_\lambda^{obs} = \frac{R_{a\lambda}(t,\theta_a)d\omega_a + R_{b\lambda}(t,\theta_b)d\omega_b + R_{c\lambda}(t,\theta_c)d\omega_c}{d\omega_a + d\omega_b + d\omega_c}$$

(2-3)

where $R_a(t_a,\theta_a)$, $R_b(t_b,\theta_b)$, $R_c(t_c,\theta_c)$ are the radiances emitted by the homogeneous parts a, b, c, of the observed surface in the direction of the detector; if one is assuming that there is no reflexion by a part of the surface (otherwise the formula becomes more complicated) one is therefore led to :

$$R_\lambda^{obs} = \tilde{\varepsilon}\, R°(\tilde{T}) =$$

$$= \frac{\varepsilon_{a\lambda}(\theta_a)R_\lambda°(t_a)d\omega_a + \ldots + \varepsilon_{c\lambda}(\theta_c)R_\lambda°(t_c)d\omega_c}{d\omega_a + d\omega_b + d\omega_c}$$

(2-4)

where t_a, t_b, t_c and ε_a, ε_b, ε_c are respectively the surface temperatures and the emissivities of the parts a, b, c, and $\tilde{T}$ is the effective surface temperature.

The problem is to define $\tilde{T}$, from this unique relation with the two unknown variables $\tilde{\varepsilon}$ and $\tilde{T}$ (even if all the emissivities and temperatures of the subpart a, b, c are known).

This is not a simple problem. It is necessary to make several independent measurements to solve it. If indeed the surface is at equilibrium and is homogeneous (the sea for instance or if the pixels are much smaller than the characteristics homogeneous structures , the problem is simply solved as said in §a). If $\varepsilon_{\lambda a} = \varepsilon_{\lambda b} = \varepsilon_{\lambda c} = \varepsilon$, and $t_a = t_b = t_c = T_S$ formula (2-4) reduces to $\varepsilon_\lambda R_\lambda°(T) = \varepsilon_\lambda R°(T_S)$

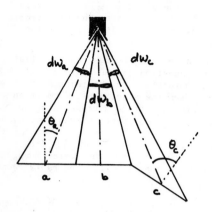

Figure 2-8. Observation of a heterogeneous surface
$$d\omega_a + d\omega_b + d\omega_c = d\omega$$

which can be given the solution
$\tilde{\varepsilon}_\lambda = \varepsilon_\lambda$ and $\tilde{T} = T_S$.

This averaging procedure may induce large errors. It depends on the angle of observation and on the altitude, as it is seen in fig. 2-9, and on the relative position of the pixel with the different contrast lines. Therefore, the observed radiance may change from one line to another one, and from one structure to another one as it is seen on fig. 2-10. Furthermore, the scanning procedure introduces errors in this configuration. Therefore, the same ground configuration may give different results according to different scanning procedures.

Fortunately the things are generally not so worse because expression (2-4) may be simplified in three cases which happen in nature :

j) If the temperature of the pixel is uniform (good conducting medium)
$\tilde{T} = T$
then

$$\tilde{\varepsilon}_\lambda = \frac{\varepsilon_{a\lambda} d\omega_a + \varepsilon_{b\lambda} d\omega_b + \varepsilon_{c\lambda} d\omega_c}{d\omega_a + d\omega_b + d\omega_c} \qquad (2-5)$$

jj) If the emissivity of all the parts are almost the same
$\tilde{\varepsilon} \simeq \varepsilon$
then

$$\tilde{T} = \frac{T_a d\omega_a + T_b d\omega_b + T_c d\omega_c}{d\omega_a + d\omega_b + d\omega_c} \qquad (2-6)$$

jjj) More generally, if the temperatures of the different part of the pixel are not very different, formula (2-4) can be simplified by expanding the Plank's formula around the mean temperature of the pixel (see formula A9). Then one is naturally led to an average emissivity given by (2-5) and by an average temperature given by

$$\tilde{T} = \frac{\varepsilon_{a\lambda} d\omega_a T_a + \varepsilon_{b\lambda} d\omega_b T_b + \varepsilon_{c\lambda} d\omega_c T_c}{\varepsilon_{a\lambda} d\omega_a + \varepsilon_{b\lambda} d\omega_b + \varepsilon_{c\lambda} d\omega_c} \qquad (2-7)$$

This is an average temperature weighted by the emissivities. The great difficulty to interpret this average temperature is that it is not this average which occurs for heat and mass transfer analysis. For instance, if one is looking for the flux of sensible heat exchanged by the pixel (see formula B 8b) one is naturally led to an average defined by :

$$\tilde{T}' = \frac{d\omega_a \frac{C_a}{r_a} T_a + d\omega_b \frac{C_b}{r_b} T_b + d\omega_c \frac{C_c}{r_c} T_c}{\frac{C_a}{r_a} d\omega_a + \frac{C_b}{r_b} d\omega_b + \frac{C_c}{r_c} d\omega_c} \qquad (2-8)$$

and there is no reason to have $\tilde{T} = \tilde{T}'$.

Fortunately $T_a \simeq T_b \simeq T_c \simeq T_m$
where

$$T_m = \frac{d\omega_a T_a + d\omega_b T_b + d\omega_c T_c}{d\omega_a + d\omega_b + d\omega_c}$$

and to first order
$\tilde{T} \simeq T_m$

which is nothing but expression 2-6. Therefore, to first order, one can use (2-5) and (2-6) to define average over a pixel of emissivities, temperature and any parameter α

$$\tilde{\alpha} = \frac{\Sigma_i \alpha_i d\omega_i}{\Sigma_i d\omega_i}$$

c) For a vegetated surface the things are more complicated because the surface is not well defined for the I.R. radiation does penetrate inside the vegetated cover. The averaging procedure will therefore depend much more on the altitude, angle of analysis, etc... than previously. Furthermore, the precise definition of a surface temperature in this case is not simple. At which distance from the ground is the "effective" surface if it exists ?

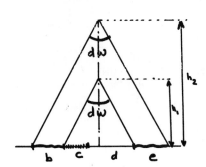

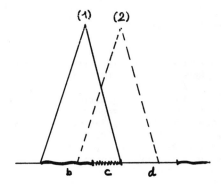

Figure 2-9. Effect of the altitude on the average observed radiance.
On this figure, the surface is flat and the different surface elements of the homogeneous part are written $d\Sigma$.
For the figure configuration :

at h_1 :

$$R_1^{obs} \simeq \frac{\varepsilon_c R°(T_c)d\Sigma_c + \varepsilon_d R°(T_d)d\Sigma_d}{d\Sigma_c + d\Sigma_d}$$

at h_2 :

$$R_2^{obs} \simeq \frac{\varepsilon_b R°(T_b)d\Sigma_b + ... + \varepsilon_e R°(T_e)d\Sigma_e}{d\Sigma_b + d\Sigma_c + d\Sigma_d + d\Sigma_e}$$

Obviously $R_1^{obs} \neq R_2^{obs}$, independently of the atmospheric effects.

Figure 2-10. Effect of the position of the pixel with respect to the inhomogeneous structure.

For line 1 :

$$R_1^{obs} \simeq \frac{\varepsilon_b R°(T_b)d\Sigma_b + \varepsilon_c R°(T_c)d\Sigma_c}{d\Sigma_b + d\Sigma_c}$$

For line 2 :

$$R_2^{obs} \simeq \frac{\varepsilon_c R°(T_c)d\Sigma_c + \varepsilon_b R°(T_b)d\Sigma'_b + \varepsilon_d R°(T_d)d\Sigma'_d}{d\Sigma'_b + d\Sigma_c + d\Sigma'_d}$$

Again the two results are different.

ii) Some problems connected with ground truth measurements

The short discussion on the signification of the emissivity or the temperature shows that one has to be careful in ground truth measurements and in the comparison between ground and remotely measured quantities. Since a session of the course is devoted to these questions, only two classes of problems will be raised :

a) What is the signification of the temperature of a vegetated surface as well as its average emissivity ?
How and where should the measurements be done to make them comparable with the remotely measured quantities ?
For instance, a thermometric measurement on a vegetated surface does not give the surface temperature which is measured from a flying platform because they do not correspond to the same physical quantity (this difficulty is less important for water or bare soils (see for instance Bonn 1977). The thermometric measurement gives the temperature of the square millimeter around the thermometer ... and even it may be a mixture of air and ground temperature. It is not the effective temperature averaged over a pixel. Since the I.R. radiation penetrates in the vegetated cover, the definition of the effective surface, if it exists, is opened. For a forest, this question is crucial.

b) A second type of problems concerns the measures which have to be done to feed the different models used in remote sensing : atmospheric corrections, geometrical corrections, boundary and initial conditions for the different time-dependent models. Some of these parameters are already shown in table 2-2. Their influence will

165

be discussed with more details in section 4 and appendix B.

2.3.3 Problems connected with theoretical models and interpretations

The main theoretical question is the following : once the spectral emissivity and the effective temperature have been extracted from the recorded output signal of the radiometer, is it possible to extract reliable informations on the ground properties ?

How to invert the system comprising a few equations (the few, measured quantities) having each of them many and many variables (the thematic parameters leading to the measured quantities). Some elements to answer these questions will be given in section 5 and more details can be found for instance in the recent book of S. Twomey (Twomey 1977).

To answer this fundamental question, very careful analysis of the relationships between thematic parameters and the observed physical parameters have to be done. To obtain these relationships it is necessary to continue both experimental and theoretical work, but having in mind the scale at which the ground is observed from space. For instance, investigation at the level of the leaves of a plant is not directly usable at the scale imposed by remote sensing.

In this evaluation the variability of the boundary conditions will play unfortunately a great part as we shall see in section 4. The boundary conditions have two types of variation :
  - regular ones which can be modelled (sun elevation, seasonal effects, ...)
  - fluctuating ones which can be analyzed on a statistical basis only (local wind, rain, clouds, ...)
How to take into account these effects by remote sensing is not a simple problem.

Many simple empirical models exist to relate observed radiance to thematic phenomena (for instance there exists a linear relationship between the net radiation flux and the evapotranspiration rate of plants, the temperature of the soil and its moisture, etc...). These models are generally fitted with constants obtained by linear regression formulae. The advantage of this procedure is that the formulae are simple and give directly interesting results in the given situation. The big disadvantage of these empirical formulae is that they are not general. The numerical coefficients depend on the geographical situation, the meteorological situation, and other parameters (for instance, the relationship between the net radiation flux and the evapotranspiration rate of plants depends on the nature of the plants, their health, ... See Perrier 1976). These formulae must therefore be used with great care for situations which are different from the original ones.

The mathematical reason for this is that the observed radiance R is a function of many thematic parameters a_1, a_2... a_n : $R(a_1 a_2 ... a_n)$. In the direct analysis one assumes all these variables to be constant but one, say a_2, the evapotranspiration ; consequently, one gets only the dependence of R on a_2 and this dependence is a function of the other parameters. This dependence may well be linear.

Conversly, when one knows the variation of R it is not possible to deduce the variation of a_2 alone because dR depends on the variation of all the variables:

$$dR = \frac{\partial R}{\partial a_1} da_1 + \frac{\partial R}{\partial a_2} da_2 + ... + \frac{\partial R}{\partial a_n} da_n$$

$$(2-9)$$

where $\frac{\partial R}{\partial a_i}$ is the partial derivative of R with respect to a_i which depends on all the parameters a_i.

To invert formula (2-9) it is necessary to have more than one equation. This fact implies the necessity of having enough measurements to deduce the interesting properties. This point will be emphasized in section 5.

Another important problem in this context is connected with the statistical models used for classifying the different parts of the ground on well-defined classes. In such classifications, one must be very confident that systematical errors can be introduced because of the systematical dependence of the remotely sensed parameters with geometry, angle of observation, etc... In the previous models, these systematical variations must be clearly defined in order to improve the accuracy of the classification.

2.4 Conclusions

In this section we have mainly raised questions to show that the extraction of the needed thematic parameters from the recorded output signals of the detector is not simple theoretically and that this inversion procedure must be looked at with care. Before we discuss the procedures to perform this extraction in section 5, we shall analyse in the next two sections whether it is worthwile and whether some relationship between the thematic parameters and the physical ones do exist.

3 PROPERTIES OF THE EMISSIVITY

3.1 Definitions and measurement

The emissivity of a natural body is the ratio of the radiance emitted by this natural body at thermal equilibrium, to the radiance emitted by a black-body at the same temperature. One difficulty is that a pixel in nature is not at thermal equilibrium and only effective emissivity can be defined as it has been briefly discussed in section 2.3.2. Another difficulty is due to the fact that the radiometer integrates the energy emitted in a band of width of wavelength which is not small (2 to 4 μ) so that the measured emissivities are in fact averaged emissivities which may depend on the radiometer properties.

3.1.1 Effective emissivity

It is well known (see for instance Fuchs and Tanner 1968) that the emissivity of a single leaf is not representative of the emissivity of a vegetative cover which is not necessarily representative of the emissivity of a pixel seen from a space platform. This is a direct consequence of the non homogeneity of the surfaces discussed in section 2.3.2. Nevertheless, it is possible to define an effective emissivity (and corollarly an effective temperature) from the reflectance properties of the cover or the pixel in a whole. By definition, we set

$$\varepsilon_\lambda(\theta,\phi) = 1 - \iint_{\text{hémisphère}} \rho_\lambda(\theta_i,\phi_i;\theta,\phi)\cos\theta_i \; d\Omega_i$$

$$(3-1)$$

where $\rho_\lambda(\theta_i\phi_i;\theta,\phi)$ is the bi-directional

effective reflectivity corresponding to the reflexion of the pixel as a whole defined in appendix A.

There is no simple relationship between the effective emissivity ε_λ and the emissivities of the constituants of the pixel, models must be built for that. Such models are largely used in the visible (Suits 1972, Verhoef and Bunnik 1976, Bunnik 1978) as well as in the microwave bands (see for instance Schanda 1976). These models are much more complicated in principle in the thermal infra-red domain because the effective emissivity depends on the temperature profile of the cover and is not therefore a specific property of the medium (for reasons analoguous to those which lead to formula (2-4)). Nevertheless, if the temperatures on the cover are not very different or if the vertical profile of temperature is not too much time dependant, the effective emissivity will not change on time too much and may be function of the medium only (see for instance Sutherland and Bartholic 1977). One expects that these effective emissivities depend on the structure of the cover, its cumulative leaf area index (LAI) and that for a vegetated cover there will exist also an asymptotic (or characteristic) spectral emissivity.

The measurement of these effective emissivities is not simple. First they cannot be done in a laboratory. The main reason is that the effective emissivity depends on structure, texture, roughness, temperature, profile, etc... of the cover, on the ratio of the cover characteristic dimensions to those of the pixel which are very different from the laboratory conditions. The measurements must be done in-situ. Secondly, if the surface is at thermal equilibrium (which is very hard to check), it could be possible in principle to measure directly the radiance of the surface and its thermometric temperature. But this is again very difficult because it is not easy to measure a surface temperature and more difficult to obtain the effective surface temperature if the surface is not at thermal equilibrium (Fuchs and Tanner 1968, Isdo, Jackson and Reginato 1975). Thirdly, the environment itself produces several perturbations which must be controlled and corrected. One of the methods to overcome these difficulties is to use a comparison between active and passive measurement with a pulsed source (Becker 1979). With this method, the effect of the environment may be eliminated as well as the measure of the temperature. Furthermore, it reduces the error of measurement because it gives directly $1-\varepsilon$ through the definition (3-1). The schematic set up of the experiment is shown on figure 3-1. The qualitative shape of the output signal of the radiometer as a function of time for a pulsed source and a non-pulsed source is shown on fig. (3-2).

3.1.2 Averaged emissivities

i) Average over the wavelength bands

The spectral emissivity is defined for one wavelength by :

$$\varepsilon_\lambda(\theta,\phi) = \frac{R_\lambda(\theta,\phi,T)}{R_\lambda^\circ(T)}$$

where T is the temperature of the emitting surface (assumed to be at thermal equilibrium), $R_\lambda^\circ(T)$ is the Planck's radiance and R_λ is the time radiance of the pixel.

The power emitted by a ground surface is measured by a radiometer which has a peculiar spectral response $f_\omega(\lambda)$. Neglecting for simplicity in this discussion the atmospheric perturbations, the signal output may be written as

$$I \simeq \iiint_{\lambda_1}^{\lambda_2} R_\lambda(\theta,\phi,T) \, f_\omega(\lambda) d\lambda \, d\omega$$

where λ_1 and λ_2 are the low and large wavelength limits of the radiometer bandwidth and $d\omega$ the I.F.O.V. solid angle. (The complete response is given in section A3.2 of appendix A). Therefore for a given target one observes

$$I(\theta,\phi,\lambda,T) = \iiint_{\lambda_1}^{\lambda_2} \varepsilon_\lambda(\theta,\phi,T) f_\omega(\lambda) R_\lambda^\circ(T) d\lambda d\omega$$

$$(3-2)$$

while for the black-body at the same temperature :

$$I^\circ(T,\lambda) = \iiint_{\lambda_1}^{\lambda_2} f_\omega(\lambda) \, R_\lambda^\circ(T) \, d\lambda d\omega$$

$$(3-3)$$

The true average emissivity is defined as

$$\varepsilon_\lambda = \frac{\int_{\lambda_1}^{\lambda_2} R_\lambda(\theta,\phi,T) d\lambda}{\int_{\lambda_1}^{\lambda_2} R_\lambda^\circ(T) d\lambda} = \frac{\int_{\lambda_1}^{\lambda_2} \varepsilon_\lambda(\theta,\phi) R_\lambda^\circ(T) d\lambda}{\int_{\lambda_1}^{\lambda_2} R_\lambda^\circ(T) d\lambda}$$

$$(3-4)$$

But from expressions (3-2) and (3-3) the measured averaged emissivity will be

$$\varepsilon'_\lambda = \frac{I(\theta,\phi,\lambda,T)}{I^\circ(T,\lambda)}$$

i.e.

$$\varepsilon'_\lambda = \frac{\iiint_{\lambda_1}^{\lambda_2} \varepsilon_\lambda(\theta,\phi,T) \, f_\omega(\lambda) \, R_\lambda^\circ(T) \, d\lambda d\omega}{\iiint_{\lambda_1}^{\lambda_2} f_\omega(\lambda) \, R_\lambda^\circ(T) \, d\lambda d\omega} \qquad (3-5)$$

This averaged ε'_λ is correct, i.e. will not depend on the spectral response of the radiometer $f_\omega(\lambda)$, if ε_λ does not vary too much in the interval λ_1, λ_2.

If contrarywise ε_λ varies very much (resonance effects as it is the case for the well known restrahlung lines in the silicate rocks) some difficulties may arise. In this case, it may happen that the average depend on the band width. For instance, ε_{8-14} may be different from $\varepsilon_{10.5-12.5}$ if ε_λ varies strongly with λ in any of the two spectral bands. This may be particularly true for the total emissivity which appears in the Stephan formula, which corresponds to an infinite band width :

$$\varepsilon_\infty \neq \varepsilon_{8-14}$$

ii) Averages over the angles

In many analysis, it is assumed that $\varepsilon_\lambda(\theta,\phi)$ does not depend on the angle.

This may lead to some errors on the edges of the picture for aerial thermography. This averaged emissivity over the angles is defined as

$$\varepsilon_\lambda = \frac{1}{\pi} \iint_{\text{hém.}} \varepsilon_\lambda(\theta,\phi) \, \cos\theta \, d\Omega$$

The same discussion as above can be directly extended to this average.

In the analysis of data, one should be aware of these difficulties and be careful in the comparison between the different available data before relevant conclusions be drawn.

168

3.2 Some properties of spectral emissivities

In spite of the difficulties just mentionned, interesting results have been obtained.
Since the emissivities of vegetated cover
will be presented by C. de Carolis in
this course, the following discussion will
be limited to soils only (Carolis 1979).

Although rocky-soils are not in the scope
of this lecture, we mention that, classification procedures, based upon the spectral properties of the emissivities of
the different silicate rocks, has been
successfully used in different geological
terrains (Vincent 1975). As for soils, the
following general properties have been
shown :

i) The value of the emissivity depends
on the mineral composition of the soils.
It varies from 0.88, or less for sands,
to about 0.98 for clay and loam according
to its texture, the grain size and indeed
its mineralogy. The values obtained in

laboratory are questionable, mainly because
the remote sensing conditions are not the
same as the laboratory measurements (see
above § 3.1).

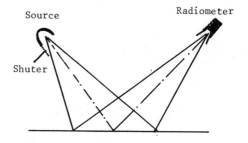

Figure 3-1. Principle of measurement of
effective emissivity based on the comparison between active and passive measurement.

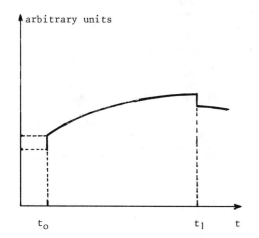

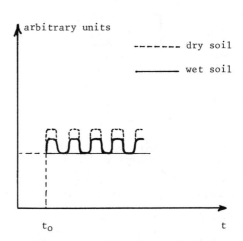

Figure 3-2. Qualitative variation of the output
signal of an infra-red radiometer when the pixel
is illuminated by an infra-red external source.
a) source turned on at time $t = t_o$
b) source periodically turned from time $t = t_o$ (Ngai 1978)

a) When the external source is turned on
at $t = t_o$, there is an instantaneous jump
due to the reflexion. Afterwards, the external source heats up the medium. From
this curve it is possible to obtain the
thermal inertia of the medium. At $t = t_1$
the source is turned off and the medium is
cooling afterwards.

b) In this case, the heating of the medium
is much lower than with a continuous source. Such a procedure is more efficient for
emissivity measurements.

169

ii) In the infra-red bands, the spectral emissivities of the different dry soils are rather flat, except when sands are mixed (see figures 3-3 et 3-4). This flatness is partly due to the fact that the most a surface is rough, the less structure there is in the spectral emissivities. This is shown on figure 3-5 for quartz and powdered quartz.

iii) The value of the emissivity increases when the water content of the soil increases. This is easily understandable since the water emissivity is larger ($\varepsilon \simeq 0.98$) than the dry sand emissivity ($\varepsilon \simeq 0.85$). This is observed on figure 3.2 b. For instance, the measures made by Fuchs and Tanner (Fuchs and Tanner 1968) gave the results shown on table 3-1. The results should depend on the water profile inside the few centimeters of the surface.

Table 3-1. Emissivities of plainfield sand and the volumetric water content of the soil layer between 0 and 2.5 cm.

ε_{8-13}	Water content %
0.94	8.4
0.92	6.2
0.92	5.8
0.90	3.5
0.90	0.7

iv) Recent measurements on saline waters confirm the dependence of the emissivity of the water on its salinity (Querry, Holland, Waring, Earls and Querry 1977). The results show a maximum of variation in the 9-10 μ band. However the variations on the emissivities are of the order of 1 to 2 % and this will induce very low variation on the soil emissivity. The analysis of the salinity of soil by a direct spectral analysis will therefore be very difficult.

3.3 Conclusions

The different variations of soil emissivities which we quoted may induce large radiance differences for soils having different compositions and structure and for soils with different moistures. It could be therefore possible, in principle, to discriminate among these soils but this is very difficult unless specific conditions on the thermal equilibrium are

respected because the emitted radiance depends on the emissivity and the surface temperature. It is therefore very difficult, without thermal model, to predict what the radiance of the different soils will be only by looking at their emissivities. This is particularly true for moisture measurements, because the moisture does modify not only the emissivity but also the thermal properties of the grounds and therefore the surface temperature.

Inversely it is not possible to neglect the different variations of the emissivities quoted above for thermal measurements because they can induce large errors. For instance, the variation of 0.90 to 0.94 of the emissivity shown in table 3-1 may induce errors of the order of 3°C on the measurement of the surface temperature.

The fact that the spectral emissivities do not depend strongly on λ (except for sandy soils) has the advantage to make the different averages quoted in §3.1.2 more accurate ... but the emissivity cannot be used to discriminate soils. On the other side, the presence of sand in the soils, may be discriminated by spectral analysis, mainly in the 9-11 μ band, provided that the surface temperature is constant (or at least has fluctuations less important than the fluctuations produced by the emissivity variations).

Generally speaking, it is more accurate in all these analysis to couple emissivity and thermal discrimination. Let us therefore discuss these properties now.

4 THE TEMPERATURE AS A TOOL FOR ANALYSING GROUND PROPERTIES

4.1 General principles

The pixel may be considered as a physical system which reacts by a surface temperature to some time dependant excitations produced by time dependant energetic fluxes coming from outside (the sun, the atmosphere, the clouds, the adjacent pixels, etc...). It is easy to understand that the surface temperature will depend on both the properties of the incident flux and the properties of the medium (the physical system).

If the temperature at a given time is interesting to know, the variation of this temperature with time is more interesting.

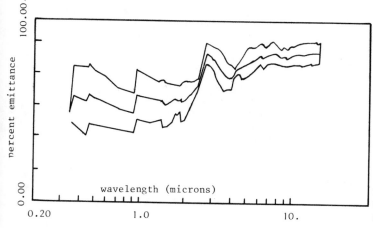

percent emittance

wavelength (microns)

0.20   1.0   10.

Figure 3-3.
Spectral emittance of all dry
soils except sans (Fligor 1974)

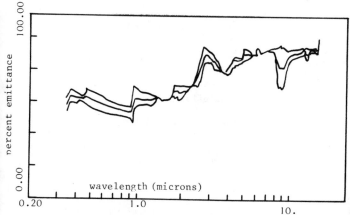

percent emittance

wavelength (microns)

0.20   1.0   10.

Figure 3-4.
Spectral emittance of all dry
soils including sans (Fligor 1974)

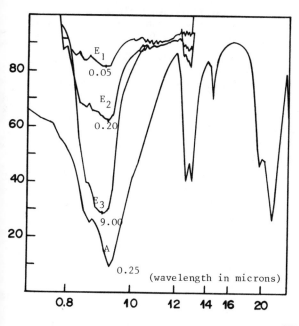

80

60

40

20

E_1
0.05

E_2
0.20

E_3
9.00

A
0.25   (wavelength in microns)

0.8   10   12   14   16   20

Figure 3-5.
Normal spectral emittance of
quartz with various surface
roughnesses.

The emittance of a polished plate
(E_3) is shown contrasted with

that of the roughned surface (E_2).

Also plotted on the diagram are
the emittance spectrum of 25 to
45 μ quartz powder (E_1) and the

absorption spectrum (A) of quartz.

(Lyon 1974)

171

Therefore repetitive, comparable measures of the temperature of the surface will bring many informations on the medium and its surface if the incoming flux is known or inversely on the fluxes if the medium has well known properties. Unfortunately, this is not an easy task and models must be developped to rely the surface tempe- rature variations to both the incoming fluxes and the medium properties.

Let us first discuss a simple model which shows the essential features and then present the different modifications which should appear on both the description of the medium and the incoming flux.

4.1.1 A simple model

The simplest model one can think of is an homogeneous material with infinite dimensions towards the bottom and in the horizontal directions. It is separated from the atmosphere by an infinite plane. The simplest flux to model the natural condition is to assume a periodic cosinus dependence.

Figure 4-1.

---

i) We assume an incoming flux

$$Q = Q_o \cos \omega t \qquad (4-1)$$

in the vertical direction and the same over the whole plane. The period is the time elapsed between two maxima of Q, it is

$$\text{\textcircled{n}} = \frac{2\pi}{\omega}$$

ii) The medium is characterized by a
- thermal conductivity λ
- volumetric heat capacity $\rho\, C_p$.
These parameters are assumed to be cons- tant through the whole medium.

iii) The average temperature is assumed to be uniform $T = T_o$.
With these very simple conditions the thermal transfer and conductibility equa- tions (see appendix B) can be solved exactly to give the temperature $T(z,t)$ at any time in the medium :

$$T(z,t)=T_o+\frac{Q_o}{\sqrt{\omega}\, P} \exp(-\sqrt{\frac{\omega}{2k}}z)\cos(\omega t-\sqrt{\frac{\omega}{2k}}z-\pi/4)$$

$$(4-2)$$

In this equation,

$P = \sqrt{\lambda\rho C_p}$ is the thermal inertia

$k = \dfrac{\lambda}{\rho\, C_p}$ is the thermal diffusivity.

Formula (4-2) learns many things which we recall briefly :

i) The amplitude of the oscillations of the surface temperature is :

$$\Delta T = \frac{2\, Q_o}{\sqrt{\omega}\, P} \qquad (4-3)$$

It is-proportional to Q_o (the incoming flux)
    -inversely proportional to P (the thermal inertia)
        - inversely proportional to the square root of ω, i.e. proportional to the square root of the period $\odot$.

ii) The phase of the surface temperature is $(\omega t - \pi/4)$; in other words, the maxi- mum of the temperature occurs at a time

$\Delta\phi = \pi/4\omega$ s after the flux is maximum (time of response) ; this phase varies with deepness.

iii) The variation of the temperature with time decreases in amplitude when the deepness increases. The penetration of the thermal wave is defined as the dis- tance L for which

$$\frac{\Delta T(L)}{\Delta T(0)} = \frac{1}{e}$$

This yields

$$L = \sqrt{\frac{2k}{\omega}}$$

172

atmosphere

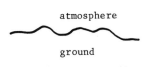

ground

— — — — — —

underground

Figure 4-2 a. Bare soil

atmosphere

- - - - - - - - - - - - - - -

cover

root

ground

— — - — - — · — - - —

underground

Figure 4-2 b. Vegetated surfaces.

Let us quote some realistic numbers (Tabbagh 1977).

For the daily wave $Q_o \simeq 40$ W/m^2

$$\textcircled{n}_j = 24 \text{ h}$$

$$\omega_j = 7.27 \ 10^{-5} \text{ s}^{-1}$$

For the annual wave $Q_o \simeq 10$ W/m^2

$$\textcircled{n}_a = 365 \ \textcircled{n}_j$$

$$\omega_a \simeq 2 \ 10^{-7} \text{ s}^{-1}$$

For the same medium $P = 1838$ W/(m^2.Ks$^{-1/2}$)

$$k = 0.5 \ 10^{-6} \text{ m}^2/\text{s}$$

$$\left.\begin{array}{l} \Delta T_j \simeq 5°C \\ \\ L_j \simeq 1,4 \text{ cm} \\ \\ \Delta\phi_j = 3 \text{ hours} \end{array}\right\} \quad \text{for the daily wave}$$

$\Delta\phi_j = 3$ hours (1/8 of a day)

$$\left.\begin{array}{l} \Delta T_a \simeq 24°C \\ \\ L_a = L_j\sqrt{365} \simeq 29 \text{ cm} \\ \\ \Delta T_a \simeq 45 \text{ days} \end{array}\right\} \begin{array}{l} \text{for the annual wave} \\ \text{(much larger than} \\ \text{the daily wave)} \end{array}$$

$\Delta T_a \simeq 45$ days (1/8 of a year)

On the principle, therefore a measure of ΔT will yield P if Q_o is known or Q_o if P is known. This measure will depend on underground properties if the wave has enough penetration in the medium. Since P depends also on the moisture, it may be possible to obtain also the moisture... but, for actual media, the relationship between T and Q is more complicated than (4-2). Let us discuss briefly what is modified both for the medium and the incoming flux.

4.1.2 More realistic description of the situation

A complete description must include three parts (cf. the points i, ii, iii) :
    - an analysis of the net incoming flux
    - an analysis of the properties of the medium
    - a definition of the initial conditions.

It must be adapted to the dimension of the analysed pixel.

i. The medium

We distinguish between bare soils and vegetated soils.
    a) Bare soils. The description is more complicated because the medium is not semi-infinite, nor homogeneous and furthermore the parameters are coupled to the temperature, moisture, etc...
We consider :

i) Geometrical parameters giving the shape and the position of the boundaries :
    - between ground surface and atmosphere (shape of the surface, roughness, etc...)
    - between ground and underground. They are defined either by the depth at which the underground temperature is known, or the geothermal flux, or the ground water table.
    - between adjacent systems. They are defined by the positions where the temperature (or the flux) is known.

ii) Constitutive parameters (see appendix B) driving the behaviour of heat and mass inside the medium. They enter directly the equations of heat and mass conductivity (eq. B-13, B-14). They have been recalled in table 2-1.
They are : the thermal conductivity λ, the volumic heat capacity ρC_p, the moisture conductivities K_ψ, the water capacity C_ℓ.

They depend on ground texture, grain size, mineralogy, moisture, etc... Some values of the thermal conductivity λ are given

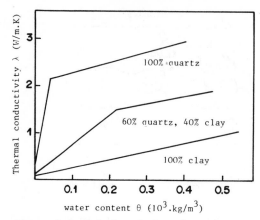

Figure 4-3. Calculated thermal conductivities for three mineral compositions
at 293 K (from Rosema b 1975)

on figure 4-3 taken from A. Rosema (Rosema b 1975). Qualitatively these curves
can be understood because the thermal
conductibility of water is larger than
the thermal conductibility of the grains
of soil and the voids between them.
The volumic heat capacity can be given by
a phenomenological expression (De Vries
1952)

$$\rho C_p \simeq 0.46 \; \theta_m + 0.6 \; \theta_o + \theta_w \qquad (4-4)$$

where θ_m, θ_o and θ_w are the volume fractions of mineral materials, organic mate-
rials and water. In the same line K_ψ and
C_ℓ do depend strongly on the moisture (cf.
for instance Philip 1975). Typical
values can be found also in A. Tabbagh
(Tabbagh 1976) or A.D. De Vries (De Vries
1966). If the medium is not homogeneous,
these parameters depend on the position,
furthermore, if there are faults or two
different terrains, they may be discontinuous.

b) Vegetated soils. The system is even
more complicated (see fig. 4-2 b) because :

i) The system must comprise three layers
with different properties :
- the vegetated cover itself
- the roots
- the soil itself.

ii) The boundary between the atmosphere
and the cover is not well defined ; it
varies with the age of the cover.

iii) The cover is transparent to the
heat and mass transfer fluxes.

iv) The roots modify the exchanges
between the soil and the cover as well
as between two adjacent systems.

It is therefore not possible to present
all the characteristics of such systems
in this lecture. Furthermore, nobody, to
my knowledge, has constructed a model
taking simultaneously all these aspects.
For instance, A. Perrier (Perrier 1976)
has analysed the two top layers, considering the two low layers as the boundary
of his model, while J. Methey (Methey
1977) has considered mainly the two lower
layers (other models will be discussed
below).

2. The boundary conditions

a) The fluxes at the atmosphere boundary

The cosine model is indeed not realistic
(see for instance, the curve quoted G in
fig. 4-4) and a detailed analysis of the
fluxes should be performed. The main fluxes
have already been given in fig. 2-1.
They are positive when directed towards
the ground. They are :

i) The radiative fluxes (discussed in
appendix A) (net radiative flux)

- the incoming flux including :
  * the direct and diffuse sun fluxes at
the ground level, R_g ; its useful part
(i.e. the flux entering the medium) is
characterized by the total albedo, a,
(the reflectivity integrated over the sun
spectrum as well as the incident and
emergent angles) ; the orientation of the
surface with respect to the sun. It will
depend therefore on the time, the latitude,
the season as well as on the slope and its
direction.

  * the atmospheric radiation flux towards
the ground, R_a (see appendix A).

- the outgoing radiation flux governed by
the total emissivity of the surface. From
the discussion of § 3, one can write

$$R_{SO} = - \; \varepsilon_\infty \; \sigma \; T_S^4$$

The net flux is

$$R_n = R_g(1 - a) + R_a(1 - a) + R_{SO} \qquad (4-5)$$

174

Generally the long wavelength R_e and the short wavelength R_s part of R_g and R_a are displayed and formula (4-5) is rewritten

$$R_n = R_s(1 - a_s) + \varepsilon\, R_e + R_{so}$$

where ε has been assumed to be $(1 - a_e)$

ii) The heat fluxes (discussed in appendix B). They are of two types :

- the convective flux or sensible heat flux H due to exchange of heat by transport it is governed by the turbulent diffusivities (depending on the wind, the roughness of the surface (aerodynamical coefficient, etc...)) , the micrometeorological parameters of the low layers of the atmosphere.

- the latent flux LE due to the vaporization or condensation of water at the boundary. It is the product of the latent heat (at the equilibrium thermodynamical conditions of the media) by the mass flux or water E. It depends on the turbulent diffusivities (generally the same as for the sensible heat flux) and the micrometeorological conditions.
These fluxes are much more complicated to analyse for vegetated covers than for bare soils. We set

$$Q = R_n + H + LE$$

b) Other boundary conditions

These boundaries are included to make the problem solvable. The most convenient boundaries are used. Except in the model of A. Tabbagh (Tabbagh a and b 1977) and of M. Raffy (Raffy 1979), the models generally assume an homogeneous medium, i.e. there is no "horizontal" fluxes of heat or of mass. When there is discontinuous features, the influence of the discontinuity on the surface temperature is of short range and asymptotic temperature values corresponding to a homogeneous medium can be postulated at distances from the discontinuity of the order of a few meters (see figure 4-18 and 4-19).

3. Initial conditions

One must impose initial conditions, i.e. an initial distribution of temperature and of mass inside the medium. They must be guessed, calculated approximately from simple models or measured. Another possibility is to assume that the surface temperature is a periodic function of time

(correction for non periodicity can be added if necessary).

4.2 The influence of the different parameters on the ground surface temperature

The surface temperature depends on several of the parameters quoted in table 2-1, but these parameters have not an equal effect. Therefore it is very interesting to find out which parameter acts principally on which part of the surface temperature. Indeed, the prediction of the quantitative influence of the different parameters on the surface temperature depends on the model. A review of these effects has been given by E. Fitzgerald (Fitzgerald 1974), K. Watson (Watson 1975) and more recently by J. Cihlar and Mc Quillan (Cihlar and Mc Quillan 1977) and K. Watson (Watson 1979).

In this section we shall discuss the effects of the different parameters on a simple model which will be presented now and show from more elaborate models the confirmation of these predictions.

4.2.1 A simple model including a more realistic description of the net incoming fluxes

The model which will be presented here is still too simple but it has the advantage to be exactly solvable and to take care of the main parameters which come into the play in natural phenomena. It will be therefore possible to analyse the particular effect of each parameter on the ground surface temperature. This model is somehow the first step of a more general model built on a Fourier analysis of the different fluxes (Becker and Hechinger 1979).

The ground is still supposed to be flat, homogeneous and infinite in all directions (one dimension problem). At a depth, L, the temperature is assumed to be a constant equal to T_L, say the temperature of the water table (the oscillations are assumed to be damped at z = L)

$$T(L,t) = T_L \qquad (4-7)$$

Instead of being described by a simple cosine function as in § 4.1.1, the net incoming fluxes will be more closely related to what is actually occuring in nature and has been briefly discussed in § 4.1.2.

The net incoming flux Q is divided into three parts.

175

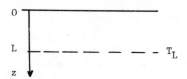

i) The global flux (short wave length)R_S

This flux is approximated by

$$R_S \simeq (1 - A)\, S\, C\, (1 + \cos(\omega t + \delta_S)) \quad (4\text{-}8)$$

where S is proportional to the solar cons-
     tant
    C takes care of the attenuation of
     the atmosphere, the seasonal effect
     the slope of the terrain in the
     N-S direction
    δ_S takes care of the slope in the
     E-W direction ($\delta_S > 0$ for slopes
     toward the East)
    A is the albedo.

ii) The long wavelength flux R_e, which
is given by (including R_{so})

$$R_e = (1 - a_e)R_a - \varepsilon\sigma\, v^4$$

which can be approximated by

$$R_e \simeq \varepsilon\sigma(T_A^4 - v^4) \qquad (4\text{-}9)$$

where ε is the average emissivity of the
ground (supposed to be equal to $(1 - a_e)$,
where a_e is the long wavelength albedo),

    T_A is an effective radiometric
temperature of the atmosphere

    v is the surface temperature of the
ground,

    σ is the Stephan constant.

iii) The non-radiative flux ϕ, which is
the sum of the latent heat flux and the
sensible heat flux. It can be written

$$\phi \simeq h(u)\{(T_a - v) + \gamma(e_a P(T_a) - e_g P(v))\} \quad (4.10)$$

where h(u) is a turbulent transfer coeffi-
     cient which depends on the wind
     speed u
    γ is the inverse of the psychro-
     metric constant
    T_a is the true air temperature

e_a is the relative humidity of the
    air which is assumed to be a cons-
    tant

e_g is the relative humidity of the
    surface of the ground which is
    assumed to be a constant

P(t) is the pressure of saturated vapor
    at temperature T.

From the simple formula (4-8) of the incoming global flux, it is logical to assume
that

$$T_A = T_M + T_D \cos(\omega t + \delta_A) \qquad (4\text{-}11)$$

$$T_a = T_m + T_d \cos(\omega t + \delta_a) \qquad (4\text{-}12)$$

T_A is indeed a function of T_a and e_a
(see for instance Staley and Jurica 1972)
where
$$T_D \ll T_M \qquad (\text{say } T_D < 0,1\ T_M)$$
$$T_d < T_m$$

With these approximations one can reasonably assume that the surface temperature
takes on the form

$$v = T_o + T_g \cos(\omega t + \delta) \quad (4\text{-}13)$$

with $T_g \ll T_o$, as it will be checked.
Inserting (4-11,12,13) into (4-9) and (4-10)
the total net incoming flux Q reads

$$Q = R_S + R_e + \phi$$

which can be written

$$Q = Q_o + Q^* \cos(\omega t + \delta Q) - \chi_g T_g \cos(\omega t + \delta) \tag{4-14}$$

where Q_o, Q^* and δ_Q are respectively the
average over a day, the amplitude and the
phase shift of the incoming flux. They
can be written

$$Q_o = (1-A)SC + h(u)\{(T_m - T_o) + \gamma(e_a P(T_m) - e_g P(T_o))$$
$$+ \varepsilon\sigma(T_M^4 - T_o^4) \qquad (4\text{-}15)$$

$$Q^* = \sqrt{Q_C^2 + Q_S^2} \qquad (4\text{-}16)$$

$$tg\ \delta_Q = \frac{Q_S}{Q_C} \qquad (4\text{-}17)$$

where

176

Models	T_O	Q^*	P^*	T_g	$\delta_Q\cdot$	$\delta_P\cdot$ (hour)	$\delta\cdot$ (hour)
Form. 4-2		202.5	1940	12.2	0	45 (3)	-45 (-3)
Form. 4-20	271.9	248	3667	7.9	5.9 (.4)	22 (1.5)	-16 (-1)

Table 4-1. Comparison between the values of the different elements of the surface temperature given by formulae 4-2 and 4-20.

The input parameters are (MKSA units)

P	A	ε	λ	L	T_L	γ
1940	.25	.936	1.82	2	284	0.015

u	e_a	T_m	T_d	δ_a	δ_A	T_M	T_D
2	.75	275	3	$\pi/6$	$\pi/6$	250	4

h	e_g	S	C	δ_S
8	1	300	0.9	0

$$Q_C = (1-A)SC \cos\delta_S + \chi_A T_D \cos\delta_A + \chi_a T_d \cos\delta_a \tag{4-18a}$$

$$Q_S = (1-A)SC \sin\delta_S + \chi_A T_D \sin\delta_A + \chi_a T_d \sin\delta_a \tag{4-18b}$$

With

$$\chi_a = h(u)\{1 + \gamma e_a \frac{\partial P}{\partial T}\big|_{T_m}\} \tag{4-19a}$$

$$\chi_A = 4\varepsilon\sigma T_M^3 \tag{4-19b}$$

$$\chi_g = +4\varepsilon\sigma T_o^3 + h(u)\{1 + \gamma e_g \frac{\partial P}{\partial T}\big|_{T_o}\} \tag{4-19c}$$

The solution of the heat equation (B-6) with the boundary condition (4-7) and (4-14) is easily derived and reads

$$T(z,t) = T_o + (T_L - T_o)\frac{z}{L} +$$

$$+ T_g \exp(-\sqrt{\frac{\omega'}{2k}} z) \cos(\omega t - \sqrt{\frac{\omega'}{2k}} z + \delta) \tag{4-20}$$

where the surface average temperature T_O is given by the solution of

$$\frac{\lambda}{L} T_O = Q_o + \frac{\lambda}{L} T_L \tag{4-21}$$

The amplitude T_g is

$$T_g = \frac{Q^*}{\sqrt{\omega} \, P^*} \tag{4-22}$$

where the effective thermal inertia P^* is given by

$$P^* = \sqrt{P^2 + \sqrt{\frac{2}{\omega}} P \chi_g + \frac{\chi_g^2}{\omega}} \tag{4-23}$$

where P is the true thermal inertia given in § 4.1.1.

Finally the phase shift δ is given by

$$\delta = \delta_Q - \delta_P \tag{4-24}$$

where δ_Q is given by (4-17) and δ_P, the phase shift due to thermal inertia, is such that

$$\text{tg } \delta_P = \frac{P}{P + \sqrt{\frac{2}{\omega}} \chi_g} = \frac{P}{\sqrt{2P^{*2} - P^2}} \tag{4-25}$$

It is interesting to notice that expression (4-20) is very close to (4-2) in which the different constants are given their value as a function of the involved parameters. The different numerical values of the different elements of the surface temperature are reported on table 4-1 for the numerical values given on that table.

4.2.2 Relative influence of the various parameters on the surface temperature

The influence of a given parameter α on a function $F(\alpha,\beta,\ldots)$ is easily measured by the relative derivative of F with respect to α, $\partial F/\partial\alpha$. For questions of homogeneity one can write :

$$dF = (\alpha \frac{\partial F}{\partial\alpha})\frac{d\alpha}{\alpha} + (\beta \frac{\partial F}{\partial B})\frac{d\beta}{\beta} + \ldots \tag{4-26}$$

These partial derivatives for the average T_o, the amplitude T_g and the phase shift δ are easily derived from equations (4-15,25). Typical numerical values are given in table 4-2. From formula (4-26), one sees that the variation dF produced by 1% relative variation of the parameter α is one hundreth of the number quoted in table 4-2. The dominant variations are underlined.

177

α ∂/∂α / α	Average T_o	Amplitude T_g	Phase δ
Physical parameters			
P	1.2	-3.86	-.21
A	-3.71	-2.11	.27
ε	-4.5	-0.7	-.03
Micrometeorological parameters			
$U^{(1)}$	1.1	-2.	-.1
$e_a^{(2)}$	2.7	0.4	.01
$T_m^{(2)}$	1.90	-0.94	-.03
$T_d^{(2)}$	0	1.47	.08
e_g	-3.77	-1.14	-.05
T_L	14.2	.16	-.01
L	-0.6	0	0

Table 4-2. Numerical values of the relative partial derivatives $\alpha \frac{\partial F}{\partial \alpha}$ of the average, amplitude and phase of the surface temperature with respect to the parameters quoted in the first column, near the typical values quoted in the caption of table 4-1.

---

(1) It has been assumed that $\frac{1}{h}\,dh = \frac{1}{u}\,du$ (linear variation of h with u)

(2) The relative variation of T_M and T_D with respect to e_a, T_m and T_d has been taken into account by

$$\frac{e_a}{T_M}\frac{\partial T_M}{\partial e_a} \simeq \frac{e_a}{T_D}\frac{\partial T_D}{\partial e_a} \simeq 0.02$$

and

$$\frac{T_m}{T_M}\frac{\partial T_M}{\partial T_m} = \frac{T_d}{T_D}\frac{\partial T_D}{\partial T_d} = 1$$

From table 4-2 the following simple conclusions can be drawn.

1. The thermal inertia acts mainly on the denominator of the amplitude. An increase of the thermal inertia will induce a decrease of the amplitude. It acts much less on the average temperature mainly through the conduction of heat from the

Input changes, effecting 1°K	absolute	relative
Albedo	0.05	
Emissivity	0.2	
Groundwater level	0.1 m	
Groundwater temperature	5 K	
Aerodynamic roughness		50 %
Wind speed	1 m/sec	
Air temperature	0.5 K	
Air specific humidity	10^{-3}	

Table 4-3. Sensibility table (after A.Rosema (Rosema 1975)). The orders of magnitude of this table are comparable with those given in table 4-2.

---

water table to the surface. It increases the average temperature if the water table is warmer than the surface temperature

$$\left(P\,\frac{\partial T_o}{\partial P} = \frac{2\lambda(T_L-T_o)}{L\chi_g + \lambda}\right)$$ (see figures 4-4a and 4-5d).

2. The albedo acts mainly on the numerator of the amplitude and on the average of the temperature which both increase when A decreases. There will be no crossing effects (see also figures 4-4b and 4-5a).

3. The emissivity acts mainly on the average temperature which decreases when ε increases, because the atmospheric radiometric temperature is less than the surface temperature

$$\left(\varepsilon\,\frac{\partial T_o}{\partial \varepsilon} = \frac{\varepsilon\sigma(T_M^4 - T_o^4)L}{L\chi_g + \lambda}\right).$$

This effect may be compared to the effect of the albedo from the ratio

$$\frac{\varepsilon\partial T_o}{\partial \varepsilon} \Big/ \frac{A\partial T_o}{\partial A} = +\frac{\varepsilon\sigma(T_o^4 - T_M^4)}{ASC}$$

Furthermore, the emissivity acts much less on the amplitude, because both the numerator and the denominator increase when the emissivity increases. Therefore the emissivity may also play a major role as far as the surface temperature is concerned. (see also figure 4-4d).

4. The wind speed U plays unfortunately a major role because it increases the exchange of sensible and latent heat. This effect depends indeed on the difference of the temperatures between the air and

178

the ground. If these temperatures are
close, the major effect occurs on the
amplitude.

5. The air humidity and temperature.
When e_a increases, the evaporation flux
decreases while the downward radiative flux
increases. Both these effects do increase
the average temperature and the amplitude.
The same things occur with the air temperature.

6. The ground relative humidity. When
e_g increases, the average temperature decreases because the upward latent heat
flux increases but the amplitude decreases
also because the apparent inertia increases.
This effect is also very important; in the
same line, an increase of e_g will decrease
the phase shift δ in absolute value.

The relative humidity, and by the way the
surface moisture, does influence the surface temperature by another mechanism,
namely the change in the thermal
inertia(see figure 4-3a), the albedo and
emissivity. For instance, the variation
of the albedo with moisture on Avondale
loam is shown on figure 4-3b (Reginato
et al. 1977) while the variation of the
emissivity is given in table 3-1 (Fuchs
and Tanner 1968).

Since the albedo decreases with moisture,
the numerator of the amplitude as well as
the average will increase by this mechanism, reducing the net effect of moisture
... Nevertheless this effect is important as it will be discussed below.

7. The temperature and position of the
water table. It acts mainly on the average
and to a less extent on the apparent inertia because the radiative emission of the
ground increases this apparent inertia
(see formula 4-19c). The effect on the
average is proportional to the difference
of the temperatures of the ground water
table and the surface
$$\left(L\frac{\partial T_O}{\partial L} = \frac{(T_O - T_L)\lambda}{(\chi_g L + \lambda)}\right).$$

The influence of the depth of the water
table on the amplitude cannot be taken
into account completely with this simple
model. Nevertheless one can understand that
a decrease of the depth will reduce the
amplitude of the surface temperature. These
effects are shown on figure 4-11 calculated
by A. Rosema (Rosema 1975) who took these
effects into account.

8. The slope of the terrain. This slope
acts on the phase shift δ_S as well as on
the coefficient C. This effect occurs main-

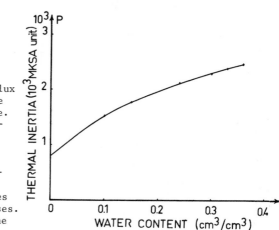

Figure 4-3a. Measured thermal inertia
for the Grenoble sand (after M. Vauclin
(Vauclin 1978)

ly in the phase shift for the (E-W) slope,
but plays almost the same role as the
albedo for the (N-S) slope. (see figures
4-5).

All these effects are confirmed by the
more elaborate calculations of A. Rosema
(Rosema 1975) who gives the sensibility
table 4-3 and many detailed analysis of the
influence of the previous parameters on
the surface temperature. In the same line,
K. Watson (Watson 1975) assuming very
dry rocks (no latent heat flux) and neglecting the sensible heat flux, calculated
the influence of the thermal inertia, albedo, temperature of the ground water table
and emissivity which are respectively shown
on figures 4-4a-d. Later on, A. Kahle
(Kahle 1977) who included the sensible and
latent heat fluxes gives in figures 4-5a-d
the effects of the albedo, the slope and
the thermal inertia. It is interesting to
see the effect of the sensible and latent
heat fluxes.

4.3 The different applications of interest
    and the corresponding models

The previous discussion, although it is
not complete and approximative, puts forward possible applications of the analysis
of the dynamics of the surface temperature
for agricultural purposes. The reliability
of these applications depends on the ability of the various models to describe the
complexe reality. Unfortunately, there is
not yet complete theoretical models. The
different existing models have been built
for specific applications allowing specific
approximations in the description of the
phenomena discussed in § 4.1. Let us first
review briefly these models.

179

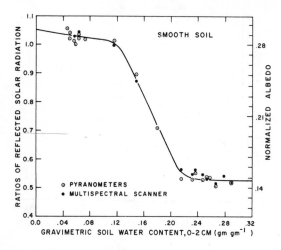

Figure 4-3b. Ratios of reflected solar radiances from continually wet and drying Avondale loam to continually dry Avondale loam for the smooth surfaces, plotted as a function of the gravimetric water content of the uppermost 2 cm of soil. The ratios were determined from the sum of wave bands 2,3,5,7,9 and 10 of the multispectral scanner and the actual albedos of the wet, drying and dry fields determined from the pyranometers.
(After Reginato et al. 1977)

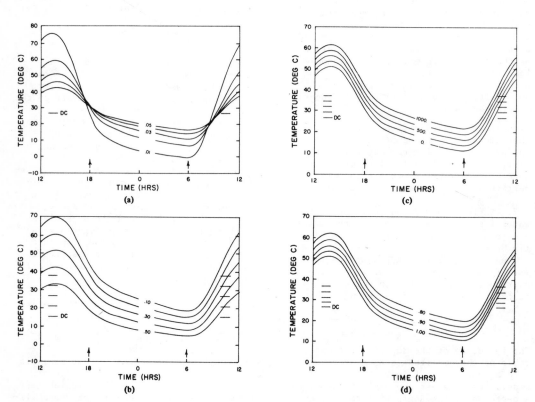

Figure 4-4.

Diurnal temperature curves for varying (a) thermal inertia (in cal/cm^2/s$^{1/2}$), (b) albedo (in fractions <1), (c) geothermal heat flux (in HFU), and (d) emissivity (in fractions <1). Times are local solar values. The mean diurnal temperature DC is indicated by a horizontal line for each curve. Note that this value is constant for varying thermal inertia [see 1(a)] and that the amplitude of each curve is reasonably independent of geothermal flux [see 1(c)] and emissivity [see 1(d)]. The vertical arrows near the time axis mark the sunset and sunrise times. Fixed parameter values used in the computation of these curves are: inertia 0.03 cal/cm^2/s$^{1/2}$, albedo 0.3, emissivity 1.0, latitude 30°, solar declination 0°, sky radiant temperature 260°, dip 0°, cloud cover 0.2. (K. Watson 1975)

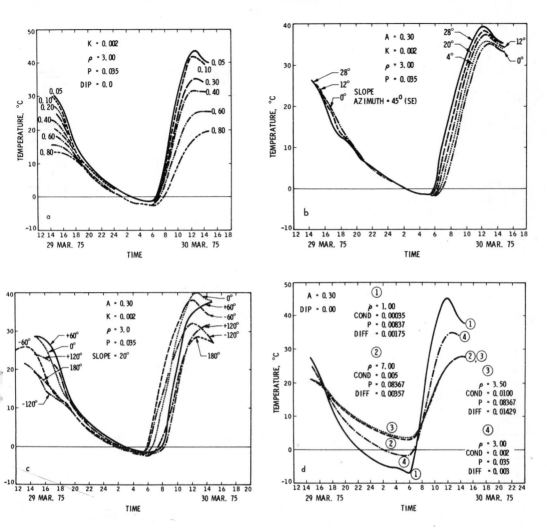

Figure 4-5. Surface temperature predicted by the model for various values of
(a) albedo, (b) slope, (c) slope azimuth and (d) thermal inertia.
(after A. Kahle (Kahle 1977)

181

Table 4-4: Characteristics of the different models

		K. WATSON (1975, 1979)	A. KAHLE (1977)	A. TABBAGH (1977 a,b)	A. ROSEMA (1975 a,b)	G.J. SOER (1977 a,b,c)
system		- geologic terrain homogeneous - bare surface - infinite medium (one dimension problem)	same as Watson can include moisture	- any bare terrain - non"homogeneous " - two dimension problems (can include faults, discontinuity ...)	- any bare terrain -"homogeneous " - include complete coupling with water content in the parameter descript.	same as Rosema plus vegetated ver
boundary conditions	air-ground	- R_S, R_a, R_g are only taken into account - L_E and ϕ neglected	L_E and ϕ included; parametrization of conductivities	the total net incoming flux is modelized by trigonometric function	modelization of all the fluxes driving the temperature. no inclusion of evapotranspiration of plants	same as Rosema inclusion of potranspiratio parametrizatio the conductiv
boundary conditions	ground – under-ground	- temperature - or inclusion of the geothermal flux	- temperature - no flux	- temperature	- ground temperature and water content	
boundary conditions	lateral conditions	one dimension model (no need)	one dimension problem (no need)	given by the calculated solution of the homogeneous medium. Two dimensions problem	one dimension problem (infinite in surface)	one dimension blem
type of equations – basic assumptions		one: conductibility of heat inside the ground -latent and convective flux neglected - geothermal flux included	only conductibility inside the ground. No coupling between moisture and thermal coefficients	only conductibility of heat inside the ground. The conductibility coefficients are assumed to be constant (the influence of moisture is included in these coeff.) Different media are connected with –continuity of temperature at boundary –continuity of fluxes	complete coupling between heat and mass transfer (only water) the values of the thermal coefficients are calculated from the calculated water content	based on the model. use crops as cators of eva transpiration
method of solution		completely analytical solution following a linearization procedure in the boundary radiative flux	numerical solution	- analytical solut. for simple geometric boundaries - computer code with finite elements and finite difference methods in other situations	completely numerical solution using finite difference methods	completely num solution using nite differenc methods
domains of application types of problem		class II and III problems for bare mineral soils (no moisture) - identification of the soils on a homogeneous structure - geothermal mapping	class II and III problems for bare soils with possible inclusion of surface moisture (without coupling)	class III problems - prospection - analysis of underground structure of bare soils - best time or prospection - transfer effects	class I problems for bare soils - moisture vegetation stress due to soil. class II problems for bare soils with homogeneous structures ; gives the coupling between thermal effects and moisture contents, influence of the water level	class I probl for vegetated evapotranspir water stress,

182

A. PERRIER 977)	M. VAUCLIN and al. (Vauclin, Hamon and Vachaud 1977)	F. BECKER and E. HECHINGER (Becker and Hechinger 1979)	DEARDOFF (Deardoff 1978)
ially the ted cover	bare terrain with inclusion of roots effects in the last version	- bare terrain - vegetated terrain as in G.J. Soer	essentially vege-tated covers
e fluxes are sed with great l nductibilities alculated	- all the fluxes are included with simple parametrization - micrometeorological parameters given (no coupling)	- all the fluxes are included with simple parametrization turbulent trans-fer coefficient calculated as in G.J. Soer and B. Seguin - micrometeorological parameters given (no coupling)	- all the fluxes are included with semi-empirical parame-trization
temperature moisture the profile the soil is lculated)	temperature of the ground at a given depth. succion and hydraulic conductivities as a function of moisture	temperature of the ground at a given depth	no need
mension pro-a homogeneous considered inite)	no need, one dimension	no need, one dimension	no need
te coupling n the diffe-henomena ntum transfer and mass er including and CO2	complete coupling between heat and mass transfer. The values of the thermal and hydraulic conductivi-ties are calculated from the calculated water content	- the water profile is not exactly calculated ; the surface moisture is simply connected to hydraulic conduc-tivities and latent heat flux - the thermal parameters are calcula-ted from moisture at the surface	only energy balance at the surface. The surface in the ground is parame-trized
ximate solu-in terms of able param. vegetated	numerical solution	exact solution by Fourier analysis and partial linearization of the fluxes	no need
I problems getated ranspira-ater stress	class I and class II as in Rosema	class I and class II for homogeneous undergrounds	class I for evapo-transpiration

4.3.1 Brief review of existing models

The different models have been built to meet applications which can be classified arbitrarily into three classes :

Class I : Applications concerning the measurement and control of heat and mass transfer between the atmosphere and the ground (soil moisture, evapotranspiration, vegetation stress, ...). In this case, the main effort is on an accurate evaluation of the heat fluxes and LE, while G_h is generally neglected or given an approximate value (because $G_h \simeq 10$ % of R_g, see for instance fig. 4-6). This class of problem is well suited to the models of A. Perrier (Perrier 1976), G.J. Soer (Soer 1977a,b,c) and J.W. Deardorff (Deardorff 1978).

Class II : Applications concerning the analysis of the structure of the surface with or without vegetated covers (type of soils mapping, type of plants, water irrigation, flows and mixing flows of water surfaces, environmental problems ...). In this case, the main effort is on R_n, the other fluxes being approximated. This class of problem can be dealt with the models of A. Kahle (Kahle 1977), K. Watson (Watson 1975 and 1979), A. Tabbagh (Tabbagh 1977 a and b), A. Rosema (Rosema 1975 a and b), M. Vauclin et al.(Vauclin, Hamon and Vachaud 1977) and F. Becker and E. Hechinger (Becker and Hechinger 1979).

Class III : Applications concerning measurements of underground properties (geological mapping, geothermal exploration, ground characteristics and moisture, prospection, archeological survey, discontinuities in the underground, etc...). In this case, the main effort is on the flux Q_h. This class can be dealt, according to their complexities with the different models, when there are discontinuities, mainly A. Tabbagh (Tabbagh 1977 a and b) and M. Raffy et al.(Raffy and Thommann 1979).

Indeed these classes are not independent and some applications appear in the three classes. Different models which may be used in the domains covered by the course are summarized in table 4-4. There are indeed other models which are not given here (cf. for instance, G. Dinelli and M. Maini (Dinelli and Maini 1977). A more general discussion is given in appendix B. The main actual limitation of these models is that the reciprocal interaction between the ground and the atmosphere is not taken into account. Progress have to be made in that direction.

4.3.2 Some applications

4.3.2.1 Measure and control of heat and mass transfer on vegetated covers- Water stress

The evapotranspiration, sensible heat and photosynthesis fluxes are not directly connected to the surface temperature and are not therefore directly measurable by remote sensing. There are several ways to proceed according as the different terms of the energetic balance equation (B-11) are taken into account.

A first possibility (see for instance B. Seguin 1978 a) is to measure the temperature, T_a, of the air, T_o, of the ground or cover surface (with all the difficulties mentionned in § 2), the various radiative fluxes giving the net flux, R_n, and G, the thermal flux into the ground. The sensible heat flux, H, is then calculated with the formula

$$H = h(T_a - T_o)$$

after a computation of the transfer coefficient h. The evaporation flux, LE, is then given by the difference

$$LE = -(G + H + R_n)$$

One can also measure H/LE (Bowen ratio) and $H + LE = -(R_n + G)$ which gives H and LE. The order of magnitude of the different fluxes occuring are shown on figure 4-6 due to M. Vauclin et al.(Vauclin 1977).

Another possibility (see for instance Van Bavel and Hillel 1976 and J.W. Deardoff 1978) is still to measure the temperature but to parametrize the various fluxes by some functions of the parameters which come into the play. The thermal flux in the ground being taken as a function of one of the other fluxes. The results are quite good but one may wonder how general the results are.

Other methods are based upon the existence of simple relations between the net radiative flux and the evapotranspiration flux. (See for instance figure 4-7a due to Idso, Reginato and Jackson 1977). A. Perrier (Perrier 1976) has shown that ETR = a + b R_n where a and b are coefficients which depend on the type of plant, the resistance of the stomates (as it is shown on figure 4-7b) but also on the wind speed and other parameters. In the same line, an earlier estimation of the possibility to measure the evapotranspiration rate over semi-aride region from the nimbus 5 satellite data has been made by R. Bossard and Y. Vuillaume (Bossard and Vuillaume 1975). Their esti-

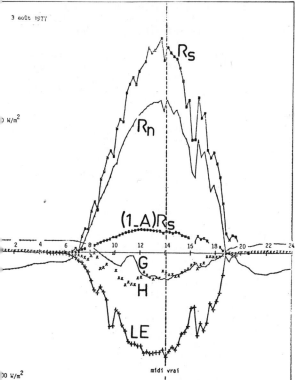

3 août 1977

R_S

R_n

$(1_-A)R_s$

G

H

LE

W/m²

midi vrai

Figure 4-6. Order of magnitude of the
various fluxes occuring in the energy
balance equation.
(After M. Vauclin et al.(Vauclin 1977)

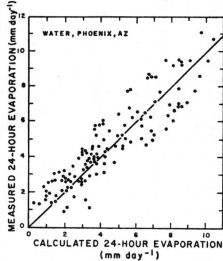

WATER, PHOENIX, AZ

Figure 4-7a
Measured daily evaporation vs. calculated daily evaporation for different crop and soil
surfaces in Arizona and California at various
times of the year. (after Idso et al (1977)

ETR (Wm^{-2})

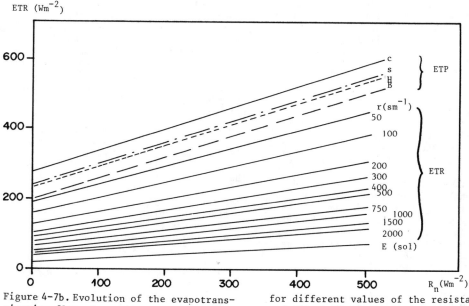

Figure 4-7b. Evolution of the evapotranspiration flux of a maize cover as a
function of the net radiative flux R_n

for different values of the resistance r
and the available water in the stomates.
(After A. Perrier (Perrier 1976))

185

mation is based upon the assumption that the evaporation rate (ETR) is equal to the net radiation flux R_n and on some approximate procedures to get this net radiation flux which reduces the accuracy of the estimation. Furthermore, the assumption that ETR = R_n will over estimate the evapotranspiration rate over vegetated covers for the conditions shown in figure 4-7 even for the potential evapotranspiration rate (ETP). This is also confirmed by the experimental results shown in figure 4-6 and by the calculations of G. Soer (Soer 1977). In this last calculation the net radiation flux is about 2 times larger than the evapotranspiration rate during the day, as in figure 4-7 and are almost equal during the night.

Finally there are several methods based upon an exact solution of the equations, discussed in appendix B (see for instance the Tellus program of A. Rosema (Rosema 1977), the code of M. Vauclin et al. (Vauclin, Hamon and Vachaud 1977) and the tergra model (Soer 1977)). As an example we quote the theoretical calculation of evapotranspiration rate on a large scale made by G.J. Soer (Soer 1977) from remote sensing measurements (thermal I.R. and visible reflectance) and complementary ground truth measurements. This calculation was compared with effective calculation from water balance estimates ; the differences were within 30 %. The method consists in evaluating the momentary evaporation and relate it to the daily evaporation through the TERGRA code. It shows the faisability of remote measurement of evapotranspiration if the adequate parameters are measured and the adequate corrections performed. There are several other tentatives to obtain directly these fluxes by remote sensing and some of them will be briefly discussed in section 5.

Since the evaporation flux depends on the water available to the plants, as it is seen for instance in figure 4-7 b(compare the curve quoted E (dry leaves) and the curves quoted ETP (saturated leaves)), the water stress will reduce this flux and consequently increase the surface temperature as it has been discussed in section 4-2. As a consequence, soil moisture stress on the plant will be detectable. Differences of the order of 2° to 3°C can be observed (see Fitzgerald 1974, 1976).

4.3.2.2 Discrimination among plants and soils - Influence of the different meteorological parameters

The difference between thermal properties and conductivities of the different covers will induce differences in the time dependence of the surface temperature, leading to a possible procedure to discriminate among types of plants or types of soils, which is complementary to the spectral analysis in the visible wavelength.

i) For vegetated covers, discrimination among plants. As an example, the results of F.J. Bonn (Bonn 1977) are shown on figure 4-8. One sees the large differences (recall that a radiometer can separate easily 0.2 to 0.3°C) between noon and 4 p.m., particularly between conifers and decideous as it has been mentioned several times.

The interesting thing to note is the evolution of these differences with the seasons, i.e. according to the leaves density (or the whole surface of the leaves). This is seen in figures 4-9a and 4-9b. These differences can be explained by the difference between the soil and leaves moistures.

ii) For soils
- Nature. A typical example is given for non-covered soils in figure 4-10. Due to different thermal inertia the two soils have different response as it has been demonstrated in section 4.2. We have also shown that the magnitude of this discrimination depends, unfortunately, on the microclimatic conditions (windspeed, air specific humidities, air temperature) and the roughness of the surfaces (which gives for known meteorological conditions the transfer coefficient h). Therefore a direct interpretation in terms of true inertia is not correct.

- Moisture. We have shown in section 4.2 how the moisture of the surface does influence the surface temperature. More elaborate calculations show, furthermore, that the moisture profile, the succion curve and the hydraulic conductivities are important. The extreme sensivity to soil moisture not only at the surface but on a whole appears to be very important between 11 a.m. and 1 p.m. according to the calculation of Rosema (Rosema 1977).

4.3.2.3 Discrimination of subsurface properties

i) Underground water
- The depth of the ground water level in-

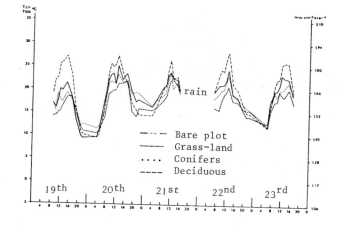

Figure 4-8. Thermal infra-red
emissions from different surfaces
in the 9.5-11.5 μm spectral range
in June 1972.
(after F.J. Bohn (Bohn 1977))

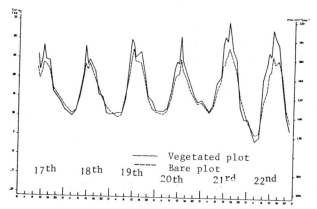

Figure 4-9a. Thermal infra-red
emission from two plots in the
9.5-11.5 μm spectral range in
May 1972.
(after F.J. Bohn (Bohn 1977))

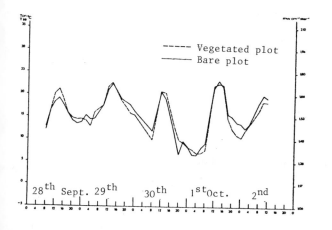

Figure 4-9b. Thermal infra-red
emission from two plots in the
9.5-11.5 μm spectral range in
September and October 1971.
(after F.J. Bohn (Bohn 1977))

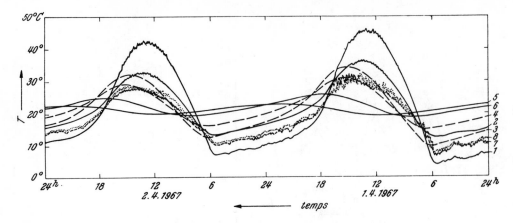

Figure 4-10. Different temperatures recorded

1. At the surface of a sand dune
2. At the surface of the sand dune
3. At a depth of 5cm in a sand dune
4. At a depth of 5cm "   "
5. At   "   " 20 cm "   "
6. At   "   " 20 cm "   "
7. The air at 5 cm above   "
8. The air at 5cm   "   "
(after P. de Felice (de Felice 1968)

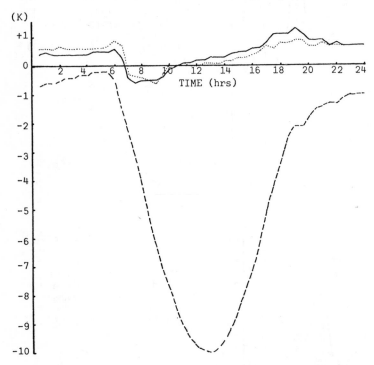

Fig. 4 -11 . The surface temperature contrast for various groundwater depths: 0.5 m (---), 1 m (—) and 2 m (...).
(From A. Rosema (Rosema 1977)

fluences the surface temperature for several reasons. First it changes the moisture profile (this effect was not included in the simple model in sect. 4.2) and consequently the thermal conductivity and the thermal inertia. Secondly, it acts directly on the average as discussed in section 4.2 and on the amplitude. Obviously the lower the ground water level is, the less sensitive the surface temperature will be to the variation of the ground water parameters. The results of Rosema for the influence of the level of the ground water on the daily thermal wave, is shown on figure 4-11.

- The temperature of the ground water level may induce variations of the absolute surface temperature of the order of 0.5 to 1°C, here the maximum change occurs during the night when the net radiative flux is inverted. The contrast between the different soils is not very much modified.

ii) Underground discontinuities
Underground discontinuities appear as contrasts on the surface because of the heterogeneities of the thermal properties. The calculations of A. Tabbagh (Tabbagh 1977) on both the daily and annual wave show that :

- The annual wave is much more sensitive ($\Delta T \simeq 0.6°$) than the daily wave ($\Delta T \simeq 0.08°C$) to the underground structure which has been analysed. Indeed these figures depend on the deepness of the discontinuity, its shape, dimensions and thermal characteristics.

- The largest contrasts appear on december and june for the annual wave and are opposite in sign, they are at noon and midnight for the daily wave.

- For faults or vertical discontinuities between two media, the temperature varies continuously from one medium to the other one, the range d of the reciprocal influence of the two adjacent media is very

short : a few decimeters. In a first approximation, it may therefore be possible to consider each medium independently of the other ones.

4.4 Conclusions

The results which we have shown are qualitatively in agreement with the very simple model discussed in § 4.2 (one reason is that any flux variation can be decomposed in a Fourier series or analysed by Fourier transform). These results give an idea of the interest of the temperature as a measure of : heat and mass exchange between the surface and the atmosphere , structures and characteristics of the surface and its underground. Nevertheless many parameters do influence these indications and may lead to strong errors (by inverting, or eracing the contrasts for instance), therefore some principles to interpret remote sensing data in the infrared bands will be discussed on the next section.

5 PRINCIPLES OF INTERPRETATION

5.1 Generalities

From the previous discussions, it appears that the measure of the radiance emitted by the surface may have interesting applications in agriculture, such as the measure and control of the ground surface temperature and soil moisture which are determinant for the evaluation of the crop yield (see Idso et al. 1977) or the evapotranspiration flux which is an important part of the water budget and may be used for plant monitoring, etc...

Unfortunately the discussion of section 2 showed that it is not simple to extract the above mentionned informations from the remotely recorded data because there are numerous parameters which distort or compose these recorded data and furthermore as we showed it in section 4.2 a given variation of the observed radiance (or surface temperature) may come from several competitive effects which are not simple to separate.

In order to overcome these difficulties the three following questions have therefore to be answered :
    i) Which data have to be recorded and when ?
    ii) How these data can be corrected for external perturbation (instrumentation, atmosphere, etc...) ?

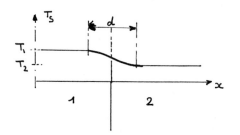

iii) How the interesting parameters can be extracted from the corrected data ?

Starting from the simple model of section 4.2 a possible scheme of inversion, which gives partial answers to the three previous questions, will be first presented in section 5.2, then from another simple model the principle of atmospheric correction in infra-red bands will be discussed in section 5.3. From the different elements gathered in these sections, a list of the data which are interesting to record can be easily established by the reader and will no longer be discussed here.

5.2 Inversion of the data

It is not possible in this short lectures to give the details of the inversion procedure. We shall only present possible approaches, knowing that they are not the only ones (see S. Twomey (Twomey 1977)).

5.2.1 General ideas

From formula (4-20), it appears that the surface temperature at a given time depends on many parameters. Consequently the knowledge of the surface temperature at a given time and only at this time is not very useful, although it may already gives interesting information if the "best time" of measurement is respected (see below some arguments). Several measurements on a day (for the daily wave) or several measurements during the year (for the yearly wave) have therefore to be done in order to extract the interesting parameters. Furthermore, the surface temperature depends on the incoming fluxes: global short wave radiations as well as atmospheric long wave radiations. These fluxes have to be measured or modelled. Generally it will be necessary to measure in several channels the visible reflected radiance (which will give the albedo or the incoming flux) as well as the thermal infra-red emitted flux. Therefore, it will be very interesting to perform a time dependent multispectral analysis.

Finally the surface temperature is very dependent on the boundary conditions which depends strongly on the local heterogeneities of the terrains. Consequently the measured radiative fluxes will reflect somehow this heterogeneity which may be characteristic of a given medium. It will be therefore interesting to perform a correlated statistical analysis of the visible or near I.R. and the thermal infra-red radiances. Let us discuss briefly these points.

5.2.2 Time dependent multispectral analysis

In order to give a general idea of the interest of such an analysis let us first analyse the temperature variation with the simple model of section 4.2 and then show what happens with more elaborate models. Let us assume also for simplicity that the ground surface is Lambertian and that we have been able to perform already all the different radiometric and atmospheric corrections.

With these assumptions, the thermal infra-red emitted radiance is (see expression A-12 in appendix A)

$$R_\nu^{IR} = \varepsilon_\nu R_\nu^{\circ}(T_S) \qquad (5-1)$$

where ε_ν is the emissivity in the band width $\Delta\nu$ around the frequence ν and T_S is the ground surface temperature which can be simply written according to (4-13)

$$T_S = T_0 + T_g \cos(\omega t + \delta)$$

Since $T_g \ll T_0$, expression 5-1 may be rewritten according to A-9

$$R_\nu^{IR} = \varepsilon_\nu \{R_\nu^{\circ}(T_0) + T_g \cos(\omega t + \delta) \left.\frac{\partial R^{\circ}}{\partial T}\right|_{T_0}\} \quad (5-2)$$

As already discussed in § 2.3.1b, this radiance depends both on the surface temperature and the emissivity.

In the same line, still assuming for simplicity that the reflectivity is independent of the angles, the bi-directional reflectivity (see A-16) is simply related to the albedo A, namely

$$\rho \simeq \frac{A}{\pi}$$

Therefore, according to (A-17), the emitted visible radiance will be

$$R^{vis} = \frac{A}{\pi} S C(\cos\omega t + 1) \qquad (5-3)$$

where A is the albedo, S and C being the same as in § 4.2 (in a more elaborate model, the angular and wavelength effects would have to be taken into account (see for instance F. Becker and E. Hechinger 1979)). In order to obtain the different interesting parameters graphically, it is interesting to follow the suggestion of J.C. Favard (Favard 1978), to plot the observed radiance in the visible part of the spectrum as a function of the radiance observed in the infra-red bands. From the expressions (5-2) and (5-3), this function is easily derived and is represented by the ellipse

$$\left(\frac{y}{a}\right)^2 + \left(\frac{x}{b}\right)^2 + \frac{2yx}{ab}\cos\delta = \sin^2\delta \quad (5-4)$$

where

$$y = R_\lambda^{IR} - \overline{R_\lambda^{IR}} = a\cos(\omega t + \delta)$$

$$x = R_\lambda^{vis} - \overline{R_\lambda^{vis}} = b\cos\omega t$$

$\overline{R_\lambda^{IR}}$ and $\overline{R_\lambda^{vis}}$ are respectively the average of the IR and visible radiance over one day,

$$a = \varepsilon\, T_g \left(\frac{\partial R_\lambda^o}{\partial T}\right)_{T_o} \quad , \quad \overline{R_\lambda^{IR}} = \varepsilon\, R_\nu(T_o)$$

$$b = \frac{A}{\pi} S\, C \quad , \quad \overline{R^{vis}} = \frac{A}{\pi} S\, C$$

and δ is the phase shift of the temperature. The different parameters a, b, δ, $\overline{R_\lambda^{IR}}$ and $\overline{R_\lambda^{vis}}$ are easily extracted from the graph of the ellipse. As it is seen on figure 5-1, one obtains :

$$\sin\delta = \frac{OE}{OI} \quad , \quad \cos\delta = \frac{I'J'}{I'H} \ i.e.\ tg\delta = \frac{OE}{I'J'} \quad (5-5-1)$$

$$\varepsilon\, R_\lambda^o(T_o) = \frac{IM + Im}{2} \quad (5-5-2)$$

$$\varepsilon \left.\frac{\partial R^o}{\partial T}\right|_{T_o} T_g = IM - Im \quad (5-5-3)$$

$$\frac{A}{\pi} S\, C = V_M - V_A \quad (5-5-4)$$

An approximative procedure to get the different interesting parameters from these equations is the following one. The ratio

$$r_\lambda = \frac{IM - Im}{IM + Im} \quad (5-6)$$

is independent of ε and may be written from formula (A-3)

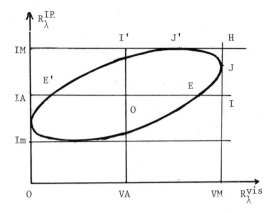

Figure 5-1. The variation of the infra-red radiance as a function of the visible radiance in the simple model of § 4.2 (equation 5-4). 0 is the center of the ellipse.

$$r_\lambda \simeq \frac{C_2}{2\lambda\, T_o}\ \frac{\exp\dfrac{C_2}{\lambda\, T_o}}{\left(\exp\dfrac{C_2}{\lambda\, T_o} - 1\right)}\ \frac{T_g}{T_o}$$

Near $T_o \simeq 280$ and in the 8–14 μm window, this expression may be rewritten to an accuracy of 0.3 %

$$r_\lambda \simeq \frac{C_2}{2\lambda\, T_o}\ \frac{T_g}{T_o} \quad (5-7)$$

It is then easy to show that an error of 0.2°C on T_g is induced by an error of 1.5°C on T_o around $T_o = 20°C$, which corresponds to an error of 2,7 % on ε (see formula 2-2). Therefore, if one assumes that for vegetated soil $\varepsilon \simeq 0.96$ and for bare soils $\varepsilon \simeq 0.90$, it is possible to calculate T_o from (5-5-2) with an accuracy of 1.5°C and to deduce T_g from (5-7) with the required accuracy. (A direct calculation of T_g from 5-5-3 assuming an error of 2.5 % on ε would lead to an error on T_g of 1.3°C instead of 0.2).

Knowing Tg, To, δ and the reflected flux, it is possible to extract the unknown ground parameters (thermal inertia, ground moisture, etc...) only if the micrometeorological parameters are known, namely the atmospheric temperature and humidity as well as the incoming short wavelength flux (see formulae 4-14 to 4-25). This discussion is left as an exercice.

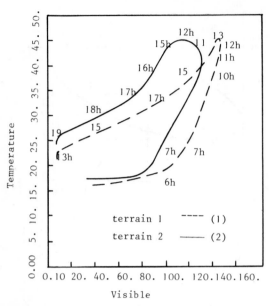

Figure 5-2. The radiance in the IR channel as a function of the radiance in the visible one. (after Favard 1978)

This simple analysis is indeed not sufficient to extract the different parameters with a good accuracy, namely the average temperature T_O and the emissivity ε. Nevertheless it shows that a day-night measurement is generally insufficient to extract the thermal inertia because

i) The maximum or minimum of emission does not occur generally at the same time for all the terrains (see figure 5-2 from Favard 1978) because the phase shift δ is not a constant.

ii) Even if it were so, it is not possible to extract the thermal inertia from equation 5-5-3 and 5-5-2 alone because even if T_g and T_O can be obtained with the previous approach, it is not possible to extract from these two parameters alone the interesting ground parameters (thermal inertia, soil moisture, etc...) because the numerator of T_g which depends on micrometeorological parameters (see 4-22, 16, 18) has to be known.
To this respect a knowledge of the phase shift δ gives very interesting complementary informations.

iii) In a more realistic model the difference between the day and night radiances are not so simply related to the amplitude T_g. However this day-night measurement gives indeed interesting qualitative contrasts which are useful for discriminating among different terrains as it has been shown for instance by K. Watson (Watson 1975) and J.C. Price (Price 1977).

Besides the fact that it is necessary to use more elaborate models such as Tergra (Soer 1977), Tellus (Rosema 1977, Watson 1978) or the Fourier analysis of the different radiances (Becker and Hechinger 1979), this discussion shows that the quantitative extraction of the ground parameters implies :

i) At least three or four measurements per day in the thermal infra-red band and in the visible band (this is possible with geostationary satellite and will be possible with the two TIROS N satellites) in order to observe the average temperature the different amplitudes and phases as well as the albedo or the incoming flux.

ii) A well equiped reliable ground truth measurement station representative of at least one pixel on the ground in order to monitor the micrometeorological parameters and the various fluxes. It may furthermore be used as a calibration point (see below) for relative measurements which are easier to make. These relative measurements allow remote measurements of fluxes, and soil moisture (with the difficulty that surface and body moistures are not equivalent). For instance, S.B. Idso et al. (Idso, Schmugge, Jackson and Reginato 1975), R.J. Reginato et al. (Reginato, Idso, Vedder, Jackson, Blanchard and Goettelman 1976) obtained good local correlations between soil moisture and measured radiative fluxes (visible and IR). Nevertheless the different parameters entering the given relationships are locally true and have to be re-estimated at different place and probably for different meteorological conditions. To avoid this local dependence, G.J. Soer (Soer 1977) and B. Seguin (Seguin 1978) have proposed more general but indeed more complicated procedures which take into account the local variation of the different parameters.

iii) Multispectral measurements both in the visible plus near infra-red and in the thermal infra-red part of the spectrum In order to improve the accuracy of the analysis. This analysis has several advantages :

j) It leads to a better knowledge of the albedo and the reflected flux.

jj) With the visible and near infra-red part of the spectrum, it is easier to discriminate the different parts of the image, say water, buildings, different soils, plants, etc... and therefore to correlate a given thermal behaviour to a given type of ground. An example of this approach has been given by L.A. Bartolucci, Ph. Swain and Ch. Wu (Bartolucci et al. 1976). Another possibility is to make correlations between the variation of the surface temperature and the spectral reflectivities, as discussed for instance by G. Nieuwenhuis (Nieuwenhuis 1979). The vegetal index

$$V \simeq \frac{R_{nir} - R_r}{R_r + R_{nir}} \qquad (5-8)$$

where R_{nir} and R_r are respectively the reflected radiances in the $(0.75-0.8)\mu m$ and $(0.65-0.7)\mu m$ bands, is a direct function of the proportion of the ground which is covered with plants. Since plants evaporate more than ground, the effective inertia p^* of plants is larger than the one of bare soils. Therefore the temperature of pixels with low vegetal index, V, will be larger than the temperature of pixels with large V. Since both the inertia and the reflectivity are furthermore a function of the moisture (see table 4-2 and figure 4-3a), anormally moisted (or dry) soils or plants will appear on a graph correlating the temperature and the vegetal index. This spectral analysis is therefore a good complement to the thermal analysis itself and leads to a better understanding of the origine of the thermal contrast observed.

jjj) A spectral analysis in the thermal infra-red bands makes first easier the separation of the respective influence of the emissivity and temperature on the emitted radiance. It gives several measures on the same pixel and allows a verification of the relationship

$$\lambda \, r_\lambda \simeq constant \qquad (see \; 5-7)$$

This spectral analysis simplifies the corrections of atmospheric effects as it will be discussed below.
Finally we recall that it gives a good complement to the spectral analysis performed in the visible part of the spectrum as it has been discussed in § 3.

All the previous discussions on the dayly thermal wave can be extended as well on the yearly wave. The varying quantity being the daily average temperature T_0 or an average of the amplitude T_g over several days. This analysis gives the seasonal variations of the surface (see for instance the discussion of § 4 and the

figures 4-8-9) from F.J. Bohn (Bohn 1977). The yearly wave has also the advantage to penetrate deeply in the ground as discussed in section 4.1. Applications of this property have been done by A. Tabbagh (Tabbagh 1977 b) for determining underground anomalies, namely archeological vestiges.

5.2.3 Analysis of the transient response of the earth to a sudden variation of the boundary conditions

Instead of using the complete cycle of variation of the surface temperature over one day or one year, it may be very interesting to analyse the fast response (transient) of the earth temperature to a sudden variation of the incoming fluxes or the boundary conditions either on a hourly scale or on a daily scale.

i) On a hourly scale, this occurs when the sun is rising, or when a cloud is suddenly shadowing a part of the ground, or when some part of the observed scene is moving, changing the local irradiation by the sun (car leaving its parking place for instance) or modifying the thermal equilibrium of the medium (a boat moving on the sea). These changes affect essentially the surface temperature.

ii) On a daily scale, this occurs for instance when good weather arises after several days of cloudy conditions, or after several rainy days. These changes affect the temperature on a larger depth. In both cases the time constant of these transients is a function of the thermal inertia and diffusivities and therefore these transients will give information on these parameters.

Some aspects of these methods have been recently discussed by M. Vieillefosse and J.C. Favard (Vieillefosse and Favard 1978). For instance, they propose to use the rapid heating of the ground surface at sunrise. This method has the advantage to reduce the number of temperature measurements which is very interesting when the weather is rapidly changing during the day but it has the disadvantage to be more sensitive to the phase shift δ (see § 4.2) which depends on the slope of the terrain, the micrometeorological conditions and the moisture of the ground during the night, and as sensitive to the sensible and latent heat fluxes (see for instance figure 4-5) where the relative variation and phase of the surface temperature and different fluxes can be qualitatively

193

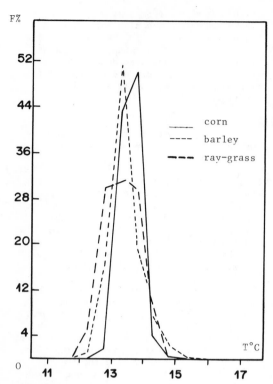

corn
---- barley
━━ ray-grass

Figure 5-3. Histogram of the frequencies of appearance of a given radiance for three different covers (after Goillot, Guyot and Malet 1976)

the effective surface temperature or of the radiances may lead to interesting discrimination among plants or to specific properties.

Such analysis have been discussed by C. Goillot et al. (Goillot, Guyot and Malet 1976). They show systematic variation of the radiance distribution for different covers (see figure 5-3) and the enlargment of the shape of the histogram with the illumination conditions. In order to quantify such an analysis one could introduce a shape parameter $s = \frac{h}{\Gamma}$ where h is the height of the histogram and Γ its half-width and correlate this parameter with the nature of the cover and its spectral properties.

From the previous discussions on the time behaviour of the surface temperature, it is expected that the evolution with time of the histograms of the temperature distributions or of the parameter s will give interesting complementary informations because not only the average temperature, but the half-width will depend on time.

observed). As an example, we have checked that the phase shift between the increase of the atmospheric temperature and the decrease of air moisture, as well as their relative values, at sunrise plays an important role on the evolution of the surface temperature (Becker and Hechinger 1979).

Since both methods (complete cycle or transient analysis) have their own advantage and disadvantage, it is interesting to use both of them in order to increase the accuracy of the determination of the ground parameters.

5.2.4 Statistical analysis of the data

In section 4 we have briefly analysed the influence of the microclimatic conditions, the orientation of leaves with respect to the sun, the wind, etc... on the surface temperature of a cover. For the same type of cover, the effective surface temperature will therefore not be uniform and a statistical analysis of the distribution of

5.2.5 Analysis based on a picture taken at one time only

It is not possible to cover the evolution of the surface temperature with time over all the year. For qualitative analysis, to check for instance the irrigated zone, a temperature measurement at a given time may be sufficient.

For the analysis of underground properties, A. Tabbagh (Tabbagh 1977) has given the best hours in the day and the best epoque in the year. He founds that the largest contrasts appear on December and June for the annual wave and at noon and midnight for the daily wave. This method is used also in geological applications to map underground faults.

For the analysis of plant properties, surface moisture, etc... a qualitative idea of what is occuring may be given by the type of correlations between the temperature and the vegetal index, or the albedo as it has been discussed in section 5.2.3. For instance, it may be interesting to classify an image according to the two parameters T/f(v) and v where v is either the red vegetal index defined by (5-8) or the green vegetal index defined by

$$v_g = \frac{R_{green} - R_r}{R_{green} + R_r}$$

where R_{green} is the reflected radiance
in the green channel and f is a function
of v defining the "normal" moisture such
that $T = \alpha f(v)$, α being a constant. Such
a classification would indicate some ano-
maly in the soil moisture. The same type
of analysis can be made with the albedo
and the temperature.

For the analysis of water mixing, thermal
water pollution, etc... these thermal maps
at a given time may be also very useful
(see for instance G. Dinelli and M. Maini
(Dinelli and Maini 1977, Becker and al 1979).

5.3 Atmospheric corrections

As it has been discussed in section 2, the
atmosphere emits and absorbs radiative
energy and consequently modifies the radiance of the surface. Since these modifications depend on the time at which the measure is made as well as on the situation
of the pixel, it is necessary to calculate
the effects of the atmosphere in all the
studies implying the comparison of temperatures measured at different time or
at different places.

There are several possible approaches to
perform these calculations. The first ones
are based on very accurate descriptions
of the phenomena ; they produce accurate
corrections but are generally time consuming (see for instance Deschamps 1977,
Scott 1974, Scott and Chedin 1977, Selly
and Mc Clatchey 1975,1976,1978 and the
general discussion by J.B. Farrow (Farrow
1975). The second type of models are based
on approximate descriptions of the phenomena leading to very simple correction
procedures which can be used on line. Such
a simple model is briefly presented in
§ A.2.4 of appendix A (F. Becker, D. Imbault
and L. Pontier 1977). This model leads to
the simple correction formula (A-29) which
has been explained in section 2. In this
section we shall discussed several ways
to use this simple model in order to perform the atmospheric corrections (Becker
1978 a).

5.3.1 One channel methods

If the measurements are done with one
channel only (say 10.5-12.5 µm), the different correction procedures are based on
formula 2-1.

i) One altitude flight. If the measurement has been done at one altitude, it is
generally necessary to perform a P.T.U.

(pressure, temperature, humidity) sounding
of the atmosphere to calculate the functions $W(z)$ and $\Theta(z)$. The constant A is
either calculated from the theory or
"measured" from a ground measurement on
a well defined site according to the relation

$$A = \frac{(T(z) - T_S)\cos\theta}{W(z)\left|\Theta(z) - T_S\right|}$$

where T_S is the radiometric temperature
of the ground, $T(z)$ the measured value
at altitude z and θ the angle of observation. If the measurements are done on a
large homogeneous surface with homogeneous
surface temperature, it is possible to
use the angular dependence of the absorption. In this case it is not necessary
to perform a P.T.U. sounding because the
surface temperature is given by

$$T_S = \frac{T(\theta_1)+T(\theta_2)}{2} + \frac{\cos\theta_1+\cos\theta_2}{2(\cos\theta_1-\cos\theta_2)}(T(\theta_1)-T(\theta_2))$$

$$(5-9)$$

This procedure has been used by the GTS and
the LMD over the mediterannean sea with
good accuracy (Becker and Pontier 1978).
Nevertheless this method must be used
very carefully because $T(\theta_1) \simeq T(\theta_2)$ and
$\cos\theta_1 \simeq \cos\theta_2$ which may lead to large
errors in the correction.

ii) Two altitude flights. In this case
it is not necessary to calculate A, nor
to have a ground truth station if a P.T.U.
sounding is performed. In this case, it
is possible to calculate A and T_S from the
two equations (5-10)

$$A = \frac{(T(h_1,\theta_1)-T_S)\cos\theta_1}{W(h_1)\left|\Theta(h_1)-T_S\right|} = \frac{(T(h_2,\theta_2)-T_S)\cos\theta_2}{W(h_2)\left|\Theta(h_2)-T_S\right|}$$

$$(5-10)$$

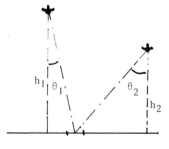

Figure 5-4. Same pixel seen at two different altitudes and with two different angles

195

where $T(h_1, \theta_1)$ and $T(h_2, \theta_2)$ are respectively the temperature of the pixel of ground temperature T_S, when it is observed at the altitude $h_1(h_2)$ and the angle of observation $\theta_1(\theta_2)$. The angles θ_1 and θ_2 are generally different because it is not possible to perform 2 flights exactly on the same projected line.

This procedure has been used successfully by Becker et al. (Becker, Bluemenroeder, Hechinger, Hourani, Ramey, Trautmann, Dechambenoy and Perrin, 1979) over the Rhine Valley. They have been able to produce on line absolute temperature image with an accuracy $< 0.4°C$.

5.3.2. Two channels methods

This method is very interesting because it avoids P.T.U. sounding. From formula 2-1 it is easy to show that

$$T_S = \frac{T_1(h,\theta) + T_2(h,\theta)}{2} + a\left[T_1(h,\theta) - T_2(h,\theta)\right]$$

$$(5-11)$$

where $\quad a = \dfrac{A_1 + A_2}{2(A_2 - A_1)}$

This equation is independent of h and θ. Therefore all the points for a given pixel must be on the same straight line whatever the altitude or the angle of observation is (see figure 5-4).

There are three ways of using it :

1. If a reference ground temperature θ_r is known (a lake, a river ...) the coefficient a can be obtained with 5-11 when the observed temperatures T_{1_r} and T_{2_r} of this reference pixel is measured. Then

$$a = \frac{2(T_{1r} - T_{2r})}{2\,\theta_r - (T_{1r} + T_{2r})}$$

2. If the measurement of one pixel (or better four pixels) with a well defined position is done at two different altitudes h_1 and h_2, a is then determined by

$$a = \frac{2(T_1(h_2) + T_2(h_1) - T_2(h_2) - T_1(h_1))}{T_1(h_1) + T_2(h_1) - T_1(h_2) - T_2(h_2)}$$

3. Over a homogeneous surface with homogeneous temperature, the angular dependence can be used. The coefficient a is given by the previous formula in which

h_i is replaced by θ_i.

Once a is known a straightforward equation is solved by any image processing system which gives a corrected image. This procedure has been applied successfully on the mediterranean sea (Becker and Pontier 1978) (figure 5-5). This model has not yet been tested over heterogeneous surfaces ; calculations are in progress at Strasbourg.

The inconvenient of formula 5-1 lies in the definition of the parameter a which appears as a difference. If a is small, formula 5-1 may be inaccurate (for the actual calculations a turned out to be of the order of 0.5 to 1.5).

6 CONCLUSIONS

Infra-red remote sensing will be a very powerful tool in the analysis, survey, control of plants, soils, etc... when the difficulties which are inherent to this method will be well challenged. The rapid survey of the possibilities of this technique, with all its imperfection, shows that interesting and useful results have already been obtained.

The main advantage of remote sensing is the possibility to have more informations, on a larger scale of space and time with the great advantage of automatic procedures for a relatively cheap cost. Here the automatic satellite plays an important role : - automatic survey
- repetitivity
- large scale.
But this is not yet without problems and limitations. For instance :

i) In the visible and infra-red band the atmosphere must be clear enough ... and this reduces strongly the capacity to repeat the measurements at some interesting periods.

ii) The recorded radiances are the synthesis of many different phenomena which must be separated in order to get reliable informations. Two interesting and complementary approaches have to be used : semi-empirical relationship obtained from a global analysis of the data and theoretical analysis giving a physical meaning to the empirical parameters as well as their dependence on neglected effects.

iii) The dimension of the pixel (at least 0.5 km x 0.5 km) on the ground integrates many local features which implies the definition of "average parameters".

196

iv) Some of the parameters modifying the radiance are not controlable (micrometeorological parameters for instance ...) and they must somehow be measured ... fortunately there exist some procedure to reduce their influence.

v) To facilitate the inversion problems recalled on point ii), it is necessary to make ground measurement (geomorphological parameters, air temperature near the ground, aerosols ...) which are not easy in order to be comparable with the remotely measured parameters (see iii)).

Fortunately the solutions of the different problems occuring in remote sensing are in fast progress thanks to the efforts on modeling, experimentation on ground with airplanes and satellites, instrumentation and image and data processing, (the most) part of the quoted bibliography is from the last three years) and the hope of having a powerful operational tool to survey our planet may become a reality in the next years ...

ACKNOWLEDGEMENTS

A part of this text has been completed at the Institut de Génie Rural of the Ecole Polytechnique Fédérale de Lausanne (Switzerland) and I would like to thank Prof. Regamay for his kind invitation and warm hospitality.

I would like to acknowledge also all my colleagues at the EPFL and at the GTS at Strasbourg for their help during the preparation of these lecture notes. Drs. A. Rosema, G.J. Soer, A. Kahne, A. Perrier are acknowledged for sending me their papers or preprints of their works, Drs. A. Tabbagh, P. Guyot, P.Y. Deschamps, N. Scott, M. Chedin, D. Imbault, L. Pontier, G. Neuwenhuis, B. Seguin for very interesting discussions and Mrs. D. Rauffer for her skillness in preparing the manuscript.

A part of this work has been completed thanks to financial support from the CNES and the DGRST.

APPENDICES : THE DIFFERENT FLUXES OCCURING IN THERMAL INFRA-RED REMOTE SENSING

In the two following appendices, the physical laws governing the different fluxes,

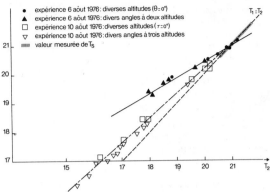

Figure 5-5. Atmospheric correction procedure using formula (5-11) and method 2. T_1 is for channel 10.5-12.5 µ, T_2 for channel 7.9-9.6 µ. The + and • points show respectively different experimental temperatures in the two channels at different altitude for the 6 August experiment, see figure 2-5, and for the 10 August experiment. The full (—) and dotted (...) lines are the result of formula (5-11) which reads

$$T_1 = (\frac{2-a}{2+a})T_2 + \frac{2a}{2+a}T_g$$

The dashed (---) line is the curve $T_1 = T_2$.

The intercept of the curves $T_1(T_2)$ with the curve $T_1 = T_2$ gives the ground temperature which is here 21° ± 0.1 for the two days. The experimental ground temperature was 20.6 < T < 21.6°.
The ▲ indicate the values obtained at different angles (after Becker 1978a).

shown in figure 2-1 of the text, occuring in thermal infra-red remote sensing are briefly reviewed. These fluxes can be divided into two parts which are not indeed independent : i) the electromagnetic radiation fluxes, ii) the heat and mass transfert fluxes.

A THE ELECTROMAGNETIC RADIATION FLUXES

A.1 The energy emitted by the scene

First of all, the surface of the natural bodies emit electromagnetic (E.M.) radiation because an accelerated electric charge emits an E.M. radiation. There are two main ways to produce such an acceleration leading to two processes of emission :

i) Thermal emission. It follows the acceleration of the charged particles present in a body due to their hazardous motions produced by their thermal agitation corresponding to the equilibrium temperature of the body. Every natural body, by the only fact that it is not at zero Kelvin does emit E.M. radiation. This process corresponds to the direct emission.

ii) Cold emission. In this case, the acceleration is produced by an external process. In passive remote sensing the incident solar irradiation induces an accelerated motion of the electric charges of the body which re-emit E.M. radiations. This process corresponds to the reflexion.

In any case, the ground will emit a radiation which has many characteristics :
an intensity, a wavelength or a spectrum (i.e. several wave length with a given intensity per unit of width of wave length), a phase, a polarization (i.e. an orientation of vibration in space), a direction of propagation.

Since the radiometers actually used in infra-red remote sensing measure mainly the intensity, i.e. the energy flux, we shall only discuss that part ... that does not mean that the other parts are not interesting.

The physical quantities measured by a radiometer is the radiance emitted by the surface. By definition, the spectral radiance $R_\mu(\theta, \phi, T_s)$ is the power emitted in the direction θ, ϕ per unit solid angle per unit area perpendicular to the beam per unit band width of wave length (unit : Watt per (m^2 sterad. μ)). The solid angle delimited by a surface $d\Sigma$ seen from the distance L, perpendicular to this surface is

$$d\Omega = \frac{d\Sigma}{L^2} \quad (A-2)$$

For a radiometer having an instantaneous field of view θ, the instantaneous solid angle is

$$d\Omega = 2\pi(1 - \cos\frac{\theta}{2}) \simeq \pi\frac{\theta^2}{4} \quad \text{if } \theta << \theta^2$$

Therefore, the power emitted by a surface of area dS, in the band width $(\lambda - \frac{d\lambda}{2}, \lambda + \frac{d\lambda}{2})$, at the angle (θ, ϕ) is

$$P(\lambda, d\lambda, \theta, \phi, d\Omega, ds, T) = R_\lambda(\theta, \phi, T)\cos\theta \, d\lambda d\Omega ds$$

$$(A-1)$$

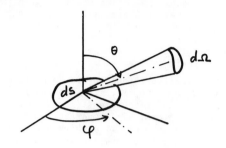

Figure A-1.

A.1.1 Spectral radiance of the black body

In order to calculate the power directly emitted by a given scene, physicists use to consider a theoretical body, called black-body because it absorbs all the energy falling on him (nothing is reflected, therefore it looks black). In the beginning of our century, Planck has shown the famous formula giving the spectral radiance emitted by such a theoretical body. The result is

$$R_\lambda^\circ(T) = \frac{C_1 \lambda^{-5}}{\exp(C_2/\lambda T)^{-1}} \quad W/(m^2 \text{sterad}.\mu)$$

$$(A-3)$$

with $C_1 = 1.19 \ 10^8 W/(m^2\mu^4)$

$C_2 = 1.439 \ 10^4 \ \mu^\circ K$

T is the temperature in Kelvin
(T = t$^\circ$C + 273)

μ is the wavelength in micrometre.

It appears that the radiance of a black-body depends only on its temperature T. It does not depend on the angles of emission nor on the intimate nature of the black-body.

The function $R_\lambda^\circ(T)$ is plotted in figure A-2 which shows the following characteristics :

i) For a given wavelength $R_\lambda^\circ(T)$ increases with T (the curves do not cross)

ii) The maximum of emission at a given temperature T occurs when

$$\lambda T \simeq 2897 \quad \mu^\circ K \quad (A-4)$$

This is Wien's law. It shows that for a

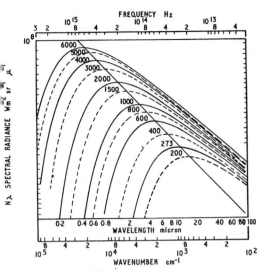

Figure A-2. Blackbody radiance (Planck's Law)

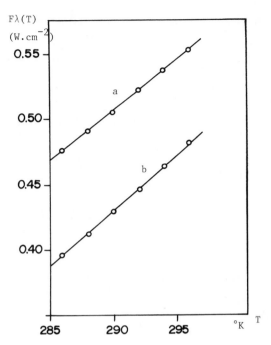

Figure A-3. Variation of $E(T) = \int_{\lambda_1}^{\lambda_2} R_\lambda^o(T) d\lambda$ with T

a) $\lambda_1 = 10.5\ \mu$ $\lambda_2 = 12.5\ \mu$

    $\lambda_1 = 7.9\ \mu$ $\lambda_2 = 9.6\ \mu$

(after P. de Felice and L. Pontier 1977)

reasonable earth temperature of 15°C, $T \simeq 285°K$, the maximum of emission occurs at

$$T \simeq \frac{2897}{285} \simeq 10.16\ \mu \qquad (A-5)$$

Fortunately it corresponds to a so-called atmospheric window, which we shall discuss below.

iii) The maximum of contrast for a given variation ΔT of the temperature occurs when the derivative $\dfrac{dR_\lambda^o(T)}{dt}$ is maximum. This occurs when

$$\lambda_m T \simeq 2400\ \mu°K \qquad (A-6)$$

For the temperature of 15°C, the maximum of contrast will occur at

$$\lambda \simeq \frac{2400}{285} \simeq 8.4\ \mu \qquad (A-7)$$

iv) The maximum of relative contrast for a given variation ΔT of the temperature occurs when the logarithmic derivative $\dfrac{1}{R_\lambda^o}\ \dfrac{dR_\lambda^o}{dt}$ is maximum. It can be shown simply that for a ground temperature of 15°C this occurs for a wavelength

$$\lambda \simeq 1.8\ \mu \qquad (A-8)$$

These figures will be very useful to choose the best channels for passive infra-

red remote sensing, compatible with the atmospheric transmission.

v) When the variation of the temperature is not too large on a given scene, the Planck's law (A-3) can be approximated by a simply linear expression. In fact, one can expand $R_\lambda^o(T)$ around the average temperature of the scene according to Taylor expansion

$$R_\lambda^o(T) \simeq R_\lambda^o(T_m) + (T-T_m)\left(\frac{dR_\lambda^o}{dT}\right)_{T_m} + \mathscr{O}\left((T-T_m)^2\right)$$

$$(A-9)$$

which can be simply calculated.

We show in figure A-3 the result of the exact calculation and the comparison with the contribution of the first two terms approximated formula (A-9).

199

From formula A-1 one sees that the power emitted by a black-body varies with the angle of emission as $\cos\theta$

$$P(\theta) = P(0) \cos\theta \qquad (A-10)$$

This law is known as Lambert's law. Surfaces emitting electromagnetic radiation according to (10) are said Lambertian. Their radiance do not depend on the angle of emission.

A.1.2 The spectral radiance directly emitted by a natural body

The theoretical black-body is generally not present in nature. It should be built with care. Nevertheless, it turns out that natural bodies emit on a way which, the most of the times, resembles the black-body behaviour. Therefore one compares the emission of natural bodies to the emission of black-bodies and define the spectral, directional emissivity as

$$\varepsilon_\lambda(\theta,\phi,T) = \frac{R\lambda(\theta,\phi,T)}{R_\lambda^\circ (T)} \qquad (A-11)$$

where $R_\lambda(\theta,\phi,T)$ is the radiance emitted by the natural body at the temperature T in the direction (θ,ϕ) and R_λ° is the radiance emitted by the black-body at the same temperature T.

$$R_\lambda(\theta,\phi,T) = \varepsilon_\lambda(\theta,\phi,T) \, R_\lambda^\circ(T) \qquad (A-12)$$

From formula A-1 one can compute the power emitted by a natural surface :

$$P(\lambda,d\lambda,\theta,\phi,d\Omega,ds,T) = \varepsilon_\lambda(\theta,\phi,T) R_\lambda^\circ(T)\cos\theta d\lambda d\Omega ds$$

$$(A-13)$$

Let us discuss somehow this formula.

i) The main point is that the power directly emitted depends on two distinct properties of the body :

- its temperature which depends on the thermal properties of the body and its environment. Its value indicates the equilibrium reached in the exchange of thermal energy (heat) and mass between the body and its environment.
- its emissivity $\varepsilon_\lambda(\theta,\phi,T)$ which characterizes the optical properties of the body.

ii) The power depends on the geometry of the experiment (through) the angles θ and ϕ (except if the surface is Lambertian. This must be checked in all experiments !)

iii) The power is emitted by the extreme surface. Roughly speaking, the penetration depth, d, is the distance travelled in the medium by the E.M. radiation after which the energy is reduced by a factor e. It is given by

$$d \simeq s\lambda \qquad (A-14)$$

where λ is the wavelength and s, a factor depending on the electric properties of the medium. For a dry ground s has a typical figure of 10, for a very wet ground, s may be as small as 0.1. Whatever the value of s is, the penetration depth is between 1 to 100 μ. Therefore only the extreme surface does participate to the emission (except for vegetated covers which are more or less transparent to the I.R. radiation according to the density of the covers). This point should be also very clear. If a thin layer of anything is on the ground the emitting surface is this thin layer, not the ground itself.

iv) Thermometric temperature, brightness temperature. The brightness temperature T_B is the temperature of a black-body which has the same radiance as the natural body at the actual thermometric temperature T_S. Therefore :

$$R_\lambda(\theta,\phi,T_S) = R_\lambda^\circ(T_B) = \varepsilon_\lambda \, R_\lambda^\circ(T_S) \qquad (A-15)$$

The brightness temperature is the temperature which one would get if he assumes that the emissivity is 1. In any situation

$$T_B < T_S$$

The emissivity properties are discussed in detail in section 3 of the text. Let us just quote some important parameters on which it depends : the nature of the surface, its roughness, the wavelength, the angle of emission and the temperature (to a less extent)(generally it should depend on the polarization of the electromagnetic radiation but there is no experimental data on this dependence).

A.1.3 The power reemitted by the surfaces

When the scene is irradiated by a source (the sun, the atmosphere, etc...) of radiance $R_{\lambda i}(\theta_i,\phi_i,T_i)$ seen from the scene through the solid angle $d\Omega_i$, the luminance reemitted in the direction $\theta_r\phi_r$ is

$$R_\lambda^r(\theta_r,\phi_r,T) = \rho_\lambda(\theta_i,\phi_i;\theta_r,\phi_r,T)R_{\lambda i}(\theta_i\phi_i T_i) \times$$
$$\times \cos\theta_i \, d\Omega_i \qquad (A-16)$$

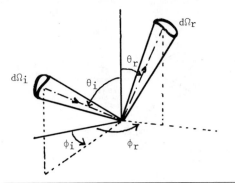

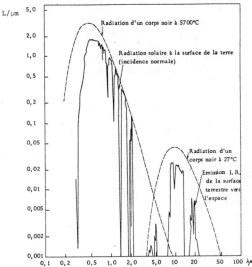

Figure A-4. Comparison between the spectral reflected radiance of the earth when it is irradiated by the sun at the zenith and the directly emitted spectral radiance by the earth at 300°K. To calculate this figure one uses : formula (A-6) with $T = 300°K$ and $\varepsilon_\lambda \simeq 1$ and formula (A-15) assuming that the sun is a black-body at $T = 5700°K$, with a diameter of $1.39196\ 10^6$ km at an average distance of the earth $d = 149.6\ 10^6$ km, and that $\rho_\lambda(\theta_i\phi_i;\theta_r\phi_r) = 1/\pi$. The absorption effects of the atmosphere have been introduced in this figure.
(After H. Flohn (Flohn 1973)).

where $\rho_\lambda(\theta_i,\phi_i;\theta_r,\phi_r,T)$ is the bi-directional reflectance at the wavelength λ. The bi-directional reflectance $\rho_\lambda(\theta_i,\phi_i;\theta_r,\phi_r,T)$ is not independent of the emissivity $\varepsilon_\lambda(\theta_r,\phi_r)$. For an opaque body in thermal equilibrium, one can show that :

$$\varepsilon_\lambda(\theta_r,\phi_r) = 1 - \iint_{hemisphere} \rho_\lambda(\theta_i\phi_i;\theta_r\phi_r,T)\cos\theta_i d\Omega_i$$

$$\text{hemisphere} \qquad (A-17)$$

In figure A-4 one compares R^d and R^r for a fictitious body for which $\varepsilon_\lambda(\theta,\phi,T) = 1$ and $\rho_\lambda(\theta_i\phi_i,\theta_r\phi_r) = 1/\pi$ and for the zenith ($\theta_i = 0$, $\phi_i = 0$).

The interesting feature is that in the range $(3-5)\mu$ there is a contribution of both the reflected and the direct radiance. We shall see later on the importance of this fact in the analysis of the data.

Formula (A-16) gives the reflected radiance due to the sun irradiation. Nevertheless, it is necessary to calculate (or measure) the direct solar irradiation at the ground level (Deschamps 1977). But as everyone knows, the atmosphere as well as the clouds do irradiate the earth surface. One must add therefore this contribution which is written

$$R^d_\lambda(\theta_r\phi_r T_a) = \iint \rho_\lambda(\theta_i\phi_i\theta_r\phi_r)R^a_\lambda(\theta_i\phi_i T_a)\cos\theta_i d\Omega_i$$

$$(A-17)$$

where $R^a_\lambda(\theta_i\phi_i)$ is the downward radiance of the atmosphere and the clouds. It is very difficult to calculate because it depends on the atmospheric conditions of the experiment. Generally, it is important to measure it from the ground. Combining formulae (11-b), (15) and (16) one has the radiance of the scene :

(direct radiance)
$$R_\lambda(\theta_r,\phi_r,T) = \varepsilon_\lambda(\theta_r,\phi_r)\ R^o_\lambda(T) +$$

(reflected radiance due to sun irradiation)
$$+ \rho_\lambda(\theta_i\phi_i;\theta_r\phi_r)\ R^{sun}_\lambda(\theta_i,\phi_i,T_i)\cos\theta_i\ d\Omega_i +$$

(reflected radiance due to atmospheric irrad.)
$$+ \iint_{hemisphere} \rho_\lambda(\theta_a\phi_a;\theta_r\phi_r)\ R^a_\lambda(\theta_a,\phi_a,T_a)\cos\theta_a\ d\Omega_a$$

$$\text{hemisphere}$$

$$(A-18)$$

A first glance at expression (A-18) shows that the ground optical properties (ε_λ and ρ_λ) are mixed with its thermal properties and the atmospheric conditions, to produce the radiance of the observed scene. One of the problems will be to find operational procedures to separate these different contributions. In this expression

- R^{sun} is the spectral irradiance of the sun at the ground level. It depends on the atmospheric conditions.
- $\varepsilon_\lambda(\theta,\phi)$ is the apparent emissivity of

the ground.
- $R_\lambda^o(T)$ is the spectral radiance of the
black-body at temperature T.
$-\rho_\lambda(\theta_i\phi_i;\theta_r\phi_r)$ is the spectral bi-directional reflectance.
- R_λ^a is the spectral downward radiance of
the atmosphere at the ground level.

A.2 Perturbation produced by the atmosphere

In the infra-red bands, the atmosphere
produces two kinds of effects : geometrical distorsions (due to refraction and
turbulence) and radiometric distorsions
(due to attenuation and emission).
The geometrical distorsions will not be
discussed here because they can be neglected to first order in passive infra-red
remote sensing with the actual resolution
of the radiometer. A general discussion
can be found in P.Y. Deschamps (Deschamps
1977), J.B. Farrow (Farrow 1975).

A.2.1 Attenuation

Due to two main effects in I.R. remote
sensing : scattering on aerosols and
molecular absorption. Ionospheric perturbations as well as Rayleigh scattering on
the molecules can be neglected as far as
one calculates the perturbations produced
on the infra-red radiation. They cannot
be neglected to calculate the total irradiance which is one of the boundary conditions for the thermal equilibrium of the
surface (see appendix B).

i) The scattering of the electromagnetic
radiation on the aerosols (the Rayleigh
scattering is negligible). It depends on
the structure and composition of the aerosols (radius of droplets, number of dro-
plets/cm3). The great difficulty is that
this structure and composition fluctuate
very much on time, leading to complicated
problems in the correction of these effects
namely the necessity to measure the aerosols composition of the atmosphere.

This is one of the biggest limitations
in remote sensing. If the visibility is
less than 5 km, the data obtained will
not be useful ... the presence of clouds
eliminates the recording from space platform (in our European regions this is
a very serious handicap) (see figure A-5).
In the following, we shall assume that
the atmosphere is clear.

ii) The absorption of the electromagnetic radiation on the molecules of the
atmosphere. This absorption is strongly

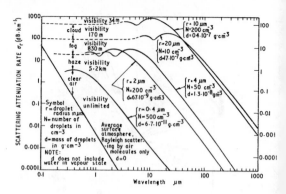

Figure A-5. Scattering coefficient versus
wavelength for Rayleigh and Mie scattering.
(quoted by J.B. Farrow (Farrow 1975))

---

dependent on the wavelength. The different
bands of absorption in the infra-red domain
are shown on figure (A-6). They define the
so-called atmospheric windows which are
well-known. As far as the correction models are concerned, there are two kinds
of molecules : those which are present in
a given proportion (say O_2, CO_2, etc...)
and those which are present in proportions
which fluctuate with time (mainly H_2O).
The main problem to take care of the absorption is to know the water content of the
atmosphere above the scene. This implies
generally a meteorological sounding giving
at each pressure level P(z), the temperature T(z) and the water vapor content
U(z) (P.T.U. profile).

This is again a limitation and some procedures to avoid such a sounding are discussed in the text. There exists several
computer programmes to calculate these
attenuation effects. Results are shown
on figures A-6.

iii) The transmitted radiance. The
transmission through the atmosphere for
the upward radiation is defined by the
transmittance $t_\lambda(z)$:

$$t_\lambda(z,\theta) = \frac{R_\lambda(z,\theta,\phi)}{R_\lambda(0,\theta,\phi)} \qquad (A-19)$$

where $R_\lambda(r,\theta,\phi)$ is the radiance transmitted
at the altitude z in the direction θ,ϕ
due to the radiance coming from the earth
at the altitude 0 in the direction (θ,ϕ)

$$t_\lambda(z) = \exp(-\tau_\lambda(z,\theta)) \qquad (A-20)$$

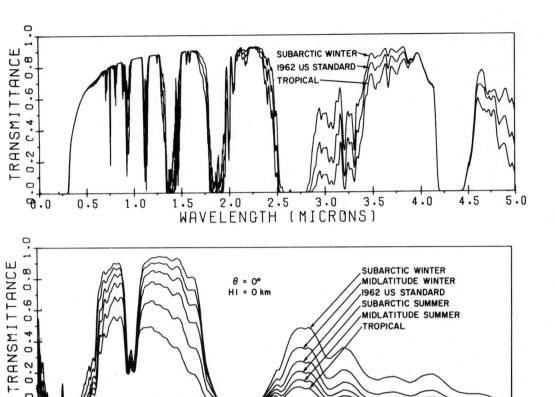

Figure A-6. Atmospheric Transmittance for
a Vertical Path to Space from Sea Level
for six Model Atmospheres
(after Selby and Mc Clatchey 1975)

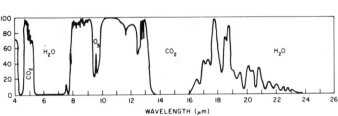

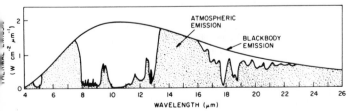

Figure A-7. (Upper) Spectral
transmissivity of the earth's
atmosphere at wavelengths in
the terrestrial radiation regime. The gases responsible for
the main atmospheric absorption
bands are indicated. (Lower)
Spectral distribution of radiation emitted by the atmosphere
compared to that emitted by
a blackbody at the same temperature.
(after K.L. Coulson 1975)

or

$$R_\lambda(z,\theta,\phi) = \exp(-\tau_\lambda(z,\theta))\ R_\lambda(0,\theta,\phi)$$

$$(A-21)$$

where $\tau_\lambda(z,\theta)$ is the optical thickness of the atmosphere defined by

$$\tau_\lambda(z,\theta) = \int_0^{z/\cos\theta} k(\lambda,x)\ dx \qquad (A-22)$$

where $k(\lambda,x)$ is the attenuation coefficient per unit length for both the scattering and the absorption at the altitude x. We can write

$$\tau_\lambda = \tau_{\lambda a} + \tau_{\lambda d} \qquad (A-23)$$

with $\tau_{\lambda a} = \int_0^{z/\cos\theta} k_a(\lambda,x)dx \qquad (A-24)$

$$\tau_{\lambda d} = \int_0^{z/\cos\theta} k_d(\lambda,x)dx \qquad (A-25)$$

where k_a and k_d are respectively the attenuation due to absorption and scattering

$$k = k_a + k_d \qquad (A-26)$$

An important fact to quote is that the attenuation will depend on the angles of observation.

On a given path, the unit of measure is the decibel :

γ_λ decibel $= 10\ \log_{10}(1/t_\lambda)$

(or $t_\lambda = 10^{-\gamma_\lambda/10}$) $\qquad (A-27)$

A.2.2 The emission of the atmosphere

The emission of electromagnetic radiation from the atmosphere comes from two main sources :

i) the direct emission by the molecule of the atmosphere and the clouds ;

ii) the reflexion by the aerosols of the radiance coming from the sun and the atmosphere itself.

This emission has not simple expressions and must be calculated with atmospheric models because it depends on the local properties of the atmosphere, mainly for remote sensing, on the distribution of the temperature of the atmosphere and the

distribution of the water content. Generally speaking, the atmosphere emits in the spatial domain in which it absorbs (cf. figure A-7).

A.2.3 The apparent radiance

The apparent radiance is the radiance actually seen by the radiometer. It is given by :

$$R_\lambda^{(a)}(h,\theta,\phi) = R_\lambda(0,\theta,\phi)\ t_\lambda(h,\theta)$$
$$+ R_\lambda^{atm}(\theta,\phi,T_{atm}) \qquad (A-28)$$

where the first term gives the contribution of the ground and the second term the contribution due to the atmosphere.

To recall the main dependences on the parameters, we quote :

 - the transmission coefficient depends on the angle of view (therefore, the correction will not be the same at the center and the adges of the image)

 - the correction is strongly wave length dependent

 - the atmosphere reduces the contrast

 - the corrections depend on the difference between the ground and atmospheric temperature. It is not a constant over the scene, it may be positive as well as negative. Order of magnitudes are given in section 5 of the text.

The following simple model will show all these features (Becker, Imbault and Pontier 1977, Becker 1978). More complete models are given by Selby and Mc Clatchey 1976 and Scott and Chedin 1977.

A.2.4 A simple model neglecting Mie scattering

If one neglects Mie scattering one can derive simple expressions to take care of both the absorption and emission of the atmosphere, when Mie scattering can be neglected.

The atmosphere is assumed to consist of a series of, n, homogeneous layers ch - racterized by a width d_i, a temperatu T_i, a coefficient of attenuation k_i. (see figure A-8).

From the previous equations, it can be shown that :

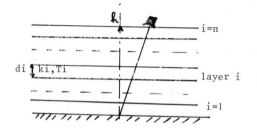

Figure A-8.

$$R_\lambda(h,\theta) = R_\lambda(0,\theta)\exp(-\Sigma_{j=1}^{n} k_j d_j/\cos\theta)$$

$$+\Sigma_{i=1}^{n}(1-\exp(-k_i d_i/\cos\theta))\exp(-\Sigma_{j=i+1}^{n} k_j d_j/\cos\theta)R_\lambda^\circ(T_i)$$

In this expression, the first term is the contribution of the ground, the second term is the contribution from the atmosphere itself.

Using linear expansions for the Planck's formula and neglecting the reflected radiation in a first approximation which is correct when $\varepsilon > 0.98$ (water, very wet soils, etc...), one can show that :

$$T(h,\theta) \simeq T_g + \frac{AW(h)}{\cos\theta}(\Theta(h) - T_g) + \varepsilon_{cor}$$

$$(A-29)$$

where h is the altitude and θ the angle of observation

T_g is the ground temperature (radiometric)

W(h) is the total water content of the atmosphere on the column of length h

A is a constant characteristic of the atmosphere and of Planck's formula parameters

ε_{cor} is a correction (which can be neglected to first order)

$\Theta(h)$ is an efficace atmosphere temperature defined by

$$\theta(h) = \frac{\Sigma_{j=1}^{n}(W_j T_j)}{\Sigma_{j=1}^{n} W_j}$$

This simple model is discussed in the text. For agricultural purposes, or dry soil, the contribution of the reflected radiation must be added. This changes the signification of the constant a, but the

general form of equation A-29 will not be changed. From this simple result, one notes that the radiometric temperature correction

$$\Delta = T_g - T \simeq \frac{AW(h)}{\cos\theta}(T_g - \Theta(h))$$

may be positive or negative according as Θ is less or more than T_g. Furthermore, it depends strongly on the water vapor content, the altitude and the angles.

A.3 The recorded signal

A.3.1 Principles of a radiometer

A radiometer is an instrument which concentrates the E.M. flux on a small detector which transforms this flux on an output intensity or voltage. The relation between the output and the input E.M. flux is not simple and must be defined by a precise calibration procedure otherwise it is impossible to use the radiometer. One can write generally for a linear relation between flux and output :

$$I = \int_{\lambda_1}^{\lambda_2} f_\omega(\lambda) R_\lambda^{(a)} d\lambda + I_0 \qquad (A-30)$$

where the spectral response of the radiometer may be written as :

$$f_\omega(\lambda) = \tau d\omega \, dA \, S(\lambda)$$

τ is the integration time,
$d\omega$ is the solid angle corresponding to the 1 FOV,
dA is the area of the entrance lens,
$S(\lambda)$ is the spectral response of the detector with its associated electronics,
$\lambda_2-\lambda_1$ is the bandwidth of the radiometer,
I_0 is a constant.
$f_\omega(\lambda)$ must be measured on flight and at the laboratory.

Note : if the scene is a homogeneous Lambertian surface at homogeneous temperature and if there is no-absorption, the recorded signal I will be the same in all the configurations of fig.A-9 although the dimension of the emitting surface is not the same.

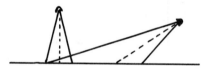

Figure A-9.

205

Radiance of the Ground

Rad. of Atm.

$$I = I_0 + \int_{\lambda_1}^{\lambda_2} d\lambda \, f_w(\lambda) \left\{ t_\lambda(\theta,\theta) \right\} \left[\varepsilon_\lambda(\theta,\varphi) R_\lambda^0(T_g) + R_\lambda(\theta_i\varphi_i;\theta,\varphi) R_\lambda^{sun}(\theta_a\varphi_a) \cos\theta_a \, d\Omega_a + \iint R_\lambda(\theta_a\varphi_a;\theta\varphi) R_\lambda^a(\theta_a\varphi_a,\bar{T}_{at}) \cos\theta_a \, d\Omega_a \right] + R_\lambda^{atm}(\theta\varphi \, T_{atm}) \right\}$$

recording ⟶ apparent Radiance at the window of the radiometer

Physical parameters	the ground	T_g extreme surface temperature (giving rise to the black-body spectral radiance measure the thermal equilibrium of the surface, function of its thermal properties and those of its environment.
		$\varepsilon_\lambda(\theta,\varphi)$ spectral emissivity in the direction θ,φ (polarization has been neglected)
		$R_\lambda(\theta_i,\varphi_i;\theta,\varphi)$ bidirectional spectral reflectivity (again polarization neglected)
	the atmosphere and meteorological parameters	- $R_\lambda^{sun}(\theta_a\varphi_a)$ irradiation of the sun at the ground level (depends on the atmospheric conditions)
		- $R_\lambda^{atm}(\theta,\varphi,T_{atm},u)$ radiance emitted by the atmosphere in the direction θ,φ at the radiometer level depends on : + Ta atmospheric temperature profile + U water content profile + optical properties of the atmosphere
		- $R_\lambda^a(\theta_a\varphi_a,\bar{T}_{at},u)$ downward radiance of the atmosphere (including the clouds) at the ground level (depends on the same parameters)
		- t_λ the spectral transmittance (depends on water content profile, U, aerosols ...) related to the spectral emissivity of the atmosphere
Technical parameters	the radiometer	$\lambda_2-\lambda_1$ wavelength band width and spectral band
		$f(\lambda)$ spectral response of the radiometer (depends on the type of radiometer and the other parameters,
Logistical parameters	conditions of experiments	h altitude of flight (depends on the relief)
		θ,φ angles of observations (depends on the relief, slopes, etc.)
		θ_a,φ_a angles of irradiation of the sun, depends on : - the hour of the flight - the season - the terrain structure (slope, directions, etc.)

Table A2 : Out-put signal of the radiometers as a function of the different parameters involved in a thermal infra-red remote sensing experiment

A.3.2 Relation between the recorded signal and the scene properties

Putting together the different expressions giving at each stage the E.M. radiation fluxes one gets the formula of table A-2. In this expression, one sees that the signal is not simply related to the ground surface parameters $\varepsilon_\lambda(\theta_r,\phi_r)$, $\rho_\lambda(\theta_i,\phi_i$; $\theta_r,\phi_r)$ and T_g.

B HEAT AND MASS TRANSFERS

B.1 Description of transport phenomena

In a medium in which neither the temperature nor the moisture are constant, transfer of heat and of mass take place inside the medium, or between two mediums.

B.1.1 Transport inside the medium

B.1.1.1 Instantaneous fluxes

The transport phenomena are quantitatively described at a given time by fluxes of heat (or of matter). They give at each time the quantity of heat (or of matter) passing through a unit surface perpendicular to the direction of propagation per unit of time.

Flux : quantity/unit surface perpendicular to the flow per unit of time.

Following A. Rosema (Rosema 1975) these fluxes can be written as it follows :

* The heat flux

$$\vec{q_h} = - \ \lambda \ \overrightarrow{grad} \ T \qquad (B-1)$$

where λ is the heat conductivity

($\overrightarrow{grad}$ T is a vector having the component $\frac{\partial T}{\partial x}$, $\frac{\partial T}{\partial y}$, $\frac{\partial T}{\partial z}$ where $\frac{\partial T}{\partial x}$ is the partial derivative with respect to x).

λ depends on the mineral composition, water content and temperature (see an example in the text).

T is the temperature at each point (x,y,z) of the medium.

q_h is in w/m^2.

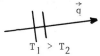

$T_1 > T_2$

* The moisture flux depends on the phase of the water:

. For liquid water (in unsaturated soils)
$$\vec{q_\ell} = - \ k_\ell \ grad(\psi - g_z) \qquad (B-2)$$
where q is the water flux (kg/m^{2}s)

ψ is the matrix potential of the water in the soil, it depends on the local temperature T

k_ℓ is the soil water conductivity under water potential gradient. It depends on the soil porosity and temperature

g is the gravitation per unit mass (9.8 N/Kg)

z is the altitude from any fixed reference

. For vapor water (in unsaturated soils)
$$\vec{q_v} = -\nu\delta D\rho_v \ \{\frac{1}{P_v} \ \overrightarrow{grad} \ P_v - \frac{M_v}{RT} \ \overrightarrow{grad} \ gz\} \qquad (B-3)$$

where q_v is the vapor flux (in Kg/m^{2}s)

ν is a mass flow factor taking into account the mixing fair and water vapor in the diffusion process

δ is a factor which takes into account the geometry of the medium

D is the vapor diffusivity in air

P_v is the vapor pressure which depends on the water potential, the temperature and the latent heat

M_v is the molecular weight of vapor

R is the perfect gas constant

T is the temperature.

If one assumes only vertical flow the total transport of moisture in the ground may be shown to be given by

$$\vec{q_{v\ell}} = \vec{q_v} + \vec{q_\ell}$$

i.e.

$$q_{v\ell} = - \ k_\psi(\frac{\partial\psi}{\partial z} - g) - k_T \frac{\partial T}{\partial z} \qquad (B-4)$$

where k_ψ is a soil moisture conductivity under water potential gradient (it can be calculated from soil properties, temperature, and water potential)

k_T is a soil moisture conductivity under temperature gradient

ψ is the water potential at a reference temperature T.

To show (B-4) it is assumed that $\psi(T)$ can be linearized according to

$$\psi(t) = \bar{\psi}(1 + \beta(T-\bar{T})) \qquad (B-5)$$

For vegetated covers the fluxes inside the medium containing the roots are discussed for instance by J. Methey (Methey 1977) and M. Vauclin (Vauclin 1979).

B.1.1.2 Variation of the fluxes with time

i) When a quantity of heat arrives into an element of volume of the medium, it changes its temperature according to the heat conservation principle at constant pressure. If one neglects the variations of volume of the medium, as well as the changes in phase of the different components of this medium, in each element of unit volume one has

$$\rho\, C_p dT = \delta Q$$

where δQ is the net incoming heat by conduction during the time dt :

$$\delta Q = - \, \text{div}(\vec{q}_h) dt$$

$$(\text{div } \vec{v} = \frac{\partial}{\partial x}\, v_x + \frac{\partial}{\partial y}\, v_y + \frac{\partial}{\partial z}\, v_z)$$

or from B-1

$$\rho\, C_p \frac{\partial T}{\partial t} = \frac{\partial}{\partial x}\left(\lambda\, \frac{\partial T}{\partial x}\right) + \frac{\partial}{\partial y}\left(\lambda\, \frac{\partial T}{\partial t}\right) + \frac{\partial}{\partial z}\left(\lambda\, \frac{\partial T}{\partial z}\right)$$

$$(B-6)$$

where C_p is the massic heat capacity at constant pressure of the soil,
ρ is the mass of the unit of volume.

Here one sees that for a heterogeneous finite medium it is not possible to factorize the heat (or thermal) conductivity λ so that the thermal diffusivity

$k = \dfrac{\lambda}{\rho C_p}$ does not appear explicitly

in the equation nor the thermal inertia

$P = \sqrt{\rho C_p \lambda}$ in the solution.

The thermal inertia gives the amplitude of the thermal response of a semi-infinite medium excited by a sinusoidal flux (see § 4.1).
To take care of the possible phase change in the water content of the medium, λ and C_p must be somehow modelized.

ii) In the same line, when a given quantity of water (or vapor) arrives in a cell of the medium, it changes its potential according to

$$C_\ell\, d\psi = \delta F$$

where δF is the net incoming moisture during the time dt : $\delta F = -\text{div}(\vec{q_{v\ell}}) dt$

C_ℓ is the liquid water capacity given by

$$C_\ell = \frac{\partial \theta}{\partial \psi}$$

where θ is the water content of the cell.

Therefore, assuming only a vertical flow, one gets from B-4

$$C_\ell \frac{\partial \psi}{\partial t} = \frac{\partial}{\partial z}\, k\psi\left(\frac{\partial \psi}{\partial z} - g\right) + \frac{\partial}{\partial z}\, k_T \frac{\partial T}{\partial z} \quad (B-7)$$

B.1.2 Transport between two media (air-ground)

These transports are more complicated to calculate and model because they strongly depend on the air turbulences, and the aerodynamic properties of the surfaces. The flux of sensible heat or convective flux ϕ is given by

$$\phi(z) = \rho\, k_h(z)\, \frac{\partial(C_p T)}{\partial z} \quad (B-8)$$

where ρ is the total air density
C_p the heat capacity (or specific heat) of moist air
k_h the turbulent diffusivity or eddy conductivity for heat.

An analogous equation may be written for the vapor flux E(z) and latent heat fluxes LE(z) :

$$E(z) = \rho\, k_v(z)\, \frac{\partial s(z)}{\partial z} \quad (B-9)$$

$$LE(z) = L_\rho\, k_v(z)\, \frac{\partial s(z)}{\partial z} \quad (B-10)$$

where L is the latent heat of vaporization
$s(z)$ the specific humidity $s = \dfrac{\rho v}{\rho}$

(ρ_v is the water vapor density ρ, the total air density $\rho = \rho_a + \rho_v$ where ρ_a is the dry air density)
k_v is the eddy conductivity for vapor.

These equations can be linearized as well when there is no vegetated cover. The coefficients k_h and k_v depend on the momentum turbulent diffusivity which depends itself on :

- the micro-meteorological parameters : wind profile, air temperature profile
- the aerodynamic properties of the surface.

Generally one assumes that $k_h = k_v$.

The relations giving $k_h (k_v)$ are very complicated and will not be given here (see for instance Rosema (Rosema 1975). Phenomenological simple expressions may be used (see for instance G.J.R. Soer (Soer 1977 a,b,c) and A.B. Kahle (Kahle 1977).

Since there is a turbulent flow, the various quantities defined above are fluctuating. What is the time average to use to define reliable average ? In the case of a vegetated cover, these equations have been discussed in great detail by A. Perrier (Perrier 1976).

In fact equations (B-8,10) are integrated and written as

$$\phi = \frac{\rho C_p}{r_a}\ (T_a(z_a) - T_S) \qquad\qquad \text{(B-8b)}$$

$$LE = \frac{\rho C_p}{\gamma(r_a+r_c)}\ (e_a(z_a) - e_s) \quad \text{(B-10b)}$$

where $T_a(z_a)$ is the air temperature at a reference altitude z_a, T_S is the ground temperature (with all the ambiguities mentionned in § 2), r_a is a turbulent transfert resistance depending on the horizontal wind velocity, the wind vertical profile, the stability conditions of the atmosphere ; $e_a(z_a)$ and e_s are respectively the pressure of the water vapor at altitude z_a and at the ground, γ is the psychrometric constant.
For a detailed discussion, see for instance J.A. Businger (Businger 1975).

B.2 Boundary conditions of the transport phenomena

B.2.1 For a non vegetated surfaces

This is a system with two boundaries, one at the interface soil-air and another one at the interface with the underground.

At the interface between air and ground, there is no accumulation of energy nor of water, therefore the fluxes coming from the ground should equal the fluxes living the ground.

The fluxes coming from the ground are from B-1 and B-4 :

heat fluxes $\quad q_h = -\ \lambda\ \dfrac{\partial T}{\partial z}\ \Big|_{\text{surface}}$

moisture $\quad q_{v\rho} = \{-k\psi(\dfrac{\partial \psi}{\partial z} -g)- k_T\ \dfrac{\partial T}{\partial z}\}_{\text{surf.}}$

The fluxes from the air are :
sensible heat flux H given by (B-8)
latent heat flux LE given by (B-10)
the vapor flux E given by (B-9)
radiative fluxes R_n discussed in appendix A.

R_n is the algebraic sum of :

R_S the sun direct and diffuse efficient fluxes (i.e. incoming flux minus the reflected flux over all the spectrum)

R_a the atmospheric downward flux

R_g the ground radiative flux.

Therefore the energy budget equation at the interface is :

$$q_h = H + LE + R_n \qquad\qquad \text{(B-11)}$$

The water budget equation is :

$$q_{v\ell} = E \qquad\qquad\qquad\qquad \text{(B-12)}$$

B.2.2 For vegetated surfaces

The complete modelization should include three coupled systems between the air and the underground : the vegetated cover, the medium with the roots, the soil.

| air |
| cover |
| roots |
| soil |
| underground |

This is a very complicated situation and there is no complete modelization for such a system. M. Vauclin (Vauclin 1979) discussed recently the influence of the root on the evaporation and on the moisture profile in the soil.

B.3 The input for model calculations without covers

B.3.1 The equation to solve

The calculation of the surface temperature and the vapor flux can be summarized as it follows for bare moisted soils.

i) Solve a system of coupled partial differential equations

$$\rho C_p(\psi,T,x,y,z)\frac{\partial T}{\partial t} = \frac{\partial}{\partial x}\ \lambda(\psi,T,x,y,z)\frac{\partial T}{\partial x} +\ldots$$

$$+ \frac{\partial}{\partial z}\ \lambda(\psi,T,x,y,z)\ \frac{\partial T}{\partial z} \qquad \text{(B-13)}$$

$$C_\ell(T,\psi,x,y,z)\frac{\partial \psi}{\partial t} = \frac{\partial}{\partial z} k_\psi(T,\psi,x,y,z)(\frac{\partial \psi}{\partial z} -g)$$

$$+ \frac{\partial}{\partial z} k_T(\psi,T,x,y,z) \frac{\partial T}{\partial z} \qquad (B-14)$$

where the functions

$$\rho C_p(\psi,T,x,y,z)$$

$$\lambda(\psi,T,x,y,z)$$

$$C_\ell(T,\psi,x,y,z)$$

$$k_\psi(T,\psi,x,y,z)$$

$$k_T(\psi,T,x,y,z)$$

are given model representations in the
different models. Some of them are shown
in the text.

ii) With the boundary conditions :
- at the air-ground interface, given
by (B-11) and (B-12) where the fluxes H, LE
and R_n are given external functions
depending both on the surface properties
and the atmospheric and microclimatic
properties ;
- at the ground-underground boundary,
given by the temperature T and the water
matrix potential ;
- at the lateral positions,
if infinite homogeneous :

$$\frac{\partial \lambda}{\partial y} \frac{\partial T}{\partial y} = \frac{\partial \lambda}{\partial x} \frac{\partial T}{\partial x} = 0$$

$$\frac{\partial}{\partial y} k_\psi \frac{\partial \psi}{\partial y} = \frac{\partial}{\partial x} k_\psi \frac{\partial \psi}{\partial x} = 0 :$$

if equal to some asymptotic values : give
these values.

iii) With the initial conditions :
$T(x,y,z;0)$ and $\psi(x,y,z;0)$ over the whole
medium. These initial conditions should
be measured on ground truth or assumed
a given value. The rapidity of the convergence of the calculation depends
generally on this choice.

B.3.2 The input parameters and the corresponding ground truth measurements
(for non-covered soils)

There are of several types.

B.3.2.1 The medium description

- Thermal properties $\lambda(\psi,T,x,y,z)$
$\rho C_p(\psi,T,x,y,z)$

- Hydrolic properties $k_\psi(T,\psi,x,y,z)$
$k_T(\psi,T,x,y,z)$
$C_\ell(\psi,T,x,y,z)$

These properties can be calculated from
more fundamental parameters such as composition, granulometry, porosity, surface
tension at the air-water inferfaces ...
These include the different discontinuities
in the medium.

B.3.2.2 The surface description

- shape to calculate a part
- optical parameters of the radiative
(emissivity ε, albedo) net flux
- aerodynamic coefficients (to calculate
the eddy conductivities k_h and k_v)
- orientation, slopes ... (to calculate
the sun irradiation).

B.3.2.3 The microclimatic condition

- wind speed profile $U(z)$ (to calculate
the eddy conductivities)
- atmospheric temperature profile
- atmospheric temperature humidity (to
calculate the flux of sensible and latent
heat , to correct the flux for absorption
in the atmosphere, to calculate the radiation emitted by the atmosphere).

B.3.2.4 The sun and atmospheric conditions

- season, latitude, longitude to calcu-
- hour late the
- cloudiness (quantity, tempe- net ra-
rature, ...) diative
flux

B.3.2.5 Underground parameters

- temperature and water potential at
a given depth
- position of the ground water level.

B.3.2.6 Initial conditions

B.3.3 Covered soils

The interesting thing is that the cover
shape, the LAI, the age, the homogeneity,
the type of the cover will influence the
effective surface temperature.

B.4 Conclusions

There are many parameters coming into the
play. The interest is that if all but one

vary, its variation may be obtained through the temperature variation. Unfortunately, it is generally not the case and several parameters vary at the same time. Furthermore, for a given contrast in the surface temperature, it is not easy to tell from which parameter the variation comes from. These points are discussed in § 5 of the text.

REFERENCES

Bartolucci, L.A., Ph.H.Swain & Ch.Wu 1976, Selective radiant temperature mapping using a layered classifier. IEE Trans. in Geoscience electr. GE 14:101.

Van Bavel, C.H.M. & D.I.Hillel 1976, Agric.Meteo. 17:453-476.

Becker, F., D.Imbault & L.Pontier 1977, Sea surface temperature and atmospheric transmission measurements by differential radiometry. Strasbourg, Proceedings of the EARSEL first general Assembly.

Becker, F. & L.Pontier 1978, Mise en oeuvre d'un radiomètre embarquable et mesures de la température superficielle de la mer. Rapport DRME (LMD et GTS).

Becker, F.1978a, Température superficielle de la mer et la transmission atmosphérique par radiométrie différentielle. Brest, Compte-rendu des Journées CNES-CNEXO.

Becker, F.1978b, Problématique de la télédétection électromagnétique et principes d'interprétation. Toulouse, Compte-rendus du Congrès OST, B1.

Becker, F., D.Blumenroeder, E.Hechinger, A.Hourani, Favel, J.Trautmann, C.Dechambenoy & Perrin 1979, Measurement and mapping of the absolute surface temperature of water surfaces by remote sensing. Ann Arbor, 13th International Symposium on Remote Sensing of Environment, ERIM. (paper C4)

Becker, F. 1979, Fundamental Physics for remote sensing. Strasbourg, Proceedings of the Int. Summer School CNES Toulouse.

Becker, F., E. Hechinger 1979, to be published.

Becker, F. (ed.) 1979, Physical and Mathematical principles of remote sensing. Strasbourg, Proceedings of the International Summer School, CNES Toulouse.

Bonn, F.J. 1977, Ground truth measurements for thermal infra-red remote sensing. Photogram.and Engineer.and Rem.Sens. 43:1001.

Bossard, D.R. & Y.Vuillaume 1975, Evaluation régionale de l'évapotranspiration et de l'humidité du sol. Rapport BRGM 75 SGN 427 AME.

Bunnick, N.J.J., 1978, The multispectral reflectance of short wave radiation by agricultural crops in relation with their mor-phological and optical properties. Agricultural University Wageningen, The Netherlands, 78, 1.

Businger, J.A. 1975, Aerodynamics of vegetated surfaces in Heat and mass transfer in the Biosphere. D.A.De Vries & N.H. Afgan (ed.), Scripta Book Cy, p. 139.

De Carolis, C. 1977, Basic problems in the reflectance and emittance properties of vegetation. Ibid. Ispra Course.

Chance, J.E. & E.W.Le Master 1977, Suits reflectance models for wheat and cotton : theoretical and experimental tests. Appl. Optics 16:407.

Cihlar, J. 1976, Thermal infra-red remote sensing : a bibliography. Canada Centre for remote sensing, Rapport n° 76-1.

Cihlar, J. & A.K.Mc Quillan 1977, Satellite thermal infra-red measurements to earth's resources. Canada Centre for remote sensing.

Collis, R.T.H. & P.B.Russell 1976, Laser applications in remote sensing. In Remote Sensing for Environmental Sciences, (loc.cit. E. Schanda Ed.)

Coulson, K.L. 1975, Solar and terrestrial radiation. Academic Press.

Deardoff,J.W.1978, Efficient production of ground surface temperature and moisture with inclusion of a layer of vegetation. Journ.Geo.Rech. 83:1889.

Deschamps, P.Y. & T.Phylpin 1976, Remote Sensing from space of the infra-red transmission of the atmosphere using polarization radiometry. Lab.Optique Atmosph. Lille.

Deschamps, P.Y. 1977, Télédétection de la surface de la mer par radiométrie infra-rouge. Lille, Thèse de doctorat.

Dinelli, G. & M.Maini 1977, Remote Sensing of thermal alteration and circulation patterns of riverine and coastal effluents. Kyoto, IFAC Symposium on Environmental System Planning.

Farrow, J.B. 1975, The influence of the atmosphere on remote sensing measurements. Summary report ESA (ESRO)CR 353.

De Felice, P. 1968, Etude des échanges de chaleur entre l'air et le sol sur deux sols de nature différente. Arch. Met.Geophys.Biokl.Ser.B16:70.

De Felice, P. & L.Pontier 1977, Sea surface temperature remote sensing with airbone infra-red radiometer. Correction of atmospheric error, Preprint LMD.

Fitzgerald, E. 1974, Multispectral scanning systems and their potential application to earth resources surveys. ESRO (ESP) n° CR 232, Spectral properties of materials.

Fligor et al. 1974, quoted by E.Fitzgerald. ESRO-ESA CR 232.

Flohn, H. 1973, quoted by J.Avias. In

La Télédétection des ressources terrestres. Tarbes, CNES p. 373.

Fuchs, M. & C.B.Tanner 1966, Infra-red thermometry of vegetation. Agro.Journ. 58:597.

Fuchs, M. & C.B.Tanner 1968, Surface temperature measurements of bare soils. J.Appl.Meteo. 7:303.

Gillespie, A.R. & A.B.Kahle 1977, Construction and interpretation of a digital thermal inertia image. Photogr.Eng. and Rem.Sens. 48:983.

Gjessing, D.T. 1978, Remote surveillance by electromagnetic waves for our water, land. Ann Arbor Science Pub. Ann Arbor.

Goillot, C., G.Guyot & Ph.Malet 1976, Premiers essais d'utilisation de la télédétection infra-rouge pour les études mésoclimatiques. Rennes ENSA, Colloque CNRS.

Hadni, A. 1967, Essential of modern physics applied to the study of the infrared. Pergamon Press, Infra-red science and Technology, 2.

Hadni, A. 1975, L'infrarouge. PUF collection Que sais-je.

Higham, A.D., B.Wilkinson & D.Kaan 1973, Multispectral scanning systems and their potential application to earth resources surveys. ESRO (ESP) n° CR 231, Basic Physics and Technology.

Higham, A.D. Multispectral scanning systems and their potential application to earth resources surveys. ESRO (ESP) n° CR 236, Summary volume.

Hodora, H. & W.H.Wells 1977, Infra-red radiometry and visible spectrometry. AGARD lecture series 88, p. 61.

Idso, S.B., T.J. Schmugge , R.D.Jackson & R.J.Reginato 1975, The utility of surface temperature measurements for the remote sensing of surface soil water status. J. of Geophys.Res. 80:3044.

Idso, S.B., R.D.Jackson & R.J.Reginato 1975, Determining emittance for use in infra-red thermometry. J.Appl.Meteo. 15:16.

Idso, S.B., R.J.Reginato & R.D.Jackson 1977, An equation for potential evaporation from soil, water and crop surfaces adaptable to use by remote sensing. Geophys.res.Lett. 4:187.

Idso, S.B., R.D. Jackson & R.J. Reginato 1977, Remote sensing of crop yields. Science, 196 : 19.

Kahle, A.B. 1977, A simple thermal model of the earth's surface for geologic mapping by remote sensing. J. of Geoph. Res. 82:1673.

Lyon, R.J.P, 1975, quoted by E.Fitzgerald, Analysis of rocks by spectral infra-red emission (8-25μ). Econ.Geol. 60:715-736.

Methey, J. 1977, Modélisation des transferts hydriques dans un versant sous prairie. Strasbourg, Thèse de 3è Cycle.

Monge, J.L., L.Pontier & F.Sirou 1976, Radiomètre ARIES. Palaiseau, Laboratoire de Météorologie Dynamique, Ecole Polytechnique. Rapport interne.

Ngai, W. 1978, Private communication.

Nieuwenshuis, G.J.A 1979, Private communication. To be published.

Perrier, R.A. 1976, Etude et essai de modélisation des échanges de masse et d'énergie au niveau des couverts végétaux. Paris VI, Thèse.

Philip, J. 1975, Water movement in soil in Heat and Mass transfer in the Biospher. D.A. De Vries & N.H. Afgan (ed.). Scripta Book cy, p. 29.

Price, J.C. 1977, Thermal inertia mapping : a new view of the earth. J. of Geophys. Res. 82:2582.

Querry, M.R., W.E.Holland, R.C.Waring, L.M.Earls & M.D.Querry 1977, Relative reflectance and complex refractive index in the infra-red for saline environmental waters. J. of Geoph.Res. 82:1425.

Raffy, M. & J.Thomann 1979, Private communication. To be published.

Reeves, R.G. (ed.) 1975, Manual of Remote Sensing. Am.Soc. of Photogr. Falls Church Virginia, Tomes 1 and 2.

Reginato, R.J., S.B.Idso, J.F.Vedder, R.D. Jackson, M.B.Blanchard & R.Goettelman 1976, Soil water content and evaporation determined by thermal parameters obtained from ground-based and remote measurements. J. of Geophys.Res. 81:1617.

Reginato, R.J., J.F.Vedder, S.B.Idso, R.D.Jackson, M.B.Blanchard & R.Goettelman 1977, An evaluation of total solar reflectance and spectral band ratoing techniques for estimating soil water content. J. of Geophys.Res. 82:2101.

Rosema, A. 1975a, Simulation of the thermal behaviour of bare soils for remote sensing purpose. Heat and mass transfer in the biosphere. Part.1 ed. by De Vries and Afgan. Scripta book Washington D.C. p. 109.

Rosema, A. 1975b, A mathematical model for simulation of the thermal behaviour of bare soils based on heat and moisture transfer. Niwars publication n° 11 p. 92.

Schanda, E. (ed.) 1976, Infra-red sensing methods. In Remote Sensing for Environmental Sciences. Spring Verlag.

Schanda, E. 1979, Interaction of E.M. radiation with the environment. In Physical and Mathematical principle of Remote Sensing. Strasbourg, Proceedings of the International Summer School. CNES Toulouse.

Scott, N. 1974, A direct method of computation of the transmission function of an inhomogeneous gaseous medium I. Description of the method. Journ. of Quant.Spect.and Rad.Transf. 14:691.

Scott, N. & A.Chedin 1977, Private communication.

Seguin, B. 1978a, Estimates of regional evapotranspiration from HEMM data. Wallingford, Second meeting of Tellus project. Working group II.

Seguin, B. 1978b, Private communication.

Selby,J.E.A. & R.E.A.Mc Clatchey 1975, Atmosphere transmittance from 0,25 to 18,5μ. Computer code LOWTRAN AFGLTR 750255.

Selby,J.E.A. & R.E.A.Mc Clatchey 1976, Atmosphere transmittance from 0,25 to 18,5μ. Computer code LOWTRAN AFGLTR 760258.

Soer, G.J.R. 1977a, Model TERGRA. Niwars Publication n° 46.

Soer, G.J.R. 1977b, Estimation of regional evapotranspiration and soil moisture conditions using remotely sensed crop temperatures.

Soer, G.J.R. 1977c, Niwars publication n° 45.

Staley, D.O. & G.M.Jurica 1972, Effective atmospheric emissivity under clear skies. J.Appl.Meteo. 11:349-356.

Suits, G. 1972, Calculation of the directional reflectance of a vegetative canopy. Rem.Sens. of Env. 2:175.

Sutherland, R.A. & J.F. Bartholic 1977, Significance of vegetation in interpreting thermal radiation from a terrestrial surface. Journ.of Appl.Meteo. 16:759.

Tabbagh, A. 1976, Les propriétés thermiques des sols. Archeophysika 6:128.

Tabbagh, A. 1977a, Sur la détermination du moment favorable de l'interprétation des résultats en prospection thermique archéologique. Ann.Géophys. 32.

Tabbagh, A. 1977b) Thèse.

Tucker, C.J. 1977, Asymptotic nature of grass canopy spectral reflectance. Appl.Opt. 16:1151.

Twomey, S. 1977, Introduction to the mathematics of inversion in remote sensing and indirect measurements. Elsevier Scientific Publishing Co, New York.

Vauclin, M., G.Hamon & G.Vachaud 1977, Simulation of coupled flow of heat and water in a partially saturated soil. Determination of the surface temperature and evaporation rate from a bare soil. München, Communication at the European Geophysical Society Conference.

Vauclin, M. 1979, ed. by F.Becker, Modèle local des transferts de masse et de chaleur entre le sol et l'atmosphère. Problèmes posés par son extension spatiale. Boulogne Billancourt, Séminaire

sur les mécanismes de transferts entre sol et atmosphère.

Verhoef, F. & N.J.J.Bunnik 1976, The spectral directional reflectance of row crops. Niwars publication n° 35.

Vieillefosse, M. & J.C.Favard 1978, "Cithare". Thermal inertia and humidity cartography over Africa by geostationary satellite. Dubrovnik, International Astronautical Federation pp. 78-127.

Vincent, R.K. 1975, The potential role of thermal infra-red multispectral scanners in geological remote sensing. Proceedings of the IEE. 63:137.

De Vries, A.D. 1952, The thermal conductivity of granular materials. Bulletin de l'Institut International du froid. Annexe I:115.

De Vries, A.D. 1966, Thermal properties of soils, Physics of plant environment. Van Wijk (ed.). North Holland pp.210-235.

Watson, K. 1975, Geologic applications of thermal infra-red images. Proceedings of IEE.63,1:128.

Watson, K. 1979, Thermal phenomena and energy exchange in the environment. in Physical and Mathematical principle of Remote Sensing. Strasbourg. Proceedings of the International Summer School. CNES Toulouse.

Wolfe, W.L. 1965, Handbook of military infra-red technology. Washington D.C. Office of Naval Research Department.

Zissis, G.J. (ed.) 1975, Special Issue on infra-red technology for remote sensing. Proceedings of the IEE. 63:1.

G. P. DE LOOR
Physics Laboratory TNO, The Hague, Netherlands

13

Survey of radar applications in agriculture

ABSTRACT

Radar is discussed as a tool in an
information system for the inventory and
the control of natural vegetations, crops
and soils. In such a system the information
is required at well determined points in
time during the growing season. Radar has
a great potential to satisfy this "all
time availability" requirement. Criteria
are drawn up which such a system must
satisfy. The available data are reviewed
and compared with these criteria. It will
be shown that radar satisfies them,
although in a different manner than we
are used to for optical systems.

1   INTRODUCTION

At several places in the world research is
going on to find out in how far remote
sensing can become a useful tool in
agriculture and forestry for:
- monitoring of the natural environment
- vegetation mapping (delineation of
  boundaries and determination of
  different plant communities)
- classification of crops and natural
  vegetation
- yield forecasting
- monitoring (health, stress)
- soil moisture determination
- soil roughness determination
  (sensitivity to slaking).
The observation and control of crops and
natural vegetation is useful only at
specific well determined points in time
during the growing season. The weather
in Western Europe and in many other parts
of the world, however, makes it often
impossible to realize this with the aid
of sensors working in the visible and
near IR. Most parts of Western Europe, for

example, have a cloud cover of 6/8 to 8/8
for at least 50% of the time. See fig.1.
It is furthermore demonstrated by the
relatively few cloud free images obtained
by LANDSAT.
  There is a need for a remote sensing
system technique which can be employed at
all times: day and night and even under
adverse weather conditions. Here radar has
great potentialities.
  These potentialities have to be
investigated. To approach such an
investigation by merely gathering and
interpreting images does not lead to the
result desired. Of course a phenomenological
use of radar imagery may already provide
some useful information for operational
purposes. However, it must never be
forgotten that radar uses cm-waves and
although the images obtained may look like
aerial photographs they are fundamentally
different. Radar images are transformations:
transformations from the microwave part
of electromagnetic (EM) spectrum (where
the radar "sees") to the visual (where our
eye can see).
  Interpretation procedures as used for the
interpretation of "ordinary" aerial
photographs must be "translated". Radar
will then prove to be a very versatile tool
for remote sensing.
This requires basic research, among others
to gain a better insight into the
interaction of microwaves with vegetation
to learn how these vegetations become
manifest in a radar image.

2. IMAGE FORMATION (metric)

The ordinary radar with rotating aerial and
imaging on a PPI screen has found wide
application, also in the air, especially
for navigational purposes. For mapping

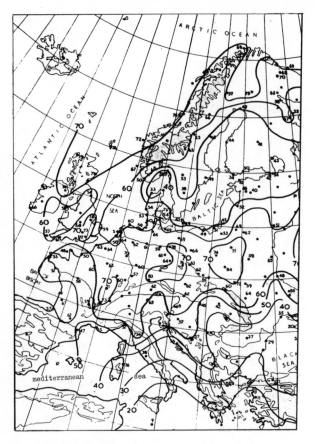

Fig.1. Percentage of time with cloud cover between 6/8 - 8/8 for June. June is best
        month with the least cloud cover.

applications, however, this type of radar
has its limitations.

With the development of the side-looking
airborne radar or SLAR, many of these
limitations have been overcome. This is
a radar with two non-moving antenna's
which are mounted in the sides of the
aircraft (or satellite). See fig.2.
Radar is an active system. It transmits
a short pulse of microwaves and records
the echoes received back in order of
arrival and puts them on a line on an
image tube e.g. in the form of small
light-dots. After the reception of the
last echo-determined by the distance over
which observation is wanted- a new pulse
is emitted, etc. Since an electromagnetic
wave propagates with the speed of light,
the turn around time is short and many
pulses can so be transmitted per second.
By this technique radar in fact measures
distances to the reflecting objects (the

time between the transmission of a pulse
and the reception of an echo). This has
its consequences on image formation. The
image so differs from the stereo-graphic
projection we are used to. This must be
taken into account when the data are to
be converted into a map.

In SLAR two antennas can be used which
concentrate the emitted energy, one on
each side of the airborne platform.
Figure 2 gives the system build-up. Only
one line is given on the image tube at a
time. By imaging this line on a film which
moves at a speed proportional to the speed
of the aircraft, a continuous image is
obtained. Scale accuracy in the flight
direction is determined by the accuracy
of maintaining the right proportionality
between the speed of the moving platform
carrying the radar and that of the film.

The mapping scales worked at are between
1: 100,000 and 1: 1,000,000 and the spatial

216

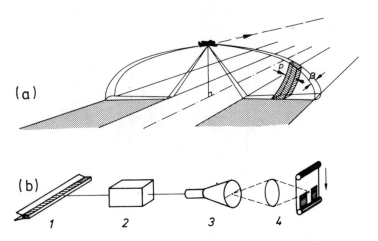

Fig.2. Side looking radar (SLAR). a. Scan configuration; b. radar system: 1 antenna, 2 transmitter/receiver, 3 image display tube on which one intensity modulated line; picture is built up to a continuous image on moving film in camera system 4.

Table 1 Average parameters of an X- and a K_a-band radar.

	X-band	K_a-band
Operating frequency (GHz)	9.5	35.0
Aerial length (m)	3	4.0
Transmitter peak power (kW)	80	50
Pulse repetition frequency (sec^{-1})	2500	3800
Pulse length (μsec)	0.2	0.1
Resolution:		
across track (m)	30	20
along track (m/km)	16	2.7
Size (without aerials) (m^3)	0.35	0.4
Weight (without aerials) (kg)	155	156

resolution of the system is in accordance with these scales. This also applies for the areas overseen. Table 1 gives the parameters of two representative systems in two common wavelength bands. The X-band system is intended for large ranges (to 100 km) and works effectively at scales smaller than 1: 250,000, where the K_a-band system is better for the shorter ranges (to 15 km) and scales between 1: 100,000 and 1: 250,000.

For further details on the metrics of image formation the reader is referred to the literature (Reeves et al. 1975, De Loor 1976).

3.   IMAGE FORMATION (Physical)

3.1   Criteria

In fig. 3 differences are given between observation with radar and with a camera. They are a consequence of the enormous difference in wavelength of the "light" used: in the order of 10^5 (light: tenths of microns, radar: cm-waves). In the visible and NIR the wavelength used is much and much smaller than the sensor (camera) and the object, where with radar both are in the same order of magnitude of the wavelength used. This has consequences for the formation of the final image and it must be clear that for such an image other criteria will apply.

	CAMERA	RADAR

optics

	lens	antenna
diameter:	to $10^4\lambda$	60 to 400 λ

resolution

order of:	μrad	mrad

object

size:	$\gg \lambda$	$\sim \lambda$

surface

smooth:	$< \frac{1}{4}\lambda$

variations:	<0.1 μm	cm's

Fig. 3 Differences in observing with a camera or a radar.

Another important aspect is the fact that radar carries its own "light"-source, which is also coherent where in the visual and NIR external sources are used —with the sun as the most important one —which are non-coherent and difficult to control.

So large differences can be indicated. For a good evaluation it becomes necessary to make a list of criteria by which a good observation system can detect, discriminate and recognize an object in this case: crops and natural vegetation. Such a list is:

1. Shape and geometry (spatial resolution)
2. Reflective properties (dynamic range and dynamic resultion)
3. Spectral properties ("color", frequency)
4. Polarization effects (horizontal, vertical, cross)
5. Temporal effects (changes in time and place; MTI).

Some of these properties are interrelated. We shall now investigate them.

3.2 Spatial resolution

The parameters determining resolution are: antenna aperture in azimuth (along track) and pulse length in range (across track). See Figs. 2 and 4. The antenna concentrates the electromagnetic energy emitted by the radar in a narrow beam with an opening in azimuth of β radians as Fig. 2 shows.

Along track resolution is then given by the well-known formula: antenna opening:

$$\beta = 1.2 \frac{\lambda}{D} \text{ radians}$$

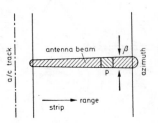

Fig. 4 Resolution in azimuth (along track) diminishes with range.

with λ the radar wavelength D the antenna aperture, both in the same units. Absolute resolution along track thus diminishes with range. Resolution in meters will never be better than the size of the aperture D.

The ratio λ/D is decisive. As fig. 3 shows this ratio is much larger than in the optical case and this means that obtainable resolution in azimuth will be much smaller. It is possible to do something about it by using the fact that we control the source of illumination and the fact that it is a coherent source. It now becomes possible to use the forward movement of the aircraft carrying the radar by treating the successive antenna positions as if they were the individual elements of a very long linear antenna array. The further away from the radar an object is, the longer it is illuminated and the longer the "synthetic" antenna array becomes. In this way it is possible to make the along track resolution independent of range with half the size of the antenna used as theoretical resolution limit.

Using satellites or very high flying aircraft such a SAR (synthetic aperture radar) is the only possibility to obtain a sufficient azimuthal resolution. From space obtainable resolutions are the same as those obtainable with camera's. Such systems, however, are very expensive: 5 to 10 x more expensive than an "ordinary" SLAR.

The ability of the radar to discriminate two objects behind each other in range (across track resolution) is determined by the pulse length p (fig. 2 and 4). Since the radar pulse travels with the speed of light (300 m/μsec) and the radar records the time for a pulse to travel to and from a target, each μsec corresponds with 150 m on the screen. Current systems use pulse lengths between 0.05 and 0.3 μsec

(which corresponds with a resolution
between 8 and 50 m) dependent on the wavelength used and the maximum range required.

So spatial resolution will always be in the
order of meters, but as already remarked
mapping scales (1: 100,000 to 1: 1,000,000)
and ranges (15 to 100 km) are in accordance
with that resolution.
    Aerial photography at the same scale
will have a comparable resolution. An
AWAR figure for a good aerial camera (with
film) is 40 1/mm (Reeves et al. 1975)
what at a scale of 1: 100,000 corresponds
with a resolution of 2.5 m.   Higher
resolution are sometimes quoted in the
literature (e.g. the Leartek system) but
this usually concerns sophisticated
equipment flown under highly controlled
conditions.
    The flying heights required for mapping
at these scales also have their influences
(even a part from visibility and cloudiness).
Even a figure of 20 m is quoted for U2
photography at a scale of 1: 100,000.
    But even in the case we would be able
to increase the resolution, say to 1 dm
(with a proportional reduction in the
area overseen), an object would look
completely different than we are used to
in the visible (fig. 3). Photograph 1
(in the back) gives an example. It shows
how an aircraft manifests itself on a
radar with such a high resolution of 1 dm.
The image would become a little "better"
when we could render the whole dynamic
range of the radar but the image would
never become the same as the photograph.
    Spatial resolution must always be seen
in relation to mapping scale and area
overseen. The combination of looking at
areas of hundreds of square kilometers in
one image with a resolution of cm's or
dm's is a contradiction in terms.
    Another key to observation, discrimination
and recognition in this context is:
texture. It will be clear from the
foregoing that texture will also be
different. Morain and Simonett (1967)
investigated texture as a key for
discriminating vegetation species. They
measured the variability in density for
specific vegetation species on radar
images (fig. 5). It is a method also used
in aerial photography (Leschack 1970).
The natural fluctuation from pixel to
pixel must be taken into account (see
par. 3.3.2).

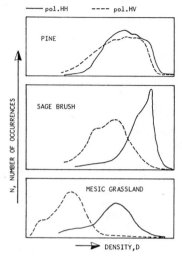

Fig. 5 Variations in density as a means
        to describe texture. After ref. 4.

3.3 Dynamic range and dynamic resolution

3.3.1 The return parameter γ

The radar-cross-section σ (in m^2) of an
object is defined with the aid of the
"radar equation" (Skolnik 1970). This
cross-section usually deviates strongly
from the geometric shape of an object
(compare photograph 1). Only for a limited
amount of objects is calculation of the
radar-cross-section possible. Examples
are the sphere and the corner reflector:
popular objects for reference targets.
    The total variation in σ that can be met
is very large: in the order of 80 dB: from
the sea and bare soils (low reflections)
to man-made objects as buildings and
bridges (high reflections). Radar can cover
them all.
    In ground returns (soils, crops and
natural vegetation) we meet objects which
are larger than the antenna beam
(distributed targets; see photograph 2).
The return must now be defined with respect
to a unit surface (1 m^2): as a reflection
coefficient. Two definitions are in common
use: σ_o and γ (Cosgriff et al. 1970;
de Loor 1976).

$$\sigma_o = \sigma/A \text{ and } \gamma = \sigma/A_i$$

with σ : the total backscatter cross-
            section
     A   : the area illuminated by the system

219

A_i: the area of the cross-section of the illuminating antenna beam at the position of the target.

The relationship between σ_o and γ is then:

$$\sigma_o = \gamma \sin\theta \text{ or } \sigma_o = \gamma\cos\varphi$$

with θ : the grazing or depression angle
and
φ : the angle of incidence (away from the vertical).

Measurements have shown that for grazing angles between 2° and 45°, many surfaces behave as diffuse scatterers and that γ is reasonably constant over this range of angles (De Loor 1976). This provides the possibility of averaging data from different sources over ranges of grazing angles. In this way we obtain a picture having a reasonable coherence from otherwise seemingly disjointed observations, and ranges for the value of γ (for grazing angles from 5° to 45°) can then be given (see table 2).

Table 2 Ranges of the return parameter γ for the X-band for grazing angles θ from 5° to 45°

	γ(dB)
Sea clutter	−40 to −15
Bare soils	−35 to −15
Vegetation and crops	−18 to 0
"Natural" surfaces	−10 to 0
Man-made structure	> 0

In this context "natural surfaces" mean: terrain with dams and hedges, rows of trees, edges of woods, etc.

3.3.2 Dynamic resolution

As may be seen from table 2, ranges can be fairly narrow for specific applications. The total variation in γ for bare soils, natural vegetation and crops is a 15 to 20 dB; for agricultural crops even less: 10-15 dB. (This is about the same range as met for the reflection coefficient in the visible and NIR). This means that the return parameter γ must be determined very accurately when we want to use it for the goals set in chapter 1. This has its consequences for the radar system to be used. Within a "window" of a 20 dB in the total dynamic range of the radar, γ must then be determined with an accuracy of

better than 1 dB. This is a severe requirement.

But there is more. Radar echo's of vegetation vary with time. Since we deal with compound reflectors (stems, leaves) moving in the wind the echo will come from continuously changing combinations and with a varying strength (coherent illumination, vector sum). See fig.6.

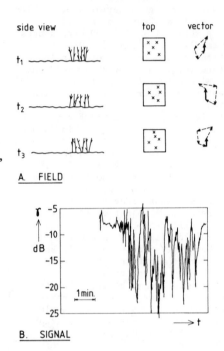

A. FIELD

B. SIGNAL

Fig.6. Variation of the reflection

The return signal will fluctuate and its strength will fluctuate according to a Rayleigh distribution (De Loor et al. 1974; Moore 1976; van Kasteren and Smit 1977). This means that single independent observations can vary considerably, even to such an extent that the whole available "window" can be covered. To find an accurate value for γ averaging over a sufficient number of independent observation will be necessary. Fig. 7 gives the number of independent samples for a specific accuracy.

For a $\pm$ 1 dB accuracy 50 independent samples are required (Moore 1976; van Kasteren and Smit 1977). Averaging can be done in space (over sufficient pixels) or in time, but since we use a moving platform it must be done in space. In radar images this effect is observed in the "spikiness" of

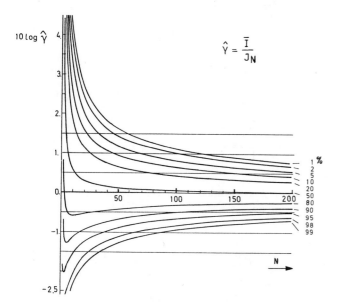

$$\hat{\gamma} = \frac{\bar{I}}{J_N}$$

Fig. 7 Confidence intervals for different percentages as a function of number of independent samples N, after van Kasteren and Smit (1977).

vegetation.
The more the number of independent observations N, the better and smoother the greytone. See photograph 3. The eye does better in this respect (Moore 1976) than a machine, since unconsciously the eye averages over groups of neigbouring pixels. An image with N = 5 to 10 (photographs 2 and 3) looks nice, but is difficult to handle for a machine ($\pm$ 2.5 dB, fig. 7).

In short we must trade spatial resolution for dynamic resolution. The choice depends on the application. When the imagery is interpreted visually we can take N smaller and so have a better spatial resolution than when data handling is done by a machine.

3.4 Spectral properties ("color")

The radarreturn of crops and natural vegetation is dependent on frequency (color). The use of more frequencies increases the discriminating possibilities. It is costly, however, since it requires a complete extra radar system for each frequency. It is known that at lower frequencies and lower incidence angles φ (to 30°) the underlying soil remains visible where at higher frequencies and higher incidence angles (φ > 30°) the

vegetation alone determines the reflection (De Loor et al. 1974; Bush and Ulaby 1977; Batliala and Ulaby 1977).

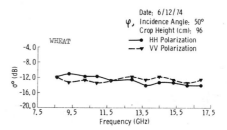

Fig. 8 Variation of the radar return with frequency. After Bush and Ulaby (1977)

3.5 Polarization effects

The radar-return is dependent on polarization (fig. 9). Knowledge of this dependency makes discrimination possible by using different combinations of polarization (fig. 9).

Different polarization combinations can be incorporated in one radar system: HH

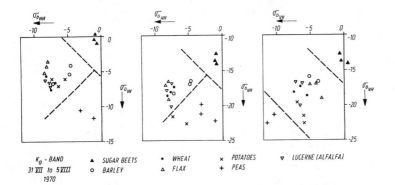

K_a - BAND	▲ SUGAR BEETS	• WHEAT	× POTATOES	▽ LUCERNE (ALFALFA)
31 VII to 5 VIII 1970	o BARLEY	△ FLAX	+ PEAS	

Fig. 9 Cluster plots at K_a-band. Vegetation discrimination on the variation of the radar return due to polarization.

(transmitting horizontally polarised waves and receiving horizontally polarised waves), VV (vertical ditto) and HV or VH (respectively transmitting horizontally polarised waves and receiving vertically polarised waves and the reverse). Most SLAR systems use HH.
Westinghouse has flown a K_a-band system using the combinations HH and HV (Morain and Simonett 1967).

3.6 Temporal effects

Crops and natural vegetation change through the growing season, see fig. 10 and photograph 4.

Being independent of the weather and the time of the day radar is eminently suited to follow these changes.

3.7 Summary: on criteria

When we now look again at the list in par. 3.1 of the criteria a good observation system must satisfy, we see that radar can meet them all. We also have seen that an effective usage of all these criteria requires a good insight in the reflection mechanisme of radar waves and the ground. Such an insight can only be built up through a well balanced measuring program. This we shall discuss in the next chapter.

4   MEASUREMENTS

4.1 The necessity of basic measurements

The current approach in dealing with remote sensing is an "a posteriori" approach: looking at and handling of remote sensing imagery. This, however, covers only part of the problem as fig. 11 shows, because it only deals with the output end of the remote sensing system.

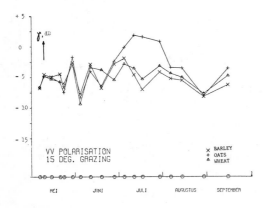

Fig. 10 Variation of the radar return of barley (×), oats (+) and wheat (△) through the growing season. After van Kasteren and Smit (1977).

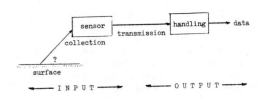

Fig. 11 The remote sensing system

222

The quality of the output (data handling and interpretation) is determined by the quality of the input. Taking wrong data (e.g. the wrong frequency, the wrong polarization) or not taking all the right data can never be compensated, neither by the largest computer nor by the most sophisticated algorithme. The alternative: taking all the data one can think of, does not solve this problem either: it only saturates the data channel.

The best approach is an "a priori" approach: putting the right questions immediately at the sensor. This requires knowledge of the input end of the remote sensing system.
This is knowledge of the object-sensor interaction: how does the object manifest itself at the sensor in the frequency range used and what are the appropriate classifiers for that specific frequency range.

This knowledge can only be obtained through a well balanced measurement program in which an interdisciplinary team does basic measurements on the ground and also carries out test flights to obtain the necessary verification and feed-back.

Such an "a priori" approach will also make it possible to use data channels optimally. Even then data handling will remain a major task, but any other approach will only cause rapid channel saturation.

4.2   Survey of the literature

Ground based and airborne measurements of the radar backscatter of vegetation crops and soils are reported in the literature. A good survey is given by Moore (1975) and Skolnik (1970). These measurements, however, mostly concern spot-checks done to collect some information on the magnitude of the radar-return of vegetation, crops and soils as a basis for radar designs. More systematic studies are only scanty available. One of the first was that by Goodyear (1959). It was done with measuring equipment mounted in an aircraft.
Major drawbacks of airborne studies are: calibration problems and the fact that in the analysis afterwards it is often very difficult to indicate precisely what area belongs to a specific measurement which makes it difficult to find the corresponding botanical parameters. A good first impression, however, of the diversity of the radar-return of different vegetation species was all the same obtained. The same technique was used later by the Naval

Research Laboratory (Daley et al. 1968) and the same problems were encountered.

Ground based measurements do not have these problems since work is done on well defined (in space) target areas but they have the drawback of a limited coverage. This last problem can be solved up to a certain extent, either by a mobile measuring system or by the use of test areas with a wide variation of plant species, e.g. on test farms. Especially the last procedure gives the possibility of obtaining botanically well controlled samples.

The first systematic series of measurements in this respect are reported by Ohio State University (Cosgriff et al. 1960; Oliver and Peake 1969). For the first time the research team was interdisciplinary and the measurements were done on botanically well defined agricultural crops.

To our knowledge at the moment the Remote Sensing Laboratory of Kansas University is the only institute in the US doing such ground based measurements on a controlled basis (Bush and Ulaby 1977; Batliala and Ulaby 1977). Their measurements now cover a wide variety of vegetations, crops and soils measured at different microwave frequencies and over several years.

In Europe a similar measurement program is carried out by the ROVE-team in the Netherlands. In this team participate: the Centre for Agrobiological Research (CABO, Wageningen) the Physics Laboratory TNO (the Hague) and the Microwave Laboratory of the Delft Technological University (Secretariat, P.O. Box 5031,2600 GA Delft). The measurements are verified by SLAR test flights carried out by the National Aerospace Laboratory (NLR, Amsterdam).

4.3   Ground based measurements, an example

Since the procedures as followed in the Netherlands are representative for such ground based measurements we shall take them as an example. (De Loor et al. 1974; Attema 1974,1978; van Kasteren and Smit 1977).

Prerequisites for a good measurement are (see also par. 3.3.2 and fig. 6):

a. The target area illuminated by the radar must be of such a size that it contains a sufficiently large collection of independent scatterers so that the envelope of the received signal is a random variable with its amplitude described by a Rayleigh distribution.

b. By spacing them in time and/or space a sufficient number of independent samples must be taken in one measurement.
(Fig. 7).

Work started in 1968 on TV-towers as measuring platform (altitude H = 75 mm) and using an X-band measuring radar of the pulse type (De Loor et al. 1974).
Its properties are given in table 3.

Table 3 Properties of the X-band measuring radar

Pulse system

Frequency	9374 MHz
PRF	1000 Hz
Antenna	1.8°, pencil beam
Polarization	HH and VV
Pulselength	0.5 μsec
Gatelength	40 nanosecs.
Output P_t	50 kW
Receiver	logarithmic; dynamic range 50 dB
P_{rec} min	-103 dB
Range variable	from 600 m to 50 km

The resolution cell was large so that the first requirement (sufficient scatterers) was easily met (fig. 12). The second requirement was met by integrating in time (100 seconds).

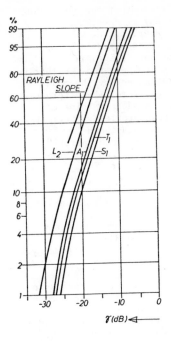

Fig. 12 Measured distributions on 4 fields

L_2 - lucerne (alfalfa)
T_1 - wheat
S_1 - suger beets
A_1 - potatoes
Polarization HH

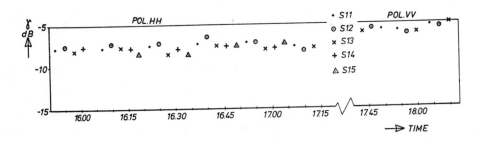

Fig. 13 Residual fluctuation of γ for 100 second measurements at 5 different places in a sugar beet field; windspeed between 1 and 4 m/sec.

By determining the fading spectra the decorrelation time could be determined and thus the integration time to obtain a sufficient number of independent samples. This decorrelation time is dependent on wind speed. As a crude rule of thumb it can be said that the windspeed in knots gives the approximate number of independent samples taken in one second (at X-band). Fig. 13 shows the residual fluctuation for 100 second measurements at 5 different places in a sugar beet field.

These measurements teached that the radar return parameter γ as a function of polarization and frequency is a useful classifier. Its range for crops and natural vegetation is limited: 20 dB. γ varies through the season and for the low grazing angles as used here ($\theta = 1^\circ - 10^\circ$) this variation is best explained by the influence of the water stored in the leaves of the top layers of the canopy. This result suggests the use of radar for in situ biomass determinations (yield forecasts).

The measuring system as described above only covers low grazing angles. This grazing angle, however, is accurately known and the two requirements: of sufficient scatterers in a measuring cell and sufficient samples in a measurement, were easily met. Working from fixed measuring platform as TV-towers, however, means that one is bound to the vegetation species available there. A short-range reflectometer, working on samples of a limited size gives a much larger freedom for experimentation in this respect. In our situation it made it possible to work on an agricultural testfarm, where a group of small well conditioned test-fields with different agricultural crops and grasses was available. It meant also that a good botanical definition of the

target could be made. These fields were very carefully prepared for this measuring program. To obtain a cell size of sufficient dimensions the antenna aperture must become fairly large at these short ranges. This means a lower accuracy in θ. The possible variation in θ, however, now becomes much larger.

Measurements were first done with an available short range X-band pulse system (De Loor 1974; Attema 1974). See table 4. Measurements were made at 3 grazing angles

Table 4 Properties of the X-band reflectometer

Frequency	9.3 GHz
PRF	15 kHz
Antenna	two horns, 17°
Polarization	VV
Pulselength	15 nanosecs.
Output P_t	70 W
Receiver sensitivity	−30 dBm
Range	5−15 m

(20°, 40° and 60°) and at 5 different places in each test field. The data, however, showed a large scatter indicating that there were problems in sampling as well in cell size (small area, small

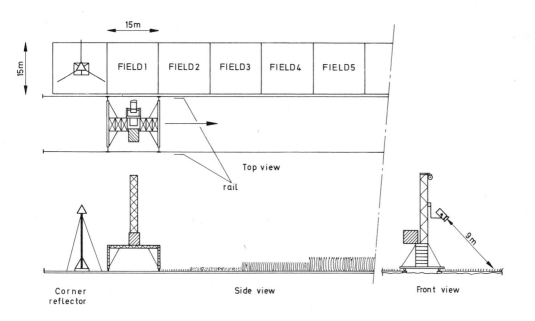

Fig. 14 The test site in Wageningen (not to scale).

number of scatterers) as in place and time (number of independent samples). So another method had to be sought.

The solution was found by building an FM/CW system (see table 5). This system is mounted on a carriage which can be moved along the test plots. Fig. 14 shows the configuration as used now by the ROVE-team.

As already remarked accuracy in grazing angle θ must be traded against cell size (sufficient number of scatterers). The variation in frequency in the FM/CW-system increases the number of independent samples (Attema 1974). Together with the variation in time and sampling in space (size of test plots) in these measurements a fair compromise was obtained (van Kasteren and Smit 1977).

Fig. 10 and fig. 15 give examples of measurements taken with this system. Grazing angles are between 15° and 85°.

Measurements were taken with this system through the growing seasons of 1975 to 1977 where in 1978 a measuring series was completed on bare soils with different roughnesses (soil moisture and sensitivity to slaking). In 1975 and 1977 the measurements were combined with SLAR flights. They confirm the measurements taken earlier (the TV-tower program). A better and larger set of data, however, taken over a wider range of grazing angles has now been obtained. Collaboration with the Remote Sensing Laboratory (Kanses University) makes it possible to compare data. Their procedure is similar to that now adopted by us, but since they have larger funds they can cover more frequencies. The comparisons envisaged are also intended to bring to light possible differences between the objects measured due to differences in climate and irrigation.

Table 5 Properties of the X-band FM-CW measuring system

Frequency	9.5 GHz (variable through X-band)
Frequency sweep	400 MHz (variable)
Modulation waveform	triangle
Modulation frequency	25-200 Hz (used at 100 Hz)
Output power	10 mW
Polarization	VV, HH, VH, HV
Range	5-15 m (used at 10 m)
Sensitivity	-27 dB (relative to 1 m² at 10 m)
Antenna's	
gain	33 dB
beamwidth	4.5°

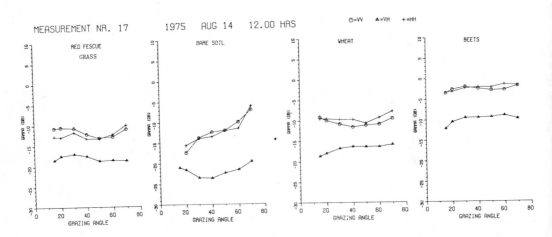

Fig. 15 Example of a measurement made at the test site in Wageningen in 1975.

5 POSITION OF AFFAIRS

5.1 Results

Where are we now ? From the literature
we know that SLAR has been used for a
number of the applications mentioned in
chapter 1. Several authors claim they have
mapped vegetation species and crops
successfully (Morain and Simonett 1967;
Haralick et al. 1970; Morain et al. 1970;
Hardy et al. 1971). Flights carried out
in the Netherlands at X- and K_a-band
(De Loor 1976) also demonstrated the
potential use of SLAR for this purpose
and in fact were the reason for our ground
based measuring program. Together with
such measurements we gained an insight
into the radarsignature of vegetation and
crops and its dependence on grazing angle,
polarization, frequency and botanical
parameters. The radar-return parameter γ
as a function of these parameters seems
to be the best classifier for
classification, monitoring and, possibly,
in situ biomass determinations. The rapid
variations in the return ("speckle") can
be explained completely by the Rayleigh
character of the target and they contain
no extra information. They must be taken
into account to obtain the required
accuracy in γ of 1 dB.

The total variation in γ due to
differences between soils, crops and
vegetation species (dynamic "window")
is small: a 20 dB (table 2).

All data, now available, make up
already a data bank of radar signatures
of a reasonable size.

5.2 Remaining questions

5.2.1 As seen from the object

Natural vegetations and crops are living
systems and depend strongly on their
natural surroundings and the weather. How
large fluctuations within a single crop
or vegetation species this may cause is
yet unknown. This is important in
relation with the high accuracy required
in the determination of γ. For example
what are the influences of: different
sowing procedures, crop rotation, soil
properties, soil moisture, etc.
Fluctuations can occur between seasons but
also within one season. 1976, for example,
was a very dry year in Europe. Are there
differences due to the place in the world,
e.g. due to different environments,
irrigation procedures, etc.? All these
fluctuations must still be investigated.
Since the observation of crops and
vegetation can only be done during the
growing season it must be clear that the
building up of insight into these phenomena
is very time consuming.

5.2.2 As seen from the sensor

All existing experience with SLAR systems
is obtained with systems that give an
image (film) as output. In most of these
systems, in particular the SAR systems,
spatial resolution was given preference
above the dynamic resolution. None of
these systems is absolute: the density and
density variations on the image film are
not directly related with the value or
variation in the return parameter γ. A
digital and absolute system with internal
calibration is required. The data stream
becomes very large when direct transmission
and recording is used, but compression at
the sensor is possible. When implementing
such compression it is necessary to take
into account the variable character of the
targets under observation (Rayleigh
scatter), which may mean deviations from
standard procedures as usually applied
in radar. Such a system is now under
construction in the Netherlands. It will
fly in 1979.

5.2.3 Some remarks about the future

Based on their experience and all available
experimental findings the Remote Sensing
Laboratory (RSL, Kansas University) -on
a request of NASA- has investigated the
possibilities of a radar system in space
for an operational organization for
classification, yield forecasting and
monitoring together with the effectiviness
of such a system (Bush and Ulaby 1977).
Especially its all weather capability makes
radar a highly effective tool. Considering
a space system they used SAR as a primary
sensor. The study, however, will also
apply to a system in an aircraft and there
the much cheaper (5 to 10 x) SLAR system
(or systems, in case more frequencies are
required) is to be preferred, not only
because of the price, but also from the
point of view of dynamic resolution. A
similar study was also undertaken recently
by the ROVE-team. First results were
reported by Smit (1978).

In their study RSL concludes that with

present-day knowledge a special system is required for soil moisture determination. (Bush and Ulaby 1977; Batlivala and Ulaby 1977). This radar must work at a fairly steep incidence angle (between $\varphi = 7^{\circ}$ to 17°) and a relatively low frequency (4 GHz). For classification, yield forecasting and monitoring a higher frequency (above the X-band) and incidence angle (around 45°) is required.
High spatial resolutions are obtainable, even from space, but the requirement of a good dynamic resolution (accuracy of γ, requiring a large number of independent observations or "looks") brings this down to a 30 to 50 m. A major problem forms the tremendous data stream, which can only be optimized with a good knowledge of the object-sensor interaction (the input, chapter 3 and 4) and can best be done in a preprocessor in the satellite.

Based on a simulation program with all the measured data they have available the RSL concludes (Bush and Ulaby 1977):

"-Data in the 13-16 GHz band seems to contain the greatest discriminating power. Moreover, VV data are better suited for a classification task than are HH data regardless of the frequency (between 8 and 18 GHz) chosen. This study found the 14.2 GHz, VV data to be the optimum data for crop classification purposes. If an additional frequency/polarization configuration is available, the 9.0 GHz, HH configuration seems most suitable (of those configurations tested) to use in combination with the 14.2 GHz, VV configuration.

-Multi-date data must be employed in the classification process if rates of correct classification in excess of 90 per cent are to be obtained. Testing data containing variance due only to scintillation, for example, it was found that the average daily rate of correct classification was only 82.7 per cent using dual frequency, dual polarized data. By using single frequency singly polarized data acquired every 10 days, however, the rate of classification rose to 97 per cent within 30 days.

-Classification results are markedly improved by removing classified categories from subsequent analyses. This and other types of ancillary information such as geographic location should be used as they become available to increase classification results".

Having an all weather system it indeed becomes possible to work with fixed

revisit periods of 5, 10 or 15 days (fig. 1 and fig. 16) and to obtain very high success percentages with only a few revisits, since each time practically all areas are covered again. This can never be realized in the same stringent

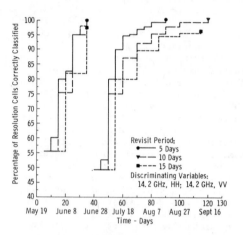

Fig. 16 Variations of cumulative classification percentages with revisit period when dual polarized, 14.2 GHz data are employed as discriminating variables. After Bush and Ulaby (1977).

manner with optical systems in the visual and NIR as is demonstrated by fig.17 from a study by General Electric and quoted by Bush and Ulaby (1977). It then takes many revisits to cover all areas at least once, where after only three revisits the chance to see the same area at all three visits has dropped to less than 3.5% (broken line in fig. 17).
Whether a combination of radar and an optical sensor is useful needs investigation in particular in view of this last disparity between these systems and the fact that the reflection mechanisms are fundamentally different for the two sensors (chapter 3).

As said before a good combination of ground based measurements and aircraft observations can be used for testing, training and evaluation. It seems useful to investigate in this way the benefits of an operational organization for classification, yield forecasting, monitoring and control: technically as well as from an organizational point of view.

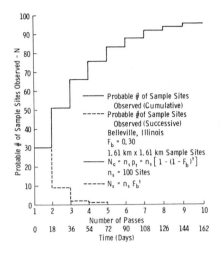

Fig. 17 Probable number of sample sites (out of 100) as a function of total number of satellite passes. Curves are shown for total observed sites (solid line) and for the number of sites observed on all passes (dashed line). From a study of General Electric (1975) as quoted by Bush and Ulaby (1977). For visual and NIR.

6.0 CONCLUSIONS

- Side looking airborne radar has a great potential in a system for classification, yield forecasting, monitoring and control. It satisfies all criteria which can be set to a sensor for the observation of natural vegetation, crops and soils. Since it can be used day and night and under practically all weather conditions strict surveying schemes can be realized.
- The radar return parameter γ (or σ_0) is the best classifier.
- The high accuracy necessary for the determination of the return parameter γ (high dynamic resolution) requires a trade-off between dynamic and spatial resolution. When the imagery is handled visually a lower dynamic resolution can be accepted (and thus a higher spatial resolution can be obtained) than in the case of data handling by computer. A combination of the two types of data handling seems to be the best approach.
- Basic research has made available a data base of such a size that it becomes possible to indicate how a radar system for the observation of vegetations, crops and soils should look like. Such a sensor must fit into an operational organization for classification, yield

prediction and monitoring. Preparations for such an organization have to start in time: technically as well as from an organizational viewpoint.
A good collaboration between radar technicians, physicists and agriculturalists is a preraquisite for the success of such preparations.

REFERENCES

Attema, E.P.W. 1974, Short Range Vegetation Scatterometry, Procs URSI Specialist Meeting on: Microwave Scattering and Emission from the Earth, pp. 177-184, Sept. 23-26, Berne.

Attema, E.P.W 1978, The Radar Signature of Natural Surfaces and its Application in Active Remote Sensing. In T. Lund (editor), Surveillance of Environmental Pollution and Resources by Electromagnetic Waves, pp. 227-252, Dordrecht, D. Reidel Publ. Cy.

Batlivala, P.P and F.T. Ulaby 1977, Feasibility of Monotoring Soil Moisture Using Active Microwave Remote Sensing, Remote Sensing Laboratory, RSL Tech. Rept. 264-12, Jan., Kansas University.

Bush, T.F. and F.T. Ulaby 1977, Cropland Inventories Using an Orbital Imaging Radar, Remote Sensing Laboratory, RSL Techn. Rept. 330-4, Jan., Kansas University.

Cosgriff, R.L., W.H. Peake and R.C. Taylor 1960, Terrain Scattering Properties for Sensor System Design (Terrain Handbook II), Engineering Exptl. Station Bull. 181, Ohio State University.

Daley, J.C., W.T. Davies, J.R. Duncan and M.B. Laing 1968, NRL Terrain Clutter Study, Phase II, NRL Report 6749, Oct. 21, Washington, Naval Research Laboratory.

Goodyear Aircraft Corp. 1959, Radar Terrain Return Study, Final report, Report GERA-463, 30 Sept.

Haralick, R.M, F. Caspall and D.S. Simonett 1970, Using Radar Imagery for Crop Discrimination: a Statistical and Conditional Probability Study, Remote Sensing of Environment 1, pp. 131-142.

Hardy, N.E., J.C. Coiner and W.O. Lockman 1971, Vegetation Mapping with Side-looking Airborne Radar: Yellowstone National Park. In AGARD Conf. Procs. no. 90: Propagation Limitations in Remote Sensing, paper 11.

Kasteren, H.W.J. van and M.K. Smit 1977, Measurements on the backscatter of X-band radiation of seven crops, throughout the growing season. ROVE report, Delft, NIWARS publication no. 74.

Leschack, L.A. 1970, ADP of Aerial
    Imagery for Forest Discrimination. Procs.
    Annual Meeting of the American Society
    of Photogrammetry, Paper 70-110,
    pp. 187-218.
Loor, G.P. de 1974, Measurements of
    Radar Ground Returns, Procs. URSI
    Specialist Meeting on: Microwave
    Scattering and Emission from the Earth,
    pp. 185-195, Sept. 23-26, Berne.
Loor, G.P. de 1976, Radar Methods. In
    E. Schanda (editor), Remote Sensing for
    Environmental Sciences, Ecological
    Series No. 18. Berlin, Heidelberg,
    Springer Verlag.
Loor, G.P. de, A.A. Jurriëns and
    H. Gravesteyn 1974, The Radar
    Backscatter from Selected Agricultural
    Crops, Trans. IEEE on Geoscience
    Electronics Ge-12, pp. 70-77.
Moore, R.K. 1975, Microwave Remote Sensors,
    Chapter 7 in Manual of Remote Sensing,
    edited by the American Society of
    Photogrammetry, Falls Church, Virginia,
    USA.
Moore R.K. 1976, SLAR Image Interpre-
    tability - Trade-Offs between Picture
    Element Dimensions and Non-Coherent
    Averaging. Remote Sensing Laboratory,
    RSL Techn. Rept. 287-2, Jan.,Kansas
    University.
Morain S.A., and D.S. Simonett 1967,
    K-band Radar in Vegetation Mapping,
    Photogramm. Eng. 32 pp. 730-740.
Morain S.A., J. Holtzman and F. Henderson
    1970, Radar Sensing in Agriculture,
    a Socio-Economic Viewpoint. EASCON
    Conv. Record, pp. 280-287.
Oliver T.L., and W.H. Peake 1969, Radar
    Backscatter from Agricultural Surfaces,
    Techn. Rept. 1903-9, 13 Febr., Electro
    Science Laboratory, Ohio State
    University.
Reeves, R.G., A. Anson and D. Landen
    (editors) 1975, Manual of Remote
    Sensing. Edited by the American Society
    of Photogrammetry, Falls Church,
    Virginia, USA.
Smit, M.K. 1978, Preliminary Results of
    an Investigation into the Potential of
    Applying X-band SLR Images for Crop
    type Inventory Purposes. Report
    Microwave Laboratory of the Delft
    University of Technology, Sept.

Photograph 1.   Photograph and high resolution radar image of an aircraft. Look at the fundamental difference between the two images.

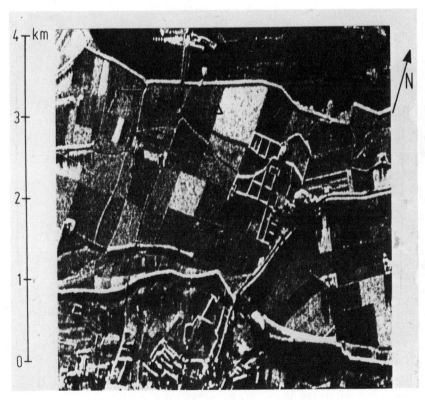

Photograph 2.   Q-band (λ = 8 mm) SLAR image of an agricultural area; July. Different gray tones for different crops. "White" are beets.

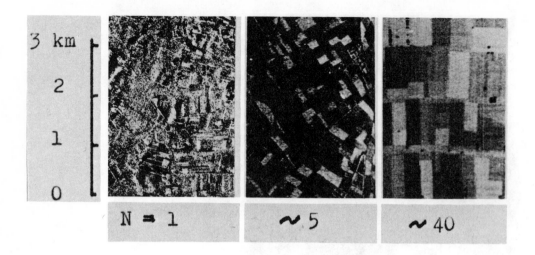

Photograph 3. Influence of the number of independent observations in one pixel on the smoothness of the reflection (gray-tone).

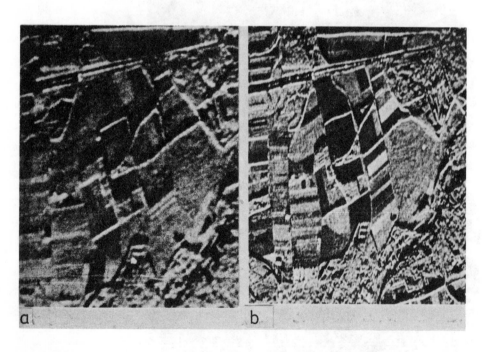

Photograph 4. Q-band (λ = 8 mm) radar images of an agricultural area. Polarization HH. Different radar signatures for different crops. Dependency on the season: a. November; b. July.

J.R.HARDY
Department of Geography, University of Reading, UK

14

Survey of methods for the determination of soil moisture content by remote sensing methods

1 THE REQUIREMENT

.What soil moisture information does the agriculturalist require? There are perhaps two main answers to this question, which correspond to the two main types of survey based on the use of remote sensing data. These are the one-off survey, and the repeated monitoring type of operation.

In the first case, the initial survey, the type of information that is required is an extension of that produced by a soil survey. How much moisture will a given soil hold between the levels of field capacity and wilting point? To what extent is artificial drainage necessary? What is the best use to which the land can be put, by virtue of the potential and limitations of its soil moisture characteristics? This is but one facet among the many required for a land capability survey.

The second, monitoring, type of survey, must be repeated at regular intervals, and must answer such questions as the level of soil moisture deficit between field capacity and wilting point for each given small area being surveyed. One must consider here the worth of the exercise: unless a cultivator can alter the soil moisture content by irrigation or any other means, the content may be only a matter of academic interest, since he can do nothing about it, and in such circumstances no money would be forthcoming to pay for a survey. The only possibility then to make such a survey beneficial beyond its costs, is if it may affect a buying and selling strategy for stocks, or in the case of water engineers, whether they need to know if any recharge of an aquifer is taking place or not, in order to decide their supply strategy. Percolation will not take place if soil moisture is below

field capacity throughout the root zone.

In every case, the use of remote sensing methods must be considered in competition and co-operation with other more conventional methods of survey, in terms of costs, benefits and accuracy. For example, can a soil moisture deficit be predicted more cheaply, quickly and accurately using remote sensing methods than by using estimates of precipitation, evaporation and transpiration, and soil moisture storage?

As in every other remote sensing operation, the rationale is generally to use the data to provide the synoptic view and information, by extrapolation from sample values determined by conventional ground methods.

Soil moisture monitoring information, to be useful, is required quickly, within a period of hours or days at most, by cultivators. Any time period of more than a few days in converting the initial data to useful information in the hands of the cultivator, means that the information is of historical rather than operational interest. Automated or interactive methods of interpretation are therefore indicated. The balancing of costs against any need for frequent monitoring also indicates the use of satellite rather than aircraft platforms for any monitoring survey.

It is reasonable to conclude this summary of requirements by stating that no operational method of determining soil moisture by remote sensing methods exists: all work in this field is in the experimental or research area at present. This paper can only indicate prospects and likely methods for development.

2 THE PROBLEM

Soil is a notoriously heterogeneous substance. This bedevils both conventional ground survey methods, and any future prospect for remote sensing operations.

For ground surveys, a statistical number of samples is necessary to give indications of the variability of soil moisture on a single site. Is it possible for the synoptic overview of remote sensing to provide some form of automatic integration over a resolution cell to reduce the effects of this heterogeneity? The heterogeneity is problem enough in a single site, but between sites it is even more marked.

For practical purposes, agriculturalists and water engineers are interested in the relationship of the actual soil moisture content at any given time to the levels of (a) saturation (the almost complete absence of air and its replacement by water in the voids), (b) field capacity (the maximum amount of water a soil will hold against gravity, corresponding to a suction of about 0.05 atmospheres), and (c) wilting point (the level of moisture at which plants wilt permanently, corresponding to a tension of about 15 atmospheres). The moisture content represented by these three levels varies with soil texture, structure and management, to name but three major influences. A generalised relationship is indicated in Fig. 1, from Hall et al. (1977), which describes the parameters and their laboratory determination.

Knowledge of moisture content is required for depths from the surface to the lower limit of the root zone. Most remote sensing methods sense only the conditions in the uppermost few micrometres of soil and these may bear little or no relationship to the conditions at root zone depths, due to the very slow upward and downward movement of water in the soil in response to hydraulic gradients. After a few days of dry weather, the soil surface may be dry, although there is ample moisture in the root zone. The converse may also apply: after a prolonged period of drought, with large soil moisture deficiencies in the root zone, the first rain will moisten the surface layer, but leave the lower layer unaffected.

A further problem is that soil, particularly in agricultural use, is frequently covered by crops or other vegetation, and it is the radiation reflected and emitted from the vegetation that provides the signal received by a remote sensor, rather than the underlying soil.

It is these signals received in typical situations that will be examined in succeeding sections, to consider the potential for remote sensing methods of determination of soil moisture. They will be examined in the various parts of the electromagnetic spectrum used for remote sensing.

3 THE VISIBLE AND NEAR REFLECTIVE INFRARED AREAS OF THE SPECTRUM

Signals received in this part of the spectrum, from about 0.3 to 3 µm wavelengths, are of reflected solar radiation. Sensors available are cameras containing film, and multispectral scanners (MSS). Space platforms include the Landsat satellites. Film is sensitive from about 0.4 µm to a practical upper limit of about 0.9 µm in the case of the 'infrared' films. Satellite MSS systems, such as that on the Landsat series, and the Heat Capacity Mapping Mission (HCMM), have an upper limit of 1.1 µm.

In this area of the spectrum, bare soils of varying kinds have a generally increasing reflectivity with increasing wavelength (Fig. 2). This diagram also shows the different reflectivities of different types of soil.

Figure 2 also shows the decrease in reflectivity in a given soil brought about by increasing moisture content. This is a matter of common experience in the visible part of the spectrum: upon moistening a soil sample, it becomes darker in colour.

This latter property could be made use of in special circumstances, but in general the differences in reflectivity between different types of soil may be as great as, or greater than, the differences caused by different soil moisture contents, and it is extremely difficult, if not impossible, to separate the two effects except when considering laboratory samples.

Differences from day to day between signals from the same site may be partly caused by soil moisture differences, but also by differences in sun elevation and atmospheric conditions, and in the case of film products, by any differences in film exposure and processing.

The conclusion seems inescapable that only rather generalised qualitative deductions are possible in this area.

A further problem is that the reflection comes entirely from the infinitesimally thin surface layer, and so no conclusions are possible directly about conditions at depth.

The agriculturalist frequently requires knowledge about soil moisture conditions in cropped areas, and in this case, as stated above, inferences will have to be made from the condition of the vegetation as revealed by the remotely sensed signal. The high reflectivity of healthy vegetation in the near infrared region from 0.7 to about 1.1µm has to be used principally for these inferences. This reflectivity is changed under stress, resulting in a changed signal, and changed appearance of film, particularly false colour infrared film. The vivid red appearance on the film of healthy vegetation is dulled. The change in infrared reflectivity may be caused either by moisture stress or disease stress, and it is difficult to distinguish the two. Again deductions have to be qualitative.

Valuable qualitative information as to relative moisture contents may be given by a different technique - skilled geomorphological interpretation of stereoscopic photography. These techniques may reveal such features as seepage lines or drainage hollows, and very small differences in relief causing moisture differences (Verstappen, 1977).

The various qualitative deductions and inferences described in this section, particularly from geomorphological interpretation, are of great value in the single survey type of operation, pointing out minor relative differences in soil moisture within a small area, and indicating likely areas requiring treatment such as drainage, rather than in any continuous monitoring operation.

Because of its wide availability, data from the Landsat satellites should be considered under this heading. Limited success has been achieved using these data. Figure 3, reproduced from Freden and Price (1977), shows typical results. These are from bare soil fields near Phoenix, Arizona. For these results, $r^2 = 0.74$. For vegetated fields, MSS brightness varies very little with moisture, except in the case of alfalfa, a lush green crop, where a r^2 of 0.88 was achieved.

Dimensionless values of satellite radiances, as plotted in Fig. 3, may be obtained by digital processing and plotting of, e.g.

$$\frac{\text{Channel 5}}{\text{Channels } (4+5+6+7)}$$

4 THE THERMAL INFRARED PART OF THE SPECTRUM

This affords considerable hope for determination of soil moisture. Thermal infrared line scan (IRLS) sensors are capable of supplying imagery or digital data, depending on the specification of the system. The magnitude of the signal recorded is a function of the emitted thermal radiation from a surface, which is proportional to $\varepsilon\sigma T^4$, where ε is the emissivity, σ the Stefan-Boltzmann constant, and T the absolute temperature. To a first approximation, it may be assumed that the emissivity of all natural surfaces is very close to one (actually around 0.9 - 0.95), and thus the signal received in the thermal wavebands (in the 'windows' at 3 - 5 µm or 8 - 14 µm at aircraft altitudes with the latter being preferred), is solely a function of the surface temperature. Because of ozone absorption in the upper atmosphere at around 9.5 µm, the satellite thermal infrared 'window' is at 10.5 - 12.5 µm.

This surface temperature is a function of the division in the outgoing part of the energy balance between evaporation and transpiration on the one hand, and on the other emitted radiation, again to a first approximation. If a large part of the incoming energy is used to supply latent heat to support evaporative processes, then both surface temperature and emitted radiation are lower (e.g. Sellers, 1965).

Thus, immediately a distinction can be made between evaporating and non-evaporating surfaces receiving solar radiation. Evaporating surfaces may be water, plant cover or moist soil, while non-evaporating surfaces may be dry soil or dry impervious surfaces such as asphalt. The dry or moist soil in this case of course refers almost entirely to the very thin surface layer, the moisture content of which may, as stated above, bear little or no relation to that of the deeper root zone below.

There is, however, also some prospect of estimating the moisture content of layers below the surface, from the thermal properties of the surface and sub-surface soil layers. This is dependent on the acquisition of day and night thermal infrared data of the area in the same 24 hours, at approximately the times of minimum and maximum temperatures, pre-dawn and early afternoon. The temperature differences between the two sets of imagery can then be used to indicate sub-surface moisture content.

A soil may be envisaged as a soil-water-air system, and if, for a given area, the soil proportion is fixed, then the voids will be filled with varying proportions of water and air according to the moisture content (Fig. 1). While soil and water have approximately similar thermal properties, air has a very low thermal conductivity and specific heat, and is thus a very effective thermal insulator. Therefore, a soil with a high proportion of air present will undergo large day and night contrasts of surface temperature, while one with a high moisture content and little air will have relatively small temperature variations. The thermal

235

property which expresses this best is the thermal inertia, the measure of the rate of heat transfer at the interface between two dissimilar media.

The thermal inertia P, is defined by

$$P = \sqrt{kC\rho}$$

where k is thermal conductivity, C is specific heat, and ρ is the density. Units are generally either (Wm^{-2} $^{\circ}$K^{-1} S$^{\frac{1}{2}}$) or (cal cm^{-2} $^{\circ}$C^{-1} S$^{-\frac{1}{2}}$). An alternative definition is,

$P = \rho C\sqrt{k}$ where k is thermal diffusivity (= k/Cρ).

The surface temperature change is governed by the radiation balance and the thermal inertia. If the radiation balance is modelled from measurements or estimates, then a thermal inertia value can be estimated. This can be fitted to estimates of bulk density, specific heat, thermal conductivity and/or thermal diffusivity derived from the varying proportions of soil, air and water present.

Figure 4 shows a typical computed surface temperature curve for different values of thermal inertia, and shows the nature of the changes to be expected.

Figure 5 shows the relationship between porosity, thermal inertia and moisture content, and points to the fact that knowledge of porosity is essential in such computations, and hence that moisture content determinations must be by % volume rather than % dry weight.

Identical thermal inertia values can be produced by different combinations of k, C and ρ, so that the thermal inertia is obviously not a unique indicator of moisture content. The method has been tested, but so far only in areas in which the evaporative flux can be ignored.

The Heat Capacity Mapping Mission (HCMM) satellite, launched in 1978, is designed to measure day and night apparent temperatures at approximate maximum and minimum temperature, in order to enable thermal inertia values to be determined.

The EEC's Joint Research Centre at Ispra has organised a Joint Flight Experiment (JFE) in which flights have been made over European test areas to collect data to simulate the HCMM data. Two models have been developed, one - the TELLUS model - for bare soil surfaces, and the other - the TERGRA model - for vegetated surfaces, in which the radiation and energy balances are modelled from micrometeorological data. These have been successfully tested using aircraft data, but so far thermal satellite data have been limited, particularly for European test areas. This shortage has been a result of a number of factors, among them European cloud cover and problems with transmission and reception of satellite data. TIROS-N thermal data are now also being evaluated for their potential for earth surface monitoring, but much of the work described in these last paragraphs is at present in an early stage of development. The resolution of these satellite systems is of the order of a kilometre, more or less, in spatial resolution, and approximately half a degree K, more or less, in thermal resolution.

A considerable problem with many aircraft borne IRLS equipment is that they have automatic gain control, and thus the grey tones recorded are indicative of relative temperatures rather than absolute temperatures. In some cases this results from the military parentage of the scanners; the military requirement is for relative images under any circumstances, and thus AGC is appropriate. Key references for further discussion of thermal remote sensing include Reeves (1975), especially pages 82-88 and 128-138, Kahle et al. (1975), Rosema (1975) and Moore (1976).

5 THE MICROWAVE PORTION OF THE SPECTRUM

This portion of the spectrum must be considered in two separate sections, since both passive and active radiation sources must be considered. In passive microwave radiometry the natural emissions of microwave radiation from the earth are detected, whereas active microwave radiation may be supplied in the form of radar energy.

5.1 The (passive) microwave portion of the spectrum

This part of the spectrum is the long 'tail' of the terrestrial emission spectrum, in which amounts of energy emitted are very small. However, this is compensated to some extent by their almost complete lack of attenuation by the atmosphere in the 'windows' due to their longer wavelength.

At any given microwave wavelength the amount of radiated energy depends linearly on both emissivity and temperature, in contrast with the thermal region, where radiated energy is much more strongly temperature dependent (proportional to T^4). Thus, while in the thermal region the signal is relatively independent of the emissivity, and in any event variations in emissivity are very small, in the microwave conditions are very different. Microwave emissivities vary widely with surface properties, including soil moisture content. As the soil moisture increases, so the microwave emissivity decreases, and this decrease in emissivity is perhaps greater than changes due to other causes, such as roughness, salinity or view angle.

Particularly with longer wavelengths, microwave radiation may be emitted both from the surface layer and from layers to limited depths of the order of tens of centimetres. However, a single microwave signal is received at a sensor and so it is impossible to separate the signal originating in different layers: it must be regarded as an integrated signal with components from different depths, decreasing with increasing depth. Figure 6 illustrates this penetration.

Because of longer wavelengths and very small amounts of radiation, the spatial resolution of most present microwave systems is limited to some tens of metres from aircraft, and kilometres from satellites. This presents problems in agricultural areas where vegetation and bare soil are mixed, since green vegetation has an emissivity near unity and soil is much lower. The two signals will be averaged by a sensor over the resolution element.

As stated above, the microwave emission from a surface is approximately linearly dependent on both temperature and emissivity. Thus microwave signals are generally presented in terms of the 'brightness temperature' B_T, where

$$B_T = \epsilon T$$

An increase of moisture content from 12% to 30% has been reported as reducing the brightness temperature of playa sediments from approximately 280° to 230° at a vertical view angle (Ohlsson, 1972). Figure 7 shows this relationship. Some knowledge of surface temperature is, therefore, required to obtain useful information on emissivity and soil moisture, and as has been show above, surface temperature can vary greatly according to evaporative state and moisture content.

All in all, because of difficulties of recording microwave signals, and low spatial resolution, passive microwave radiometry for the determination of soil moisture is still very much a research and development area rather than an operational one.

5.2 Radar

This is an 'active' microwave remote sensing system, in that pulses of microwave energy are emitted from an antenna and reflected or scattered at the ground surface. A portion of this energy is returned to the antenna, the radar return, and it is this signal which is interpreted for ground properties.

The radar return from a ground 'target' will depend primarily upon the following ground properties: its roughness in terms of the radar wavelength, the slope and aspect of the surface in relation to the sensor, and the complex dielectric constant of the surface.

The roughness in terms of the radar wavelength determines whether incident energy is reflected in the specular or diffuse mode. Commonly used radar wavelengths are around 8mm or 3 cm, although the total radar waveband is from 3 mm to 3 m. Hence some level soil surfaces with a fine tilth may be near specular reflectors, from which no useful information on surface characteristics can be obtained.

The influence of slope and aspect of the surface relative to the antenna has a profound effect on the returned signal, as can be seen from the relief and shadow effects on any SLAR imagery.

The complex dielectric constant of a surface varies considerably with its moisture content. The value for water is about 80, while dry soils are typically at about 5. Absorbed water has a relatively lower dielectric constant than free water, but even so the moisture content of a surface will tend to dominate its reflection/absorption properties, rather than its composition. Thus qualitative indications at least of soil moisture content may be available from radar data.

Penetration of radar energy below the surface is only possible in the driest of materials. Figure 7 can be used to illustrate the penetrative capability of radar energy as well as passive microwave, but it must be noted that the commonly used radar wavelengths, around 8 mm and 3 cm, are both found at the extreme right hand edge of that diagram.

Working in the same wavelengths, but involving higher quantities of energy, radar has a better all-weather capability than passive microwave, and does not have the same limitations on resolution if synthetic aperture or focussed synthetic aperture systems are used. The only spaceborne imaging radar whose data are available to non-military users was that aboard SEASAT-1, launched in 1978. The sensors aboard this satellite had a short life of about three months, but data were collected and archived, a quantity of which were obtained over land areas. Results of analyses of these data for soil moisture and other terrain properties are awaited.

6 CONCLUSION: SUMMARY OF METHODS AND POTENTIAL, COSTS AND BENEFITS

As stated above, there is as yet no operational method of determining soil moisture for large areas from remotely sensed data. Every piece of work reported is in the experimental and research stage.

A single survey may be used to detect relative differences over an area, which may be useful for such applications as drainage requirements. The question at issue here is whether this can provide any information which is not already available, at a comparative cost. The answer may be negative in agriculturally developed areas, although useful information may be obtained for relatively unknown areas.

A monitoring survey could, if successful, provide information which is not already available by other means, although estimates of regional soil moisture deficiency are made routinely using estimates of rainfall and evapotranspiration. The problem here is whether the value of the information obtained (the benefit) could be greater than its cost; the only areas where this could be undoubtedly true are those where irrigation is available.

The costs of providing such a monitoring survey using aircraft as platforms would be high, and so a spaceborne system is indicated, with automated or semi-automated data processing.

In the visible and reflective infrared portions of the spectrum, differences in reflectivity are relatively small and masked by changes due to other factors, as mentioned above. Thus this is not a promising area for soil moisture monitoring, although some correlations have been found between surface soil moisture and signals recorded by Landsat, particularly in Channel 5, and in some areas where the surface is normally dry, the areal extent of surface wetting due to local rain has been determined. This is of value principally in arid and semi-arid areas.

The thermal infrared portion of the spectrum offers perhaps the best chance of determination of soil moisture content on a regular basis, provided good enough radiation balance information is available. Data would be required in thermal inertia form, although much of the effort in this area so far has not been directed towards soil moisture (e.g. Kahle et al., 1975). The results of the HCMM experiment will be of great interest in this respect when they are available, for this satellite has the potential for recording day and night temperature differences, and other satellites such as TIROS-N offer possibilities.

In the thermal area, satellite images are already available such as those from the SCMR (Surface Composition Mapping Radiometer aboard Nimbus 5, with a resolution of about 600 m. Particularly at night there are differences in temperature between moist and dry areas, with moist areas having higher surface temperatures under radiation conditions.

Landsat C, launched in 1978, had a thermal channel with a resolution of approximately 240 m, but the sun synchronous orbit time of about 0930 restricted its usefulness for moisture determination, since at this time temperature differences are relatively small (Fig. 4). This sensor was also short-lived.

Spaceborne passive microwave recording suffers from defects of low resolution, although brightness temperatures are available at this resolution, e.g. from the ESMR (Electrically Scanning Microwave Radiometer) aboard Nimbus 5. With a knowledge of land surface temperature, the microwave emissivity of land areas can be determined, but these will be a compound of the signals from vegetated, bare soil and built up areas in any resolution cell While vegetation has a microwave emissivity of near unity, water is relatively low at about 0.4. Soil values lie between the two, near unity for rough dry soil and reducing with increasing moisture content.

As already stated, no imaging radar system is available for land areas in space although data were collected from SEASAT-1 during its short life. Aircraft imagery can be used to give qualitative differences, but the expense of such surveys is almost certain to debar their regular use.

In conclusion, it seems that the most likely area for development of soil moisture measurement by remote sensing is in the thermal infrared, by thermal inertia methods. Since visible and reflective infrared measurements are already available (e.g. Landsat) these must be explored to their fullest (limited) potential. Passive microwave may be a possible method, but is beset by reolution and other problems. Radar seems ruled out for the present on grounds of cost of aircraft surveys and non-availability of spaceborne sensors.

7 REFERENCES

Barrett, E.C. and L.F. Curtis 1976, Introduction to Environmental Remote Sensing. London, Chapman and Hall.

Carslaw, H.S. and J.C. Jaeger 1959, Conduction of heat in solids, 2nd edn. Oxford, Oxford University Press.

Condit, H.R. 1970, The spectral reflectance of American soils, Photogr. engg. 36:955.

Curtis, L.F. 1977, Remote sensing of soil moisture: user requirements and present prospects. In R.F. Peel, L.F. Curtis and E.C. Barrett (eds.), Remote sensing of the terrestrial environment, Colston Papers 28, p.143-158. London, Butterworth.

Farrow, J.B. 1975, The influence of the atmosphere on remote sensing measurements: summary report, ESA Contractor Report CR-353. Neuilly, France, ESA.

Freden, S.C. and R.D. Price (ed.) 1977, Report on significant results and suggested future work, obtained from Landsat follow-on principal investigator interviews, NASA document X-902-77-117. Greenbelt, Md., U.S.A., Goddard Space Flight Center.

Hall, D.G.M., M.J. Reeve, A.J. Thomassen and V.F. Wright 1977, Water retention, porosity and density of field soils, Technical Monograph No.9. Soil Survey of England and Wales, Harpenden, England.

Kahle, A.B., A.R. Gillespie, A.F.H. Goetz and J.D. Addington 1975, Thermal inertia mapping. In Proc. 10th International Symposium on remote sensing of environment, p.985-994. Ann Arbor, Michigan, U.S.A., Research Institute of Michigan.

MacDowall, J. et al. 1972, Simulation studies of ERTS-A and B data for hydrologic studies in the Lake Ontario basin: 4th Annual Earth Resources Program Review. Houston, Texas, MSC.

Moore, D.G. 1976, Thermal infrared remote sensing as a measure of soil moisture and evapotranspiration. Brookings, South Dakota, U.S.A., Report SDSU RS176-06, Remote Sensing Institute, South Dakota State University.

Ohlsson, E. 1972, Summary report on a study of passive microwave radiometry and its potential application to earth resources surveys, ESRO Contractor Report CR-116. Neuilly, France, European Space Research Organisation (now European Space Agency ESA).

Reeves, R.G. et al. (ed.) 1975, Manual of remote sensing, 2 vols. Falls Church, Va., U.S.A., American Photogrammetric Society.

Rosema, A. 1975, Heat capacity mapping, is it feasible? In Proc. 10th International Symposium on remote sensing of environment p. 571-582. Ann Arbor, Michigan, U.S.A., Environmental Research Institute of Michigan.

Sellers, W.D. 1965, Physical Climatology. Chicago, University of Chicago Press.

Verstappen, H.Th. 1977, Remote sensing in geomorphology. Amsterdam, Elsevier.

The figures for this paper will be found on succeeding pages.

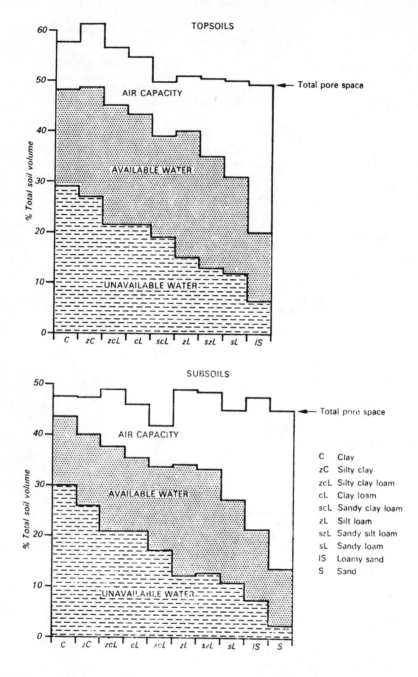

Fig.1   Total porosity, air capacity and water retention for certain
        particle-size classes.   (Reproduced from Hall et al., 1977).

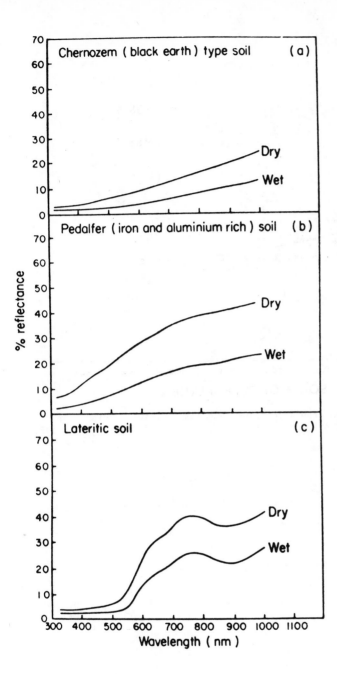

Fig.2 Spectral characteristics of soils in the visible
       and near infrared parts of the spectrum.
       (Reproduced from Barrett and Curtis, 1976, after
       Condit, 1970).

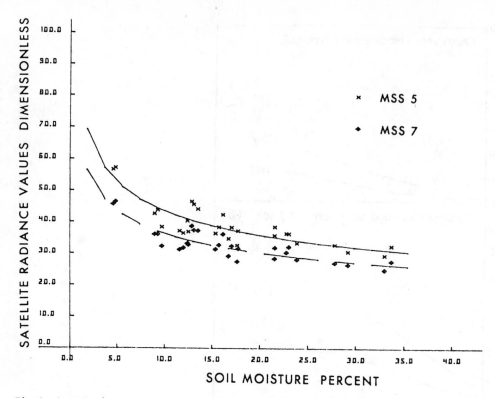

Fig.3 Correlation between Landsat radiance values and soil moisture content for bare fields near Phoenix, Arizona. (Reproduced from Freden and Price, 1977).

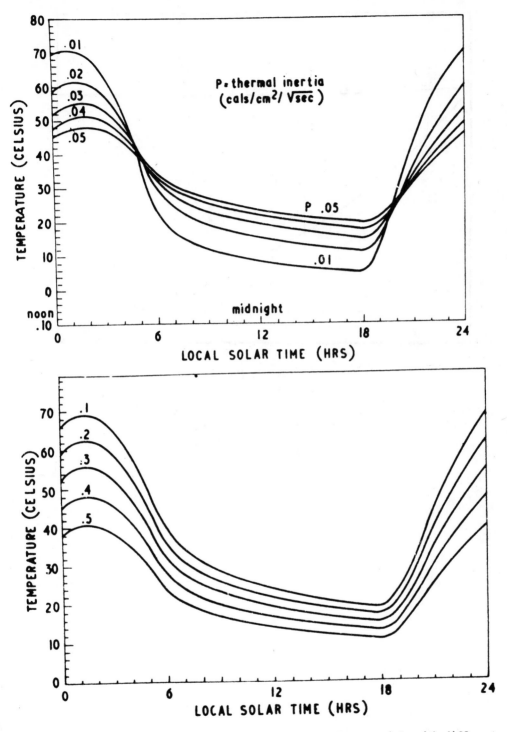

Fig.4 Diurnal surface temperature variations computed for materials with different
thermal inertias (above) and albedos (below).   (Reproduced from Farrow, 1975).

243

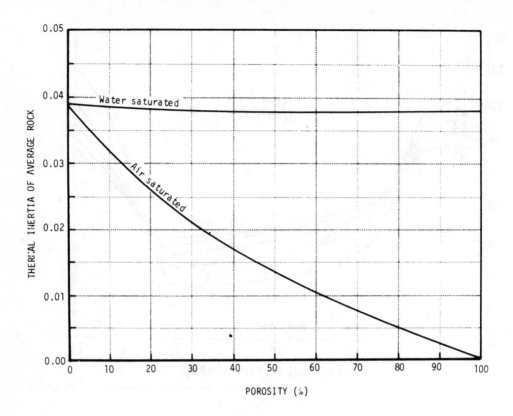

Fig. 5 Thermal inertia variations of average rock (soil) as a function of porosity.
(Reproduced from Reeves et al., 1975, after Carslaw, 1959).

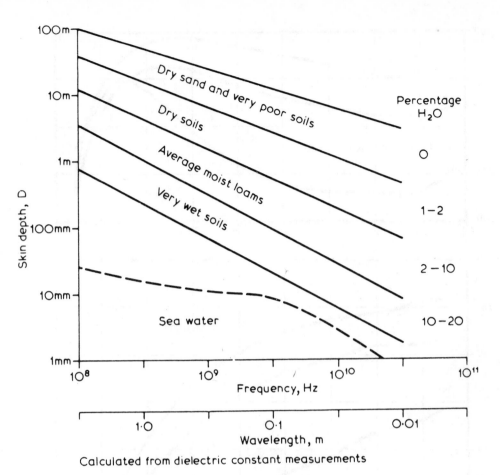

Calculated from dielectric constant measurements

Fig.6   Microwave penetration of soils with varying wavelengths and moisture contents, calculated from dielectric constants.   (Reproduced from Barrett and Curtis, 1976).

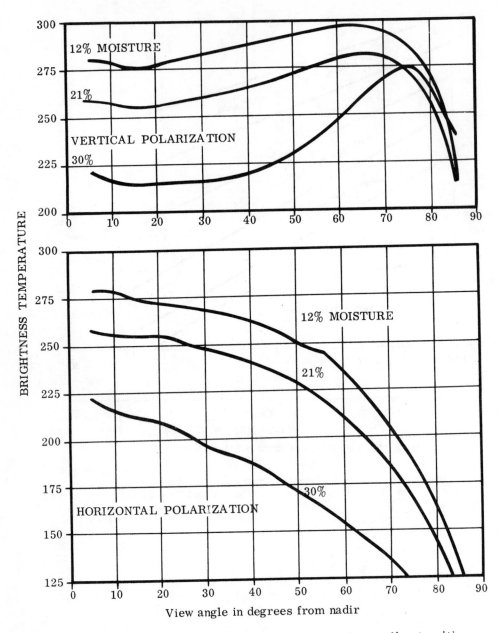

Fig.7 Variation in microwave brightness temperature of playa sediments with moisture content. (Reproduced from Ohlsson, 1972).

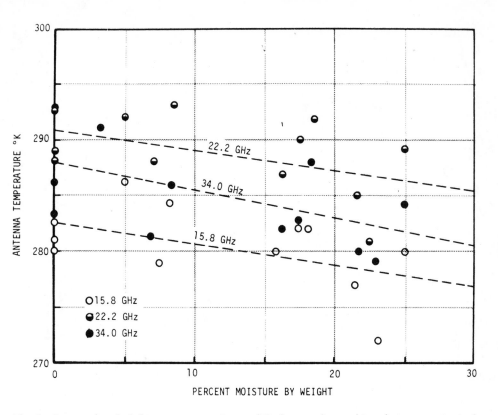

Fig. 8 Decreasing brightness temperature with increasing soil moisture content for three microwave wavelengths. (Reproduced from Reeves et al., 1975).

J. R. HARDY
Department of Geography, University of Reading, UK

15

The acquisition of ground data
for surveys based on remotely sensed data

1 TERMINOLOGY

The title for this lecture was given as
'Ground truth methods', but I wish to
begin these notes by taking issue with
this title. 'Ground truth' is a term
which is by now fairly firmly entrenched
in the literature, but 'ground truth' very
rarely is 'truth'. Truth is an absolute
term, and only the simplest of information
can be absolutely true. For example, the
simple statements that 'the crop sown in
this field is spring wheat' or ' this
parcel of land is mainly woodland' could
be true, but usually much more complex
data than these are required as will be
evident below. The terms 'ground data'
or 'ground information' are much to be
preferred to 'ground truth', and will be
used below. Arguments have also been put
forward objecting reasonably to the
word 'ground', since water, ice or air
may be involved, and arguing for 'surface
data' (Barrett and Curtis, 1976).

2 THE PURPOSE OF GROUND DATA COLLECTION

Remote sensors can be used to acquire a
very large quantity of data for a very large
area in a very short time (for example a
Landsat 'scene' is recorded in approximat-
ly 25 seconds), and it is this synoptic
view that provides much of the strength of
surveys based on remotely sensed data.
The fundamental purpose of ground data
in such surveys is to allow extrapolations
to be made, using the remotely sensed
data, from small known areas to larger
unknown areas, thus producing information
for the total area.
Ground data are used for two main
purposes to permit such extrapolations.
The first is to provide a key for inter-

pretation or calibration, using a known
area for which ground or other information
is available: this may be termed the
training function. The second is to
enable a check to be made of the accuracy
of the survey based on the remotely
sensed data. This is termed the test
function. Both are vital: it is of
little use having information if its
reliability is unknown.
The same methods of obtaining ground
data may, in some cases, be used for
acquiring both training and test data:
indeed at the time of collection, it may
frequently not be known for which purpose
the ground data are to be used.

3 INFORMATION REQUIRED BEFORE GROUND DATA COLLECTION

For the ground data and resulting survey to
be most effective, some information is
ideally required before the whole project
begins. Those concerned should be aware
of the objectives, scale and specification
of the final survey product, the area to
be covered, and the details of the remote
sensor system or systems to be used.
This information will permit the most
efficient and economic system of ground
data collection to be designed.
For example, there are widely different
ground data requirements for a survey
designed to determine weed and disease
infestation of a small area using aircraft
photography, or for a survey designed to
determine soil moisture deficits over a
number of sample areas using thermal imagery,
or for a survey designed to determine the
area under wheat in a very large area
using Landsat data.

4 SCALE CONSIDERATIONS IN GROUND DATA COLLECTION

A number of the major factors which influence the ground data collection methods will now be discussed. Perhaps the first of these is the scale effect, and even this can be interpreted in at least two ways. There is the size of the resolution element of the remotely sensed data on the one hand, and on the other the size of the area to be surveyed using remotely sensed data and the scale of the final product required.

Most remotely sensed data used in agricultural applications consist of records in some form of electromagnetic radiation reflected or emitted from the ground surface. These records can be broken down into resolution elements, and the size of a resolution element will vary according to the nature of the data. In the case of low level air photographs, elements of a few centimetres in extent can be distinguished. A Landsat pixel as recorded on computer compatible tapes, represents the signal from an area approximately 80 metres square, while sensors aboard atmospheric observation satellites such as the Nimbus series may image single elements of hundreds of metres or kilometres in extent.

Whatever the size of this element, the remotely sensed data will record the aggregated radiation from the whole element. For example, for a cultivated field with 50% of the surface covered by the crop, 30% by bare soil, and 20% by weeds, on a large scale aircraft photograph each of these components may be separable, whereas Landsat will record a signal which will be a combination of the radiation from the three components in their respective proportions, and perhaps from neighbouring fields as well if they are covered by the 'pixel'. The proportions must, therefore, be recorded as ground data. If a ground survey is designed specifically for use with data from a particular platform/sensor combination, this scale effect must be taken into consideration in the design of the survey.

This scale effect must be considered in another way. There will be different requirements, for example, if the remotely sensed data are to be used to produce areas under given crops for a whole state, or if detection of incipient crop disease sources over a small area is required. The first case requires relatively simple data for a large number of points throughout the area, while the second may require much more detailed information from a small number of fields. This is concerned with the scale and purpose of the final product.

5 MULTI-LEVEL DATA COLLECTION

Differing scales of the remotely sensed data and of the final product may introduce another concept: that of multi-level data collection. If space imagery is to be used for the main survey, and the scale of this survey is small with rather generalised data as for example crop inventory, then a multi-level sampling procedure using say ground and aircraft data may be useful. If the case of a crop inventory is considered, most crops can be identified with confidence by a skilled interpreter from aircraft photographs. At a relatively low level of skill, classes such as bare soil, pasture, small grain (wheat, barley, oats, rye), maize and root crops (sugar beet, potatoes etc.) can be identified. Higher levels of skill are required to separate crops within these general classes, for example, potatoes from sugar beet, but this can be done with good reliability, although ground data are useful to check or provide identification in difficult cases.

The possibility thus exists of supplementing ground data in these cases by information derived from air photographs, to extend the training and test data. For example, one of the largest ground data surveys undertaken by the Reading University team was for the evaluation of Skylark rocket photography of Argentina (Townshend, 1977). This survey was undertaken in co-operation with staff of the Instituto Nacional de Tecnologia Agropecuaria (INTA) for the ground data collection, and aircraft were provided by the Argentine Air Force for aerial surveys. Some 330 000 km^2 were photographed from the rocket of which approximately 92 000 km^2 were cultivated. For the crop inventory part of the project, detailed information was collected for 744 fields, with summary data for a further 1411 fields, within two to five days of the rocket launch. In addition, 33 sample strips with a total length of over 2000 km^2 were photographed from an aircraft. These two sources of data were integrated to form a large body of information in combination.

Depending on circumstances, some map or other archival material may be able to be used as ground data. This applies if relatively static elements of the environment are to be sensed, such as areas under forest or urban use. Here the cover may remain permanent for a period of years and, subject to checking in the field or from air photographs, map data may be acceptable.

6 THE TIME DIMENSION IN GROUND DATA COLLECTION

This introduces consideration of the time scale in ground data acquisition. Examples above are of ground data which may not change substantially for years, while crop cover may change substantially in a period of days at some stages of growth. Surface moisture content, however, may change substantially in an hour or more.

The time scale for ground data may thus conveniently be considered in three divisions. The first is those data which change on a time scale of minutes or hours, such as radiation balance and soil moisture. These data must be collected simultaneously with remote sensing data acquisition, or the ground data are of very little or no use. This will be referred to as 'hour scale data'.

The second concerns effects which may change over a time scale of days or a week or two at most, such as crop condition and leaf cover. The remotely sensed appearance of a crop can change drastically in such a period. These effects will be referred to as 'day scale data'. There are effects which are intermediate between these first two scales, mostly in the nature of cultivation practices such as ploughing or harvesting, which can change the appearance of an area in hours.

The third time scale will be referred to as 'year scale data', since it is concerned with almost invariant parameters, such as morphological characters, soil texture and structure (subject to cultivation practices) and mature trees, urban and water cover (water cover may of course vary rapidly at time of floods, and these changes are of great interest in the 'day' or 'hour' scales). Whatever the purpose of the survey, the slope and aspect of a ground data area should always be recorded, since if this varies appreciably from the horizontal, the incoming radiation is affected. This effect may be allowed for quantitatively for direct sunlight, although diffuse radiation cannot be simply allowed for (e.g. Monteith, 1973, Sellers, 1965).

These varying time scales can pose a dilemma for the earth scientist. It is obvious that data such as soil moisture must be synchronous with the overflight of the sensor. If clear weather is needed for a mission, then, if aircraft are used, there will almost certainly be standby delays for both the aircraft and the ground data team, with increasing costs and logistical problems. A sudden weather change may involve the abortion of a mission at the last minute. If a satellite overpass is involved the problems are even more acute, in that the time is fixed regardless of weather conditions, and a late decision will have to be taken, based on weather forecasts, as to whether to collect data for a particular pass or not.

For efficient collection of ground data, the rapid processing of the remotely sensed data is highly desirable, together with its speedy transmission to the ground team. The processing will indicate whether or not the remotely sensed data have been acquired correctly. If this is not the case, then time and costs can be saved if the ground team can be quickly informed. If a copy of the remotely sensed data can be passed to the ground team quickly, critical points may be checked in the field while there is still time.

As an illustration of this, the experience with the recent Joint Flight Experiment in U.K. may be quoted. Since the prime objectives were soil moisture content and terrain properties, data were obtained at three stations by continuous manning during the 24 hours of the flight period. These data were radiation and energy balances, ground heat flux and soil temperature and moisture samples. This is the 'hour' scale of operations. For extrapolation of these results, a land cover survey was required of the area imaged by the sensors. Arrangements were greatly facilitated by the availability of 'quick look' images. Within three hours of the aircraft landing after the flight, images were collected for the ground team and were in their hands a few hours later. This enabled a check of data quality to be made, and the complete land cover survey was made directly on prints of the images rather than maps, in the following three days.

This may be compared with the situation confronting the ground team for a survey based on satellite data. Since the time is fixed for the overpass, cloudy conditions may prevent any data being obtained for that pass. If sky conditions are clear, there is always a delay period which may run into weeks, before the team will know whether data have been successfully obtained, and have data available for interpretation. In this situation, the collection of synchronous data is always something of a gamble, and the advantages and disadvantages of postponing ground data collection until imagery is available will have to be considered.

The disadvantages of postponing data collection are obvious: in some cases data collection later may be of little or no use (hour scale data). Ploughing, mowing or harvesting will present problems with day scale data, and no knowledge of the ratio of leaf cover to soil and/or weed cover will be available for crops.

The advantages of ground data collection postponement are the certainty that the cost in time and labour is only expended when the remotely sensed data are available, the easier location of sample points or fields on imagery (particularly in poorly mapped areas), and the possibility of being able to design a better sampling programme, considered below.

7 THE MAP BASE FOR GROUND DATA COLLECTION

The question raised above of the quality of mapping of an area may next be considered. A map may almost invariably be considered as a reliable representation of the country depicted, at the time at which it was surveyed or photographed, and within the limitations of its scale. Maps are nearly always more or less out of date: development and changing field boundaries usually make field mapping difficult, even in countries such as Britain that are considered well mapped.

Problems of this kind were acutely illustrated for the Reading University team in the work in Argentina decribed above. For much of the uncultivated area, the best maps available were at a scale of 1/500 000 and decades out of date. The road network had changed beyond recognition, so that visual air navigation was difficult as well as ground navigation. In such cases a preliminary survey may have to be made, even if only based on a road traverse and car odometer. Consideration must be given in such cases as to whether the landmarks chosen to mark the beginning, route and end of the traverses will be visible on the imagery.

8 THE SAMPLING STRATEGY FOR GROUND DATA COLLECTION, AND ACCURACY ASSESSMENT

The last points raised bring up the question of the sampling strategy to be adopted. This is affected by a number of considerations other than statistical sampling procedures. These include the problem of cost in money and time: how much is available, should it affect the procedure, or is it so limited as to make meaningful data collection impossible? These are vital considerations at the planning stage.

There are also problems of access. If an ideal sampling system is designed, can the points be reached at all, and perhaps more pertinent in many cases, can the work be carried out in a time scale such that the data are of use?

These considerations, and the necessity for positive location on the imagery, may dictate the use of road traverses for the collection of data. These are likely to be statistically unsatisfactory, but their use may be forced upon the surveyor by practical considerations of time and logistics.

Alternative systems are the selection of sample points or plots by defining a grid net of points, or by a selection of points using random numbers to define the point co-ordinates in the two axes. These methods may be combined if a random sample point is selected in each grid square. Another possibility is to survey the whole of a block of ground if such variables as land cover are being assessed. These methods frequently pose problems of access, location, cost and time taken.

In any situation, however, the sample should ideally be stratified in some way. If separation into a number of distinct classes is the objective, these classes should be such that the within sample variance is much less than the between sample variance. The total sample should contain an adequate sample of each class. This is particularly true of any test data.

Van Genderen and Lock (1976) have pointed out that in a small sample, the probability of finding no errors may be quite high even if classification accuracy is low, and they propose a method of accuracy assessment based on probability statistics

In general, a sample of around 30 in each class seems adequate for accuracy assessment in most cases, but it may prove impracticable to obtain sufficiently large samples of a class if that class occupies only a very small proportion of the total area, or population. In this case, less reliable estimates of accuracy will have to be accepted for these minority classes, which is probably an acceptable solution in normal usage.

Accuracies may be expressed either in probability terms as proposed by Van Genderen and Lock, or by a simple proportion of the points correctly classified. The latter is common practice in the literature, and the method is described for example by Zalensky and Scherck (1975). In either case a classification matrix is informative and should be made up. One axis of the matrix is provided by the ground classes, and the other by the classes derived from the remotely sensed imagery. A perfect classification will then have all the points on the diagonal. Errors will be shown in the appropriate off-diagonal elements.

9 THE DATA TO BE COLLECTED

Decisions as to the data to be collected will involve consideration of all the above factors. In general, experience of the Reading University team has been that we tend to collect too much data at each point rather than too little, but this may not be a general fault. This is frequently because sensors or research methods are being evaluated, rather than an operational survey carried out, and also that cost considerations do not perhaps have to rank so highly with a research team as they would with a commercial group.

The data collected may be divided in general into two groups, visual and instrumental, with some data in an inter-mediate class.

9.1 Visual data collection

A high proportion of data can be estimated visually. Crop or vegetation type, stage, height, row spacing and orientation, percentage ground cover, with the same information for any weeds present, is one group. Soil variables such as percentage of total view covered by soil, proportion of this area covered by stones, field texture, description of structure and roughness (agricultural practice) is another. Even soil moisture can be estimated by feel and appearance. Micro-relief can be described. Soil colour is commonly measured using Munsell colour charts. Vegetation colour can also be estimated, but less satisfactorily, due to such effects as light and shade, and variability. A range of colour samples is available: we use Munsell colour charts for soil, but prefer the Methuen colour charts (Kornerup and Wanscher, 1967) to full Munsell charts, the ISCC-NBS Centroid color system (Kelly and Judd, 1968), or the Royal Horticultural Society colour charts, for vegetation.

Considerations of generalisation and coding of the above data will arise in any project. Is it simpler for example, for an observer to write (1) if crop height is less than 10cm, (2) if it is between 10 and 20 cm, and so on, rather than measuring or estimating and recording the actual crop height? Loss of exact information has to be balanced against ease of recording and coding.

Skill and experience are of value in visual collection of ground data, but whatever the standard of the personnel, estimates of some of the above variables may differ from observer to observer and training and standardisation may be necessary, for example in estimating percentage ground cover. Members of the Reading team have frequently estimated values for the same area and compared results. If these estimates are compared with accurate values obtained by quadrats, then individuals' estimates can be corrected. A training and standardisation scheme was part of the preparation for the Argentinian project described above, in which British and Argentinian personnel were involved.

9.2 Instrumental data collection

Some of the above data may be improved by some form of mechanical or instrumental supplementation. For example, the accuracy of soil and vegetation percentage cover may be improved by using quadrats, but the time taken is increased. This has to be a conscious decision.

Ground photography can be of great value as supplemental information, but some caution must be observed. It is recommended that a board describing the number or location of the site is photo-graphed as part of the scene to avoid possible later error. The photography may be an ordinary horizontal view, a downward oblique, or a vertical photo-graph, and true colour or false colour infrared film, or both, may be used. Reversal (transparency) film is normally recommended.

The vertical photographs may, with certain provisos, be used at a later stage for cover estimates instead of quadratting in the field, thus speeding up the field data collection. They may also be used to estimate the proportion of the scene in shadow, which has an important influence on the signal, but is extremely difficult to estimate visually. The major drawback is the limited area covered at near vertical view angles, if this is the view angle of the remote sensor. A hand-held camera can have a maximum height above ground of about 1.5 metres. If a normal lens of 50 mm focal length is used, the area viewed is 72 x 108 cm, and extreme view angles are 23.4° from the vertical. If a wide angle lens of 35 mm focal length is used, the area covered will be extended to 103 x 154 cm, with viewing angles up to 31.7° from the vertical. Only the small central portions of such images should be used for quadratting for comparison with a vertical looking remote sensor such as Landsat.

This area covered may be improved by standing on top of a vehicle to take the photograph, but this requires access to the cropped area for the vehicle, and resultant crop damage.

Slope, aspect and other morphological information requires the use of such instruments as a compass and clinometer.

Soil moisture may be measured *in situ* by tensiometers or neutron probe apparatus, and subsurface soil temperatures have to be measured *in situ*. Soil and vegetation samples may be removed for subsequent standard laboratory analysis, if the survey budget will stand it. Ground heat flux measurements must be taken *in situ* if required.

The major remaining class of instrumental measurements involve estimates of radiation and energy balance, and ground measurements at point samples of the signal values that will be recorded by a remote sensor after degradation by the atmosphere.

The measurement of the radiation and energy balances at a point are standard procedures in microclimatology (e.g. Sellers, 1965). Reflected solar radiation in narrow wavebands may be measured by instruments such as an ISCO spectroradiometer. While these are suitable for calibration purposes, it has been our experience that they are bulky and present difficulties in field use. As a result of these problems, small, cheap and portable bandpass radiometers have been constructed at Reading University (E.J. Milton). So far the instrument has been designed to display on a meter immediate integrated outputs of reflectances corresponding to the channels of Landsat MSS, together with ratios of any signals.

The sensors used in the field radiometer are selenium photovoltaic cells for Channels 4, 5, and 6, and a silicon photodiode for Channel 7, in combination with appropriate Wratten gelatine filters. The overall dimensions of the instrument are 20 x 15 x 15 cms, and the weight is 1.6 kgs. Development is proceeding further.

In the thermal band, such instruments as the Barnes Instatherm remote reading thermometer may be used to sense the surface equivalent black body temperature.

In the microwave band, a 3 cm radar has been mounted on a vehicle and used by the Bristol University team to measure radar backscatter for various surfaces.

Atmospheric degradation may be estimated by correlation with local horizontal visibility readings taken at a nearby meteorological station or in the field, although this method is not accurate. A better estimate may be made in the field using a spectroradiometer with a narrow field of view pointed directly at the sun to record only the incoming direct radiation. This must be repeated with varying solar altitudes during a day to deduce a measure of the atmospheric extinction coefficient. This is unlikely to be feasible in a normal agricultural project (Hulstrom, Adams and Oldham, 1977).

10 CONCLUSIONS

These notes pose a wide range of questions and offer only generalised solutions. The nature of the ground data collected must vary for every project according to the various factors outlined. It is difficult, considering the range of these factors, to suggest any standardised scheme of ground data collection which could meet the requirements of a variety of institutions and programmes.

The use of such ground data in conjunction with the remotely sensed data for training, testing and extrapolation purposes is another major field of consideration, posing a further range of problems.

11 REFERENCES

Anderson, J.R., E.E. Hardy and J.T. Roach 1972, A land use classification system for use with remote sensor data, U.S. Geological Circular 671.

Barrett, E.C. and L.F. Curtis 1976, Introduction to environmental remote sensing. London, Chapman and Hall.

Beckett, P.H.T. 1974, The statistical assessment of resource surveys by remote sensors. In E.C. Barrett and L.F. Curtis (Eds), Environmental Remote Sensing. London, Edward Arnold.

Benson, A.S., W.C. Draeger and L.R. Pettinger 1971, Ground data collection and use, Photogramm. Engg. 37: 1159.

Curtis, L.F. 1977, Remote sensing of soil moisture: user requirements and present prospects. In R.F. Peel, L.F. Curtis and E.C. Barrett (eds.), Remote sensing of

the terrestrial environment, Colston
Papers 28, p. 143-158. London,
Butterworth.

Curtis, L.F. and A.J. Hooper 1974, Ground
truth measurements in relation to aircraft and satellite studies of
agricultural land use and land
classification in Britain. In European
Earth Resources Satellite experiments,
ESRO SP-100. Neuilly, France, ESRO,
(now European Space Agency).

Hord, R.M. and W. Broomer 1976, Land use
map accuracy criteria, Photogramm. Engg.
42: 671-677.

Hulstrom, R., L. Adams and L. Oldham 1977,
Inflight performance evaluation of
satellite remote sensor data by ground-
based measurements, Jour. Brit. Inter-
plan. Soc. 30: 172-177.

Kalensky, Z. and L.R. Scherck, 1975,
Accuracy of forest mapping from Landsat
CCT's. In Proceedings, Tenth Inter-
national Symposium on Remote Sensing of
Environment, p. 1159. Ann Arbor,
Michigan, ERIM.

Kelly, B.L. and D.B. Judd 1968, ISCC-NBS
Centroid color system. In J.T. Smith
(ed.) Manual of color aerial photography, p.523. Falls Church, Va.
American Society of Photogrammetry.

Kelly, B.W. 1970, Sampling and statistical
problems. In Remote sensing for
agriculture and forestry, p.329-353.
Washington, D.C., National Academy of
Sciences.

Kornerup, A. and J.H. Wanscher 1967,
Methuen handbook of colour, 2nd Ed.
London, Methuen.

Mitchell, C.W. 1973, Terrain Evaluation.
London, Longmans.

Monteith, J.L. 1973, Principles of
environmental physics. London, Edward
Arnold.

Sellers, W.D. 1965, Physical climatology.
Chicago, University of Chicago Press.

Stobbs, A.R. 1968, Some problems of
measuring land use in underdeveloped
countries: the Land Use Survey of Malawi,
Cartographic Jour. 5: 107-110.

Thaman, P.R. 1974, Remote sensing of
agricultural resources. In J.E. Estes
and L.W. Senger (eds.) Remote Sensing:
techniques for environmental analysis,
p. 189-223. Santa Barbara, California,
Hamilton Pub. Co.

Townshend, J.R.G. 1976, Ground information
for the earth-resources Skylark. In
E.C. Barrett and L.F. Curtis (eds.),
Environmental remote sensing II, p.217-
245. London, Edward Arnold.

Van Genderen, J.L. and B.F. Lock 1976, A
methodology for producing small scale
rural land use maps in semi-arid
developing countries using orbital MSS

imagery - (NASA-CR). London, Department
of Industry.

Webster, P. and P.H.T. Beckett 1968, Nature,
219: 680.

S. SCHNEIDER
Federal Research Institute for Regional Geography and Regional Planning
Bonn, Germany

16

Land inventory, land use and regionalization

The growing experience gained from a variety of civil applications of remote sensing techniques during the last decade have demonstrated new ways of solving many problems in the geographical and planning fields. They have opened a broader view of the interrelationships between various natural phenomena and human activities. But the fuller utilization of remote sensing technology by many public services and managements, may be expected to bring even more important achievements during the next decades through the qualitative and quantitative improvement of the remotely sensed information and the increased effectiveness of its procurement. This can be achieved through the agency of quick, large to medium scale, multi-sensor observation of special features within small areas, or by less detailed, small to global scale but repititive and synoptic surveying.

As experience has shown, remote sensing, including aerial photography can provide the data for information in many fields of research and practical application, such as:
- Natural landscape division
- Land-use (inventory and planning), land-use changes, land-use conflicts
- Soil classification and conservation (agricultural crops, irrigation, soil erosion)
- Plant diseases (crop and forest protection)
- Mineral and oil exploration
- Disaster assessment and prediction (volcanoes, earthquakes, landslides, inundations)
- Global weather mapping
- Horizontal and vertical temperature distribution
- Surveying of changing amphibian zones
- Variations in the humidity of the soil
- Water resources inventory and management for agricultural and industrial use
- Water quality (pollution)
- Thermal plumes, thermal streams
- salinity
- Air pollution monitoring

- Traffic surveying and control

"Remote sensing systems have to offer various levels of the sensing capabilities, various degrees of reliability, various stages of information processing, dependent on the areas of application and on the degree of development from experimental to operational utilization." (Schanda, 1976).

In Germany the conventional official land-use data are limited to certain traditional land-use classes reflecting the cultural landscape of several decades ago. Consequently they are based on expanding units. In other words, the territories of our communities as the smallest statistical reference units are increasing in size through administrative centralization.
In Germany there existed in 1961 a total of 24 503 communities; in 1975 the total number of communities was only 1o 914 without any change of the total spatial dimension of the state area. The official statistical information based on communities is therefore no longer so detailed as before.

Remote sensing data are complementary to those statistical data which present detailed information such as the areal use of types of buildings, the number of floors and the floor space.

Moreover remote sensing data expand the available information to cover

ecological, socio-economic and environmental problems such as: the monitoring of air-and water-pollution, of the different sorts of fallow-land, of the green spots in the cities or of the progress in reclaiming land in opencast-mining districts. To all these reasons, the topicality and the reliability of remote sensing data demonstrate their applicability in regional planning. Our experience shows that the task of land-use survey by remote sensing techniques could be solved with an interpretation accuracy about 9o to 95 %.

The systematical development of airphoto interpretation in combination with computerized data handling has led to the use of aerial photography being favoured for land-use analysis as a first step to a regional information system.

In the Rhine-Neckar Region, the Frankfurt Institute for Planning Data has detected remarkable differences of areal dimension of up to 25-3o% between the size of the same land use units in the old cadastral classification and in the remote sensing data. Moreover, the land use data of this region have been digitized in a very short time and recorded in regional handbook of land-use statistics.

A balance sheet as well as thematic maps to the scale of 1:5o 000 and 1:25 ooo respectively have been printed out. For the first time it was possible to present a complete information on the land-use structure of the non built-up areas of the region.

Because of the world-wide coverage by landsat imagery, several countries have used this multispectral data material for an inventory of their territory, for monitoring land-use changes, land-use conflicts and environmental damages. Last year the Federal Department for Regional Planning initiated a research project to examine the feasibility of transposing the Landsat data into a thematic map of land-use for planning purpose to the scale of 1:2oo 000. The sample-sheet covers the Rhine-Neckar region and the Upper Rhine Valley between Mannheim and Speyer.

Test areas of the imagery are enlarged to the interpretation scale of 1:5o 000; the selected features are generalized to the final scale of 1:2oo 000.

The land-use categories and features are classified by a digital image handling system on a screen; although it may be difficult in some instances to separate real land use categories from signals without additional evidence, the results of the object-classification are promising. We hope this map will be the first sheet of a new thematic series of land-use map.

As far as the application of thermographs over a large area such as the Ruhr District, is concerned it may be advisable to consider multispectral satellite imagery. Over the Ruhr District a haze cover had been detected after filtering the four Landsat spectral images by a multispectral viewer. The same effect could be achieved by digital image processing.

Similar studies on the thermal conditions in the area of the Main Basin near Frankfurt and in the Rhine-Neckar Region near Mannheim led to the establishment of fresh air corridors which are to be kept free of buildings or trees.

It has been proved that it will no longer be possible in future to change the use of a large area without at the same time making a statement on the changes of the environmental conditions with the help of remote sensing technology.

The aim of creating a map series with particular geodetic grid required the rectification and mounting of several satellite images which may belong to different orbits. This is a problem and task of general interest. Some first steps have been made in this direction as we know by the Institute of Applied Geodesy in Frankfurt/M. (This institute as well as the State survey of NRW have a long experience in creating air-photo maps (1:25 000 and 1:5 000).

The technological and geometrical difficulties accompanying the evaluation of line-scanner imagery will be eliminated by developing an optic-electronic sensor for the visible and the near infrared spectral bands. This new device has been developed in the Federal Republic of Germany by Hofmann (MBB), using Fairchild photoelements. It is based on the use of solid state image sensors. The advantages are manifold, they include
- no mechanical scanning
- built-in geometric accuracy
- high quantum efficiency of silicon for visible and near IR-wavebands
- high resolution
- good stability
- low voltage operation.

These charged coupled devices have 5oo, 1o24 and 1728 single cells. One sensor measures 32 x 17 x 5 mm. The resolution is 3o - 4o linepairs per mm. On a projected Space Shuttle mission the CCD's will have a ground resolution of 7.5 m from 92o km of altitude over a 185 km cross track scene. (Dr.Ing. Hofmann, MBB, 28.6.1976 and R.C. Heller, Helsinki, July 1976).

A new compilation of Landsat and Radar imagery was presented at the ISP-Congress of Helsinki 1976 by Harris and Graham (USGS and Goodyear Aerospace Corp.).

The simultaneously view retains all the information available from each sensor system and provides additional detailed data. Radar views terrain at a low grazing angle, apparently emphasizing terrain relief. Details presented in the Radar imagery have a resolution five to ten times finer than that available in the Landsat imagery. Landsat imagery, on the other hand, is not subject to the target shadowing effect of Radar and provides unique signatures of water siltage, soils and vegetation type and condition through the differential reflectance at the four wavelengths commonly used in analysis.

Combining imagery from the two sensor systems is facilitated by planning radar flights so that the viewing direction corresponds to the Landsat orbital path across the earth's surface and the sun angle, minimizing differences in shadows and highlights.

The Landsat imagery is sensitive to water and vegetation type and condition, so that repeated coverage is desirable. On the other hand, radar imagery is principally affected by terrain slope and general type of cover. It is therefore well suited to aircraft operation.

Another problem in connection with the land-use inventory is the monitoring of land-use conflicts, such as between big chemical plants and recreation grounds or housing projects or between gravel-pits and water protection zones. This last problem has been studied in the Upper Rhine Valley where a large series of gravelpits follows the course of the river. Some of these gravelpits have been transformed into waste dumps.

Because of the delicate conditions for obtaining drinkingwater, the relation between the groundwater-level and the water level in the gravel pits has been studied last year using aerial photography and Landsat imagery.

The monitoring of recreational functions within the gravel-pit zone of the valley by color photography and by thermography has revealed the existence of potential land-use conflicts. It is just in the neighbourhood of urban agglomerations that the water in the gravel-pits and old river branches is the subject of considerable conflict.

A remote sensed documentation of the actual land-use situation in order to avoid future land-use conflicts seems to be increasingly employed by planning boards as well as by industries. The regional planner has the responsibility of providing a balanced level for various types of land-use with the aid of reliable data in order to answer the economic and environmental requirements.

The successful development of infrared technology - photography and radiometry - has led to its application in thermal studies of different planning projects. Nevertheless it must be admitted that some data obtained from a single sortie have been applied prematurely and too enthusiastically; and they have led to misinterpretation and wrong actions.

Two thermal infrared images taken within a period of several hours may yield completely inverse results. Therefore the success of infrared measuring and imaging depends to a high degree on the point in time and the meteorological conditions. Caused by the wide range of variations the results of one sortie of thermal taken only once are not or only little representative.

Several planning boards of the Ruhr-District, the Frankfurt-Area and the Rhine-Neckar District have been encouraged to study this problem. The most interesting results were presented in the polycentric Ruhr District with several large cities and between them buffet zones (green belts) which have so far been preserved from development.

These buffet zones represent predominantly farm land with small woodland areas, public parks, cemeteries, allotments, playing and sports grounds in between. With the foundation of the Ruhr planning board in 192o, this board was made responsible inter alia for the configuration and limitation of green areas and zones, which are important to all housing estates.

It has been proved that regional green areas filter and renew the air. The green areas filter the dust particles out and cause air currents, which scatter pollution widely, thus reducing fall-out.

Even noise can be reduced by a corresponding configuration of the green areas. Because of its close neighbourhood to the residential areas, the regional green zone system for the central part of the Ruhr district is to a large extent suitable as a local recreation area. The thermographs made by line-scanner during day and night and different seasons present clearly the regional green zones in colored equidensites. This gives an actual presentation of the differences in temperature around the areas co-vered with buildings of differing density as well as for the relati-vely sharp demarcation line around the built-up areas. The thermographs of the central Ruhr District are an impressive proof of the fact that continuous open areas (green belts) have a cooling effect on cities. Generally speaking, the infrared thermograph corresponds with the test result of the city climate, namely that the surrounding tempe-rature also rises with an incre-asing density of development. By differentiating the radiation tem-perature in stages of o.5°C, not only the exact position but also

the geometrical pattern of these objects can be accurately deduced from the heat evaluation. The urban thermographs allowed a differenti-ation between areas covered with 3-floor closed building blocks and those covered with separate 2-floor buildings. At the same time, the exposures give a statement on the insulation of the building elements (from Ruhrsiedlungsverband).

The same problem of the influence of the large industrial plants on the adjacent residential areas by high emissions of heat leads back to the problems of city climate. The temperatures of the open areas, their sequence in the course of a night and their registration by a thermography offer the best means of attacking this problem.

Using remote sensing methods, we have extreme valuable techniques for viewing the whole, regardless of how large it may be. If our objective is general information over very large areas, we have at our disposal satellite data. At the other extreme, we can observe phenomena just a few centimeters in size with very large scale aerial photography. It is of ex-treme importance that right tech-nique to the level of analysis and decision-making should be applied.

To realize the full potential of remote sensing earth observation systems, these systems must move out of the experimental stage into an operational mode. Experimental systems cannot provide data on a continuous basis which is important to the users in regional planning.

It is also important that we recognize the limitations of our remote sensing tools. With remote sensing techniques we can view the whole but we cannot necessarily analyze the whole in detail. With very few exceptions, some amount of ground truth information is required primarily in the form of aerial surveys and analyses, of airphoto information to supplement satellite surveys.

Features observable through remote sensing are generally one of two types: those whose values decay rapidly with time after observation, such as inundation, discharge plumes, soil moisture etc. and those which have a long-term value such as geological structure, land use, ocean colour, forest type and cover etc.

We need organizational structures to analyse and disseminate information and to make management decisions for the two types of data.

Data and information which has enduring value can be put in the form of published books, maps, images etc. and can be handled by library-type dissemination centres and archives. On the other hand data

where value decays rapidly with time, must be telemetred in digital or graphical form to analysis centers, analysed in near-real time and then telemetred immediately to end-users and decision makers.

One of the greatest advantages of remote sensing techniques is the ability to extend very limited ground-acquired information over very large areas, thus adding the synoptic and synchronous aspect to the detection of change.

263

S. SCHNEIDER
Federal Research Institute for Regional Geography and Regional Planning
Bonn, Germany

17

Environmental monitoring by remote sensing methods

The problem of environmental monitoring by remote sensing methods can be studied particularly on water pollution. Though most European countries do not have quantitative problems of water supply, the serious problem of water qualities becomes increasingly apparent. The future assurance of sufficient supply of good drinking water demands the highest priority.

The Department of the Interior has asked the Federal Institute for Regional Geography and Planning to build models of an operational remote sensing system for monitoring and estimating water quality. Two test areas have been chosen: the industry district of the small Saar river between Saarbrücken & Saarlouis with an extreme width of 30 - 40 m and the middle Upper Rhine valley between Karlsruhe & Mannheim with a width of about 240 - 280 m. We have used different types of line-scanners with and without blackbodies and have preferred a combination of infrared radiometer-thermometer and infrared line-scanner in order to avoid atmospheric influences as far as possible and to get a distinct differentiation of surface temperature across the river. We used a helicopter and a simple oneengine aircraft. The helicopter equipped with an infrared radiometer for measuring the temperature above the water surface was flown in a longitudinal axis along the narrow Saar and in 38 transverse axes over the Rhine. Over small rivers a single central temperature curve may be sufficient for representing the temperature above the water surface. Over big rivers and estuaries it will be necessary to get the data of the water surface temperature in at least three curves, i.e. along the centre of the river and along both sides. This situation is shown in the diagram of the Rhine near Mannheim-Ludwigshafen. Another possibility would be the continuous print-out of measured computerized data, as e.g. one number for 16 measured data (Lower Elbe near Hamburg) in a tidal river. (Thermocomp).

The radiation temperature-profiles across the river as well as the temperature imagery won by line-scanning demonstrate that the distribution of temperature over the width of a big river may differ considerably. One of the last results using these methods combined with digital evaluation is the so-called "Thermocomp"-map of the lower Elbe region which will now be compared with a "Thermocomp"-map of the Upper Rhine.

The conventional methods of measuring the water surface temperature are helpful but not sufficient in comparison with the areal overview given by remote sensing methods. All essential discharges into the Saar river as well as into the Rhine river have been detected. In both cases we had to combine the results of multispectral photography and of infrared scanning. In one case of a big cellulose plant the heavy polluted and heated discharge plume had been represented by both sensors: colour camera and infrared scanner.

The form and spread of the pollution plumes depend on the water quantity and the flow velocity, so that the discharge plumes of the Saar and the Rhine have been usually quite different.

After flowing for a short time the plumes of the Saar cover the whole river surface; the plumes of the Rhine are pressed along the banks as small and long bands extending in some cases, more than 30 km in length. The midstream surface temperature of the Rhine is seldom influenced by the heated discharge plumes near the banks, which means that the water surface temperature of big rivers in industrial and often densely populated regions is usually quite different close to the bank and in mid-stream.

An overheating of the river water would probably have the following consequences: as the water temperature rises, the oxygen demand of fish and other organism is increased. An overheating of the Rhine would probably result in the danger that the biological balance in the water would be severely disturbed, causing a loss of the remaining selfpurifying power of the river. Ships would be troubled by the heavier development of mist. The recreational value of the river landscape would be lowered by increasing turbidity and visible algae development.

In extreme cases, mass formation of algae leads to intensive decomposition processes with decay and odour problems. Dangerous influences on the production, preparation and distribution of drinking water from surface water can be predicted, although not yet quantitatively determined today.

The water quality of the small Saar river could be characterized by interpretation of the existence or non-existence of aquatic plants and plant-communities as indicators of pollution. In the case the canalized waterway of the Upper Rhine the observation of aquatic plants must be confined to the zone of old branches and tributaries - 6-8 km of width - where the interpretation is based on certain plant communities as indicators of eutrophication.

It is precisely this wetland zone of old waters and tributaries, which is of great natural interest, which is now more endangered by human activities than the waterway itself. We identified more than 30 discharge points in the Upper Rhine Valley, where the sewage flew into the old waters and not immediatly into the main river. The land use conflicts arising in this zone include industrial location, location of power plants, refuse pits, pipe lines, gravel pits as against recreation and sporting grounds, forestry, agriculture and areas of nature conservation.

Because of the many possibilities of land-use conflicts in this wetland zone, a landscape inventory by airphoto interpretation has been carried out. First of all, infrared colour aerial photography, combined with ground checking, gives clear differentiations of plant societies.

Infrared represents smallest waters as well as waterlines, eutrophication near river banks, detergents, refuse pits, big industrial plants in spite of the mist and smoke in the valley.

As long as experiments and experiences of water turbidity and colour must be continued, we prefer the indirect way of water quality interpretation by studying the plant societies, espec. the hydroflora.

The best platform for getting good information on the hydroflora was the small one-engine aircraft, taking the airphotos from a low height above ground.

The radiometric temperature curves as well as the line-scanner imagery of these rivers have now been used by administration and planning authorities in solving difficult location problems such as the siting of power plants.

All data received by remote sensors and their digital picture processing such as the type of water, characteristics of flow, sort, source and colour of discharge plumes, distribution of surface temperature as well as the kind and degree of eutrophication make it evident that remote sensing methods can be used for environmental monitoring.

If we consider the value and efficiency of remote sensing technolo-

gy, there are manifold possibilities of application dependent upon
the problems concerned.

One of the first results of remote
sensing application in W-Germany
are the maps on the water quality
of the Saar river and on the water
protection zones and the land use
conflicts in the Upper Rhine district.

M. TAILLADE-CARRIERE

Service ARGOS, Centre National d'Etudes Spatiales, Toulouse, France

18

Satellite data collection system – Agricultural application

1 DATA COLLECTION DESCRIPTION

In a satellite data collection system, parameters such as temperature, pressure or other variables are sensed at the platform (in situ or direct sensing) and the data encoded, formatted and transmitted to processing facilities via a satellite communications link.

Satellite data collection can be described by several characteristics :
- the technique is used to monitor unattended sensory platforms via telecommunications
- in situ data is collected data rather than remotely sensed data of the type obtained from radiometers or multispectral scanners
- the systems employ large quantities (hundreds...) of low cost ($5,000) platforms
- the platform message durations are short (32 to 1000 bits)
- position location can be accomplished by the measurement of Doppler effect or the measurement of range
- the systems are compatible with both low and geostationnary orbits ; thus possibly offering near real time or real time data as needed

Many of the applications involve both the tracking of moving platforms (weather balloons, buoys, ice islands, wild animals) as well as the collection of data from fixed sites (moored buoys, water survey stations, volcano surveillance stations...)

In operation, data is relayed from the remote platforms to a satellite either randomly or upon interrogation command.
Data can be stored aboard the satellite for readout over a central receiving site or can be relayed directly to a control processing facility or to a local regional user terminal. Once data is received on the ground, information is formatted and disseminated to users. In the case of position location the coordinates are computed and disseminated to the user along with the collected sensory information.

2 REVIEW OF EARLY AND PRESENT SYSTEMS

2.1 Various data collection systems

Several systems and techniques have been developed and tested, and others are planned for future launch. Figure 2.1 shows a decision tree for distinguishing various data collection systems.

2.2 Interrogations, recording and location system (IRLS)

The major elements of the system (figure 2.2.) are :
- a central ground processing facility
- a satellite carrying receiving equipment
- the remote platforms

In operation unique addresses or codes identifying the platforms were programmed into the satellite from a ground acquisition facility at the beginning of each orbit.
As orbital time elapsed the platforms were interrogated at predetermined times and the received data was stored aboard the satellite for retransmission to the ground facility at the end of each orbit. Position location was accomplished by a ranging technique with a minimum of two interrogations or range measurements required for each platform.

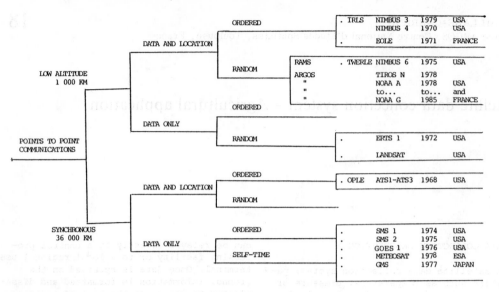

NOTE : AN ORDERED SYSTEM RESPONDS TO AN INTERROGATION FROM THE SATELLITE WHEREAS A RANDOM SYSTEM TRANSMIT AT PREDETERMINED INTERVAL.

IRLS   Interrogation, Recording and Location System

RAMS   Random Access Measurement System

TWERLE Tropical Wind Energy Conversion and Reference Experiment

NOAA   National Oceanic and Atmospheric Administration

ERTS   Earth Research Technology Satellite

ATS   Applications Technology Satellite

OPLE   Omega Position Location Equipment System

SMS   Synchronous Meteorological Satellite

GOES   Geostationnary Operational Environmental Satellite

GMS   Geostationnary Meteorological Satellite System

FIG. 2.1. Various data collection systems

The IRLS was the first global satellite system that demonstrated the worldwide capabilities of satellite data collection. The IRLS and EOLE were ordered system that utilized <u>receivers on the platform</u> to initiate the platform transmission to the satellites. This resulted in substantial cost, size and power consumption requirements for these platforms.

2.3 Random access data collection system

The data collection system developped for the Landsat satellite series was the first random access system. In this system, platforms transmit their sensory information to the satellite randomly. The short duration transmission initiated by platform timers allows multiple platforms to be ser-

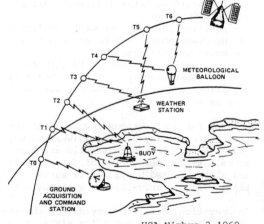

FIG. 2.2. <u>IRLS Concept</u>

USA Nimbus 3 1969
Nimbus 4 1970
FRANCE Eole 1971-
1974

270

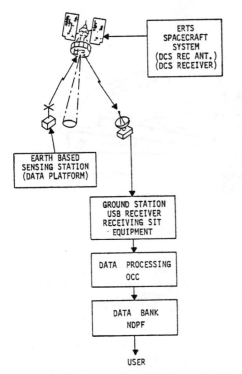

FIG. 2.3. Landsat data Collection System

viced (capacity for 2000 platforms simultaneously in the satellite field of view with a 95% probability of data collection from each platform). The Landsat DCS uses an UHF uplink from platforms to the satellite while the data received by the satellite is retransmitted immediately on a S band downlink to the ground terminal.

2.4 Random access data collection and location system

"RAMS" system using the multiple access technique was developped to support the Tropical Wind Energy conversion and Reference Level Experiment (Meteorological Experiment with approximately 400 balloons). RAMS carried by the Nimbus F satellite was launched in 1975 and is still in operation. RAMS permits global scale experiments to be performed utilizing low cost, simple data collection platform equipment.

The platforms transmit a one-second message to the satellite at random times at the rate of once per minute.
The satellite records a Doppler frequency measurement and a time tag and formats the received data. This information is stored aboard the satellite for readout over the Fairbanks, Alaska, ground station and transmission for processing.
The position location coordinates of each platform are computed and the data is transmitted to investigators.
Another application, as shown in fig. 2.4. used the OMEGA system to derive vertical profiles. The Carrier Balloon System (tested in 1975) involved large balloons carrying dropsondes commanded via SMS satellite.

As the sondes were released, they received and relayed the OMEGA signals to the lar-

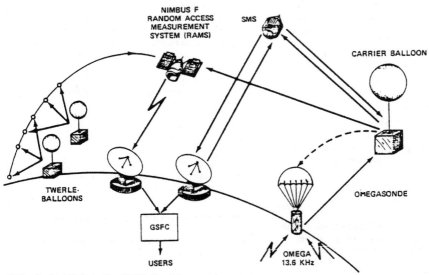

FIG. 2.4. Nimbus F RAMS System

271

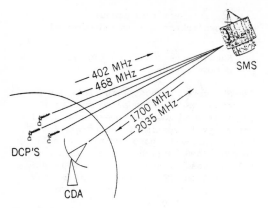

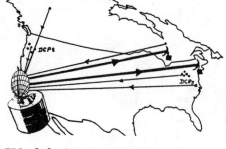

FIG. 2.5 SMS System

FIG. 2.6 Comsat experiment

ge balloon for retransmission to a processing center via SMS satellite (geosynchronous).
The phase measurements derived from the
sondes during descent were used to compute
the vertical wind profile from 20 mb levels
to the surface.
The position of large balloons were known
using RAMS/Nimbus F low altitude satellite.

2.5 Ordered system with self time capability

The SMS data collection system is principally an ordered system with the interrogations initiated by command.
Interrogated DCP'S transmit their data upon
receipt of a unique address command transmitted via the satellite from the ground
station.
Alarm DCPs transmit an alarm signal when
their sensors exceed a specified threshold
value. Upon receipt of this alarm signal,
the ground station will interrogate the
platforms to obtain the actual data.
A self-timed capability allows transmissions
at pre-established intervals under control of
an internal clock (self-timed DCPs).

The system has a UHF platform to the satellite link and data is received at the
ground through an S band downlink. The system is intended for non-moving applications
and can handle 10,000 platforms.

2.6 Commercial synchronous satellite data
collection project

In 1977, Comsat General Corporation (USA)
engaged in a joint development program
with the US geological survey (USGS) and
Telesat CANADA to collect environmental
data via satellite.

This program represents the first use of
present-day commercial satellite communications technology to improve the management
of resources on earth.
Under a six-month evaluation program,
Comsat will erect 15 self-timed data collection platforms.
The DCPs will automatically transmit the
data (mainly hydrological) in random bursts
via Anik satellite operated by Telesat CANADA
to control receiving and distribution points.

2.7 Launch schedule of data collection
missions

In table 2.7 a launch schedule of early,
present and up-coming data collection missions is shown.
The systems are grouped on the basis of
data collection only and data collection
with position location.
Assuming a three-to-five-year lifetime for
the synchronous missions and two years for
the low altitude, two major operational
programmes appeared after 1978.

- GOES programme with eight spacecraft
which have been approved to date :
  - 5 launched successfully
  - 3 under construction.
- TIROS N/NOAA/ARGOS programme with eight
spacecraft which have been approved to
date :
  - the first one launched in October 1978,
  - the second planned for April 1979
  - 6 others under construction.

2.8 Earth coverage of the main systems

2.8.1 Data collection worldwide geosynchronous system

Three GOES satellite will form a part of an

272

TABLE : 2.7

D A T A   C O L L E C T I O N   M I S S I O N S

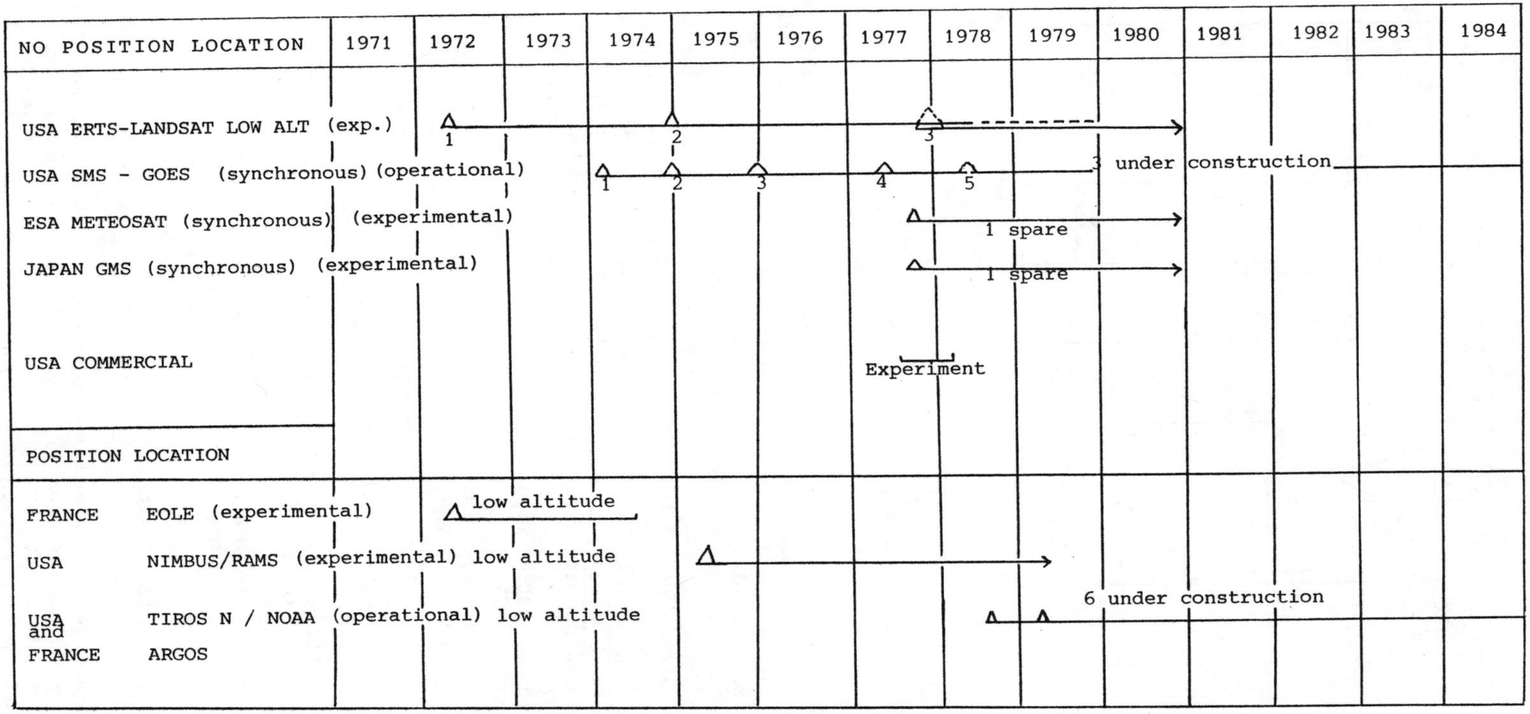

NO POSITION LOCATION	1971	1972	1973	1974	1975	1976	1977	1978	1979	1980	1981	1982	1983	1984
USA ERTS–LANDSAT LOW ALT (exp.)		△1		△1	△2			△3		under construction →				
USA SMS – GOES (synchronous) (operational)				△1	△2	△3	△4	△5		3 under construction →				
ESA METEOSAT (synchronous) (experimental)							△	1 spare →						
JAPAN GMS (synchronous) (experimental)							△	1 spare →						
USA COMMERCIAL								Experiment						
POSITION LOCATION														
FRANCE EOLE (experimental)		△ low altitude												
USA NIMBUS/RAMS (experimental) low altitude					△				→					
USA and FRANCE TIROS N / NOAA (operational) low altitude ARGOS								△	→		6 under construction →			

TWO SATELLITE DATA COLLECTION SYSTEMS WILL BE OPERATIONAL AFTER 1978

. GOES SYSTEM (still operational from 1976)

. TIROS N / ARGOS SYSTEM

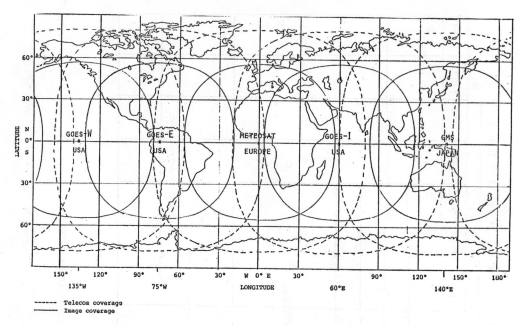

------ Telecom coverage
——— Image coverage

FIG. 2.8.1. Worldwide geosynchronous system coverage

international array of five geostationary spacecraft , targeted for the first GARP global experiment, a project under the auspices of the World Meteorological Organisation (WMO) and the International Council of Scientific Unions (ICSU).
The other two satellites, added to the GOES satellites, will provide GLOBAL coverage from approximately 60°N to 60°S.
Meteosat (European Space Agency) centered over Greenwich Meridian (Eastern Atlantic) and launched in 1977.
GMS (JAPAN) centered over 140°E (Western Pacific) and launched in 1977.
Figure 2.8.1. shows the coverage of the worldwide geosynchronous Meteorological Satellite System.

2.8.2 NOAA/ARGOS low orbit operational system

The eight satellites will be launched successively in such a way as to keep two of them simultaneously on orbit.
The two satellites will provide a worldwide coverage for data collection and location.

2.8.3 LANDSAT low orbit (data collection system)

This experimental system provides a global coverage. However, as the data received by the satellite is retransmitted immediately (without on-board storage) the system can

be used only whenever the satellite is in mutual view of a receiving station and a transmitting data collection platform (see fig. 2.3).

2.8.4 Nimbus/RAMS low orbit satellite

This experimental satellite provides a global coverage with location and data collection capability (see fig.§ 2.4).
A unique satellite was launched in July 1975 and will operate until failure.

2.9 Conclusion

Since 1978, two major data collection satellite systems providing global coverage will be available.

- Geosynchronous system (data collection ONLY)
with five satellites simultaneously on orbit
  2 operational GOES - Figure 2.9 (1)
  3 other satellites :
    METEOSAT
    GMS - Figure 2.9 (2)
    GOES I (operated by ESA)

- Low Polar Orbiting System (data collection and location)
with eight TIROS-N. NOAA/ARGOS satellites providing an operational system from the end of 1978 to 1986 at least.
Figure 2.9 (3) and 2.9 (4).

GOES DATA COLLECTION SYSTEM

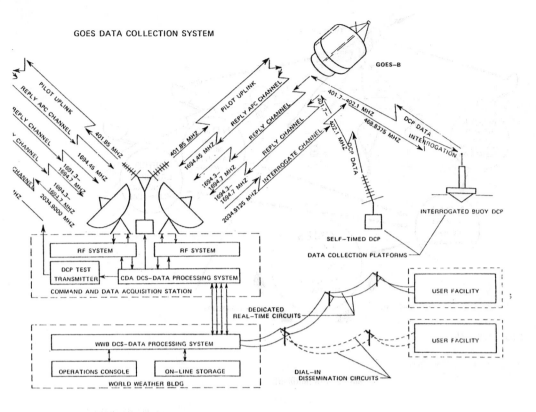

GOES/DCS SYSTEM DESCRIPTION

FIG. 2.9. (1)

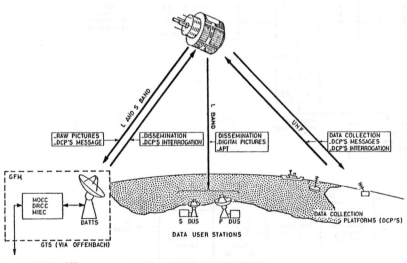

FIG. 2.9. (2)   METEOSAT SYSTEM DESCRIPTION

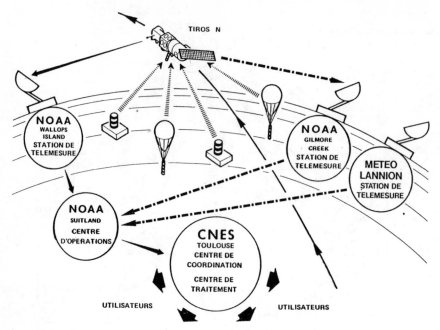

FIG. 2.9. (3)

ARGOS DATA HANDLING

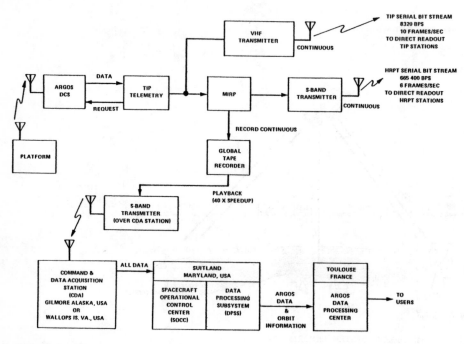

FIG. 2.9. (4)

3 COMPARISON OF THE TWO MAJOR SYSTEMS

The two major systems :
  - synchronous system
  - low orbit system
are compared in their various parts :
  - satellite system
  - data collection platform
  - data processing
  - data distribution
  - charges

3.1. SATELLITE SYSTEM

GEOSYNCHRONOUS SYSTEM   LOW ORBIT SYSTEM

- Name and authority in charge

			NASA (USA)
- GOES	NOAA (USA)	- TIROS N/ARGOS	NOAA (USA)
- METEOSAT	ESA		CNES (FRANCE)
- GMS	IMA and NASDA		Service Argos
			Toulouse, France

- Satellite orbit

Geostationnary, equatorial orbit
Altitude 36.000 km   Near polar   altitude 830 km

- In orbit redundancy

GOES System : operational with two space-   Operational system (1978-1985)
craft on orbit
    75° W longitude   Eight satellites programme
    135° W longitude   2 satellites simultaneously on orbit
3 others satellites planned
                                                 (each one able to assume the data collec-
METEOSAT : experimental   tion operation)
  One flight unit : 0° longitude
  One spare

GMS : experimental
  One flight unit : 140° E longitude
  One spare

  GOES satellite
  One fligth unit : 70°E longitude

- Earth coverage

Global from 60°N to 60°S with the whole   Each satellite provides a global coverage
system. See figure 2.8.1. for individual
satellite coverage.

- Capability

Data collection   Data collection and location (with 2 to
                                               3 km RMS accuracy)

277

GEOSYNCHRONOUS SYSTEM	LOW ORBIT SYSTEM

- DCP capacity in data collection

10,000 per satellite   16,000 for the system

- DCP type

Ordered Access   Random access
- interrogate
- alarm
- self-timed

- DCP reporting

On demand (interrogate)   When   data collection platform is visi-
Per 3-6 hours (self-timed) or less   ble from one satellite 7 times a day at
                                        the equator to 28 times per day at the
                                        poles.

- DCP radio frequency report channel

Bandwiths are divided in 3KHz channels 401,650 MHz
- International DCPs   Unique frequency for all DCPs
  402 to 402,1 MHz
- Domestic DCPs
  GOES   : 401,7 to 402
  METEOSAT : 402,1 to 402,2
  GMS   : 402,2 to 402,4

- Interrogation channel

International DCPs   No interrogation
  468,765 MHz   Random access
Domestic DCPs
  GOES   : 468,825
           : 468,837
  METEOSAT : 468,925
  GMS   : 468,924

- DCP antenna type

Directional, 15° pointing accuracy   Omnidirectional

Gain 10 db - 13 db   Linear or circular polarization
Circular polarization

- DCP transmitter power

5 Watts   600 mW to 3 Watts depending on antenna
                                        polarization and mission

GEOSYNCHRONOUS SYSTEM	LOW ORBIT SYSTEM

- DCP Report message

1. Total duration

10 to 60 sec.	0,3 to 0,9 sec.

2. Bit rate

100 bits/sec.	400 bits/sec

3. Number of bits for sensor data

2000 to 5200 bits	Less than 256 bits i.e. 32 coded 8 bits analog sensors

- DCP power consumption

Higher than 600 mW depending on transmitting rate and duration (400 Watt during transmission)	About 100 mW for data collection mission (including transmission periods)
5 kg	1 kg

- DCP sensor inputs

- Parallel digital - serial digital - analog	- parallel digital - serial digital - analog

- Time delay for disponibility of the results
(at the processing center)

Near real time between measurement and distribution	Between 2h to 6h measurement to distribution delay

- Locations for the data processing centers and results dissemination outputs

GOES (SUITLAND) USA METEOSAT (DARMSTAD) GERMANY GMS JAPAN	NOAA/ARGOS Service Argos - Centre Spatial de Toulouse 18 av. Edouard Belin 31055 TOULOUSE CEDEX FRANCE
Each center processes and disseminates the data of the DCPs in view of the corresponding satellite.	Unique ARGOS center (Toulouse) processes and disseminates the data of all the DCPs in operation wherever on Earth.

- Charges

Data processing
. GOES : Free of charge
	Data processing
	. Data collection
	   - raw data (i.e. coding in decimal, octal, hexadecimal , free of charge
	   - results in physical parameters : 1$ per platform per day
	. Location : 20$ per platform per day special contracts possible

Data dissemination At USER'S charge	Data dissemination At USER'S charge

| GEOSYNCHRONOUS SYSTEM | LOW ORBIT SYSTEM |

- Data processing

GEOSYNCHRONOUS SYSTEM

The sensor outputs are converted to an ASCII coded serial message by the DCP for transmission to the satellite.

Once the message is received the data processing system :
- checks for the correctness of the platforms address
- examines for error conditions
- stores the message on disk in an area allocated to the User (or owner) platform.

LOW ORBIT SYSTEM

The sensor outputs are converted into a binary coded serial message by the DCP for transmission to the satellite.

Once the message is received the data processing system :
- checks for the correctness of the platform address and errors conditions
- processes the data according to the User's need

Three types of processing are available in increasing order of complexity :
- coding of sensor data in decimal, octal hexadecimal, BCD
- conversion into physical parameters using the calibration curve of each sensor with possibility of inside-outside limits checking
- special processing (each one must be discussed separately)

The results of sensor data processing (and also location if needed) are broken down into "experimenters files" and stored.

- Dissemination of the results to the USERS

GEOSYNCHRONOUS SYSTEM

Once the data is stored, it can be accessed by a USER "dialling in" to the system.

Alternatively, data can be immediately routed directly to the USER via one of the dedicated lines, or stored on magnetic tapes to be sent by mail.

LOW ORBIT SYSTEM

Two cases must be considered :

. Real time distribution using :
- telephone (call from the USER and direct access to the computer)
- International or private telex
- Computer to computer dedicated lines
- Global Telecommunications System of the World Weather Watch (automatic connection in Paris) with agreement of French Meteorological Office (see Fig. 3.4)

. Differed distribution by mail :
It applies to listings, punched cards, magnetic tapes obtained through the weekly ending of Data Bank.

3.2. CONCLUSION

Major advantages of each system

GEOSYNCHRONOUS SYSTEM

- Operational for GOES System
- Data available in real time (alarm possibility at any time)
- Data available on interrogation. The rate of reports can be variable on request
- Self-timed DCP capability. Allows DCP to be used without command receiver - lower cost
- 60°N to 60°S global coverage - using international DCP

LOW ORBIT SYSTEM

- Operational and redundant
- Global coverage
- Simple data collection platform
- Low cost (2000$)
- Low weight (1000 gr for electronics)
- Low consumption : 100 mw in average for a 8 sensors data collection platform
- 32 analogic sensor inputs capability
- Easy to operate
  . Unique frequency for all the DCPs

GEOSYNCHRONOUS SYSTEM	LOW ORBIT SYSTEM
- Length of sensor data DCP message 2000 to 5000 bits	. omni-directional antenna

Major parameters choice between the two systems

GEOSYNCHRONOUS SYSTEM	LOW ORBIT SYSTEM
- Low latitudes (60°N to 60°S) DCP operation - Real time needs - Alarm need without delay - Large messages (2000 to 5000 bits) - Operational (GOES system)	- Operational 1978 - 1986 - High latitudes (data collection every 50 minutes in the polar areas and only 7 times per day at the Equator) - Two to six hours measurement to distribution delay acceptable - Measurement only when satellite in visibility acceptable - Global scale data collection experiment - Location needed (moving DCP) - Less than 256 bits message (i.e. 32 8bits analogic sensors). - Lower cost of DCPs.

4 AGRICULTURAL DATA COLLECTION

4.1 General

Today the agricultural economy is faced with increasing challenges due to :
 - dwindling areas of prime soils suitable for intensive crop production,
 - increasing competition for good soil areas by industry and housing,
 - increased fertilizer and operating costs,
 - environmental constraints on chemical control of pests and weeds.
Ever increasing demands for food and fiber to meet current and future needs necessitate a close scrutiny of techniques that can be used to utilize existing agricultural facilities more efficiently.
Most effort involves management practices to maximize production without creating undesirable conditions as a result of man-environmental interaction.

4.2 Satellite data collection capabilities

Data collection satellite systems can collect pertinent data via instrumentation that can be utilized in an operational mode. They can provide :
 - advanced ground computerization techniques that enable the data to be quickly processed and interpreted
 - integration of the data collection platforms (ground-based), satellite, computer and data dissemination in a real time mode (synchronous satellite) or near real time (less than 6 hours for TIROS N-ARGOS low

altitude satellite).
They can make a contribution to provide integrated data for the improvement of real time (or near real time) inventory, monitoring and management of agricultural systems.
The development of agricultural data monitoring and processing systems on a near real time basis can greatly enhance and complement the multispectral remote sensing satellite imagery.

4.3 Applications

The applications of satellite data collection systems to agriculture can include the following areas :

4.3.1 Production of crops

Crop growth in an integration of daily atmospheric and phytological events. Conditions that prevent growth include disease, insects, drought, severe storms (floods, hail, wind, frost), low fertility, weeds, salinity, air pollution...
Some of these phenomena can be measured by in situ data collection observations. For example, temperature, relative humidity and duration of free moisture can be used in models to predict damage by disease and insects.
Soil moisture and incident crop response to

various moisture levels are important parts of any predictive growth model. Evapotranspiration models use much of the data which can be obtained via DCS (initial soil moisture, temperature, solar radiation...)
As an application of an evapotranspiration model, the advisability of supplemental crop irrigation to maximise yields can be determined on a short time basis.

Crop yield prediction models also include inputs (meteorological variables...) which can be given by DCP on a near real time basis.
DCS in situ measurements for agronomic production can be :

| Planting date | - soil temperature (0-5 cm depth) |
| | - Soil moisture (0-30 cm depth) |

Daily growth	- Air temperature (maximum-minimum)
	- Relative humidity
	- Soil moisture vertical profile
	- Solar radiation

Disease	- Temperature (maximum-minimum)
	- Relative humidity
	- Hours of free moisture

Insects	- Temperature (maximum-minimum)
	- Hours of free moisture
	- Soil moisture (surface)

4.3.2 Forest resources

DCS can make contributions to operational monitoring of forests by giving data from remote sites in near real time.

4.3.2.1 Forest fire

DCS can improve efficiency and accuracy of forecast data by giving information from sites not currently covered and quicker access to the data.
Fire measurements can be :
 - Wind speed and direction
 - Temperature
 - Relative humidity
 - Radiation

4.3.2.2 Forest hydrology and snow monitoring

DCS measurements in the high elevation forests or inaccessible sites can give data to be used for water release measurements from the snow-pack.
Measurements can provide data :
 - on water quality (effect of a forest on water quality)
 Measurements : organic nitrogen, oxygen, temperature fluxes.
 - on pest management
 Measurements : temperature, humidity, precipitation, net radiation, soil moisture.

5 GOES APPLICATION EXAMPLE

5.1 a) the three DCP's allocated to investigators at Michigan State University through NASA-AMES will be utilized in an environmental monitoring system of surface parameters to be integrated into pest management control system. This system is experimental. The DCP phase of this system is funded by NASA-AMES.
b) the principal investigator for implementing the DCS program at MSU is :
Dr. Robert Boling Jr., Department of Electrical Engineering and System Science, Michigan State University, East Lansing, MI 48824, telephone :(517) 353-6490.
c) data will be collected by the DCS beginning approximately Nov.10,1976, to continue for approximately two years. The data, collected at three sites sites distributed about the lower peninsula of Michigan will be utilized to interpolate weather conditions (primarly soil temperatures) across the prime grain crop growing regions.
d) During this experimental phase (the first eight months of operations),data is not required immediatly, but rather will be used to calibrate the interpolation equation and to verify the instrumentation system. Accordingly, initial data perishability will be approximately two weeks.
e) the final user of the data will be an NSF funded research program conducted through the Entomology Dept. of Michigan State University.
5.2 The system will be self-timed.

5.3 There will be three DCP's, all self-timed, none having the emergency alarm provision. The first DCP will be deployed November 10,1976 ; the other two DCP's will be deployed approximately one month later.

5.4 The initial deployment of a DCP will be at East Lansing, Michigan, co-ordinates 42428428, for testing and initialization purposes. Shortly thereafter, that platform will be moved to the vicinity of Gull Lake, Michigan (42248524). The other two stations will be initially installed in the vicinities of Sandusky, Michigan (43258250) and

Gladwin, Michigan (45398430). All sites
will be fixed.

5.5 Each DCP will sample six analog sensors
yielding an eight bit number per sensor.
The sensors will be scanned at twelve minute intervals. There will be 15 scans incorporated per transmission, for a total
data set of 720 bits. The format of the
scanned data frame will be : the first five
eight-bit numbers will correspond to temperature measurements and the sixth-number
will monitor battery condition. This six-
input scan frame will be repeated fifteen
times per transmission. Each sensor message will be expressed by an eight-bit word.

5.6 Reporting times will be on three hour
intervals on the self-times schedule in
accordance with the NOAA-NESS published
authorization schedule.

5.7 The data is to be supplied on magnetic
tape, and on 1200 bauds dial up to be delivered to :
Dr. Dean Haynes, Department of Entomology
Michigan State University, East Lansing
MI 48824.
Data is required during the experimental
phase (i.e. the period from November 15
to March 15) approximately once per week.

5.8 Commercial services cannot meet our
program needs for the following reasons :
  1) the DCP's will be eventually deployed
in areas inaccessible to commercial telephone lines,
  2) microwave service cost is prohibitive
  3) Conventional data loggers, recording
on cassette tape, or similar storage devices have far too long lag times relative
to data perishability in the operating
system.
  4) an earlier study has shown that for
the eventual planned monitoring system,
satellite communication is by far the most
economical communication service.

6 CONCLUSION

Agriculture involves a spatial domaine that
is enormous in size. The number of platforms
required should be very large to supply
areas of intensive agriculture, rangeland
or forestry.

To-day, utilizations of satellite DCP systems in agriculture are limited (compared
to remote sensing techniques or to the use
of DCPs by people of other disciplines -
hydrology for example-).

Satellite DCP systems, however, can be a
good and cost competitive solution to provide data for agriculture :
  - when there is a real lack in the existing meteorological network (remote areas)
  - to provide ground truth and complement
satellite remote sensing measurements
  - when an interdisciplinary platform can
be operated, i.e. when people of two or
several disciplines agree on the location
and the common use of a platform (example
agriculture, hydrology, meteorology)
  - for a time limited experiment (mobili-
ty of a platform).

7 REFERENCES

- Satellite Data Collection User Require-
ments Workshop, Final Report June 1975,
Edward A. Wolff, Charles E. Cote, J. Earle
Painter.

- Satellite Data Collection Newsletter,
edited by Dr. Enrico P. Mercanti, Code 952,
NASA-GSFC, Greenbelt Md - USA.

- GOES DCP User's Guide NOAA/NESS (November 1976).

- METEOSAT DCP User's Guide ESA - MPO
(March 1977).

- ARGOS DCP User's Guide - Service ARGOS,
CNES 18 av. E. Belin, 31055 TOULOUSE CEDEX

- IInd ARGOS User's Meeting, November 2-3,
1977.

- New Ventures in Satellite Communications,
Comsat General Corp. 950 l'Enfant Plaza,
S.W. Washington DC 20024

- A European Yield Forecasting Agromet System , (A proposal to use Meteosat and Landsat 3 data in correlation with a Meteorological DCP network), 1977, Georges Fraysse,
Joint Research, European Communities Center,
Ispra, Italie.

- Le Système ARGOS utilisé en télédétection
comme aide à la collecte de données (vérité
terrain), Novembre 1977, Laidet (GDTA),
Toulouse, France.

- J.D. Oberholtzer, NASA, Wallops flight
Center, Wallops Island, Virginia 23337, USA
(Agrometeorological network over Telephone
lines - soil temperature - air temperature-
wind speed and direction solar intensity
and precipitation).

- GOES Application example.

- Florida Fish & Wildlife Office USGS -
Corps of Engineers, US. Forest Service,
A DCP network (20 platforms) and ERTS imagery are being used to monitor and make
environmental assessments and to manage the
water resources of 3600 sq kilometers in the
Everglades. Proved particularly useful during the 1973-74 winter spring drought
where damage was minimized by planning based upon accurate knowledge of the water
stored in the various lakes, canals, and
conservation areas. Soil moisture sensors
on the DCP's are also being used to warn
of potential fire hazards.
David Cox (305) 724-1571,
Edward Vosatka, Florida Game and Fresh
Water Fish commission, 7630 Coral Drive,
West Melbourne, Florida 32901.

R. BRYAN ERB
National Aeronautics and Space Administration
Lyndon B. Johnson Space Center, Houston, Texas, USA

19

The large area crop inventory experiment (LACIE)
Methodology for area, yield and production estimation
Results and perspectives

1. INTRODUCTION

1.1 The Agro-Economic Situation

Mankind is becoming increasingly
aware of the need to manage better the
resources of planet Earth - its atmosphere,
vegetation, oceans, fresh water, soils,
minerals and petroleum supplies. As the
world's populations increase and a higher
standard of living is sought for all, more
careful planning is required to make
effective use of these resources to produce adequate food supplies. Agricultural
production is highly dynamic and depends
on complicated interactions of prices,
weather, soils, and technology over the
planet. Each day the outlook changes as
these ingredients are altered due either
to natural events or as a result of man's
decisions. Wheat, the most important
internationally-traded crop, is being
planted, harvested, and grown throughout
the year in various regions of the world.
Much wheat is grown in semi-arid conditions
with marginal weather and cultivated with
a wide range of technology levels and thus
production is subject to extreme variations.
The world's wheat supply has fluctuated
from the critical deficiencies of the 1972,
1974 years to the apparent oversupplies of
the current period. Both of these conditions have had severe economic impact.

1.2 Current Inventory Information and the Need for Improvement

Current world food supply estimates
are a compilation of estimates generated
for the most part by the various national
agricultural information systems. The
quality of the information systems in the
various countries and the estimates range
from timely and reliable to virtually

non-existent. Frequently, estimates are
based on past trends, sometimes adjusted
by subjective judgment, rather than on
objective information. The primary
properties of an effective, world agricultural information system are objectivity,
reliability, timeliness, adequacy in terms
of coverage, and efficiency.

In the light of the manifest need for the
nations of the world to manage better the
planet's agricultural production, improved
information throughout the year on crop
prospects for the important producing
regions is extremely critical. This need,
coupled with the state of development of
remote sensing, brought into focus the
feasibility of applying remote sensing,
together with related technology, to the
task of developing a technology for global
agricultural monitoring.

1.3 The Components of Production

Wheat production is the product of area
and yield. Figure 1 shows wheat production
variability for the years 1965 to 1975 for
the U.S. Both yield and area have fluctuated significantly from year to year.
Weather is extremely variable over a geographic region and to get an acceptably
accurate forecast of production, it is
critical to associate the right weather
with the actual area being affected. Where
the effects are so severe as to remove area
from production, this reduced area must be
estimated. The estimate of yield at a
country level is directly dependent upon
the geographic distribution of area actually planted and then harvested. Therefore,
not only must a survey system monitor the
total area harvested, but it must also monitor the geographic distribution of the area
harvested as well as the associated

geographic pattern of weather and other growing conditions.

U.S.A. TOTAL WHEAT PRODUCTION COMPONENTS VARIABILITY

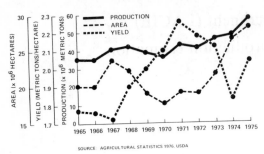

SOURCE   AGRICULTURAL STATISTICS 1976, USDA

Figure 1.— Variability in Area, Yield, and Production in the U.S.A.

1.4   The Background of LACIE

The roots for LACIE were intentionally and carefully established in 1960 by the Agricultural Board of the National Research Council in the U.S. when experiments were conceived to examine the feasibility of using multispectral remote sensing for agricultural crop monitoring. An organized research program was established in 1965 by the U.S. Department of Agriculture (USDA) and the National Aeronautics and Space Administration (NASA). This program led, in an orderly fashion, from the first successful computer recognition of wheat in 1966 using multispectral measurements collected with aircraft to follow-on development and testing of satellite capability in 1972. The success of several feasibility investigations in 1972-73 (Erb, 1974) conducted with the Earth Resources Technology Satellite, then known as ERTS 1 and later renamed Landsat 1, led to the design and initiation of LACIE in 1973-74. LACIE was a logical next step in the chain of research and development activities and was designed to test on a regional or national basis the technology developed over the previous decade, and to establish the technical feasibility of a global agricultural monitoring system.

1.5   The Roles of the Agencies Participating in LACIE

Each of the three agencies of the U.S. Government (the National Aeronautics and Space Administration (NASA), the U.S.

Department of Agriculture (USDA) and the National Oceanic and Atmospheric Administration (NOAA) of the U.S. Department of Commerce) that conducted LACIE brought particular expertise to the experiment. NASA was responsible for the overall technical design and management of the experiment, for the acquisition of Landsat data for the area analysis and for all data handling and logistics. NOAA was responsible for the development and operation of the yield models and weather summaries and for the acquisition and handling of meteorological data. The USDA was responsible for acquisition of historical agricultural data, for the acquisition of current-year ground data for accuracy assessment and for the compilation of the production reports. There was substantial involvement of a number of research establishments at universities and in industry and the major support contractor for NASA, the Lockheed Electronics Company, was responsible for much of the implementation of the experiment.

2.   THE DESIGN OF THE EXPERIMENT

LACIE was initiated in 1974 to demonstrate the forerunner of an operational global wheat monitoring system. The experiment objectives were:

o To demonstrate an economically important application of repetitive multispectral remote sensing from space.

o To test the capability of the Landsat together with climatological, meteorological and conventional data sources to estimate the production of an important world crop – wheat.

o Commencing in 1975, validate technology which could provide timely estimates of crop (wheat) production.

Performance goals, based on both an analysis of the capabilities of existing conventional survey systems and a projection of future needs, were established for the experiment to be utilized as evaluation criteria. These included:

o An accuracy goal for estimates at harvest to be within ±10% of true country production 90% of the time (referred to as the 90/90 criterion). An additional goal was to establish the accuracy of these estimates (made on a monthly basis from early season through harvest period) prior to harvest.

286

o A timeliness goal to demonstrate that the Landsat data could be reduced to acreage information within 14 days after acquisision in an operational environment.

o All estimates to be based on objective and repeatable procedures and not adjusted within the experiment utilizing outside information sources.

The LACIE was focused on monitoring production in selected, major wheat-producing regions of the world. The experiment extended over three global crop seasons and was designed for expansion up to eight regions (Figure 2). The early phases of the experiment concentrated primarily on a "yardstick" wheat-growing region of the U.S., the nine-state, wheat region in the U.S. Great Plains (USGP), where current information relative to wheat production and the components of production were available to permit quantitative evaluation of the technology. As the experiment progressed, a combination of programmatic policy decisions, availability of resources, and the LACIE experimental design permitted an orderly expansion of the initial scope to include the monitoring of wheat production in two additional major producing regions, (Canada and the USSR). This expansion included exploratory studies for monitoring wheat production in five other major-producing regions (India, Peoples Republic of China, Australia, Argentina and Brazil). In addition, at the end of Phase I, key USDA management decisions resulted in the incorporation of a USDA-User System within the USDA-LACIE effort.

The experiment extended over three overlapping global crop seasons, each of which was considered an experiment phase. Phase I of LACIE, global crop year 1974-75, focused on the integration and implementation of technology components into a system to estimate the proportion of wheat in selected study segments within the major producing region and the development and feasibility testing of yield and production estimation systems. An end-of-season report for area estimates of wheat/small grains in the U.S. Great Plains was generated.

In Phase II, global crop year 1975-76, the technology, as modified during Phase I, was evaluated for monitoring wheat production for the U.S. Great Plains and Canada, and "indicator regions" in the USSR. Monthly reports of area, yield, and production of wheat for these three major-producing regions were generated. A substantial level of effort was expended to identify significant problem areas and to incorporate recommended technology components into the LACIE analysis systems for use during Phase III.

During Phase III, global crop year 1976-77, new technology, developed during Phase II, was implemented and evaluated for monitoring wheat production for the U.S. Great Plains and the USSR. Monthly reports of

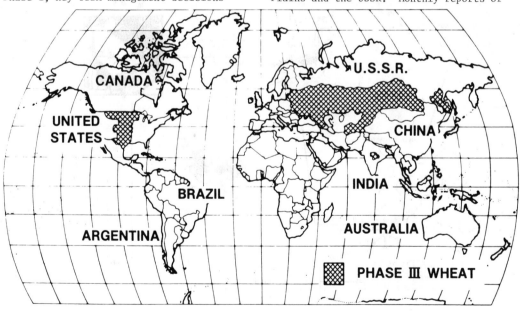

Figure 2.— LACIE Study Areas.

area, yield, and production estimates of wheat for these major producing regions were generated.

3. THE LACIE TECHNICAL APPROACH

The technical approach (Figure 3) to the LACIE was to estimate production of wheat on a region-by-region basis where production is the product of area and yield. Area was derived by classification and mensuration of Landsat multispectral scanner (MSS) data and yield estimates were obtained from statistical regression models which relate wheat yield to local meteorological conditions, notably precipitation and temperature. The integrating factor for the area and yield estimates was a sampling and aggregation strategy which efficiently allocated sample segments (5x6 nautical mile) to be acquired by Landsat and analyzed for wheat percentage, defined the strata boundaries for the wheat yield models, and formulated the upward expansion (aggregation) of the area and yield estimates to regional and country estimates of production. These aggregations resulted in experimental commodity reports of wheat area, yield and production for user evaluation and accuracy assessment. The performance evaluations provided the mechanism both for verifying where the LACIE technology was performing adequately and for isolating and identifying problems.

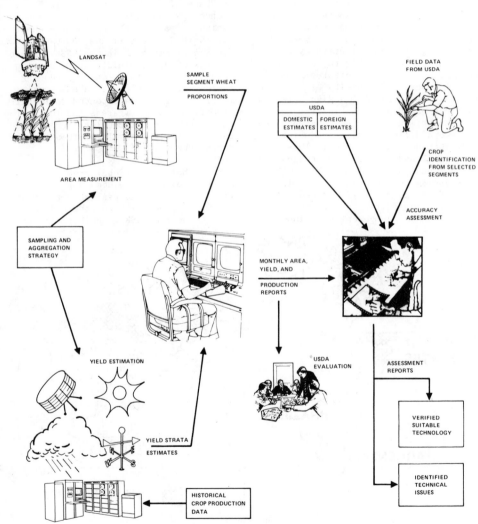

Figure 3.— LACIE Technical Approach.

288

3.1 Landsat Data Acquisition

The initiation of Landsat data acquisition (Figure 4) is at the Johnson Space Center (JSC) where the sampling strategy defined the locations of the segments to be acquired. The Landsat acquisition information was transmitted via existing Apollo communication lines to the Goddard Space Flight Center (GSFC) which commanded the Landsat for multispectral scanner acquisition each 18 days during the crop season. Data was, for the majority of LACIE, transmitted to ground receiving stations at Maryland, Alaska, or California either in real-time or by use of the on-board tape recorders. During the latter parts of LACIE, ground stations in Italy and Pakistan were utilized to conserve the on-board recorders. Data from the ground stations were shipped to the GSFC where the Landsat preprocessing was performed. The data was screened for cloud cover, registered to previous acquisitions, and the sample segment data extracted and transmitted in digital computer compatible format to JSC where it was entered into an electronic data base. In addition, electronically regenerated full-frame (100 n.m. x 100 n.m.) film in 70mm black-and-white format for each MSS band was shipped to

the USDA Aerial Photography Field Office in Salt Lake City which converted it to 9-inch color infrared (IR) film composites and shipped them to JSC. The 9-inch composites were prepared four times per crop season.

3.2 Analysis for Area Estimation

The analysis of the Landsat data was performed at the JSC (Figure 5) where procedures were designed and personnel were trained to perform computer-oriented crop identification and mensuration without the availability of ground truth. The analysis was basically a four-step process. In the first step, the Landsat and ancillary data was prepared and assembled so that a trained analyst could perform crop identification. The assembled Landsat data products included full-frame color IR film, segment level color IR film products, and graphical plots of MSS response. Ancillary data included historical agronomic practices, crop growth stage information based on historical data and current year weather and summaries of the meteorological conditions for the current crop year. The second step was the labeling by the analyst, based on established procedures of a small percentage of the segment data elements

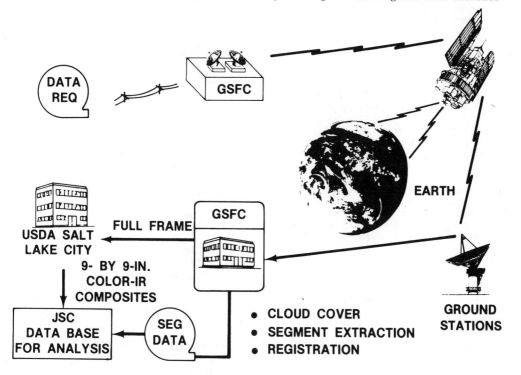

Figure 4.— Landsat Data Acquisition for Area.

289

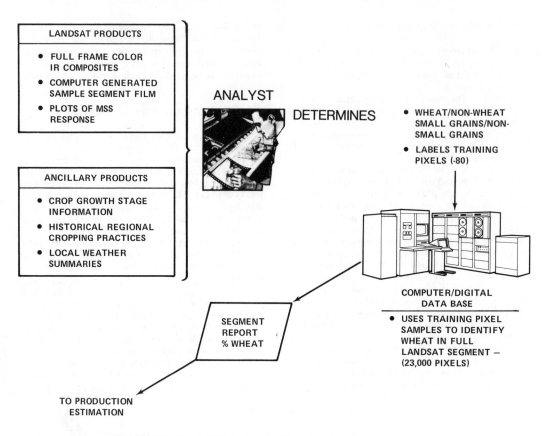

LANDSAT PRODUCTS

- FULL FRAME COLOR IR COMPOSITES
- COMPUTER GENERATED SAMPLE SEGMENT FILM
- PLOTS OF MSS RESPONSE

ANCILLARY PRODUCTS

- CROP GROWTH STAGE INFORMATION
- HISTORICAL REGIONAL CROPPING PRACTICES
- LOCAL WEATHER SUMMARIES

ANALYST

DETERMINES

- WHEAT/NON-WHEAT SMALL GRAINS/NON-SMALL GRAINS
- LABELS TRAINING PIXELS (-80)

COMPUTER/DIGITAL DATA BASE

- USES TRAINING PIXEL SAMPLES TO IDENTIFY WHEAT IN FULL LANDSAT SEGMENT — (23,000 PIXELS)

SEGMENT REPORT % WHEAT

TO PRODUCTION ESTIMATION

Figure 5.— LACIE Analysis for Wheat Area Determination.

(pixels) as being either wheat or non-wheat or small grains or non-small grains. This labeling was strongly based on the variability (Figure 6) in the multitemporal (over time) crop appearance of ground cover types afforded by the sequential Landsat coverage. In the third step, the analyst labels were used in a computer to train a multivariate pattern recognition algorithm to identify wheat or non-wheat for all the data elements (approximately 23,000 pixels) of the Landsat segment, and to tabulate the results as a percentage of wheat for the segment. The final step was the evaluation by the analyst of the result as acceptable before submitting the result for wheat production estimation. It should be noted that early attempts by the analysts to discriminate between wheat and other small grains such as barley were not generally successful and labeling was primarily either small grains or non-small grains. Historically derived ratios were then applied to the resultant segment level estimates of small grains to estimate

wheat. A procedure for analyst discrimination between spring small grains based on subtle differences in crop growth stages was tested in North Dakota during LACIE Phase III and shows promise.

3.3 Meteorological Data Acquisition

The overall implementation and operation of the applications involving meteorological data were under the direction of NOAA's Center for Climatic and Environmental Assessment (CCEA). This included global meteorological data acquisition for use in wheat yield models, in wheat growth stage models (crop calendars), and in the weather summaries used by the area estimation analysts. In Washington, DC, weather data was routinely acquired through the World Meteorological Organization's (WMO) Global Telecommunications System and was augmented by foreign data from the U.S. Air Force's Environmental Technical Applications Center (ETAC) and domestic data from the National Weather Service (NWS), the Federal Aviation

290

WINTER WHEAT
GROWTH STAGES

| Seedbed Prep and Planting | Emergence | --------------- Dormant --------------- | Tillering and Jointing | Booting and Head Emergence | Ripening and Maturity | Harvesting |

| SEP | OCT | NOV | DEC | JAN | FEB | MAR | APR | MAY | JUN | JUL | AUG |

PHOTOGRAPHS

LANDSAT IMAGES

W-WINTER WHEAT, P-PASTURE

Figure 6.— Variability in the Appearance of Wheat.

Agency (FAA), and by imagery of cloud cover and type acquired by the National Environmental Satellite Service. Preprocessing of this data for the project was assisted through the NOAA Center for Experimental Design and Data Analysis. This primarily involved preparation of temperature and precipitation at individual meteorological stations and representative values over the yield model strata. This data was transmitted to the computers of the National Meteorological Center (NMC) in Suitland, Maryland, (Figure 7).

3.4 Yield Estimation

The wheat yield models utilized in LACIE were statistical regression models based upon recorded historical wheat yields and weather. These regression models forecast wheat yield for fairly broad geographic regions (yield strata) using calendar monthly values of average temperature and cumulative precipitation over the strata, thereby providing monthly updated yield estimates during the growing season. Figure 8 illustrates the factors which influence wheat yields. Along with the required meteorological data, the yield models for each of the model strata were stored on the NMC computers. Operation of

the yield models was under the control of the NOAA-CCEA Modeling Division at Columbia Missouri, (Figure 7). After the yield estimates were generated, they were transmitted to the NASA-JSC for input to the wheat production estimation.

3.5 Crop Calendar Models

Models which estimated the current year's growth stage for wheat utilizing meteorological data as input were also implemented on the NMC computers and under the operational control of the NOAA, Columbia, MO, personnel. These models utilized daily values of meteorological data and were run on a biweekly basis for selected meteorological stations in the regions of interest. At JSC, the crop calendar model results were input to a program which interpolated to define a wheat growth stage at the location of the sample segments at the times of Landsat acquisition for utilization by the analysts performing the crop identification and labeling.

291

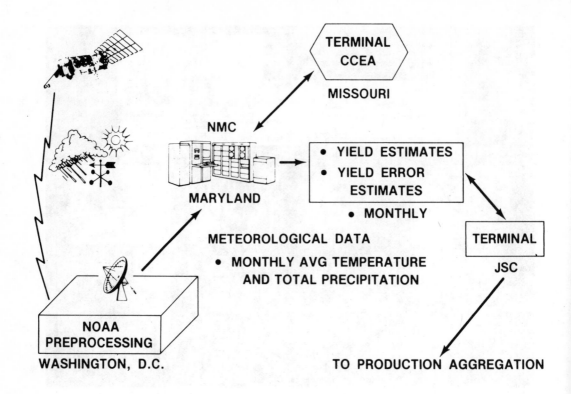

Figure 7.— Meteorological Data Acquisition and Data Flow for Wheat Field Estimation.

MAJOR FACTORS:	SOILS	TECHNOLOGY	WEATHER
	DIFFERS BY REGION, AND COUNTRY	-WHEAT VARIETIES, -IRRIGATION -CROPPING PRACTICE -FERTILIZERS	-SOIL MOISTURE, TEMPERATURE
VARIABILITY	STABLE	CHANGES SLOWLY	FLUCTUATES DRASTICALLY

Figure 8.— Sources of Variability in Wheat Yield.

3.6 Production Estimation

The wheat production estimation process (Figure 9) involves the upward expansion (aggregation) of the segment level wheat percentages to the yield strata regions where the aggregated area estimates and yield model estimates were multiplied to provide estimates of production. Estimates of production for larger regions are the sum of the appropriate strata level production estimates. The statistical sampling approaches on which the production estimation procedure was designed allow country level production accuracies to be within a few percent while requiring analysis of only 2 to 5% of the total area using the Landsat 5x6 nautical mile sampling segments. Confidence limits on the area, yield, and production estimates were also estimated.

3.7 Accuracy Assessment

The LACIE accuracy assessment effort (Figure 10) was designed to determine the accuracy of the LACIE area, yield, and production results. This assessment was performed both at the large area level (i.e., state, region, country) and at the detailed level (i.e., segment, yield model and lower) in order to isolate problem areas and identify factors to be addressed for potential resolution. Although comparison to USDA and foreign country estimates were made, the primary assessments were made over the USGP "yardstick" region where reliable USDA estimates are available at the state and higher levels, and where collection programs provided information down to the field level for detailed evaluations. This field level data was acquired during Phase II and III for accuracy assessment sample segments representing approximately one-third of the total USGP sample segments. Field data for some selected Canadian segments were also provided. From accuracy assessment results, LACIE was able to identify the sources of error and prioritize issues for further research, as well as to verify procedures and approaches used.

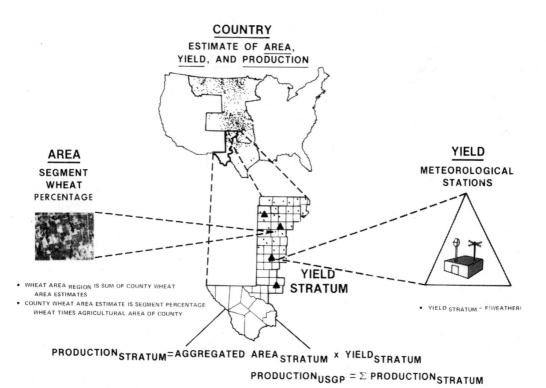

COUNTRY
ESTIMATE OF AREA, YIELD, AND PRODUCTION

AREA
SEGMENT
WHEAT
PERCENTAGE

YIELD
METEOROLOGICAL
STATIONS

• WHEAT AREA REGION IS SUM OF COUNTY WHEAT AREA ESTIMATES
• COUNTY WHEAT AREA ESTIMATE IS SEGMENT PERCENTAGE WHEAT TIMES AGRICULTURAL AREA OF COUNTY

YIELD STRATUM

• YIELD STRATUM = F(WEATHER)

$$PRODUCTION_{STRATUM} = AGGREGATED\ AREA_{STRATUM} \times YIELD_{STRATUM}$$

$$PRODUCTION_{USGP} = \Sigma\ PRODUCTION_{STRATUM}$$

Figure 9.— Production Estimation Procedures

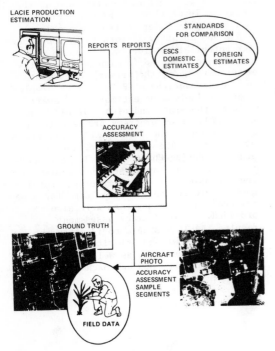

LACIE PRODUCTION
ESTIMATION

STANDARDS FOR COMPARISON

REPORTS REPORTS

ESCS DOMESTIC ESTIMATES

FOREIGN ESTIMATES

ACCURACY ASSESSMENT

GROUND TRUTH

AIRCRAFT PHOTO

ACCURACY ASSESSMENT SAMPLE SEGMENTS

FIELD DATA

Figure 10.— LACIE Accuracy Assessment.

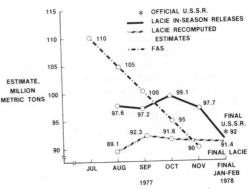

Figure 11.— Phase III USSR Wheat Production.

4. RESULTS

4.1 Accuracy of the Estimates

The experiment established that the technology developed for LACIE met the performance goals for wheat production inventory in important cases. Notably LACIE produced, in August 1977, what proved to be an accurate indication of the USSR spring wheat shortfall. This was well before more definitive information was released by the USSR.

The 1977 Soviet final production estimate released in January 1978 was 92 million metric tons and the LACIE final estimate was 91.4 million metric tons, a difference within 1% as shown in Figure 11. Additionally, two crop years of study in both spring and winter wheat regions of the Soviet Union resulted in estimates that support the experiment performance goals. Compared to historical information, this LACIE achievement represents a significant advance in acquiring an accurate and timely wheat production estimate in an area of great significance.

For comparison, Figure 11 shows USDA projections and LACIE initial and recomputed results. The recomputation involved a simulation of what the LACIE results could have been in a truly operational situation with timely (30 day delay) analyses. These results are extremely encouraging, indicating that USSR results could be within 3% in August, about one and one half months prior to harvest.

The accurate performance of the LACIE estimate in the USSR situation was validated by more intensive evaluation in the U.S. yardstick area. Phase III results in this region support a conclusion that the technical modifications incorporated into the experiment had indeed led to significant improvement from previous Phase II technology in the results from the analysis of Landsat data. The production estimates for the region are compared throughout the season to the "true value" as represented by the USDA Statistical Reporting Service. The LACIE estimates marginally met the 90/90 accuracy goal at harvest and even achieved this one and one half to two months prior to harvest. The results of the area and yield components for the region are shown in Figure 12. It can be noted that, on the average, the acreage estimates were quite good while the yield forecasts tended to be under those of the Statistical Reporting Service. The models were developed with data for the 45 years prior to each of the test years and, when tested on 10 years of historic data, were supportive of the 90/90 accuracy goal. An analysis of the yield model behavior indicates that they generally perform adequately if no significant changes in trend occur and if the average weather conditions for a region are not drastically different from the historic data used in their

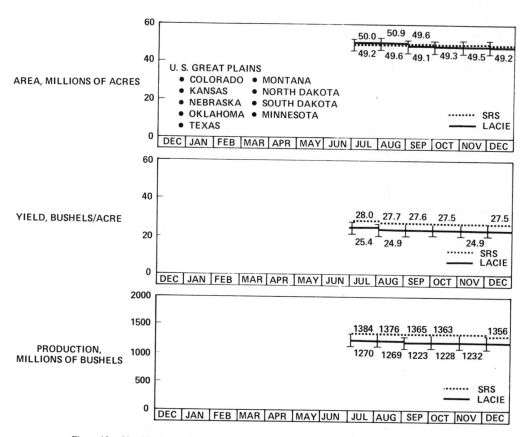

Figure 12.— Monthly Comparisons of LACIE and SRS Estimates, U.S. Great Plains, Phase III.

development. Where extreme departures from normal weather do occur, the models tend to respond in the right direction but do not capture the extent of the excursion. However, as could be seen in the Phase III USSR spring wheat regions these models did perform adequately in a departure from normal which, while not extreme, was of great importance to the U.S. and other countries. The Phase III results for production, area and yield in the "yardstick" winter wheat region of the U.S. generally support the results achieved in the USSR.

The results in the strip/fallow areas of spring wheat regions of the U.S. exhibited a tendency to underestimate the spring small grains. Econometric ratio models, developed in Phase II and used to estimate the spring wheat from the LACIE estimates of small grains, worked well for the region. As indicated above, the yield models tended to underestimate the expected values of the yields at harvest. The area estimates were less than 1% under as

compared to the 10.7% underage experienced in Phase I and the 14% underage of Phase II. Figure 13 displays the results for Phase III spring wheat. If the major differences between the spring wheat regions of the yardstick area and the USSR are taken into consideration, the yardstick results are supportive of what was observed in the USSR results in Phases II and III. That is there is nothing inherently difficult about spring wheat and it can be estimated accurately under the right conditions.

In general, if the yield models had performed as they did in Phases I and II, and on the average in the 10 year test, the 90/90 accuracy goal would have been exceeded in the yardstick region. It is also concluded that in regions where the minimum field dimension tends to be similar to the Landsat spatial resolution, the estimates tend to be low. More recent results are indicating that spring wheat can be differentiated from spring barley during the wheat soft dough stage.

Considerably more research will be required to accomplish this reliably. However, LACIE investigators are optimistic that with Landsat D considerable improvement will be possible in these more difficult regions.

As an example of the progress achieved in obtaining improved wheat estimates,

Figure 14 compares the LACIE segment wheat proportion estimates with ground truth for Phases II and III. This data indicates a significant improvement in the proportion estimates derived from Landsat using the Phase III procedures and supports the improved aggregated results previously described for the total region.

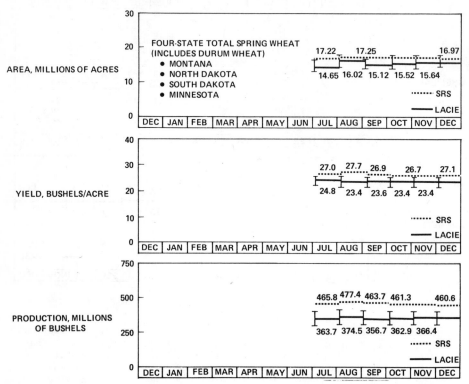

Figure 13.— Monthly Comparisons of LACIE and SRS Estimates, U.S. Spring Wheat, Phase III.

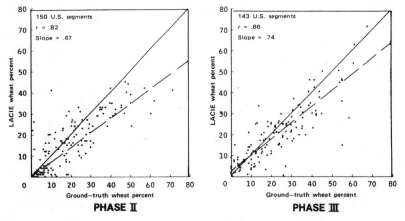

Figure 14.— Comparison at Phase II and Phase III U.S. Estimates with Ground Truth.

4.2 Transfer of the Technology

A decision was made by USDA early in 1976 to initiate an additional activity to develop a data analysis system to transfer and exploit the emerging LACIE technology for USDA use. This prototype was approved in January of 1976 to serve as the vehicle for the transfer of technology from applied research to an application test within USDA.

The goal of this activity was to develop the basic analytical capabilities, hardware and software to support the testing and evaluation for USDA use of the technology developed during LACIE. This USDA-led effort within the LACIE involved the active participation by NASA and NOAA in providing assistance in the transfer of technology from LACIE to the USDA user system.

5. CONCLUSIONS AND OUTLOOK

LACIE was a carefully conducted experiment designed to research, develop, apply and evaluate a technology to monitor wheat production in important regions throughout the world. LACIE utilized quantitative multispectral data collected by Landsat in concert with current weather data and historical information. The experiment exploited high-speed digital computer processing of data and mathematical models to extract information in a timely and objective manner.

The totality of results from the three crop years of focused experimentation strongly indicated that:

o The current technology can successfully monitor wheat production in regions having similar characteristics to those of the USSR wheat areas and the U.S. hard red winter wheat area.

o With additional applied research, significant improvements in capabilities to monitor wheat in these and other important production regions can be expected in the near future.

o The remote sensing and weather-effects modeling approach followed in LACIE is generally applicable to other major crops and producing regions of the world.

Note: Between the delivery of these lectures at the Ispra Advanced Seminar on Applications of Remote Sensing in

Agriculture and Hydrology in November/December 1977, and the preparation of this paper for publication, some additional analysis of results was conducted and, in January of 1978, the official USSR production figures were released. Accordingly, these final Phase III results were included for the sake of completeness. Further, extensive documentation of the experiment was accomplished and papers thereon presented at a LACIE Symposium. References to the symposium documentation are included for the reader who may wish more detail.

6. REFERENCES

Erb, R. Bryan, Nov. 1974, ERTS-1 Agricultural Analysis, Vol. I of a series of analyses on the utility of ERTS-1 Data for Land Resources Management, NASA TMX 58117.

7. BIBLIOGRAPHY

LACIE: Independent Peer Evaluation of the Large Area Crop Inventory Experiment, Oct. 1978, The LACIE Symposium, JSC-14550

LACIE: Proceedings of Plenary Session, Oct. 1978, The LACIE Symposium, JSC-14551

Large Area Crop Inventory Experiment (LACIE) Executive Summary, August 15, 1978, JSC-13749

Large Area Crop Inventory Experiment (LACIE) Symposium Proceedings - To be published in 1979.

R. A. PACHECO
Food and Agriculture Organization of the United Nations, Rome, Italy

20

Applications of remote sensing to agriculture development in tropical countries

1 INTRODUCTION

The rapidly expanding global population and the increasing world demand for food places strong stress on the limited land, mineral, vegetation and water resources of the world. Industrialization during the past decades has added to the stress on those resources and the environment. Never before in human history has the need been greater for sound planning and management of these resources and for environmental protection. One of the vital elements of any effort to meet these needs is the capability to make inventories of the renewable and non-renewable resources of the earth and to monitor environmental changes.

The primary basis for the economic development of most developing nations, particularly those within the tropical belt, lies in their natural resources. Yet these nations, with few exceptions, do not have thorough enough knowledge about the nature, quantity and location of their resources, to harness them effectively for the welfare and progress of their people. Most tropical countries have yet to determine the extent and condition of their arable land, forest, rangeland and water resources and to identify promising areas for mineral exploration. One of the crucial elements, at the present time of world food and energy shortages and of spreading environmental deterioration, is the need to monitor the changing condition of their natural domain – to forecast crop yields, to detect land degradation and pollution of water, to recognize alterations in land use, to give early warning and assess damage of natural disasters and to observe many other aspects of environmental change.

As we know, remote sensing offers the feasibility of monitoring agricultural resources for rapid and continuous assessment of plant, soil and water resources and interrelated problems. A successful remote sensing operations programme must be tailored to solve or manage the maximum variety and complexity of applications. This paper considers the role of remote sensing and attempts to present a brief analysis of the problems of utilization of remote sensing data. In what follows the case studies are mostly taken from the developing, mainly tropical, countries with which FAO is principally concerned as an organization. There are a wide range of data collection systems and remote sensors available which no doubt have been described in other lectures. We will deal with some of them that have been used in developing countries.

2 FROM AERIAL PHOTOGRAPHY TO ORBITAL MULTISPECTRAL SENSING

2.1 Aerial photography

FAO's field programmes have made use of remote sensing techniques for over twenty-five years. Over one hundred of the technical assistance projects currently being operated by FAO involve the use of aerial photographs, at scales ranging from 1:10,000 to 75,000 on a range of black-and-white and colour films. Photography is employed extensively for land inventory, land use, soil and water resources surveys, vegetation, forestry, etc.

At the present time aerial photography is by far the most widely used remote sensing technique. It can provide a basis for height information, planimetry, size and shape.

The direct application of remote sensing to food production depends on assessing from imagery the crop areas and the state and condition of crops and pasture. The aerial photographic techniques used now-a-days have been developed and proved technically and economically sound over several decades. Careful choice of the film-filter combination, season of photography in relation to plant growth and photographic resolution (expressed in scale in terms of the focal length of the camera lens and flying height) enables crops including cereals to be identified, their areas measured (Howard, 1976), their condition and yields assessed (e.g. Howard and Price, 1973); insect infestations located (Spurr, 1960, Wenderoth et al, 1975) and the presence, absence and development of disease determined (e.g. Brenchley and Dadd, 1962). Frequently, photographic scales between about 1:8,000 and 1:20,000 are used, (i.e. resolution in practice better than about $1\frac{1}{2}$ metres). Much smaller scales are satisfactory for land use and land capability mapping (e.g. 1: 60,000).

The major advantages of conventional aerial photography are the high resolution; the wide choice of methods, instruments and trained personnel. The major disadvantages of aerial photos applied to agriculture result from uneconomic repetitive coverage (within a year); non-uniform and uncalibrated intensity measurements that obstruct automatic density processing, and the relatively high cost per km^2, especially for large scale aerial surveys.

The introduction in the 1960's of reliable colour photography and, more recently, infrared colour photography, has led to new applications in natural resource surveys. High flight infrared false colour and panchromatic imagery was simultaneously flown for the FAO/UNDP Land Resources Survey Project, Sierra Leone in 9 days (end of 1975 and beginning of 1976). This was the first time such imagery had been specifically ordered for a nation-wide project of this type and it has subsequently proved highly suitable for both land system and vegetation/land use mapping at reconnaissance level, leading to the production of derived maps of soil and land suitability. The small scale of the photography and its generally high quality has enabled rapid mapping the two mentioned subjects. An outstanding feature of the CIR film is the ease with which cultivated and fallow crop land could be mapped. CIR permits a ready distinction to be made between vegetation and soil that is not always possible when using black-and-white photographs. A comparative study of the characteristics and costs of various types of remote sensing products is given in Table 1 (Schwaar, 1976).

2.2 Rocket photography

Rockets have been used in land-use studies in northern Argentina (Hardy, 1974). Experience in this short lived camera-capsule sensor has shown that it is operationally expensive. Difficulties in providing ground control and using the circular shaped photographs have restricted its operational use.

2.3 Orbital photography

The imagery from spacecraft platforms, i.e. Gemini, Apollo, and Skylab, owing to the scanty and irregular coverage prevents its extensive use by developing countries. Also, manned space remote sensing experiments cost considerably more than near-polar orbiting unmanned satellites. Recently the USSR Soyuz Salut satellites have been used to cover extensive areas of several countries.

2.4 Airborne multispectral scanners

Multispectral scanners (MSS) have certain advantages and disadvantages when compared with photography. There is no question that the advantages that are particularly important are having the capability to provide spectral data in wavelengths not available from photography and being able to provide precision radiometric data directly on computer compatible tape. The major disadvantages are that at current costs one hour airborne MSS data is at present about five times that of aerial photography and at the same scale and resolution is much lower. After this investment, the actual production of imagery, interpretation and/or the automatic processing has still to be executed. Therefore, it is personally envisaged that multispectral scanning will not readily become widely operational in the developing countries.

Table 1 Comparative acquisition costs of various types of remote sensing products (prepared by Schwaar 1976)

SENSOR	Original scale	Maximum enlargement	Minimum economy survey area	Cost/Km.2 at original scale
ERTS/ LANDSAT	1/3,369,000	1/60,000 to 1/125,000 (1)	n.a.	0,000068 US (2)
SLAR	1/250,000 to 1/500,000 (3)	1/50,000 to 1/25,000 (1)	250,000 Km.2	8.00 to 15.00 US
SKYLARK	1/3,000,000 to 1/1,000,000(4)	1,100,000 to 1/50,000 (1)	400,000 Km.2	Unknown
High altitude photo IRC and Colour B&W and IR B&W	1/80,00 to 1/125,000	1/15,000 to 1/20,000	50,000 Km.2	1.50–3.00 US 1.20–3.00 US
Med. scale Photo IRC and COL B&W and IR B&W	1/35,000 to 1/50,000	1/7,500 to 1/10,000	10,000 Km.2 to 15,000 Km.2(5)	7.00–15.00 US 5.00–6.50 US
Large scale photo IRC and COL B&W and IR B&W	1/7,500 to 1/15,000	1/1,500 to 1/2,000	2,000 to 2,500 Km.2(5)	15.00–29.00 US

(1) Subject to geometric patterns size and other characteristics of ground features.
(2) Excluding costs of surveys/acquisition (for costs of enlargements of various products, re: EROS Data Centre).
(3) According to SLAR system used.
(4) According to type of camera and depression angle used as well as altitude of sensor.
(5) Extremely variable owing to mobilization and demobilization costs bearing on overall photography survey.

3 SIDE-LOOKING AIRBORNE RADAR (SLAR)

Radar has not had broad application in developing countries, because of the higher cost, lower planimetric accuracy and lower resolution compared with aerial photographs. The advantages of this active system are that it can penetrate cloud-cover and, if necessary, it can be flown at night. Other advantages of radar are rapid data acquisition, penetration of smoke or haze, and real-time in-flight display of the developed film for on-the-spot examination of the imagery for quality. In circumstances where speed is essential, radar with its cloud penetrating capability, is the only current system suited for agricultural purposes in much of the tropical rain forest areas of the world, (Howard, 1976). Examples of radar missions for quick regional surveying of cloudy and inaccessible areas are Brazil, Venezuela, part of Colombia and Ecuador, Nigeria and Indonesia. In some areas radar has been used for oil and mineral exploration and for forest reconnaissance surveys.

4 THE TECHNOLOGY OF REMOTE SENSING FROM OUTER SPACE

In the long evolution of techniques for resource information gathering, the earth resource survey satellite series Landsat-1 and -2 represent a major technological advance. The new element is the combination of two outstanding innovations - the use of space platforms instead of aircraft, and the use of the multispectral sensor in place of the conventional photographic camera.

As we know satellites have the capability of providing man with repetitive imagery of the same ground areas. However, different field disciplines have markedly different time-frequency requirements.

Table 2 — Satellites suited to providing inputs to food production (Howard 1976)

Satellite	Spectrum	Resolution (metres)	Time interval between repetitive coverage of same ground area
* Landsat-1	Visible and near infrared	80 (0.4 ha)	18 days (since 1972)
** Landsat-2	Visible/near IR	80 (0.4 ha)	18 days (since 1975) (combined with Landsat-1 every nine days)
DMSP (VHRR)	Visible/thermal IR	600	12 hours
** NOAA (VHRR)	Visible/thermal IR	900	12 hours
NOAA (SR)	Visible/thermal	4,000	12 hours
Landsat-C (1977)	Visible/IR/thermal	0.2 ha, 0.4 ha	18 days
Nimbus-G (1978)	Visible 5 Narrow bands ($\pm$ 10 nm)	800	
Seasat-A (1978)	Microwave (synthetic Aperture RNDMR)	25	72, 151 days
GOES (by 1980)	Visible/therma	2,500/5,000	$\frac{1}{2}$ hour

* Used in agricultural studies. Data receiving stations are located in United States, Canada, Brazil and Italy with a range of up to 3,000 km. Satellite has telemetry capability for transmitting data from ground control platforms (DCPs). Stations are planned in Iran, Japan, Zaire and Argentina.

** Receiving stations are located in United States, Brazil, Canada, France, West Germany, Japan, China and are planned in Belgium, Italy, Sweden and the United Kingdom.

Table 3 - Some uses of remote sensing in soil studies

Country/project	Kind of project. Use made of remote sensing.

Conventional photogrammetric photography (panchromatic film)

Many	Many projects use black and white air photographs and only occasionally are they not available.

High altitude, air photography: panchromatic black and white film

Sierra Leone	Land systems mapping, land suitability mapping, present land use.

Infra-red photography

Sierra Leone	High altitude black and white and colour.

SLAR

Nigeria	Land systems, present land use.
Indonesia	Soil and land use mapping (small part of Sumatra).
Nicaragua	Sub-contract.
Columbia	Trial area.

Thermal sensing

Lebanon	Coastal zone study.

LANDSAT imagery

Sudan	Soil survey project - exploratory surveys.
Sudan	Land Development project. Exploratory and reconnaissance surveys. Drainage.
Sudan	The Sudd. Multitemporal studies.
Sudan	Savanna development project reconnaissance soil and land use surveys.
Jordan	Exploratory survey of soil regions and erosion hazards.
Indonesia	Soil survey project. Exploratory studies soils and land suitability.
Indonesia	Transmigration project. Land classification for setteement.
Philippines	Soil survey.

For example, the geologist and soil surveyor require few repeated satellite coverage of the same ground area (Howard, 1976). Amongst other natural resource managers, the hydrologist and forester will probably be satisfied by annual seasonal coverage, other than in periods of natural disasters (e.g. drought, flood, fire, biotic damage). Thus agriculturists, including agricultural economists, fishery officers and rangeland managers, are potentially the major natural resources users of repetitive satellite imagery.

It is unfortunate that cloud cover (see Figure 1 - Landsat) resolution of the imagery (see Table 2) and the global availability of ground receiving station (Figure 2) restrict the immediate wider use of satellite data (Howard, 1976a). It is also unfortunate that over a considerable time only one of the four tape recorders on the two polar orbiting Landsats was functioning, since regions without receiving stations cannot benefit continuously from the nine days coverage of Landsat-1 and Landsat-2. The launching of Landsat-C in 1977, Seasat in 1978 and the GOES programme with its five geostationary satellites (to be in orbit by 1980 to provide primarily global low resolution weather data at $\frac{1}{2}$-hour intervals) may partly alleviate the situation. It should be noted that weather satellite data receiving stations are much cheaper to establish than Landsat stations and at commercial prices Landsat imagery and tapes are relatively expensive, being many times the NASA/EROS Data Center prices. These are obviously important decision making factors in planning a satellite application system for food production.

Whilst Landsat is classed as an experimental system in the United States, Landsat data is increasingly being used operationally in developing countries (e.g. for thematic mapping at 1:250,000). It is also considered useful, as pointed out by Howard (1977), to recognize that remote sensing as an aid to food production is applied either directly or indirectly to agricultural crops or to other earth's resources which influence food production. Concerning the latter, remote sensing applications related to forestry, hydrology, geomorphology, etc. must be included in the overall application of remote sensing to aid food production. In this context, however, it needs noting that there has been a marked lack of action in applying the synoptic approach by the remote sensing of 'other resources' to the improvement of food production.

As indicated by the increasing range of published works on the applicational uses of satellite imagery (vide Manual of Remote Sensing; NASA Proceedings, 1973, 1974; Abstracts NASA Survey Symposium, 1975; Proceedings, Remote Sensing of Environment 1973, 1974; Canadian Remote Sensing Symposia), Landsat imagery is proving increasingly useful in different parts of the world in providing direct and indirect in-puts in programmes associated with food production. Temporal imagery composite, combining two or more bands in time sequences are proving useful in the study of seasonal changes occurring over large areas (e.g. pastoral savanna burning in Africa). Spectral band combinations have been found useful in showing up hitherto unknown large-scale natural patterns tn the Earth's surface, the significance of which are yet to be determined. In Thailand, Landsat-1 imagery was used to determine the rate of de-afforestation - between 1961 and 1973 the forest cover decreased from 50% to 37%; and in the Sudan the seasonal inundation of the Sudd is being examined using a temporal sequence of Landsat imagery.

Landsat imagery has certain advantages when compared with aerial photography and radar. The advantages are the synoptic view (Landsat scenes cover areas of about 185 by 185 Km.), the repetitive cover, the multispectral capabilities (four spectral bands), and the near-orthographic projection. The major limitations of Landsat data include: (1) the much lower resolution than aircraft imagery, (e.g. aircraft 3m or better); (2) the effect of cloud cover, since the data are obtained at fixed intervals in time (i.e. mid-morning), there is no chance of avoiding clouds (e.g. rain forest areas - about 55% not adequately covered, Howard 1976a); (3) the lack of steroscopy although available in research; and (4) sometimes atmospheric attenuation. Due to these constraints, studies conducted in various parts of the world show varying results on the practical utility of Landsat data. In many instances, however, where the data do not have to meet highly demanding user specifications, they may do very well, i.e. where user requirements are less stringent and where the resource data base is meager.

Because of its spatial resolution, Landsat data alone will not suffice for large-scale earth resource mapping, but is marginally acceptable for thematic mapping at scale of 250,000. Cartographically, Landsat is acceptable at 1:500,000 or smaller and can be used for up-dating maps at larger scales (e.g. 1:100,000). In these cases,

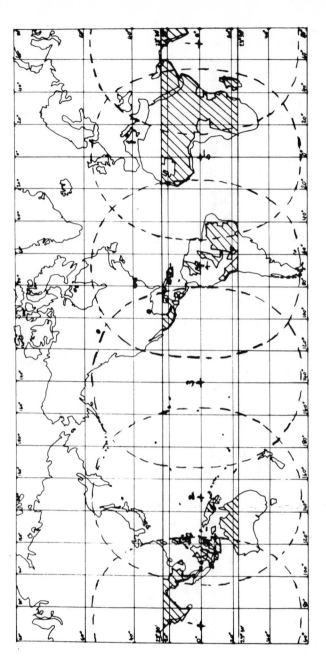

Figure 1 - Hatched areas (///) indicate the approximate ground coverage with 10% cloud cover or less of tropical countries by the polar orbiting Earth resources satellites, Landsat-1 and Landsat-2 (July 1972 to April 1975). Also shown are the locations of the operational Landsat stations (·) The broken lines indicate approximate coverage of the GOES geostationary weather satellites to be in operation before 1980 (1:USSR; 2:CMS - Japan; 3:SMS-B - USA; 4:SMS-A - USA; 5: METEOSAT - Western Eur pe). These geo-stationary satellites may have an important impact on the development of crop forecasting.

305

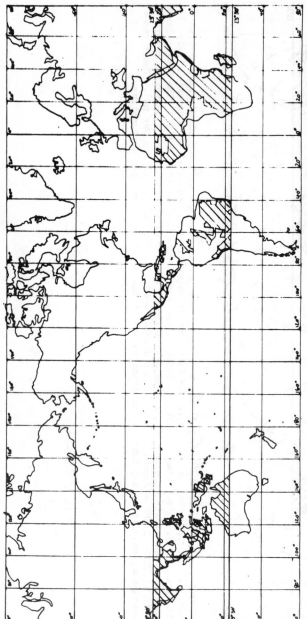

Figure 2 – Map showing the coverage of the tropics by Landsat imagery between August 1972 and April 1975 (10% cloud cover and less: ///)

Landsat's advantages are not on a scene resolution but in the all inclusive seasonal view. Furthermore, Landsat data add an important new level to multistage data collection, which includes sensing by low flying aircraft and ground observation. Field investigations to obtain ground data (i.e. so-called "ground truth") are essential for the interpretation of remote sensing data. Aerial photos remain vitally important for the acquisition of data of higher spatial resolution, including the provision of data to supplement information obtained from Landsat and to reduce sampling on the ground.

5 FAO CASE STUDIES RELEVANT TO AGRICULTURE DEVELOPMENT AND MONITORING

The following provide examples from FAO's work using satellite imagery:

(a) Tropical forest cover monitoring (Cameroon, Benin, Togo). This programme is being carried out using both Landsat and radar imagery. Initial results are highly satisfactory.

(b) Satellite remote sensing techniques for improving desert locust survey. It was found in N.W. Africa that the operational use of Landsat data for vegetation assessment may be undertaken by monitoring on a seasonal basis the adjusted and complemented by the information from the weather satellite data (Hielkema and Howard, 1976). By applying different Landsat data processing and analysis techniques (i.e. visual, analogue and digital analysis) different levels of detail, both qualitative and quantitative can be obtained on the location, size and vegetal cover of ground favourable to the desert locust.

(c) Flood monitoring. FAO has undertaken two pilot investigations using Landsat imagery. The first covered the catastrophic Indus floods in Pakistan in 1973 (Ruggles and Howard, 1974) and the second involved a multi-temporal Landsat interpretation of the flood region of the Sudd Basin in Southern Sudan (De Pauw and Spiers, 1977).

(d) Soil survey. A summary of FAO's main activities on the use of remote sensing for the soil studies is given in Table 3.

6. DISCUSSIONS AND CONCLUSIONS

The several FAO case studies are considered indicative of the level of information to be obtained from satellite imagery for operational purposes. What is obvious is that Landsat imagery can be used for

providing a range of thematic maps at a scale of about 1:250,000 or smaller. This is important, since in many developing countries there is incomplete national mapping at a constant scale, that in many cases the existing maps are well out of date and urgently require revision and that in some countries Landsat imagery will provide the first opportunity of nation-wide mapping at uniform scale. The provision of several types of thematic maps based on Landsat imagery, may be possibly further improved by the temporal study of Landsat scenes. The value of combining imagery of the same ground area but of two different dates when conditions were markedly different is well illustrated by the flood monitoring in Pakistan (see also Howard, 1976b, 1977).

In the scenario of possible future world-wide traumas, there should be increasing opportunities for regional satellite sensing along the lines of these case histories. Attention has been drawn in this paper to the application, in combination, of relatively low resolution imagery for monitoring weather and climate as affecting hydrological resources and the relatively high resolution Landsat imagery for mapping some terrestrial resources and for monitoring terrestrial change. The desert locust satellite programme, initiated by J.A.Howard, illustrates the value of using simultaneously both low resolution imagery (i.e. NOAA) and high resolution imagery (i.e. NOAA) and the use of the thermal mapping of terrestrial resources is provided by the US/Mexico screw-worm control programme in which areas of increased screw-worm activity were mapped thermally using satellite data transformations (Barnes, 1974).

At the present time, satellite sensing is still very much at the applied research stage of development as compared with aerial photography; but its operational uses are rapidly increasing. Based on the experience of the life of Landsat-1 and -2 and the launching in 1978 of Landsat-3, regular global coverage by these satellites seems secure until about 1981. Further, Landsat-3 is equipped with a thermal scanner and the RBV sensor, should have a ground resolution about twice that of Landsat-1. Seasat, to be launched in 1978, should provide experimental data useful to the management of coastal resources; and the and the GOES programme, with its complete low resolution coverage of the earth's resources management related to food and hydrology. The future role of high spectral resolution (as distinct from high spatial

307

resolution) will be more fully evaluated through the Nimbus-G satellite with its narrow-band coastal zone scanner. The future role of low resolution imagery, which is now available daily is seen as providing data on cloud cover, cloud type, cloud temperature and windforce, wind-direction and transformations of these data into information related to hydrological processes, crop development and crop yield. Transformations include probability of rainfall, rainfall intensity, net short-wave (solar) radiation near the earth's surface, net long-wave (terrestrial) radiation, and possibly evapo-transpiration, which is a critical parameter of soil moisture, plant stress and crop yield.

7 REFERENCES

American Society of Photogrammetry (1975) - Manual of remote sensing Ch.19 - Water resources assessment 1479-1545, Am. Soc. Photogram., Falls University, Va. USA.

Barnes, C.M. 1974, Basics of the screw-worm Personal communication.

Barrett, E. 1977, Rainfall monitoring in the region of the North-West African Desert Locust Commission 1976-77. Consultant's report (AGD(RS)4/77), FAO, Rome.

Barrett, E. 1977, The assessment of rain-fall in North-Eastern Oman, through the integration of observations from rain-gauges and meteorological satellites. Consultant's report, FAO, Rome.

Culler, R.C., Jones, J.E. and Turner, R.M. 1972, Quantitative relationship between reflectance and transpiration of phreatophytes - Gilar river. NASA Ann. Earth Resources Programme, Rev. 4.

Curry, D.T. 1977, Identifying flood water. Remote sensing of environment 6: 51 - 61.

De Pauw, E.F. and Spiers, Berna 1977, Multispectral landsat imagery inter-pretation of the flood region draining to the Sudd, Southern Sudan (AG:DP/SUDD/71/553), FAO, Rome.

Everett, J.E., Berberman and Cuellar, J.A. 1977, Distinguishing saline and non-saline rangelands with Skylab imagery, Photogram. Eng. 1041-48.

Geary, E.L. 1967, Coastal hydrology. Photogram. Eng., 44-50.

Hardy, J.A. 1974, Skylark rocket photo-graphy as an aid to developing countries COSPAR Working group 6 report, 2: 86-95.

Hielkema, J.U. and Howard, J.A. 1976, Application of remote sensing techniques for improving desert locust survey and control (AGP:LCC/76/4)

Howard, J.A. 1970, Stereoscopic profiling of land units from aerial photographs,

Austr. Geagr. XI, 259-68.

Howard, J.A. 1973, Concepts of integrated satellite surveys for developing countries, FAO, Rome (AGD(RS)3/73).

Howard, J.A. and Lanly, J.P. 1975, Remote sensing for tropical forest surveys. Unasylva 27 (108): 32-37.

Howard, J.A. 1976, Remote sensing of tropical forests with special reference to satellite images, IUFRO, Oslo.

Howard, J.A. 1976, Satellite remote sensing of agricultural resources for developing countries - present and future - an international perspective, FAO, Rome (AGD(RS)2/76).

Howard, J.A. 1977, Concepts of remote sensing applications for food production in developing countries (in Earth observation systems for resources management and environmental control, D.J. Clough and L.M. Morley, plenum Press, London, part 3 chapter 13 (Ed.).

Lyon, R.J.P. and Lee, K. 1968, Infrared exploration for coastal and shoreline springs. Tech. Report 68-1, Stanford University, USA.

McNeil, H. 1973, Personal communication. Brazilian Coffee Board, Rio de Janeiro.

Myers, V.I. 1970, Remote sensing for defining aquifers. NASA Ann. Earth resources program Rev: 3.

Olson, C.E. 1970, Remote sensing of broad-based tree species under stress. Third international symposium in photo interpretation, ISP, Dresden PDR: 689-96.

Pacheco, R.A. 1977, The use of Landsat imagery for assessing soil degradation in Morocco. International Society of Soil Science, FAO, Rome.

Solomon, S.I. 1977, Personal communication. Hydrological and Water Resources Department, WMO, Geneva.

J. DEJACE & J. MÉGIER
Commission of the European Communities
Joint Research Centre, Ispra Establishment, Ispra (Varese), Italy

21

Agreste project: Experience gained in data processing, main results on rice, poplar and beech inventories

This text is an extract from the final report of the Agreste Project edited by A. Berg, G. Flouzat and S. Galli de Paratesi (Nasa Landsat Satellites Investigations Nr. 28790, June 1978)

1 INTRODUCTION

The Agreste Project [1] has been sponsored by the Commission of the European Communities to study the range of applications of remote sensing to agriculture and forestry under European conditions and particularly in France and Italy.
The data used are mainly data from the LANDSAT-1 and LANDSAT-2 satellites launched by NASA. Data from airborne scanners are also used but to a minor extent.

One of the major objectives of Agreste is to study the possibility of performing inventories of irrigated crops like rice, planted forests like poplars and natural growth forests like beeches by computer-aided analysis of digital data from LANDSAT. The studies presented thereafter deal thus with inventory investigations over the Italian test areas of Agreste, in Northern Italy.

A further task is considered by using aircraft data: the possibility of discriminating varieties of rice.

The data processing work has been done at JRC Ispra, but the ground data have been collected, and the ground truth maps have been set up, in tight collaboration with: Ente Nazionale Risi - Montara (ENR) for rice; Istituto per la Sperimentazione della Pioppicoltura, Ente Cellulosa e Carta - Casale Monferrato (ISP) for poplar forests; Istituto Nazionale Piante da Legno - Torino (INPL) for beech forests.

2 RICE INVESTIGATIONS

2.1 Description of rice fields

Most of the Italian rice fields (91.4% in 1975) are concentrated in the Northern part of the country, along the Po Valley, in the provinces of Milan, Pavia, Novara and Vercelli. Rice cultivation is very extended in this region with a global surface of rice fields of 1620 km² in 1975. Its concentration increases from East to West, with rice fields of progressively larger

size. The rice fields are organized in continuous parcels separated by little dams from 10 to 80 cm thick for the main dams. The parcels are flooded at sowing time with a layer of water of an average depth of 10 cm (± 3 cm). The average dimension of the parcel is about 1 ha (10,000 m²) with wide variations from 0.1 to 5 ha and a general rectangular shape. Inside a large rice area which is "seen" by LANDSAT as a continuous water body during the sowing period, dams and ditches which bring water to the parcels, constitute 5 to 6% of the total surface.

The main rice varieties cultivated in Northern Italy are Balilla and Balilla GG (28.2% of the total cultivation area in 1975), Padano (12.7%), Ribe (10.6%), Arborio (8.7%), Roma (8.3%) and Ringo (7.5%). Other less important varieties are present, including Europa (3.3%) and Romeo (2.6%). Four of the mentioned varieties, together with two others, have been investigated in section 4. The other main cultivations present in the test area (situated immediately to the South of the town Mortara, 40 km to the South-West of Milan) are, in order of decreasing importance: corn, wheat, oat and simil, poplars, beets, potatoes, melons and water melons.

Water bodies are not present in the test area apart from the Po river and some of its tributaries.

2.2 Ground data collection

Table 1 presents the phenological stages and soil conditions in the test area of Mortara, collected by ENR in correspondence to LANDSAT-2 passages in 1975, for the most important varieties Balilla and Balilla GG.

Ground truth maps providing the location of the rice fields were established for two test areas in collaboration with ENR. A test area of 120 km² was chosen for the year 1973 around the village of Ferrera, between the town Mortara and the Po river (test area "Ferrera"); for the year 1975, an area of about 250 km² was considered, including the town Mortara and the test rice fields of ENR (test area "Mortara"). For a limited part (44 km²) of the test

Table 1 : Development stages of rice plants and soil conditions at LANDSAT-2 passages, 1975, for varieties Balilla and Balilla GG

Date of passage	Development stage	Soil conditions
April, 22	sowing (April, 17–18)	under water
May, 10	no information collected	" "
May, 28	emergence	" "
June, 15	beg. of tillering	" "
July, 3	tillering - field coverage not complete	" "
July, 21	tillering - complete field coverage - beg. of booting	" "
Aug., 8	late flowering (beg. Aug., 5 – 6)	" "
Aug., 26	late milk stage	drained
Sept., 13	late dough stage	"
Oct., 1	harvest (from Sept., 29 – Oct., 6)	"
Oct., 19	harvested	"

area "Mortara", a ground truth map of some rice varieties considered in 1975 was established.

2.3 Data processing for rice inventory with LANDSAT data

2.3.1 Processing of 1973 LANDSAT-1 data [2]

A LANDSAT-1 scene acquired on May 10, 1973 over the Po Valley was made available, with no cloud cover, at a time where almost all the rice fields were under water after sowing, before emergence of rice plants above the water level. A single level slicing on channel 7 of LANDSAT allowed to localize the water bodies by mapping the lower values of the channel

response. The threshold value for the class "water" was easily determined by studying the histogram of channel 7 values over the investigated test area "Ferrera". Figs. 1 and 2 allow the comparison between reference map and classification results for a part of the test area "Ferrera" (about 45 km^2). The shapes of the rice fields, although rather complex, are fairly well reported and the smaller fields (some pixels in size) are recognized. A few rice fields, not yet flooded on May 10, are obviously not found and, on the contrary, some flooded meadows are mapped as water bodies. The comparison of rice areas between reference and classification maps results in an overall underevaluation by less than 3% of the rice covered surface. Of course, by this method, other water bodies like large streams (not present in the above test area) are mapped together with rice fields; however, even ignoring the fact that the relative surface occupied by streams is very limited in the Agreste test site no. 1, a solution would be to map the streams by using a LANDSAT scene acquired before sowing stage or after harvest, when rice fields are drained from water. This kind of correction has not yet been performed.

2.3.2 Processing of 1975 LANDSAT-2 data

Among the LANDSAT-2 passages listed in section 2.2. , only three resulted usable over the test area "Mortara" which was chosen for the study of rice in the year 1975 (Table 2). No early LANDSAT data were available at the end of May before emergence of the plants above water level. The tillering stage prevailing in the June and early July scenes introduces a great heterogeneity in the data due to a) variation of

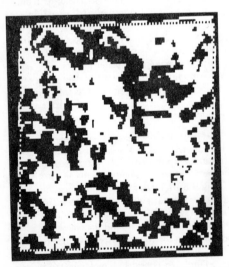

Fig. 1 : Ground truth map for rice fields (in black) 1973

Fig. 2 : Classification result by level slicing on channel 7 of LANDSAT. Rice is black.

Table 2 : LANDSAT-2 available scenes over rice test-
area, 1975

ID Nbr.	Date	Growth stage
2144 − 09331	June, 15	Begin of tillering
2162 − 09331	July, 3	Tillering
2234 − 09320	Sept., 13	Late dough stage

vegetation coverage from field to field and b) variation of aspect of the rice plants from variety to variety. The scene of September 13 has not yet been processed.

The concept of supervised classification was
used. The heterogeneity of the data convinced us, on
the basis of previous studies, to use a quadratic rather
than a linear classifier, but with a special care to define
and choose the training sets. As a matter of fact, the
maximum likelihood method (ML) was chosen,
with supervised classes described by normal parameters. If the class distributions taken into account are
not normal, as is generally the case, the ML scheme
does not provide any more the best theoretical classifier. It results, however, in a very reasonable and efficient quadratic classifier, provided the class distributions are unimodal or nearly so. A further question
arises from the evaluation of the a priori probability
$P(C_k)$ for class C_k but the supervised context allows
here to use equal existence probabilities for the
classes of interest and thus to ignore them in the
classification process.

2.3.2.1 Processing of data from June 15, 1975

The study was made on the reduced test area "Mortara" (see section 2.2). In a first stage the training
sets for rice were chosen by combining a clustering
method with a uniformity mapping procedure , both
using euclidean distance as similarity measure between two points in order to put in evidence clusters

of points suitable in the sense of uniformity and of
geographic location. The same was done for non rice
classes, mainly poplar groves and corn fields. A first
classification and mapping provided a basis to identify
- in connection with the ground truth map - reliable
sampling areas to define training sets for unimodal
rice classes. No care was purposely devoted to the
fact that the new classes were largely overlapping
(owing to the large variance values) in order to keep
the entire variance of the rice data. The mean and
standard deviations of the rice training classes are
given in Table 3. The training sets for rice represented
less than 4% of the total area processed.

The classification and mapping results using this
second set of five classes together with three non rice
classes, showed that a trade-off was needed between
non-recognition of rice zones and misclassification as
rice of non rice zones by setting a proper "membership threshold" on rice classes. The classification results are mapped in Fig. 4 and can be compared to
the ground truth displayed in Fig. 3. They were obtained by setting on each rice class a rejection threshold of 3% of the maximum probability of the corresponding class distribution.

The global rice percentage area was found to be
43%, whereas the ground truth value is 35.5%, thus
an overestimation by 21% of the rice covered areas.

2.3.2.2 Processing of data from July 3, 1975

At the time this work was done, a procedure had been
set up to construct a digitized version of ground truth
maps by discrete elements following the scanning
grid of LANDSAT (see section 3.4.1) for more details. A discrete version of the ground truth map with
rice varieties for the limited test area "Mortara"
was thus realized. It was then possible to localize
training samples within the different varieties with a
fair accuracy. Due to the reduced dimensions of the
fields, each training sample was rather small (from
4 to 20 pixels) and several samples were merged for
each variety. The training sets represented about 5%
of the total processed area. The statistical parameters
of some retained varieties are presented in Table 4.

Table 3 : Mean and standard deviation values
for rice classes

		LANDSAT-2 Channels			
		4	5	6	7
R_1	μ	24.3	21.7	46.9	19.1
	σ	1.1	1.0	3.0	1.8
R_2	μ	28.8	20.8	43.2	17.2
	σ	1.1	0.9	6.0	3.7
R_3	μ	25.3	22.8	45.2	18.3
	σ	1.2	1.5	3.4	2.2
R_4	μ	27.8	26.1	48.0	19.1
	σ	1.4	1.8	3.5	1.8
R_5	μ	25.1	23.4	45.3	17.9
	σ	1.4	1.7	4.6	3.0

Table 4: Mean and standard deviation for various
rice varieties, July 3, 1975

		LANDSAT-2 Channels			
		4	5	6	7
Arborio	μ	23.7	21.2	65.6	31.8
	σ	1.3	0.63	5.5	3.9
Balilla GG	μ	21.7	19.2	48.9	21.7
	σ	1.3	1.5	5.1	3.3
Gritna	μ	22.1	20.4	58.7	28.3
	σ	1.2	1.0	4.5	2.7
Carnaroli	μ	22.6	20.1	48.7	21.0
	σ	1.5	1.5	4.6	2.5

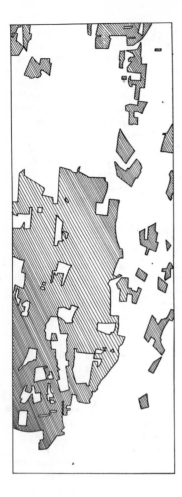

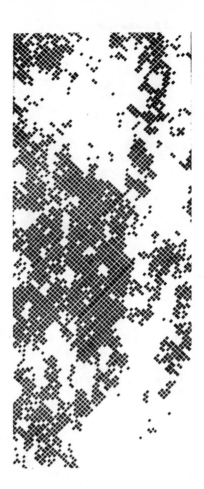

Fig. 3 : Ground truth map for rice fields in test area "Montara", 1975. Rice fields are hatched areas. Approx. scale 1/80,000

Fig. 4 : Classification results by ML method. Rice fields are black

The classification work was then performed with the ML method, starting from 4 rice classes corresponding to the 4 varieties and from some other non rice classes. This time, no rejection threshold was set for the classes.

A pixel-by-pixel comparison between discrete ground truth and classification results was performed under the form of a confusion matrix (see Appendix CFMAT) shown in Table 5. It is organized in one row for each ground truth category (R = rice, NR = non rice) and gives the percentages of pixels correctly classified into that category and misclassified into the other category.

On the righthand side of the R column, one has indicated in each row the percentage of pixels classified as "rice" which actually belong to the ground truth category R (row R), but also to category NR

(row NR). The same is done on the righthand side of column NR. The global result RCL/RGT (total number of pixels classified as "rice"/total number of pixels belonging to the category rice on the ground truth) is also given. This presentation allows the most complete evaluation of the classification performances.

Table 5 : Confusion matrix (expressed in %) for data of July 3, 1975, ML method. RCL/RGT = 109%

Ground truth	Class	
	R	NR
R	90/82	10/20
NR	33/18	67/80

From the detailed results of Table 5 it is seen that 90% of rice present on the ground truth was correctly recognized, but with 18% misclassifications within the pixels classified as rice.

As a global result (RCL/RGT), it is seen that rice covered areas are over estimated, this time, by 9%. Two circumstances may explain the improvement of the results, as compared to those obtained by processing the data of June 15. On the one hand, the development stage (more advanced tillering) may be more favourable to the discrimination of rice from non rice; on the other hand, sampling over the actual varieties may improve the representation of the intrinsic reflectance features of rice. The last assumption seems to be reinforced by the fact that sampling rice independently of the varieties location brings a degradation of the results.

It must be added, however, that, although the consideration of rice varieties as training classes improved the accuracy of rice classification as a whole, the problem of discriminating varieties with LANDSAT data is not yet solved. Preliminary results in this sense show that two groups of varieties can be rather well discriminated from each other, Carnaroli and Balilla GG from Arborio and Gritna.

2.3.3 Conclusions for rice inventory in Italy with LANDSAT data

The generality of any conclusion is naturally restricted by the fact that, up to now, the test areas studied are limited as compared to the extension of the rice cultivated areas in Italy, although the test areas were chosen as typical of rice cultivation. In this respect, a comparison should be made over a wide region between recognized rice areas at the flooded stage (corrected from other water bodies) and coverage data given by ENR from cadastral maps updated every year.

Extrapolation from results obtained for a 4 km² test zone allows to expect good performances in the inventory of rice areas over wide regions, such as Northern Italy, with a global accuracy not inferior to 95%. Such an accuracy is obtained by using LANDSAT data during the period between sowing and emergence of rice. The suitable time interval extends over 3 to 4 weeks; a cloudfree LANDSAT scene should therefore be obtained within this short period. This might prove to be difficult to realize; actually we did not succeed for the year 1975, even by considering LANDSAT-1 as well as LANDSAT-2 passages, thus with a coverage period of 9 days.

Using data acquired at later development stages, does not bring the same precision in inventory. A global accuracy of 90% has nevertheless been reached. This conclusion is also limited by the reduced extension of the studied area (44 km²). The last result was obtained by using the scenes acquired on June 15 and July 3; however, an exhaustive multi-temporal analysis combining these scenes with the last scene available, acquired on September 13, is still to be done.

In a preliminary estimation, one could evaluate to a minimum of 5 or 6 the number of successive LANDSAT scenes necessary for an exhaustive rice inventory. The actual availability of LANDSAT data over the test area chosen did not allow to fulfill this requirement. One of the reasons for this was that the test area for the year 1975 is situated toward the edge of one of the two overlapping LANDSAT frames (in the sense East-West). This unfavourable circumstance was not given sufficient importance when choosing the test area.

2.4 Processing for airborne scanner data for identification of rice varieties [3]

2.4.1 Characteristics of the data

The data were acquired on August 7, 1975, between 9.11 and 10.32 a.m. by a Bendix M²S scanner at a date on which the various rice varieties should be in the flowering stage (Table 1) considered to be favourable for the scope of the operation.

The technical data of the scanner are reported in Table 6. The data from channels 1, 2 and 10 were

Table 6 : BENDIX M²S scanner technical data

	Chan. No.	Center (µm)	Width (µm)	Chan. No.	Center (µm)	Width (µm)
Scan angle 100°	1	0.410	0.06	7	0.680	0.04
	2	0.465	0.05	8	0.720	0.04
Roll-compensation ± 10°	3	0.515	0.05	9	0.815	0.09
	4	0.560	0.04	10	1.015	0.09
Geometric resolution 2.5 · 10⁻³ rad	5	0.600	0.04	11	11.0	6.0
	6	0.640	0.04			

not used, however, due to an unacceptable level of noise. The resolution of the data at the altitude of 1500 m is 3.8 m. The flight was made over a part of the test area "Mortara"; the strip of data processed here covers a zone approximately 3 km² in area, where six rice varieties are present, labelled in the following way on the ground truth document (Fig. 5): Gritna (G), Balilla GG (B), Arborio (A), Carnaroli (C), Rocca (RC), Romeo (RO). The other symbols correspond to Corn (M) and wet meadows (MR).

2.4.2 Ground truth preparation

The work was done in collaboration with ENR (see section 2.2). The final document (Fig. 5) was prepared following the procedure described in section

3.4.1, but in a much easier context owing to the resolution of the data. Not all the rice fields present in the strip were characterized from the point of view of rice variety. Fields containing unidentified varieties or mixtures of varieties - what often occurs in the studied zone - are black on the ground truth, together with roads, field ways and some inhabited areas.

2.4.3 Classification methodology

The classification was exhaustive, including also the zones uncharacterized on the ground truth. The ML and MED methods were used . No rejection threshold was applied to any of the classes and no point was the refore left unclassified. The training of the algorithms was done on little portions of the ground truth classes as seen in Fig. 5. Care was taken that the statistical distributions in the training sets be nearly unimodal although training sets of the same ground truth class could present noticeable differences between the respective distributions. Better global results were obtained, however, with the ML method by merging together all the sub-distributions of the same class, than by processing such "sub-classes" separately. The opposite was true, on the average, with the MED method, particularly class A and class M were divided respectively into three sub-classes.

2.4.4 Atmospheric corrections

As the total scan angle of the Bendix M^2S scanner is $100°$, it is well known that the variation of the thickness of the atmosphere between the scanner and the ground along a scan line has a systematic effect on the data acquired and may then cause a degradation of the classification results. The "long track averaging" procedure used here to correct this effect, assumes that, in absence of the above mentioned atmospheric effect, the mean and variance for each channel along a column of data (i.e. following the flight axis) would have the same value for all the columns (i.e. from edge to edge in the strip of data acquired). The routine set up calculates then in a first step the mean and the variance of the columns from edge to edge of the strip and successively transforms in a second step the data of each column in order that all the columns have the same mean and variance. This procedure is repeated separately for each channel. The correction on the variance has proved to be useful to correct what appears as a lack of contrast towards the edges of the strip in a somewhat accurate visualization of the raw data. This correction method has been applied to all data.

2.4.5 Outline of the results

The results are presented in the confusion matrices displayed in Tables 7 to 9, where UC means uncharacterized areas on the ground truth from the point of view of rice varieties, and O stands for roads, inhabited zones and other non vegetal items. In the classification process, UC and O are grouped under the label O. In the last row, labelled "total classified/ total ground truth", the global classification results for each category are indicated.

The best result was obtained, as expected, with the ML method using all the 8 channels available (Table 7). It is seen that the discrimination is on the average very good between rice varieties on the one hand, between rice and other vegetal species on the other hand. The percentage of ground truth classified as such varies from 65 to 83%, apart from variety RO. The data compression by principal components analysis (P.C.) from 8 to 5 dimensions affects very little the results (Table 8), but the calculation time is decreased from 29 to 18 min (CPU time with IBM 370/165, 134320 pixels processed); the fraction of the total variance conserved in the transformation is 99.2%. The classification results are displayed on Fig. 6 for this last case. It is seen that the uncharacterized rice fields (in black) on the ground truth are mainly classified as mixtures of varieties with the exception of zones where rice was not recognized; it must be said in this respect, that other varieties, not identified on the ground truth, are present in the same region and were not sampled here.

Table 9 contains the results obtained with the MED method after a linear transformation on the data defined by P.C. analysis, but retaining the dimensionality of the space (i.e. 8 dimensions). The transformation improves the results because it has the tendency to decouple the variables as the global variance-covariance matrix of the whole area becomes diagonal.

2.4.6 Conclusions

Recognition of rice varieties at the flowering stage by using airborne scanner data at rather low altitude (1500 m) seems to be feasible. The accuracies obtained on a reduced test area (3 km²) range from 65% to 83%, apart from one case. Further studies are planned, however, first to check the reliability of the results on a larger zone and second to use flights at higher altitudes in order to study the variation of the discrimination accuracy with the resolution of the data.

Table 7 :   Rice varieties confusion matrix for ML 8 channels

	C	B	A	MR	M	C	RC	RO	O
G	71.4	7.9	9.8	0.0	2.4	0.2	2.5	0.2	5.4
B	2.9	76.4	1.2	0.1	3.1	1.7	0.4	0.3	13.9
A	11.3	1.3	66.8	0.0	3.9	0.2	3.2	0.9	12.3
MR	0.1	0.9	0.4	83.5	2.2	0.0	0.1	0.0	13.8
M	0.8	0.0	0.2	0.1	80.1	0.0	0.1	0.0	19.2
C	0.8	10.3	0.3	0.0	1.0	64.7	14.7	2.5	5.6
RC	0.9	1.3	4.5	0.0	1.3	0.5	78.3	10.2	3.0
RO	0.2	1.7	8.8	0.0	1.2	0.1	26.0	51.2	10.8
O	0.5	1.2	0.1	1.6	2.8	0.1	0.1	0.1	93.5
UC	2.0	12.0	3.4	0.4	13.8	3.6	1.9	0.6	62.3
Tot. cl./tot GT	105	97	83	89	108	68	210	84	

Table 8 :   Rice varieties - confusion matrix for ML after data compression from 8 to 5 dimensions by P.C. analysis

	C	B	A	MR	M	C	RC	RO	O
G	69.1	7.4	11.5	0.0	3.0	0.6	3.1	0.2	5.0
B	2.9	74.4	1.2	0.0	4.4	2.7	0.4	0.3	13.6
A	15.2	1.0	63.7	0.0	6.4	0.2	2.8	0.9	9.7
MR	0.0	0.0	0.5	85.4	3.9	0.0	0.0	0.0	10.1
M	0.3	0.0	0.3	0.2	78.0	0.0	0.1	0.0	21.1
C	1.1	11.7	0.4	0.0	1.2	62.4	14.6	2.8	5.7
RC	2.1	1.3	5.4	0.0	1.3	0.6	76.3	10.3	2.5
RO	0.2	1.5	11.9	0.0	1.2	0.2	27.0	47.9	9.7
O	1.1	1.4	0.4	1.4	5.4	0.2	0.2	0.1	89.7
UC	2.5	11.9	3.6	0.4	17.6	4.0	2.0	0.5	57.4
Tot. cl./tot GT	111	96	82	90	117	67	204	83	

Table 9 :   Rice varieties - confusion matrix for MED 8 dimensions after P.C. transformation

	C	B	A	MR	M	C	RC	RO	O
G	61.1	6.6	10.0	0.0	3.5	0.1	0.8	0.1	14.9
B	1.9	65.8	0.5	0.0	4.5	1.1	0.3	0.1	25.8
A	13.4	1.8	60.4	0.0	3.9	0.0	2.0	0.4	16.1
MR	0.3	0.0	0.2	68.2	0.7	0.0	0.0	0.0	30.6
M	0.2	3.2	1.9	0.0	69.3	0.0	0.0	0.0	25.4
C	1.4	23.5	0.5	0.0	2.2	43.4	10.9	0.8	17.2
RC	1.5	2.7	20.8	0.0	0.7	0.1	56.7	9.9	7.5
RO	1.1	5.4	26.8	0.0	0.7	0.0	20.1	29.8	16.0
O	0.2	1.8	0.3	0.4	1.6	0.0	0.0	0.0	95.5
UC	2.3	15.1	3.4	0.2	10.5	2.3	0.7	0.3	65.3
Tot. cl./tot GT	98	104	89	70	100	45	150	48	

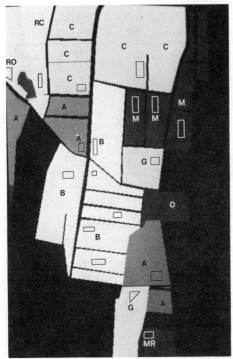

Fig. 5 : Ground truth map for rice varieties, 1975

Fig. 6 : Classification results by ML method after data reduction by PC analysis. The grey levels correspond to those displayed on the ground truth

- Class 3 : 25 — 75% ground coverage (groves of 4 — 6 years)
- Class 4 : more than 75% ground coverage (groves of 7 years and more)

Poplar groves were then grouped in two comprehensive classes:
- Class A : young groves with ground coverage less than 25%
- Class B : intermediate and adult groves with ground coverage over 25%

and were reported on the available UTM map, scale 1 : 25,000.

3.2 Particular conditions of the study

The dimensions of the poplar groves in the test areas considered, range from some hectares to 100 — 200 hectares, so that the training sets for classification had to be defined on reduced numbers of pixels (from some pixels to some tens of pixels) in order to avoid sampling on mixed pixels at the edges of the groves.

Another characteristic of the problem was that the retained training sets belong only to the forementioned class B; in fact, it was not possible to

3 POPLAR INVENTORY WITH LANDSAT DATA

3.1 Ground truth preparation

Three test areas have been investigated in the Po valley region (TA_1, TA_2 and TA_3); they cover surfaces of 40, 30 and 45 km^2 respectively. These areas represent typical zones for poplar cultivation in Italy, which assumes a great importance for packing and wood pulp industries.

The relevant ground truth data were collected by a conventional aircraft flight which delivered infrared black and white photographs at a 1 : 10,000 scale. This flight was performed in August 1975 over the Po Valley.

An exhaustive conventional photo-interpretation of all the poplar groves was made by research workers of the ISP.

Following the criterion established by ISP [4], the poplar plantations were subdivided into 4 classes corresponding to various ranges of ground coverage percentage, and age, i.e.:
- Class 1 : 0 — 5% ground coverage (groves of one year)
- Class 2 : 5 — 25% ground coverage (groves of 2 — 3 years)

retain training sets for class A because the insufficient ground coverage causes too much dispersion in the data and hence in the classification results, owing to the great variability of the ground conditions. It must be noted, however, that, although recognition of class A would provide a valuable production forecasting, recognition of class B is sufficient for an evaluation of the amount of wood available for industrial needs.

Actually class B contains about 90% of the total amount of wood present in the two classes and the whole amount of timber available each year for industry.

The last point to be stressed is that the ground truth document did not provide reliable information on ground coverage, other than poplar groves and water bodies. Three ways are open to overcome this difficulty:

- setting membership thresholds for the classes of interest in order to leave as unclassified the pixels which actually do not belong to these classes [5];
- using auxiliary classes defined by an unsupervised clustering process on the uncharacterized parts of the ground truth [6];
- using some "pseudo-classes" with wide variance, defined also on the uncharacterized parts of the ground truth, when the clustering process is not successful. In this last case, the pixels which actually do not belong to the classes of interest, are likely to be attributed to the "pseudo-classes" in the course of the classification process [2].

For the task of identifying poplar groves in flat areas, the last two approaches were found feasible (unlike what happened for inventory of natural forests in mountainous regions, see Chapter 4). They were therefore applied preferentially to the first approach in order to avoid the problem of defining thresholds for the various classes. Moreover, the second approach, which defines unidentified but homogeneous clusters of pixels outside the characterized zones of the ground truth, was found suitable for most of the cases.

3.3 Analysis of 1972 – 1973 data (preliminary results)

Some preliminary results have been obtained by using two LANDSAT-1 scenes of October 7, 1972 and May 10, 1973. The 1972 ground truth document was made for TA_1 where poplar groves are located along the Sesia river, close to the confluence with the Po river. For the year 1973, the area TA_2 was considered where poplar groves are situated in the vicinity of the Po river, about 60 km downstream from the first area.

A comparison was made between various classi-

fication methods and it was concluded that the maximum likelihood scheme (ML) used in a "supervised" context with normal parameters to describe the training classes, gave on the whole better and more reliable results.

The training sets were situated in both cases outside the part of the ground truth document utilized in the study.

The comparison between classification results and ground truth was done in a rough way in the case of data from October 7, 1972. It allowed, however, to ascertain that the bulk of intermediate and adult groves (class B) was actually recognized.

The various surfaces in the ground truth for the data of May 10, 1973 were evaluated and the surface percentage of ground truth document covered by poplars was compared to the classification results. Since the problem of superposition of small areas between ground truth and classification results was not yet solved, it was difficult to evaluate the amount of pixels wrongly classified as "poplar". The global result indicated, however, that about 70% of intermediate and adult groves (class B) were correctly recognized.

3.4 Analysis of 1975 data [7]

3.4.1 Discrete ground truth construction

The ground truth document was established (according to the procedures described in section 3.1) on the TA_3 area situated also along the Po river, immediately to the North of the town Valenza. Such a reference map was then gridded into discrete elements corresponding to the LANDSAT pixels and stored on magnetic tape to be compared afterwards with the classification results. The gridding was done by following an iterative process between reference map and classification results for a given LANDSAT scene in order to ascertain the direction of the satellite scanning on the map and the location of some "one pixel" marks both on map and LANDSAT data. The interpreted areas were then obtained from the corresponding discrete contours (given as input) by running a contour-follower routine. Obviously, as three different LANDSAT scenes over the same zone were utilized, the same pixel-to-pixel correspondence had to be established between the three individual scenes.

The overall accuracy of the superposition procedure is crucial for the reliability of the classification results, on account of the reduced dimensions of the individual areas studied, which can even be one pixel in size in extreme cases. In this situation it was decided to discretize the continuous document (ground truth on U.T.M. map) and deform it in order to achieve the superposition with the LANDSAT scanning grid (i.e. the mapped classified results). Fig. 7 shows the discrete ground truth obtained for the studied zone, photographed on the video screen digital display unit. It is seen that some poplar groves are one pixel in size, whereas most of them range from tens to a few hundreds.

3.4.2 Monotemporal analysis of the LANDSAT data

Three LANDSAT-2 scenes were available on the test area TA_3. They were acquired on June 15, July 3 and September 13, 1975, respectively. The ML method of classification was mainly used and training sets were defined only for poplar groves of class B, within the zone to be classified (see Fig. 7). The number of training pixels used represents less than 3% of the total number of pixels. The training sets were determined by a preliminary clustering of the data of the whole area using Euclidian distance between points as similarity measure, in order to put into evidence clusters which make sense from the points of view of geographical location and uniformity. They were then cleaned from marginal or anomalous points in order to exhibit unimodal or nearly unimodal distributions.

The data from each of the three scenes were processed separately and a pixel-by-pixel comparison between discrete ground truth and classified results was done under the form of confusion matrices (see Appendix CFMAT).

The results obtained are given in Tables 10, 11 and 12. These confusion matrices are organized in one row for each ground truth category and they give for each category the percentages of pixels correctly classified into that category and misclassified into the other categories. For the reasons stated above, the poplar classes A and B were classified in a single poplar class P. In the P column, one has indicated, on the right, the percentages of pixels classified as "poplar" which actually belong to the ground truth categories P_A and P_B, but also to categories W and R. This allows a complete evaluation of the classification performance. The ratios PCL/PGT (total number of pixels classified as poplars/total number of pixels belonging to the two poplar classes

on the ground truth) are also given.

The best overall results could be those given for the September 13 scene since 78.4% of poplars P_B and 52.2% of poplars P_A were correctly recognized. But on the other hand, it should be noted that, although 57.7% (40.7 + 17.0) of the pixels classified as poplar actually belong to this category, 42.3% (4.4 + 37.9) do not. Some of these results are displayed in Fig. 8. Similar comments apply to the results of the other scenes.

The MED method provided very similar results, although lower by 1 to 4 units on the percentage figures, but with computing times reduced by a factor around 3.

3.4.3 Multitemporal analysis of the LANDSAT data

Three approaches were proved to combine the single processings at the three different dates (using the ML method) in order to increase the accuracy and the reliability of the results [7].

Improvements were obtained by combining the data of June 15 and September 13. No further improvement was achieved by invorporating also the data of July 3, probably because they are too much correlated to those of June 15 from the phenological point of view.

The first approach proves to minimize the misclassification in both directions (i.e. poplar classified as "non poplar" and "non poplar" classified as "poplar") by taking into account, in the combined processing, only the pixels classified in the same class in the separate processings. The resul represents then the intersection of the successive classifications for the various classes (Table 13).

The second approach considers that the total amount of information available for the discrimination of the classes of interest, is present and integrated in the superposition, pixel-by-pixel, of the single data frames. The set of frames is then considered and processed as a single frame with 8 channel data from two LANDSAT scenes (Table 14).

The last method considered uses a probability

Table 10 : Results (%) for scene 6/15			
PCL/PGT = 134%			
W = water; P = poplar of			
classes P_A and P_B; R = rest			

| Ground-truth | Class | | |
	W	P	R
W	68.4	26.4/ 5.5	5.3
P_A	1.9	46.9/13.5	51.2
P_B	0.5	78.5/35.9	21.1
R	3.9	23.9/45.1	72.2

Table 11 : Results (%) for scene 7/3			
PCL/PGT = 94%			
W = water; P = poplar of			
classes P_A and P_B; R = rest			

| Ground-truth | Class | | |
	W	P	R
W	67.3	6.5/ 1.9	26.2
P_A	1.8	34.8/14.3	63.5
P_B	1.0	61.9/40.5	37.1
R	2.3	16.1/43.3	81.7

Table 12 : Results (%) for scene 9/13			
PCL/PGT = 118%			
W = water; P = poplar of			
classes P_A and P_B; R = rest			

| Ground-truth | Class | | |
	W	P	R
W	61.9	18.7/ 4.4	19.5
P_A	2.9	52.2/17.0	44.9
P_B	1.2	78.4/40.7	20.4
R	2.1	17.7/37.9	80.1

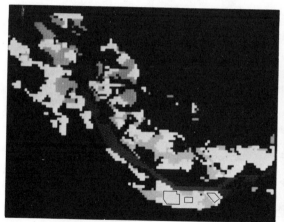

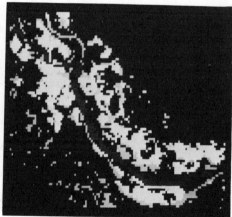

Fig. 7 : Ground truth map for poplar groves, 1975. The training sets are drawn with black lines.

Fig. 8 : Classification results by ML method. Data of September 13, 1975. White: poplars from class A and B well recognized as poplars; light grey: non poplar items misclassified as poplars; dark grey: Po river recognized as such.

Table 13 : Results (%) for intersection between scene 6/15 and scene 9/13
PCL/PGT = 75%
W = water; P = poplar of classes P_A and P_B; R = rest

Ground-truth	Class W	P	R
W	54.4	9.7/ 3.6	35.9
P_A	0.9	28.9/14.9	70.2
P_B	0.3	65.8/54.0	33.9
R	1.4	8.1/27.4	90.5

Table 14 : Results (%) for superposition of scenes 6/15 and 9/13
PCL/PGT = 113%
W = water; P = poplar of classes P_A and P_B; R = rest

Ground-truth	Class W	P	R
W	61.7	27.0/ 6.6	11.4
P_A	1.8	49.1/16.7	49.1
P_B	0.4	79.0/42.7	20.6
R	2.0	15.3/34.0	82.7

Table 15 : Results (%) for probability products for scenes 6/15 and 9/13
PCL/PGT = 108%
W = water; P = poplar of classes P_A and P_B; R = rest

Ground-truth	Class W	P	R
W	70.8	16.4/ 4.2	12.8
P_A	2.1	49.4/17.7	48.5
P_B	0.5	80.8/45.9	18.8
R	3.3	13.7/32.2	83.0

scheme which attributes the pixel to be classified to the class for which the product of the membership probabilities for that class in each separate frame, is the highest (Table 15, Fig. 9).

Results of the three forementioned approaches can be summarized as follows:

-   The "intersection" method reduces very much the number of misclassifications in the sense "non target" classified as "target", but at the same time it increases too much the misclassifications "target" classified as "non target".
-   The "superposition" method reduces, on the whole, the misclassifications in both senses.
-   The "probability product" method has the same effects, with slightly better results, than the superposition method; it is moreover better adapted in a "supervised" context because of the higher degree of freedom allowed in the choice of the training sets for each individual scene.

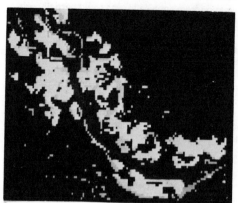

Fig. 9 : Multitemporal analysis combining data of June 15 and Sept. 13, 1975. Classification results ML method using the probability product scheme. Grey levels as in Fig. 8

319

3.5 Conclusions

Among the various approaches used in the LANDSAT data analysis on poplar groves in the Po region, the multitemporal analysis on three 1975 scenes led to the most satisfactory results:

- 81% of poplar groves belonging to class B (adult and intermediate groves over 3 years old) and 49% of poplar groves belonging to class A (young groves up to 3 years old) were correctly recognized in a unique class "poplar".
- On the other hand, 4% of the areas classified as poplar groves are actually water bodies and 32% are actually ground objects outside the poplar groves.
- A careful examination of the ground truth document shows that the first type of misclassification is due to mixed pixels in the cases of groves located along the Po river; the second type of misclassification is mainly due to natural woodlands containing poplars together with willows and robinias in variable proportions.

From an operative point of view, it would be desireable to separate class A from class B on the classification results in order to properly evaluate the amount of timber presently available for industry and the expected production in the successive years. In order to prove to which extent the last objective can be reached, a further effort should be made in the sampling procedure within class A, by considering in particular the detailed classes 1 and 2 (see section 3.1).

4 IDENTIFICATION OF BEECH FORESTS

The inventory of beech forests in Northern Italy is considered as an essential part of a broader inventory of natural renewable resources. Actually, beech wood originating from coppice forests could partially substitute poplars for the production of wood pulp. These coppice forests constitute in Northern Italy an important unutilized resource, especially in Piemont where they represent a significant fraction of the total forested areas. The identification of these beech forests by remote sensing techniques has to take into account various objective difficulties arising mainly from the mountainous relief of the investigated areas and from the typical mixed composition of the natural deciduous forests.

4.1 Ground truth data

The area under investigation is situated in the Italian part of the Maritime Alps, south-west of Cuneo (test site no. 3). It covers part of the valley of Stura di Demonte and an affluent called Vallone dell'Arma.

The ground truth, prepared by INPL from I.R. aerial photographs (scale 1 : 17,000), considered the following types of vegetation:

- Chestnut forest on north-oriented slopes at medium altitudes;
- Beech forests at higher altitudes and on the edge between the two valleys;
- Deciduous forest with closely mixed species;
- Crops and meadows at the bottom of the valleys.

A significant portion of the area (rocks, bare soils, villages, etc.) is not fully characterized.

The ground truth has been reported manually on the 1 : 25,000 map and then digitalized following the method described in section 3.4.1, Fig. 10 gives the display of digitalized ground truth.

4.2 MSS data and items of classification

Two LANDSAT scenes, free of clouds over the studied area, were available for this investigation, namely those of July 3rd, 1975 and of September 13th, 1975.

The two scenes were used first separately and later merged into a unique 8-channel scene.

Because of the mountainous relief of the area under investigation, the angle of incident sunlight to the ground varies from 10 to 65°, thus leading to extreme variations in direct irradiance. A complete digitalization of the relief which would have taken account of this relief effect was beyond the scope of the study. A simple solution was therefore adopted by defining two beech classes, one in sunlight (beech 1) and the other in shadow (beech 2).

Spectral signatures, i.e. mean and variance-covariance matrix were obtained for each class on homogeneous training samples covering 40 to 200 pixels.

4.3 Results

4.3.1 Data from July 3rd, 1975

The best results were obtained, on the July data, with the maximum likelihood method, using probability thresholds for each class (see Appendix). From the confusion matrix given in Table 16, one notes the following features:

a) The percentage of correct beech classification (i.e. the ratio of the number of pixels of real beech recognized as beech to the total number of beech pixels on the ground truth) amounts to 63%.

b) If beech, chestnut and mixed deciduous forests were considered as a unique class, the percentage of correct classification would be 76%.

c) Discrimination between beech and chestnut is illustrated by the following figures:
   - 74% (11 + 63) of beech area are classified as deciduous forest (either beech or chestnut);
   - among these pixels, 85% (63/74) are correctly classified as beech and thus discriminated from chestnut;
   - 67% (50 + 17) of chestnut are classified as deciduous forest;

- among these pixels, 75% (50/67) are correctly classified as chestnut.

d) Finally, the precision on surface estimation (ratio of total number of pixels classified to the number of pixels in the ground truth for a single class) ranges from 81 to 118%.

Because of the limited precision of discrete ground truth, it has seemed unreasonable to still refine the parameters of classification. Fig. 11 shows the results of classification.

4.3.2 Data from September 13th, 1975

The scene of September 13th (Table 17) gives satisfactory results as far as surface estimation (precision between 82 and 135%) and global classification of deciduous forest (75% of correct classification) are concerned. But discrimination between beech and chestnut is poorer. This is probably to be related to the plant phenology, the spectral signature of chestnut in early July being influenced by flowering.

4.3.3 Use of a background class

The use of a background class should, in principle, appear more satisfactory since it does not need any arbitrary fixation of threshold. Actually, in this case, this method does not give as good results as the threshold method, even if the number of pixels classified in this background class is artificially reduced by the introduction of a low "a priori" probability.

4.3.4 Multitemporal analysis

Multitemporal analysis starting from an eight-channel scene gives poorer results than the analysis of each single scene, in spite of an increase in computing time (by a factor of 2.8 in this case).

4.3.5 M.E.D. method

Modified Euclidean Distance (M.E.D.) method brings a relevant reduction in computing time (a factor 4 in this case). This method was applied to the July scene in the three different ways indicated in Table 18,

Fig. 10 : Ground truth map for beech and chestnut forests. White: beech; Light grey: chestnut; Dark grey: meadows

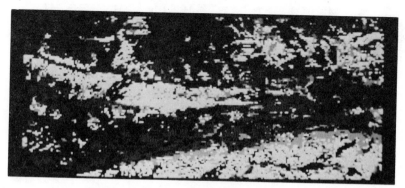

Fig. 11 : Results of classification by ML method on LANDSAT data of July 3, 1975. Same grey levels as in Fig. 10

Table 16 : Results of classification (M.L. with thresholds) on the 7/3/75 scene. (a) in number of pixels, (b) in percentages of ground truth.
Legend: M.C.: meadows and crops; chest.: chestnut; n.cl.: not classified; m.d.f.: mixed deciduous forest; n.ch.: zone not characterized on ground truth; ratio: ratio of classified area to effective area

	Classification					Classification			
G. truth	M.C.	chest.	beech	n.cl.	G. truth	M.C.	chest.	beech	n.cl.
M.C.	917	116	220	639	M.C.	48	6	12	34
chest.	144	407	140	127	chest.	17	50	17	16
beech	229	354	1924	489	beech	10	11	63	16
m.d.f.	15	32	96	23	m.d.f.	9	19	58	14
n.ch.	463	56	103	292	n.ch.	51	6	11	32
					ratio	.97	1.18	0.81	
	(a)					(b)			

Table 17 : Results of classification (M.L. with thresholds) on the 9/13/75 scene (for legend, see Table 16)

	Classification					Classification			
G. truth	M.C.	chest.	beech	n.cl.	G. truth	M.C.	chest.	beech	n.cl.
M.C.	1076	118	284	414	M.C.	57	6	15	22
chest.	112	354	267	85	chest.	14	43	33	10
beech	448	552	1748	318	beech	15	18	57	10
m.d.f.	40	50	73	3	m.d.f.	24	30	44	2
n.ch.	238	33	127	516	n.ch.	26	4	14	57
					ratio	1.01	1.35	.82	
	(a)					(b)			

Table 18 : Results of classification (M.E.D.) on the 7/3/75 scene. (a) on original data without limiting distance, (b) on data after principal component transformation, without limiting distance, (c) on transformed data with limiting distance. All results are in percentages of ground truth. (For legend see Table 16)

Ground-truth	Class M.C.	chest.	beech	n.cl.	M.C.	chest.	beech	n.cl.	M.C.	chest.	beech	n.cl.
M.C.	46	13	6	35	58	13	7	21	59	6	12	23
chest.	16	63	9	12	15	67	9	9	20	42	23	15
beech	7	24	46	23	12	26	47	16	13	11	62	14
m.d.f.	2	39	39	21	8	38	39	14	10	14	63	13
n.ch.	55	12	6	27	63	11	8	18	62	5	13	20
ratio	1.15	2.12	0.58		0.91	2.06	0.56		1.19	1.04	0.83	
	(a)				(b)				(c)			

first on the original data without limiting distance for classes, then on the same data after a principal component transformation, and finally on these transformed data by using a limiting distance for each class. Numerical results are given in Table 18.

One notes that the principal component transformation improves the results of the classification. This can be explained theoretically by the following remarks. In the channel space, reported to principal axes, the overall variance-covariance matrix is diagonal and the variance-covariance matrices for individual classes are tending towards diagonal matrices, the non-diagonal terms of covariance being small relative to those of variance. From a geometrical point of view, this means that the clusters of pixels for individual classes are tending towards spheres (taking into account the particular distance used in M.E.D.). This fact facilitates the distinction between classes and thus improves the classification.

Here again, the use of limiting distances, corresponding to the probability thresholds in M.L. method, significantly improves the classification (part (c) of Table 18.). The results are comparable to those of the M.L. method, with an important reduction in computing time.

4.3.6 Extended ratio approach

Another preliminary approach was attempted by using classification methods based on ratio of channels or, more generally, on ratio of linear combinations of channels ("extended ratio"). This work was done at the JRC Ispra by H. Gulinck, now at Centrum voor Landbouwkundige en Ecologische Interpretatie van Teledetektiegegevens, Leuven (B). In this case, corrections for atmospheric absorption which were useless in methods based on distances between pixel and class, strongly influenced the classification results.

Path radiances for the LANDSAT channels were measured on July 3rd and September 13th, in connection with rice investigation, in the zone of Mortara situated about 150 km apart from the beech test site. Corrected values were obtained by subtracting these path radiance values (transformed to bit scale) from the channel values, thus defining a new origin in channel space.

The best separation of beech forest from other categories was obtained applying to corrected values the following extended ratio:

$$\frac{3 \times \text{chann. } 4 + \text{chann. } 6}{\text{chann. } 7}$$

and subsequently a simple monodimensional level slicing. Classification was performed on a limited area (21 lines, 66 elements, i.e. 21 x 66 pixels).

The numerical results (confusion matrix) are given in Table 19, considering only two categories "beech" and "non-beech".

From these first results, this approach appears

Table 19 : Confusion matrix, scene of July 5th, extended ratio method with path radiance corrections, results in percentages on ground truth

	Classification	
G. truth	Beech	Non-beech
Beech	83.6	16.4
Non-beech	16.2	83.8
Ratio	104.6	96.4

as promising and should be studied more in detail. Particularly, values of path radiance should be deduced from the satellite data rather than from ground measurements.

4.4 Conclusions

Particular difficulties arise in the application to mountainous regions of remote sensing techniques from satellite with the aim of vegetation classification. Nevertheless, this classification remains possible by such techniques if the algorithms are adapted in such a way to create a class of rejects, which gathers all objects which are not sufficiently characterized on the ground truth. The test used for checking the goodness of classification, the so-called confusion matrix, is much more severe than the simple comparison of surfaces where the two types of classification errors, omissions and comissions, always partly compensate each other. It should be emphasized that the confusion matrix takes into account also possible errors in superposition of LANDSAT scanning grid and ground truth document (map or aerial photograph). Under European conditions, where fields or forests are often a few pixels in area, this superposition problem acquires a crucial importance.

The classification methods used so far allow a satisfactory recognition of beech and even its discrimination from other deciduous species (such as chestnut). The better discrimination achieved in July shows the importance of the observation date, which should be carefully studied a priori in collaboration with specialists (in this case foresters).

APPENDIX - OUTLINE OF DATA PROCESSING METHODS USED

Unsupervised classification methods

Two clustering methods have been used, both utilizing the euclidean distance as similarity measure between data points, and membership threshold to define the clusters.

The first method [8] constructs the clusters from a random subset of points and attempts to grow the cluster (class) to which a point is attributed by examining also the neighbouring points on the same line. All the points (pixels) of the studied zone are then classified into the various classes defined before, by using euclidean distance between point and mean of the classes. There is no iteration process in this method.

The second method uses all the points of the studied zone in the clustering process. A rejection criterion is also introduced within the points already clustered in order to take into account the evolution of the classes. Several iterations are generally performed in order to ensure a reasonable stability of the results.

Supervised classification methods

The maximum likelihood method (ML), with normal parameters to describe the training classes, has been used [9]. The following decision law is implemented for vector (point) x:

$$h_k(x) \geqslant h_\ell(x), \ \forall \ell \neq k \rightarrow x \in \text{ class k}$$

where

$$h_k(x) = \ln P(C_k) - \tfrac{1}{2} \ln | \Sigma_k |$$
$$- \tfrac{1}{2} [(x-M_k)^\mathsf{T} \Sigma_k^{-1} (x-M_k)]$$

M_k mean vector of class k
Σ_k covariance matrix of class k

In the present context, $\ln P(C_k)$ is cancelled from the expression of $h_k(x)$ (see section 2.3.2 for discussion).

The modified euclidean distance method (MED)

This method can be considered as formally identical to the ML scheme, but with the assumption that all the covariance matrices describing the class distributions are diagonal. From another point of view it can be seen that the method calculates the euclidean distances between any vector to be classified and the mean vectors of the classes of interest, it then weighs the distances by the variance of the classes; lastly it constructs the following distance between the vector and class C:

$$\sum_{i=1}^{n} \frac{di^2}{v_i} + \ln \prod_{i=1}^{n} v_i$$

where d_i is the euclidean distance for channel i, v_i is the variance for channel i, n is the number of channels (dimensions) and the first term is the weighted euclidean distance. The unknown vector is consequently assigned to the class for which the distance just described is minimum. This second method is simpler and much less time-consuming than the ML method.

Uniformity mapping procedure

This procedure [9] computes the average euclidean distance between a given point and its immediate neighbours. This distance is a measure of the uniformity of the data around the point. A uniformity map is then set up by starting from the uniformity value attributed to each point.

Determination of confusion matrices (CFMAT)

This program simply performs a pixel-by-pixel comparison between the discrete version of the ground truth document and the classification results.

If m ground truth categories have been classified into n different classes, then m x n cases are considered in the comparison process and a m x n confusion matrix is built and organized, as described in sections 2.3.2.2 and 3.4.2 for instance.

REFERENCES

[1] Cassinis R., Galli de Paratesi S., Guyader M., "Agricultural resources investigations in Northern Italy, Southern France and Madagascar", Symp. on European Earth Resources Satellite Experiments, Frascati (Italy), Jan. 28 – Feb. 1, 1974

[2] Mégier J., "Classification automatique des données du satellite LANDSAT appliquée à agriculture et sylviculture", Cinquièmes Journées d'Optique Spatiale, Marseille, 14 – 17 Oct., 1975

[3] Dejace J., Mégier J., Mehl W., "Computer-aided classification for remote sensing in agriculture and forestry in Northern Italy", XIth Int. Symp. on Remote Sensing of Environment, Ann Arbor, Michigan, USA, April 15–19, 1977

[4] Cellerino G.P., Lapietra G., "Interpretazione di aerofotografie in piccola scala per il censimento della pioppicoltura", Cellulosa e Carta, no. 3, Marzo 1977

[5] Dejace J., "Inventaire des hetraies dans une région des Alpes Maritimes italiennes à partir des données du satellite LANDSAT", Coll. sur l'Utilisation des Satellites en Télédétection, Paris, Sept. 1977

[6] Lapietra G., Mégier J., "Acreage estimation of poplar planted areas from LANDSAT satellite

data in Northern Italy", Proceedings of the
Symp. held during the XVI IUFRO World Con-
gress - Subject group S6.05, Remote Sensing,
Oslo, 21—26 June, 1976

[7] Mégier J., "Multi-temporal digital analysis of
LANDSAT data for inventory of poplar planted
groves in Northern Italy", Int. Symp. Image
Processing, Graz, Austria, October 1977

[8] Turner B.J., "Cluster analysis of multispectral
scanner remote sensor data", Remote Sensing
of Earth Resources, vol.1, ed. by Shahroki,
The University of Tennessee, Tullahoma,
Tenn., 1972

[9] Borden F.Y., Applegate D.N., Turner B.I.,
Lachowski H.M., Merembeck B.F., Hoosty J.R.,
"Satellite and aircraft multispectral scanner
digital data users manual", ORSER-SSEC
Technical Report 1.75 (1975), Office for Re-
mote Sensing of Earth Resources. The Penn.
State University.

ACKNOWLEDGEMENTS

The authors want to express their appreciation for
the efficient help they obtained in setting up the
ground truth documents, from G. Baldi and R.
Malagoni, "Ente Nazionale Risi", Mortara;
C. De Carolis, "Istituto di Patologia Vegetale",
Milano; G. Cellerino and G. Lapietra, "Istituto per
la Sperimentazione della Pioppicoltura, Ente
Cellulosa e Carta", Casala Monferrato; G. Mondino
and R. Salandin, "Istituto Nazionale Piante da
Legno", Torino.

Part 2   Hydrology

B.MARCOLONGO
Laboratory of Applied Geology, CNR, Padua, Italy

22

Photointerpretation applied to hydrogeological problems:
Soil moisture content, soil permeability, hydrogeological structures

1 INTRODUCTION

Photointerpretation is an extensive term which can be applied to the a-nalysis of images taken in different bands of the electromagnetic spectrum, presented as picture-like photographic documents; therefore it includes the study of panchromatic W&B photos, IR false colour photos, MSS images, thermal IR images, Side Looking Radar images (SLR), etc.

This interpretative process has essentially inductive character, starting from certain effects or consequences and then going back to the determinant causes; to do that the researcher needs keys of interpretation which are founded, among other things, on some parameters common to all the images, such as 1. tone, 2. texture, 3. structure and 4. morphology. Tone, strictly speaking, is the optical density of the film at any point. Obviously in almost all cases it is the differen-ce in tone between objects or bet-ween an object and its background which is important. Texture is the space variation of density across the film; the general texture of i-mages is created by tonal repetition in groups of objects which are often too small to be discerned as indi-vidual objects. It gives the visual impression of roughness or smooth-ness. Structure is the group of structural elements, defined by re-petitive relationships existing a-mong the component textural ele-ments; the structural elements are then characterized by: a. kind of relationship among the texture ele-ments; b. relative tone of element; c. total tone of the zone. Morpho-logy (or shape) is the distinctive morphologic aspect of an unit ele-ment or of a group of unit elements.

In this introductory lecture I will particularly speak about photointer-pretation applied to hydrogeologi-cal problems, done on photographic and non-photographic images, taken in the visible band of the electro-magnetic spectrum, and on thermal infrared images. Such an interpre-tation sees as highly developed the last and the most elevated phase of a normal interpretative process, called deduction; it relates an ima-ge not directly seen on the photo with an object not directly perce-ptible, following the previous sta-ges of individuation, identification and classification. In fact the main hydrogeological structures are al-most always buried and so they are recognized and described only by me-ans of interposed and suitable "in-dicators", identified on the images.

For easiness of interpretation and treatment the aquifers can be sche-matically subdivided into:

1. aquifers contained in loose ma-terials permeable through (gravels porosity and sands);

2. aquifers lodged in lapideous rocks permeable through porosity (such as sandstones) and through fracturing and/or karst (such as calcareous rocks).

The first can be unconfined, with upper boundary formed by a free water-table or phreatic level under atmospheric pressure, or confined, that is completely saturated aquifers whose upper and lower boundaries are impervious layers.

They are typical of alluvial **plains and** fairly wide valley floors.

The "indicators" are directly **linked with the hydrogeological structures, recharge zones,** groundwater discharge ways, storage areas, in the case of phreatic aquifers, whereas for confined (or artesian) aquifers these interpretative elements offer information on the consistence of the water resources only in indirect or deductive way.

The aquifers lodged in lapideous rocks are fundamentally characteristic of mountainous or hilly zones; of basic importance to their description are tectonic-structural studies, owing to the strict correlation between fracturing-faulting systems and the potential existence or the possibility of formation of aquifers themselves.

Moreover drainage density analysis, performed with the help of remote sensing images, is extremely significant for the correlation betwen stream flow and ground-water bodies. The run-off from a natural watershed consists of water from various sources, and usually it is separated into three components (direct surface run-off, inter-flow, base-flow), of which the base-flow is the principal one from the hydrogeological point of view. Besides the drainage density there are many other watershed parameters, determinable with image interpretation, which are of use in quantitative evaluation of water supplies.

Later on we will see some typical examples illustrating the concepts now exposed.

1.1 PHOTOINTERPRETATION IN PLAIN ZONES AND WIDE VALLEY FLOORS: parameters ("hydrogeological indicators"), methodology, examples

The plain zones and the wide valley floors are formed by alluvial deposits more or less coarse carried and washed down the water courses.

A quite characteristic hydrogeological element of plain zones is the so called "rising ground-water band or line", which indicates the demarcation between the upper and lower plain itself. The two parts consist respectively of draining sandy and gravelly formations, bearing unique undifferentiated aquifer, and alternating silt clay beds and sand deposits; the latter normally form multistratum aquifers, the uppermost of which is always phreatic. In consequence of the decrease of particle size the water finds an obstacle to its own underflow, and because of the reduction of topographic slope, too, we have the rising of ground water forming various springs.

Other very important hydrogeological elements are the paleoriver beds, these being ancient and abandoned water courses now filled with coarse loose materials (gravel and sand), which are often seats of conspicuous water-bearing strata.

To identify and to describe these and other hydrogeological aspects, the specific photointerpretative parameters are:

1. the mosaic of agrarian parcels (subdivision of the soil into plots of land) in case of cultivated areas;

2. the paleoalluvial and alluvial morphological aspects (such as "sunken meanders, alluvial terraces, etc.);

3. areas with different soil moisture content in the unsaturated zone of the subsoil.

The parcel subdivision represents a reflex of the human adaptation to the environment. When we analyze such an agrarian partition, we must consider the global structure of the whole, namely the dimensions and the orientation of the single parcels. The distribution can be in fact equally balanced or random, being essentially linked either to certain types of soils, or to certain ground-water situations: coarse and draining alluvial soils are often covered by elongated and isoorientated parcels; soils relatively rich in humus on the contrary show polyhedric or draught-board parcels; the most poor soils frequently have large and badly delimited parcels; phreatic currents often isoorientate the parcels; high ground-water table, or high soil moisture content, with consequent more available quantity of water for agricultural purposes, determine intensive exploitation with dense unorientated agrarian mosaic.

The first slide presents a typical example of agrarian parcel mosaic; it is a W&B panchromatic photo uncontrolled mosaic, showing two kinds of parcel distribution, the former belonging to an area of draining coarse alluvial materials (sandy soils), the latter to a calcareous zone with initial karst phenomena (red iron soils) (Fig. 1 Mosaic of Crocetta del Montello Piave river North Italy).

Starting from these principles, we must carefully analyse and mark all these bands of agrarian parcels which, either for their meandering form, or for their prevalent extension direction unconformable with those of the limitrophe plots of land, are evident elements of rupture of the general agrarian texture. These bands correspond to paleoriver beds and to abandoned meanders, important to define the main groundwater flows of the phreatic aquifers.

To accomplish such an analysis of parcel structures simple W&B panchromatic photos are sufficient.

As an example we report the results of a photointerpretation study done on W&B photos, to individuate the principal hydrogeological structures

of the alluvial plain between the rivers Astico and Brenta (North Italy).

The second slide shows the aspect of parcel structure of a paleofluvial course and a series of abandoned meanders (Fig. 2).

The third and fourth slides show respectively a circumscribed area of clay and silt soils (which represent the outcrop of a pleistocene paleoplain), and a few relict meanders (Fig. 3, 4).

Lastly the fifth slide displays the global results (hydrogeological and paleohydrographic map) obtained all over the investigated area (western part of the Venetian plain), based almost uniquely on the parcel mosaic analysis and the morphology of paleoalluvial traces (Fig. 5).

The sixth slide shows the stereogram (or block diagram) of the eastern portion of the same area, a part of wider stereogram built up analysing more than five hundres stratigraphic columns from mechanic and geophysic boreholes. It represents the "ground truth" to control the methodology applied in the photointerpretation; it also emphasized the relationships betwen the superficial hydrogeological aspects (as identified by photos) and the deeper hydrogeological structures (drawn up by "deduction" in the interpretative process) (Fig. 6).

The interpretation of the fluviomorphological aspects can be done on various remote sensing images, but it is greatly improved by the stereoscopic view, because of the tridimensional relief perception.

Among the principal fluvio-morphological structures, significant for hydrogeological interpretation, we have to mention: alluvial terraces; sunken relict meanders; boundaries of flood plain; boundaries of actual river beds; etc.

Slide n. 7 shows an outstanding example of alluvial terrace, delimitating the flood plain of the Brenta river, which be seen as steep escarp under stereoscope. The terrain on the right of the terrace is higher than on the left (Fig. 7).

Soil moisture content (as percentage = $\frac{\text{vol. water}}{\text{vol. soil}}$ x 100), that is the amount of water stored near terrain surface, is of considerable significance from both theoretical and practical viewpoint, being a basic component in the hydrogeologic cycle, energy exchange near the surface, and in modelling various ecosystem processes. Moreover soil moisture in the unsaturated zone of the subsoil gives interesting indications as to particle size (seiving curves) and then on the porosity and permeability of the upper sediments (empirical laws of Eckis, 1934 and Lane-Washburn, 1946); generally speaking the soil moisture, other conditions being equal, is higher in fine particle soils* (fine sand, silt, clay) than in the coarser ones (medium sand, coarse sand, gravel and cobbles).

Reflectance at photographic wavelengths (0.4-0.9   ) is sensitive to differencesin the surface characteristics at the soil-plantair interface.

A densitometric analysis done on W&B photos can be very useful to delimitate areas with equal soil moisture content. But we have to consider that the moisture-reflectance relationship is influenced by even small variations in soil texture; with increasing content of fine soil particles, integral reflectance of the soil increases for all moisture levels.

The addition of water to soil also changes the soil thermal properties and in general it increases the conductive capacity of the soil (C_c = C.   . K, where C is the specific heat,   = density, K = tempe-

---

* Here the term soil is used with extensive meaning, designating the pedogenetic substratum as well the "illuvial" and "eluvial" horizons.

rature conductivity), thereby causing an attendant decrease in the temperature range at the soil-air interface. Neverthless, factors such as surface vegetation tend to mask the temperature range at the soil-air interface, so that attempts to utilize thermal remote-sensor data in order to map soil-moisture distribution are not very valid (Mac Dowal et al., 1972, Burge et al., 1971).

Multispectral photography in both the visible and near infrared can on the contrary give successful results.

Specific photointerpretative works done the alluvial Venetian plain for hydrogeological purposes, have shown that the use of bands 0.5-0.6 (green-yellow) and 0.8-0.9 (near infrared) and their successive analogous elaborations (product, ratio, additive synthesis of product and ratio) should be routine procedure in describing soil moisture content and other hydrogeological features.

The presence of humidity in the unsaturated region of the subsoil, or water-bearing layers extremely close to the surface of the terrain, brings about a sharp drop of reflection in photographic infrared (band $0.8-0.9\mu$), together with normal green-yellow reflection (band $0.5-0.6\mu$).

Finding the "ratio" betwen these bands thus makes it possible to bring into evidence zones characterized by the same medium soil moisture content.

Moreover plant cover presents considerable reflection both in the photographic infrared band and in the green-yellow band. Thus the determination of the product leads to an exaltation of the outlines of plant coverage.

To synthetize graphically all my lecture I would like to show you the slides n. 8 and 9 (respectiveli: Fig. 8 ratio of Skylab bands $0.5-0.6\mu$ and $0.8-0.9\mu$; Fig. 9 product of Skylab bands $0.5-0.6\mu$

and 0.8–0.9μ) and slide n. 10 (Fig. 10 results of photointerpretation of the two combined treatments), concerning the eastern part of the Venetian plain.

Here the "ground-water rising line" (or "spring line"), hydrogeological element common to almost all plains, is identified by tracing the borders of the areas characterized by the prevalence of one of these two parameters: soil moisture content "high", soil moisture content "low".

The paleochannels in the upper plain can be mapped by looking at the simultaneous presence of the following parameters (or situations): "low humidity", "small herbaceous crops", "isoorientated parcels".

The existence and extension of alluvial conoids (or alluvial fans) is described by marking and comparing the borders of the areas where, on the average, parameters such as "isoorientated parcels" and "irregular distribution of parcels" are preponderant.

At the end of this chapter dealing with the hydrogeological photointerpretation on plain zones, I would like to present the synthesis of a study on IR false colour Skylab images, regarding the delta of the Po (Fig. 11).

The hydrogeological and geopedological (types of soils) situation are connected and compared through performing mixed photointerpretation.

Furthermore the area is strongly marked by the human activity (agriculture), and so can draw quite interesting consideration on the interferences betwen man and environment; for instance the relationship between areal distributions of land reclamations and phreatic water levels. Lands with very shallow or quite surfacing ground-water tables are frequently reclaimed.

The thematic map of the hydrogeological, geopedological, and mor-

phological aspects of the delta of the Po shows all this (Fig. 12).

1.2 PHOTOINTERPRETATION IN MOUNTAINOUS AND HILLY AREAS: parameters ("drainage pattern", "lineations"), methodology, examples

In the mountainous and hilly areas the aquifers are principally found in rocks permeable by fracturing and karst (limestone, calcareous rocks and intrusive or effusive igneous rocks) and only locally permeable by porosity (sandstones).

There, morphologic and morphometric analyses of the "drainage pattern" and other watershed parameters (basin shape and area, "relief ratio", etc.) and structural or tectonic analyses of fractures and faults are of primary importance.

The former analyses provide in fact significant information on the degree of permeability of the rocky substratum and the stream flow dimension, from which the "base flow" can be derived (assessment of the aquifer's potentiality).

The latter analyses give suggestions on the underground water-circuits (where we have unidirectional permeability controlled by fracture and fault systems), aquifer's recharge areas and ground-water storage areas.

The system of channels, called a "drainage pattern" (or "erosion design"), is always the starting point of a hydrogeological analysis of remote sensing images.

In general the simple W&B panchromatic photos provide information, concerning the channel networks, often far superior to topographic maps.

The interpretative principle that a pattern depends, among other things (tectonics), upon the lithology and hence the permeability of rocks leads to the conclusion that common characteristics such as density, ramifications or orientation indicate materials with the same degree of permeability or hydrogeological conditions.

The drainage pattern may be studied from a. a qualitative and b. quantitative point of view.

a. Several authors (von Engeln, Lobeck, Smith, Parvis, Zernitz et al.) have described certain types of drainage pattern which are regerded as fundamental or basic. These types are (Fig. 13):

1. dentritic (or treelike) the most common erosional pattern, often developed on impermeable materials (as clayshale);

2. angular (or trellis) structurally controlled by fault lines, fractures or joints, on hard resistant rocks of granular texture (massive igneous rocks, massive sandstones, crystalline rocks, etc.) permeable through fracturing;

3. radial an arrangement of consequent streams flowing radially from an uplift dome or volcano, on fine and granular tuffs permeable through porosity;

4. parallel, peculiar of impermeable fine-textured materials (shales or clay) with tilted surfaces. Parallel pattern is rather common on coastal plains and valley fills, permeable through porosity;

5. annular, appearing on annular structures like domes or basins, formed by rocks of different erosional resistence, and therefore of different hydrogeological behaviour (possible multistratum aquifer).

Patterns of internal drainage occur in soluble rocks like limestone or gypsum (permeable through erosion) or in insoluble porous materials like sandstone or conglomerates (permeable through porosity). The "sinkhole pattern" (Longwell and Knopf) should be regarded as basic.

The sinkholes on soluble rocks, termed locally as "dolines", "uvalas", "poljes", have a roundish outline.

The sinkholes on insoluble granu-lar rocks, are roundish or irregular shaped depressions.

In both the cases they represent infiltration centres where the rainwater or the superficial water percolates into the rock beds beneath. The analysis of their distribution is quite important to determine the recharge area of an aquifer.

Besides the basic drainage types described so far, special drainage patterns are known (Fig. 14).

Among the common or significant types we have to mentionare:

1. deranged pattern, combination of surface and subsurface drainage of glacial drift regions (glacial till or moraine with "knob and kettle topography", typical of certain regions of the Alps). The surface channels are without destination erratically contorted and haphazard on a semipervious terrain;

2. dichotomic pattern, found on alluvial fans or on deltas (built up with very permeable coarse granular sediments, with streams radiating from a common point ("gap");

3. phantomatic or illusory pattern, indicating shallow subsurface drainage with no channel development. It becomes visible due to soil shading or vegetation changes (tonal or densitometric analysis).

An outstanding example is given by the buried channel network of the Cellina-Meduna alluvial fan, visible on slide n. 8 (Fig. 8), and interpreted on map, Fig. 10, previously shown.

For hydrogeological studies to be done with considerable detail (photographs at 1/20.000 or larger scale, preferably examined with magnification), quite important features are the "gullies"; these are the drainageways of the fourth order or initial running water cut into consolidated or unconsolidated materials.

According to D.R. Lueder (1959) young gullies show the following characteristics related to the hydrogeological properties of the sediments:

334

1. short gullies, strainght narrow V shaped, with rather shallow depth and steep gradient: in non-cohesive, permeable through porosity materials, like sand and gravel;

2. long gullies, deep, wide, gently sloping smooth curves, with many branches, and uniform gradient: in cohesive impervious sediments, like shales, clays, marls;

3. rather long gullies, extremely variable in depth, not shallow, with moderate width and all types of cross section like U or V: in intermediate semipervious materials, as sandy shale, silt, interbedded deposits like flysch.

b. The quantitative description of a drainage basin and its channel network is normally subdivided into information concerned with the linear aspects of the channel system, the areal aspects of the basin and the relief aspects of the drainage basin and pattern.

The procedure introduced by R.E. Horton (1945) and modified by Strahler and Schumm (1964), to assign to each tributary an order number, is still the starting point of such an analysis. Stream order number has been shown to be directly related to the size of the contributing watershed, to channel dimensions and to stream discharge (Strahler, 1964).

There are relationships codified as laws among number, length and slope of streams of each order and "bifurcation ratio R_b" and "length ratio R_1" (number and mean length of streams of i order to number or mean length of streams of next higher order).

In establishing some of the areal aspects of drainage basin, the area of individual channel segments of a given order is mapped and the relationship of area to length and other factors can be established. Once areas are defined, the relationship of basin area to stream discharge can be examined ($Q = j\ A^m$,

where j and m are constants derived by fitting a regression line to available data between Q and A).

Another item of interest is "drainage density". From the base map delineating the entire channel network of streams of all orders in a given watershed basin, the ratio of the cumulative total of channel segment lengths for all orders to the total basin area can be established; this ratio is referred to as drainage density.

Controlled values of drainage density range from lows around 4.8 km/km^2 to highs of about 2100 km/km^2.

Low values of drainage density have been shown to correlate to regions of highly permeable subsoil materials (or highly resistant) under dense vegetative cover. Relief generally is low.

High drainage density values are associated with regions of impermeable subsurface materials (or weak zones), with sparse vegetation and high relief.

For interpretative purposes the drainage density is qualitatively classified as fine, medium, coarse. The coarser the density is, the more permeable the substratum is, other conditions being equal.

Therefore analysis of drainage density is very significant to evaluate the base flow, owing to the inverse correlation between the two entities (higher the drainage density, smaller the quantity of base flow). This may be explained by the fact that within an area of homogeneous climatic conditions, the drainage density is strongly influenced by the permeability of the rocks. On impermeable rocks direct surface run-off is high, but recharge of the ground water is insignificant.

Drainage density is also valuable for predicting mean annual run-off, especially in semi-arid climate areas.

A second approach to hydrogeological interpretation in mountainous areas is the structural analysis of lineation patterns, the lineations being potential ways of infiltration and percolation of ground-water. This is particulary true for aquifers contained in rocks permeable through fractures, were often the permeability is unidirectional and governed by the fault and joint system. Therefore the lineation pattern is essential to describe recharge areas, ground-water flow directions and storage areas; in other words, the wole hydro-morphological model of the aquifer.

In this respect it seems appropriate to go deep into the concept of "lineations".

With the term lineations, whose synonyms could be "linears", "lineaments" or "linear features, we usually mean a rock discontinuity with prevalently rectilinear development, to which a tectonic nature (fractures, joints, faults) is generally attributed. Nevertheless in the photointerpretative analysis the physical objects are observed through an intermediate passage (images), or through a certain manipulation, which surely modify their external appearances.

A substantial alteration occurs when one presents lineations having different natures, as if theywere similar and homogeneous. For example a linear hydrogeological structure (namely almost rectilinear paleoriver bed, weathered and soaked rocky band included in more impervious materials, etc.) can reveal itself with the same photographic characteristics of a fault.

Therefore lineations often represent the "superficial reflection" of deeper physical situation and can also underline the presence of faults and fractures through an intermediate system of "indicators".

The lineation patterns can be very well analysed, on satellite images thanks to their characteristics

of synthesis, (for scales not smaller than 1:100.000 - regional studies).

As an example of the concepts illustrated so far, we show the total fields of lineations drawn from band (0.8-1.0μ) of Lansat-1 image and from derivative of the same band 7 (Fig. 15 and 16), covering a test-area of the central Alps (Bolzano - Innsbruck).

The lineations have been subdivided, through preliminary screening with geological and topographic maps, into homogeneous groups characterized by a prevalent geo-interpretative aspect (morphological, lithological, tectonic and antropic), these groups are distinct from another group of lineations not yet classified, which can be considered as the new information obtained from the satellite image interpretation.

The "lineation pattern" of band 7 depicted on slide 17 (Fig. 17) shows how often the same lineation, unitary from the geometric point of view, is not equally homogeneous as photointerpretative nature along all its development; it can fact present a morphologic stretch, followed by a lithologic or tectonic segment, etc.

The illumination influence on the interpretation of linear features in mountainous areas (like those here considered, whose Landsat-1 images have sun azimut and elevation respectively ranging from 147º to 157º and 22º to 30º) is sometimes very strong. In fact we can have the percentage increase or decrease of the presence of certain lineation classes.

On this subject I elaborated a logic model, which can be quantified after statistical analysis, for the qualitative compensation of the unlike data obtained from the visual interpretation of satallite images. The model is shown in slide 18 (Fig. 18). It gives the relationships among the enhancements caused to different lineation groups by shades at various year periods.

To reduce the influence of sun azimut and elevation on tilter surfaces, we used also the analogous treatment of "autocorrelation"(or "derivative") of band 7, suitably done throughan optical process.

The slide 19 (Fig. 19) shows the total field of lineations interpreted on autocorrelated band 7.

Rose diagrams respectively of band 7 and its autocorrelation prove the improvement attained with the treatment (Fig. 20). Above all it is important the counter-clockwise shift of the maximum percentage values from the N 30°-60° E class (band 7) to the N 0°-30° E class (derivative of band 7).

In fact the shadow effect jointed with the particular topographic situation of the area rotates the lineation directions, as seen on the images, towards the sun direction (clockwise distortion). The autocorrelation process permits to avoid that.

At the end we present an example of structural analysis for hydrogeological purposes done on thermal images (9-11$\mu\nu$) and harmonic analysis of the same images (this treatment, which gives results similar to those obtained by derivative or autocorrelation, was electronically processed).

Slides n. 21 and 22 show respectively the band 9÷11μtogether with its harmonic analysis and the derived interpretation (coastal area in South Italy) (Fig. 21 and 22).

The linear discontinuities are marked taking into account every fine texture variation, or every boundary betwen two areas with a different textural aspect.

Here the rocks are principally intrusive igneous "gabbros" or metamorphized types of the same chemins and calcareous sandstones. Therefore the recognition of the principal lineation systems serves to guide more detailed studies,

to be done on aerial photos, with the purpose to find out the model of the underground water-circuits.

2 REFERENCES

American Society of photogrammetry 1975 - Manual of Remote Sensing. Vol. I and II. A.S.P., Falls Church

Baggio P., Marcolongo B., Sottani N. 1975 - Il bilancio idrogeologico degli acquiferi nella pianura a nord di Vicenza. Quaderno 1. "Studi Trentini di Scienze Naturali", 52/3, Trento.

von Bandat H. F. 1962 - Aerogeology. Gulf Publ., Houston.

Barret E.C., Curtis L.F. 1974 - Environmental remote sensing. Vol.I and II. Edw. Arnold, London.

Cardamone P., Casnedi R., Cassinis G., Cassinis R., Marcolongo B., Tonelli A.M. 1976 - Study of regional linears in Central Sicily by Satellite imagery. "Tectonophysics", 33, 81÷96, Amsterdam.

Cassinis G., Marcolongo B. 1976 Variation in alignment patterns obtained by satellite imagery with respect to scale and enhancement techniques: application in two central areas of the Alps. Round Table on "Geology and Artificial Satellites", E. Majorana Int. Centre for Scientific Culture, Erice.

von Genderen J.L. 1969 - Technical exercises for the Sub-Department of Geography - Vol. II, Special exercises. I.T.C., Delft.

Girard C.M., Girard M.C. 1975 Applications de la télédétection a l'étude de la Biosphère. Masson et Cie, Paris.

Kruseman G.P., De Ridder N.A. 1970 Analysis evaluation of pumping test data. Int. Inst. Land Reclamation and Improvement, Wapeningen.

Lambe T.W., Whitman R.V. 1969 Soil Mechanics. J. Wiley & S., London.

Lechi G.M., Marcolongo B., Tonelli

A.M. 1974 – Applicazioni delle immagini multispettrali da satellite alla individuazione di mate-
riali da costruzione nella pianura veneta. Atti del 1° Conve-
gno Int. Pietre e Minerali Lito-
idi, Torino.

Lechi G.M., Marcolongo B., Tonelli
A.M. 1976 – The principal hydrogeological aspects of the eastern
Venetian plain as identified by
multispectral photography from
satellites. XXI Congress of the
Italian Society of Photogrammetry, (I.S.P.), Bologna.

Marcolongo B. 1973 – Fotointerpre-
tazione sulla pianura alluviona-
le tra i fiumi Astico e Brenta,
in rapporto alle variazioni del
sistema idrografico principale.
"Studi Trentini di Scienze Naturali", Trento.

Marcolongo B., Marini C., Semenza E.
1976 – Lineation's pattern by
satellite images in a test area
of the Alps. Round table on "Geology and Artificial Satellites",
E. Majorana Int. Centre for Scientific Culture, Erice.

Marcolongo B., Mascellani M. 1977
Satellite images and their treatments applied to the identification of the "Roman Reticulum" in
the Venetian Plain. XXXVI Photogrammetric week, Stuttgart.

Marcolongo B., Matteotti E. 1976
Utilizzazioni di immagini da satellite nella programmazione ter-
ritoriale regionale. XXI Congres-
so Naz. Soc. It. di Fotogramme-
tria e Topografia (S.I.F.E.T.),
Bologna.

Meyerink A.M.J. 1970 – Photo-interpretation in hydrology, a geomorphological approach. I.T.C.,Delft.

Miller V.C. 1961 – Photogeology.
Mc Graw- Hill, London.

23

H.HAEFNER
Department of Geography, University of Zurich, Switzerland

Snowcover monitoring from satellite data
under European conditions

From any standpoint involving man's dependence on water,
it is difficult to overemphasize the importance of snow
in the hydrologic cycle. However, there appear to be no
reliable estimates of the amount of water which is deposited annually in the form of snow and removed by melt on a
continental or worldwide scale.

Perhaps the primary research need in this area at the present
time is a more thorough study of the contribution of meltwater from the alpine belt of major mountain ranges.

> Donald Alford (in: Ives&Barry: Arctic and
> Alpine Environments, London 1974, p. 107)

1 INTRODUCTION

Freshwater is one of the critical natural
resources, which affects men directly and
indirectly. Consequently there is a great
need for detailed and rapid data for a
better monitoring and management of water
resources on a global basis.

Freshwater in many parts of the world,
in particular in middle latitude and polar
climates and in high mountain environments
derives primarely or to a substantial part
from the melting of the seasonal snowpack
(Fig. 1).

Therefore snow is a very important natural resource which needs continuous monitoring. This may be undertaken best by
remote sensing techniques, especially by
high resolution satellite systems.

Snow suveys with aerial photographs have
been amongst the first and obvious applications of photointerpretation. Already in
the 1930-ies ground photography has been
used by Potts in the Rocky Mountains of
Colorado, USA, and a first comprehensive
report on applications of air photos was
published by the US Army Corps of Engineers in 1948 (Hall 1952).

Snow as a research topic has been taken
up again in the initial phase of weather
satellites. TIROS-imagery of April 1960
was used in a survey on Eastern Canada

(Barnes & Bowley 1974). The first systematic study was published by Fritz in 1962.
Since then a vast number of projects has
been undertaken using various satellite
systems, as reviewed in WMO-Report No 19
(1973), by Barnes & Bowley (1974), and by
Rango & Itten (1976). The start of the
Landsats' increased the activities to an
even greater rate. These results are summarized best in the proceedings of a workshop
held in South Lake Tahoe, California, sponsored by NASA and the University of Nevada
(Rango, Edit. 1975).

Snowmapping with remote sensing techniques has several advantages:
- Large areas can be observed simultaneously, allowing a truely regional comparison (which is impossible on the ground).
- Areas which are remote and inaccessible
can be surveyed.
- The reflectance characteristics of snow
are distinct and - in general - clearly
separable from other features.
- Only a single feature - snow - has to
be classified and separated from all other
objects (whilst for other purposes, e.g.
land use, there are always numerous categories to be distinguished).

But inspite of these extensive activities
and the great number of research papers on
the subject, it has to be mentioned that

the development of operational snowmapping
systems is rather slow. Snowcover monitoring from satellite is not yet undertaken
on a routine basis. This may be due to
various reasons such as cloudcover problems, non-availability of data in almost
"real-time", difficulties in dealing with
very different types of snow, influence
of relief in high mountains and vegetation
cover (forests). In particular in Europe,
where the problems are somewhat different
than in other parts of the world, these
facts call for special attention.

Therefore the purpose of this paper is
to review snowcover monitoring by satellite techniques as a basic information for
water management and run-off forecasting
under the specific demands and conditions
of Europe. In this respect the paper is
connected with the one by Dr. Martinec,
with whom our research activities are
closely coordinated.

2 EUROPEAN NEEDS AND OBJECTIVES

2.1 Purpose for snow surveys

Surveying the snowcover with remote sensing techniques may be applied for quite
a many different purposes, such as:
- water management
  = freshwater supplies and reserves
  = run-off prediction
  = irrigation practices
  = hydroelectric power production
  = flood control
  = sedimentation
  = navigation
- hazard
  = avalanches
  = floods
  = in urban areas
  = to traffic and communication systems
- planning
  = settlements
  = recreation
  = transportation
  = agriculture
  = land evaluation in general
- scientific demands
  = course of transidient and climatic
    snowline
  = seasonal and regional distribution
    and variation of snowline
  = snowmelt pattern
  = influence on radiation balance and
    atmospheric conditions
  = geoecological impacts
  = influence on man-made features

2.2 Types of surveys

To gain the necessary background information for the different applications, various snowparameters have to be measured,
such as:
- areal extent of snowcover in its seasonal variation
- snowdepth
- water equivalent
- surface conditions
  = snowtype
  = temperature (especially freezing-
    thawing line)
  = wetness
  = age
  = pollution
  = spectral characteristics
- separation of snow and clouds
  etc.

Sofar existing satellite systems provide
primarily information on the areal extent
of the snowcover. In this respect we always have to keep in mind, that what we
receive by a sensor system is the spectral
response (be it reflection or emission)
from a two-dimensional surface. What we
want to know are the identity and the characteristics of a three-dimensional object.
To obtain this information proper interpretation methods have to be developed. For
efficient water monitoring data on snowdepth or - even better - on the water equivalent are essential. There was great hope
that these data may be obtained from MW-
devices. Numerous experiments were carried
out in this direction (they are summarized
best by Moore 1974), but up to now no definite solution could be presented.

2.3 Demands for Europe

The Remote Sensing Working Group (RSWG) of
the European Space Agency (ESA) clearly
pointed out, that "the main European emphasis is placed on the management and conservation of known resources, rather than
on the exploration and exploitation of new
resources at national level (ESA 1977, p.2).

In addition to these requests there are
specific constraints to be taken into consideration, such as:
- climatic conditions
- statistical requirements (very high
accuracy etc.)
- smallness of individual areal units involved (field pattern etc.)
- intermixing and heterogenity of units
(land use / settlement pattern etc.)

Therefore the following recommendations were established for a European earth observation satellite program:
- all weather capability of sensing system
  - repetition rate of 6 to 9 days
  - 6 to 10 spectral bands
  - ground resolution of 15 to 30 m

2.4 Objectives

Even so these recommended specifications cannot be met by present days spaceborne data acquisition systems, methods of snow-cover monitoring shall be reviewed regarding these European demands. Examples will be presented which - hopefully - may lead to an operational classification system suitable for Europe.

Up to now the most urgent needs are in the field of run-off monitoring from snow-melt and run-off prediction. For a day to day calculcation or the use of mathematical models the most important parameters are the extent and the changes of the snow-covered areas. Therefore main attention is given to the development of an accurate and fast method to classify digital Landsat-data, as the presently most suitable earth observation satellite system. But other methods and data acquisition systems shall be considered, too.

3 BACKGROUND INFORMATION ON SATELLITE SYSTEMS THAT HAVE APPLICATIONS TO SNOW SURVEYS

No detailed technical discussion of existing and future satellite systems useful to snow survey shall be presented here. Reference is made to the corresponding handbooks. Fig. 2 lists the most important systems.

From the existing systems Landsat and NOAA-VHRR are of particular interest. Additional useful data is expected from HCMM and NIMBUS-G.

It has to be underlined, that the proper selection of the data acquisition source is of great importance especially regarding the development of an adequate and economical operational interpretation system. There always has to be a realistic relation between:
- size and structure of the study area (watershed)
- expected accuracy of the results
- remote sensing system (ground resolution/repetition rate)
- method of data processing and information retrieval.

Only the selection of an appropriate combination will allow to set up an economic interpretation system, capable of producing the needed information within a very short time lapse.

If very detailed results of a limited watershed are needed, only aerial photography can provide a sufficient data base. Fig. 3 represents a high altitude b&w aerial underflight to verify the results from satellite classification. Small scale photography produces an extremely detailed picture and is useful for such purposes.

But already catchment areas bigger than approx. 100 km^2 are difficult and unecono-mical to cover continuously from airplanes and too many photos have to be interpreted. Landsat data may already be appropriate from 10 km^2 on upwards and are best suited for areas of hundreds of km^2. A convenient relation between study area, sensor systems and ground resolution can be summarized as shown in Table 1.

It is obvious that with increasing acreage there has to be a tradeoff in accuracy. What never can be given up is time to deliver the vital information. Consequently the developed interpretation method - as well as the number of frames or tapes to be treated - has to be of a practical and economical order.

An interesting aspect could occur by combining two different data acquisition systems. For example most recent studies in the small watershed of Dischma-valley (43 km^2) by Martinec showed, that Landsat data provided as accurate (or better) results as the measurements from orthophoto-graphs. Satellite data could substitute aerial photography regarding accuracy, but not regarding repetition rate. An 18-day cycle is by far too long for detailed studies.

The combination of a high resolution earth resources satellite system with relative low repetitation rate together with a weather satellite system of high repetivity and low ground resolution should be tested. But sofar no systematic research has been conducted along this line.

4 SPECTRAL PROPERTIES OF SNOW

As already mentioned, snow represents a well defined and clearly distinguishable feature. But on the other hand it has to be emphasized strongly that no such general feature as "snow" exists. The physical characteristics of snow and especially the condition of its surface vary greatly in different climatic regions of the world according to weather condition, season, aging process etc. Consequently the spectral response of snow differs considerably between different snowtypes and of the same snowcover during the diurnal and seasonal cycle. Besides snow is always covering an underlying object which may influence its spectral reflectance. The question arises how thick a snowcover has to be so that its spectral reflectance is not influenced anymore by the underlying stratum. Barnes & Bowley (1968, p. 106) estimated that "from an analysis of the brightness level alone, snowdephts greater than about 4 inches (approx. 10 cm) cannot be differentiated in satellite photography". Staenz (1976, p. 113) observed from his ground measurements on the spectral characteristics of snow that the influence of the snowdepth varies with the snowtype. For example, measuring slightly wet snow, an influence could be noticed up to 14 cm, measuring very wet snow it amounted up to 28 cm.

A specific problem occurs in forested areas. It is almost impossible to classify snow in dense tree stands. Only a comparison with the surrounding open areas may lead to definite conclusions.

Until recently little data on the spectral characteristics of snow were collected under natural conditions. For many years a single generalized spectral curve (Fig. 4) was known and widely used in many publications.

Measuring the solar energy reflected from snow poses quite a many problems. A good example is the study by O'Brian & Munis (1973), which was conducted under controlled lab conditions simulating various conditions of natural aging such as melting and refreezing (Fig. 5).

For our purpose it was found more appropriate to measure in the field under various meteorological conditions. Since the study should support our Landsat‑classification and only portable equipment was feasible, an EXOTECH-100 was used (Staenz 1976).Here the measured spectral wavelengths are limited to the wellknown four bands in the visible and near IR. The reflected radiance of six snowtypes was measured at different slope angles, daytimes, in sun and in shadow, as well as the same sample during thawing of the surface from frozen condition in the morning to a wet one in the afternoon. Additional measurements from a helicopter served as means for investigating atmospheric radiation attenuation of the lower atmosphere.

Up to now the results show that two main groups of snow are presumably discernible using the four MSS bands (Fig. 6). The most important snow paramteres (as observed in the field and described by Staenz, 1976), modifying the reflected radiances are: grain form, snowdepth (up to 30 cm), snowdensity, wetness, pollution of surface, roughness of surface, age of snowpack.

The influence of the snow wetness on the brightness values was measured during about two hours at the same place. The snow surface changed from frozen to wet during the measuring period. From the target reflectance values measured in all four channels, ratios were compiled. They are presented in Fig. 7 in function of time. Ratio 4/7 and to a lower amount 5/7 show a distinct modification from the frozen and dry to the wet snow conditions. The combination of these two channels consequently will provide good information on the moisture conditions of the snow surface.

A comparison of the reflectance measurements in sun and shadow is given in Fig. 8a and 8b. In Fig. 8a the results (mean values) of brittle snow, determined with a reference panel, show a good correspondance. In contrast to the conformity of the target reflectance stands the reflected radiance (mean values $\bar{E}$), as shown in Fig. 8b with a clear separation between measurements in sun and in shadow. Similar results were achieved from other snowtypes.

It is obvious that such measurements are time consuming and depend very much on the weather and snow conditions. Therefore not all snowtypes could be covered sufficiently to gain the necessary statistical basis. The measurements have to be continued.

Just recently a longterm observation program of the microwave emission and scattering behaviour of snow was started by Schanda using a test site at Weissfluhjoch/Davos. First results could already be published (Hofer & Schanda 1977).

The purpose of our study was to get a better understanding of the variability of the spectral properties of snow and to use the results for a correlation with Landsat brightness values. Here we have to deal with another problem, namely the saturation of the pixels.

In Landsat digital output snow as the brightest feature is often "cut off". Fig. 9 shows an example representing snow in sun and in shadow in the four channels in a block-diagram-form.The saturation of the sun-exposed snow in the channels 4 and 5 is shown clearly. Therefore the "Int. Working Seminar on Snow Studies by Satellites" organized by WMO in Geneva in 1976 recommended that "future Landsat-MSS and similar satellite sensors be designed to measure snow radiance through the entire range of expected values and not be cut off or 'saturated' as they do in Landsat-1 and -2".

4.1 Spectral bands for snowmapping

Judging from our experience it has to be concluded that for efficient snow studies there are at least three channels needed, one in the visible (Landsat-band 5 is preferred to 4 to determine the areal extent), one in the near IR (Landsat-band 7 to separate dry and wet snow) and one in the 1.55-1.75 μm band (to separate snow and clouds).

5 GROUND TRUTH

In an initial experimental phase extensive ground measurements and field observations should be carried out in addition to remote sensing studies for:
- a better understanding of the characteristics and importance of the various snow parameters and its influence on the spectral signatures,
- a correct selection of the test samples to be used in a supervised classification (chap. 7),
- the verification of the classification results.

Therefore a routine ground investigation program was set up observing the important snow parameters during all Landsat-passes on more or less cloudfree days (Fig. 10). The goals were to determine the significant parameters which should be observed furthermore as a minimum requirement (Staenz 1976, Stirnemann 1976).

As a result of this field campaign we learned that for a description of the physical properties of the snow and in particular of the snow surface all mentioned

parameters are needed. Only air temperature and density are highly correlated.

High altitude aerial underflights serve as excellent references to verify the accuracy of the classification (Fig. 3).

In an operational system ground measurements and observations will have to be reduced to a minimum. For the time being they cannot be omitted completely primarely for the proper selection of the test samples. But with the built-up of data banks on the spectral behaviour of different snowtypes, the task can be minimized.

6 DATA INTERPRETATION

A good review on the state of the art - especially on the operational aspects in the various countries - was presented at the "WMO Int. Working Seminar on Snow Studies by Satellites" in Geneva, 1976. The latest synopsis on satellite data analysis techniques was presented by Rango & Itten (1976) considering the following selected methods:
- photointerpretation (from single to complex methods)
  = snowline discrimination and planimetering of snowcovered area
  = mean snowline altitude conversion to snowcovered area
  = interpretation of gridded imagery
  = optical enhancement devices - density slicing
  = electronic interactive analysis
- digital image processing
  = unsupervised ⟩ batch or interactive
  = supervised

It has to be emphasized that for quite a many purposes analog techniques are more feasible and economical, especially for mapping the areal extent of the snowcover. Photointerpretation is by far the most simple and easiest approach and does not require costly equipment. The methods may be facilitated and improved by using various optical and electronical enhancement and density slicing devices, in particular for producing better photo products and for a fast areal measurement. Only when continuous, routine snowmapping is desired in a longterm project, the interpretation procedure should be automated and quickened as much as possible, which can only be achieved by machine-aided processing.

6.1 Analog processing

Since the analog techniques are rather well-known, only short notice shall be given regarding the degree of detailness and accuracy to be achieved.

The DISCHMA-valley test site (Fig. 18) is studied in detail regarding the snowmelting process. The interpretation was carried out from aerial photographs onto topo-maps 1:10,000 using seven coverage categories (100-99 % / 98-88 % / 87-63 % / 62-38 % / 37-13 % / 12-2 % / 1-0 % snowcovered area). The situation was mapped on three different dates during the most important melting period (May to June).

From these basemaps a multiplicity of information can be gained on the distribution and dynamics of the snowcover regarding exposure, slope angle, altitude, vegetation type etc. in its seasonal variations. This allows a qualitative and quantitative evaluation of the changes of the snowcover and of its geoecological impacts. The results of this comprehensive study will be published soon. In Fig. 11 a very generalized example (only totally snowcovered areas, e.g. more than 87 %; map-scale reduced) is given of the many-folded graphical and tabular possibilities resulting from such an accurate and detailed mapping.

The accuracy achieved with high resolution satellite photography could be tested with Skylab S-190-A and -B photos (Haefner & Geiser 1975). Optical enlargers were used for the interpretation and the transference onto a topo-map (Fig. 12). It could be concluded that the map-scales as indicated in Table 2 were best suited and that the course of the transidient snowlines (in particular on glaciers) could be evaluated with the given accuracy.

For local studies the use of larger map-scales up to 1:25,000 (contour intervals of 10 m) is possible with good enlargement equipment, which again would improve the accuracy considerably. But for these purposes aerial photography is recommended.

6.2 Digital processing

For a combined classification of multi-spectral scanner data on a pixel by pixel basis digital methods have to be applied, leading to an automated evaluation. Three major aspects should be kept apart (Fig. 13):

- preprocessing
- feature extraction
- presentation of the results

Before the actual feature extraction takes place, several preprocessing procedures may be applied. Two different tasks have to be considered here. The first one deals with the reformating and organization of the digital MSS-data for the own hardware components and the specific objectives. The second one contains all corrections of the data in accordance to certain criteria to improve the classification. These programs include geometric as well as radiometric corrections and rearrange the data for an output in a maplike form.

Fig. 14 gives a theoretical approach to decision making techniques. It is based on the fact that the spectral signatures of each ground element represents a four-dimensional "feature space", where each element is characterized by a single point. In order to classify pixels the feature space must be subdivided into different "sub-volumes" or categories, either by a supervised technique based on training samples or by a non-supervised one, applying clustering.

Sofar mainly supervised classification systems based on training samples are used. Since the accuracy of the classification depends heavily on the correct choice of these training samples, they have to be delineated and statistically tested with specific care. The training samples are actually subject to statistical fluctuations. The larger the number of pixels and the closer together they are, the better their definition. Using the HP-9830 with a plotting device it is possible to receive a two-dimensional graphical presentation of the distribution of the samples in the feature space. Combining always two different variables it is possible to get a good idea of the position and the separability of the training groups within the feature space.

To avoid an inclusion of pixels which are too far away from any training sample to make a sensible assignment, thresholding-techniques are used and the corresponding pixels classified as unassignable.

Various algorithms exist and were tested (Muri 1976) regarding their usefulness (accuracy versus expenditure):
- D-Class (distance in feature space)
- A-Class (angular position in feature space)
- maximum likelihood
- stepwise linear discriminant analysis
- PPD (parallel epiped method) etc.

The most economical algorithm with satisfactory accuracy for mapping the extent of
the snowcover was the PPD. If additional
features are involved such as clouds, or a
differentiation of the snowcover is re-
quested, more sophisticated algorithms have
to be applied. At present two systems are
operational at Zurich, one for the CDC
6400/6500 at the Swiss Federal Institute of
Technology (Fig. 15) and one for the IBM
370/155, based on a PPD-algorithm, developed for an easy interpretation by un-
trained people (Fig. 16). Here an interactive system (IBIS) and a batch mode (BIS)
are available.

A comparison of different available image
interpretation systems was carried out by
Itten (1975). The main results are compre-
hended in Fig. 17. From these studies he
concluded that "a highly interactive, specially designed system together with a
skilled applications specialist can, for
the future bring maximum operational use in
satellite snowcover observations " (Itten
1975, p. 245).

Present-day computers in general lack an
adequate output device for the presentation
of the classification results. Special
equipment is required for high-quality
outputs. At Zurich we dispose of the Photo-
mation System P 1700 by Optronics Inc.,
which records the information pixel by
pixel on a film (Fig. 26) in a maplike
mode.

7 EXAMPLES OF SNOWMAPPING IN THE ALPS

The concept as outlined in the previous
chapters was realized step by step in the
various projects carried out during our
Landsat-1 and -2 and Skylab-projects with
NASA (Haefner 1975, 1975, 1977). The most
important factors for snowmelt run-off prediction by a day to day calculation or for
the use of the model by Martinec are always the extent and the changes of the
snowcovered areas. It is the aim to not
only monitor the changes of the snowcover
but also to gain evidences on the snowpack
and the conditions of the snow surface.

The methods applied were improved and
modified with the advancement of our research and according to the specific objectives of the different studies. The
test sites varied also regarding location
and size, but were mainly in Grisons
(Fig. 18).

7.1 Areal extent under different relief and illumination conditions

The objective was to classify snow and ice
(glaciers) considering all manifestations
such as sun position, sun angle, slope
angle, shadows, exposure etc. (Gfeller 1975).
Consequently quite a number of learning
groups for snow and background features had
to be located. They were lateron combined
together into four main categories:
  - snow and ice in sun
  - snow and ice in shadow
  - background (snowfree) in sun
  - background (snowfree) in shadow
All four MSS-channels were used as variables and the data classified by linear discriminant analysis. Two different sets of
learning groups were established, one for
the "normal" situation with uniform, dry,
fresh snow, the second one for situations
with partly melting and/or refrozen snow.
The results showed only a very small number
of incorrectly classified pixels and are represented in tabular and graphical form
(Fig. 19). For a comparison the results
with both classification sets are presented
for the same day. Since one pixel covers
4,514 m^2, it is easy to calculate the
areal extent from this pixel by pixel
classification.

From this example we can conclude that the
objective of classifying the extent of the
snowcover in sun and in shadow can be
achieved with high accuracy.

An open problem - we learned - is the
calculation of the true surface. Even the
best geometric correction gives us only the
orthogonal projection of the true surface.
In high mountain areas - such as our test
site - there may be a difference up to
20 % between these two surfaces. Therefore
to overcome this problem, a digital terrain
model has to be correlated with the satellite data to improve the areal measurements.

7.2 Exact location of the transition zone

For run-off predictions and studies of the
melting process it is especially important
to locate the exact position of the transidient snowline. In reality this is not a
line but a smaller or larger transition
zone of melting snow patches between the
totally snowcovered and snowfree area. In
a study in Central Grisons (test site
approx. 2,700 km^2, Fig. 18) special attention was given to this problem.

The classification was done with a PPD-algorithm using band 5 and 7. This allowed the most efficient and economic classification. Careful ground truth and aerial underflights (Fig. 3) were used to exactly locate the training samples in particular in and around the transition zone. The selected categories and the statistics of the corresponding training samples are presented in Fig. 20 and their classification matrix in Fig. 21. The total area was then classified and mapped into three main categories:
- snowcover
- transition zone
- snowfree area

Fig. 22a and b show the extent of these three categories for April 22nd, 1975 of the total test site (a) and an enlarged section (b). The areal measurement is summarized in Fig. 23, with 9 % belonging to the transition zone, whereas 65 % were still totally snowcovered, and 26 % completely snowfree. The accuracy of the classification was tested in four smaller sample areas, where detailed measurements from aerial photography were taken independently. The results indicate that the snowcover was always classified with 93 % accuracy or better (Fig. 24). In average the snowcover was mapped 5.6 % too large, the snowfree background 2.6 % too small, compared with the total size of the test area. The achieved accuracy is quite satisfactory.

Consequently the objective of a very precise location of the transition zone can be fulfilled with Landsat-data. Open problems remain regarding the border of the selected test sites. Sofar we used rectangular areas. But for practical applications watersheds should be classified. This can be achieved by using masquing techniques. A better and "direct" method is discussed in Chap. 7.4.

7.3 Separation of snow and clouds

As already mentioned the cloud problem is one of the most obstructive problems in dealing with satellite imagery. Nevertheless partly cloudcovered images have to be used, too. A separation of snow and clouds is relatively easy with analog techniques, due to the different textures. But a separation based on the spectral signatures from Landsat is not possible. Much better changes were offered by the Skylab-EREP-S-192 multispectral scanner with 13

different channels. Of particular interest is the channel between 1.55 μm and 1.75 μm. In addition to a band in the blue region (0.46 to 0.51 μm) and one in the near IR (0.78 to 0.88 μm) a clear separation of snow and clouds could be gained (Muri 1976).

The test area was in the Valais around Zermatt. Fig. 25 summarizes the learning groups which had to be set up to finally reach a separation into the following five main categories:
- snow and ice in sun
- snow and ice in shadow
- clouds
- background (snowfree area) in sun
- background (snowfree area) in shadow

Of the various algorithms tested the D-Class, based on the euclidean distance provided the best results. Fig. 26 gives the output in geometric corrected, e.g. maplike form. It has to be noticed that some errors arise especially due to the forming of mixed pixels, e.g. at the edges of the clouds and around the lakeshore (reservoir of Mattmark etc.).

Consequently the integration of a band 1.55 to 1.75 μm in future satellites is a "must" for snowmapping under European conditions. It could be clearly demonstrated that this channel enables an automated separation of clouds from snow.

An open problem is to interpolate the extent of the snowcover in a partly cloudy scene. This may be achieved when having a longterm and sufficient knowledge on the changes of the snowlines. Then it should be possible to extrapolate from the situation in the cloudfree parts the course of the snowline where it is cloudcovered.

7.4 Delineation of watersheds

To determinate the areal extent of the snowcover in areas of undefined shape such as watersheds of various sizes, with an automated classification system using multispectral and multitemporal data, a new method was developed recently (Urfer 1978). The results should be in such a form, that they serve as input for the run-off model by Martinec. Therefore not only the extent of his test site, the DISCHMA-valley (Fig. 18) was delineated but in addition separated into various altitudinal zones.

To achieve a combination of the satellite data with the ones from the terrain model a fifth "artificial" channel was constructed in addition to the four Landsat-bands. The data of the different altitudinal zones (three zones between 1,500 and 3,000 m) were digitized from a topo-map in pixel units. Well defined and easy recognizable reference points have to be used which can exactly be located in the image as well as on the map. Using these points as reference system it is possible to adjust the digitizer in such a way, that he devides the corresponding distance into the same number of units as there are Landsat-pixels. This presumes that the distance between the pixels along a line was equalized in a preprocessing step. In doing so the terrain can be arranged into the same number of "terrain elements" as Landsat-pixels, each one belonging to one of the three altitudinal zones of the watershed. But obviously much more zones or other grouping criteria could be introduced. For a new Landsat-scene, with a slightly different central point, the two different data sets just have to be superimposed correctly by locating the same reference points again.

Since the DISCHMA-valley is included in overlapping Landsat-passes of two following days data from both orbits were used. As the two orbits are not parallel to each other, but shifted with an angle of about 1.4° toward each other, two different sets of "terrain elements" had to be digitized, one for each orbit.

The method was tested for two different watersheds in Grisons (Fig. 18). The smaller one, the DISCHMA-valley, is the test site of Dr. Martinec's run-off model. The results are presented in his paper. The second, larger one, corresponds to the watershed of the Landwasser in the region of Davos (Fig. 27). Fig. 27 shows the latter test area devided into the three different altitudinal zones. Fig. 28 presents the results of the snow classification for June 8th, 1976. For the snow classification the same procedure was used as described in Chap. 7.2. From the combined data sets it is easy to retrive the areal extent of the snowcover for each altitudinal zone. The comparison with the results from aerial photography, gained in the DISCHMA-valley, prove that Landsat-data are of equal accuracy or - in the highest belt - even better.

Therefore the objective of an automated classification of test areas of any size and shape could be solved by introducing a fifth "artificial" channel containing the same number of "terrain elements" as Landsat-pixels.

A remaining problem is a better characterization of the surface conditions.

HCMM- and Nimbus-G-data will be of great interest in this respect. A next step has to deal with a separation of the snowcover into different snowtypes. Additional information should be furnished on the snowdepth and the water equivalent.

8   CORRELATION WITH OTHER DATA

Output systems should always be set up in such a way, that they allow a combination and unification of data from different acquisition sources. This may be achieved best by spacially referencing or geocoding the data, to perform spacially-oriented processing. A possible method has been demonstrated in the previous chapter (7.4).

One of the best solutions is to incorporate and organize all data in a geographic information system. This enables not only the combined utilization of a variety of data sets of different origin but also the fast retrieval of the needed information for specific applications.

9   CONCLUSIONS

As shown from the various examples mapping of the areal extent of the snowcover and an exact delineation of its transition zone can be undertaken from satellite data in an automated procedure with high accuracy for areas of any irregular shape.

The remaining problem concerns primarely the cloud problem, which cannot be overcome directly with optical scanner devices. A channel in the 1.55 to 1.75 μm region of the spectrum is essential. Then, with sufficient background information, also partially clouded scenes could be interpreted and the missing parts interpolated.

Regarding a further evaluation of the conditions of the snow surface, additional studies are needed.

For further mapping activities a combination of a low resolution data acquisition system with high repetition rate (NOAA-VHRR) together with a high resolution, low repetivity system (Landsat) should be carefully tested.

For studies on the snowdepth and the water equivalent further systematic researches with active microwave systems are indispensible.

10 REFERENCES

Barnes J.C. & Bowley C.J. 1968, Operational guide for mapping snow cover from satellite photography. Allied Research Ass. Inc., Concord, Mass.

Barnes J.C. & Bowley C.J. 1974, Handbook of techniques for satellite snow mapping. ERTS Doc 0407-A, Env. Res. & Techn. Inc., Concord, Mass.

ESA 1977, European remote sensing space programme - mission objectives and measurement requirements. ESA/EXEC(77)3, Paris.

Fritz S. 1962, Snow survey from satellite pictures. Proc. 1st Int. Symp. on Rocket and Satellite Meteorology, Washington.

Gfeller R. 1975, Untersuchungen zur automatisierten Schneeflächenbestimmung mit Multispektralaufnahmen des Erderkundungssatelliten ERTS-1. PhD-Thesis, Univ. of Zurich.

Haefner H. 1974, Remote sensing applications for geoecological studies in the high mountain environment. ISP Com. VII, Proc. Symp. on Remote Sensing and Photo Interpretation, Banff, Alberta.

Haefner H. 1975a, Snow survey and vegetation growth in high mountains (Swiss Alps) and additional ERTS-investigations in Switzerland. ERTS-1 Final Report, Zurich.

Haefner H. 1975b, Snow survey and vegetation growth in the Swiss Alps. EREP-Final Report, Zurich.

Haefner H. 1977, Snow mapping and land use studies in Switzerland. Landsat-2 Final Report, Zurich.

Haefner H. & Itten K.I. 1977, Snow studies by satellites in Switzerland. In WMO Publ.-Project on Snow Studies by Satellites, Phase II, Geneva.

Haefner H. & Geiser U. 1975, Kartierung von Höhengrenzen zwischen Mt. Blanc und Gotthard Massiv mit Skylab-EREP-Aufnahmen. Geographica Helvetica, No 3.

Hall W. 1954, Report on the use of air photographs in snow surveys. Arch. Int. de Photogramm., XI-III, Amsterdam.

Hofer R. & Schanda E. 1977, Signatures of snow in the 5 to 94 GHZ range. Radio Science.

Itten K.I. 1971, The determination of snow-lines from weather satellite pictures. Berichte III. Int. Symp. für Photointerpretation, Dresden, Int. Archive of Photogrammetry, Leipzig.

Itten K.I. 1975, Approaches to digital snow mapping with Landsat-1 data. Proc. Workshop on Oper. Applic. of Sat. Snowcover Obs., South Lake Tahoe, Calif.

Martinec J. 1977, Possibilities and obstacles in the hydrological exploitation of satellite data. In WMO Publ., Geneva.

Meier M.F. 1974, New ways to monitor the mass and areal extent of snowcover. In Beck, Baker, Ruthenberg (Edit.), Approaches to earth survey problems through use of space techniques.

Moore R. 1974, Snow cover. Paper, Active Microwave Workshop, Houston.

Muri R. 1976, Automatisierte Trennung von Schnee und Wolken im Satellitenbild (Skylab S-192). M.S. Thesis (unpublished), Dept. of Geography, Univ. of Zurich.

O'Brian H.W. & Munis R.H. 1973, Red and near-infrared spectral reflectance of snow. Manus. for USACRREL Res. Report, Hannover, U.S.

Rango A. (Edit.) 1975, Operational applications of satellite snowcover observations. Proc. Workshop, South Lake Tahoe, Calif.

Rango A. & Itten K.I. 1976, Satellite potentials in snowcover monitoring and runoff prediction. In Nordic Hydrology 7.

Seidel K. 1976, Digitale Bildverarbeitung. Techn. Report, Dept. of Photography, Swiss Fed. Inst. of Technology, Zurich.

Staenz K. 1976, Radiometrische Untersuchungen über das Reflexionsverhalten von Schnee. M.S. Thesis (unpublished), Dept. of Geography, Univ. of Zurich.

Stirnemann H.P. 1976, Kartierung der temporären Schneegrenze mit Hilfe der Landsat-2 Aufnahmen. M.S. Thesis (unpublished), Dept. of Geography, Univ. of Zurich.

Urfer H.P. 1978, Routinemässige Schneekartierungen in hydrologischen Einzugsgebieten mit Landsat-Daten. M.S. Thesis (unpublished), Dept. of Geography, Univ. of Zurich.

WMO 1973, Snow survey from earth satellites - a technical review of methods. WMO/IHD Report No 19, Geneva.

WMO 1976, Int. working seminar on snow studies by satellites - Draft Report, Geneva.

Tab. 1: Correlation between size of study area and satellite system

Size of study area	Sensor resolution	Sensor
≤ 10 squ.km	≤ 10 m	airborne photographic camera (various flight altitudes)
10 - 1,000 squ.km	approx. 100 m	Landsat
> 1,000 squ.km	approx. 1,000 m	NOAA-VHRR

Tab. 2: Results of mapping the location of the transidient snowline from Skylab-EREP camera system

Photographic system Skylab	Map-scale	Contour intervals	Accuracy
S-190-A	1:200,000	100 m	± 50 m
S-190-B	1:100,000	50 m	± 20 m

SATELLITE	ORBIT			STARTING DATE	EARTH RES. SENSORS	NO OF SPECTRAL BANDS	GROUND RESOLUTION (in m)
	ALTITUDE (km)	INCLINATION (degrees)	REPETITION RATE				
LANDSAT-1 + 2	920	99	18 days	1972,1975	MSS RBV	4 3	80 80
SKYLAB	435	50	5 days	1973	MSS Multib.ph.	13 10	60 15 - 60
NOAA 5	1520	102.1	12 hours	1976	VHRR	4	1000/4000
LANDSAT- 3	920	99	18 days	1978	MSS RBV	4 + 1 1 (Stereo)	80 + 240 40
LANDSAT-D	917	99	18 days	1981	Thematic Mapper(MSS)	5 + 1	30 - 40 + 90 - 120
HCMM	600	98	12 hours	1978	HCM RADIO-METER	2	500
NIMBUS-G	925	99	2 - 3 days	1978	CZCS SMMR	6 4	800 16 - 144 km
PREOP. EUROP. EARTH RES. SATELLITE	?	?	?	1985 - 1986	MSS RADAR	8 - 10 2 Frequ.	30 20 - 50

Legend: MSS Multispectralscanner   SMMR Scanning Multichannel Microwave Radiometer
        RBV Return Beam Vidicon Camera   CZCS Coastal Zone Colour Scanner
        SAR Synthetic Aperture Radar   VHRR Very High Resolution Radiometer

Fig. 2: Existing and planned satellite systems suitable for snowmapping

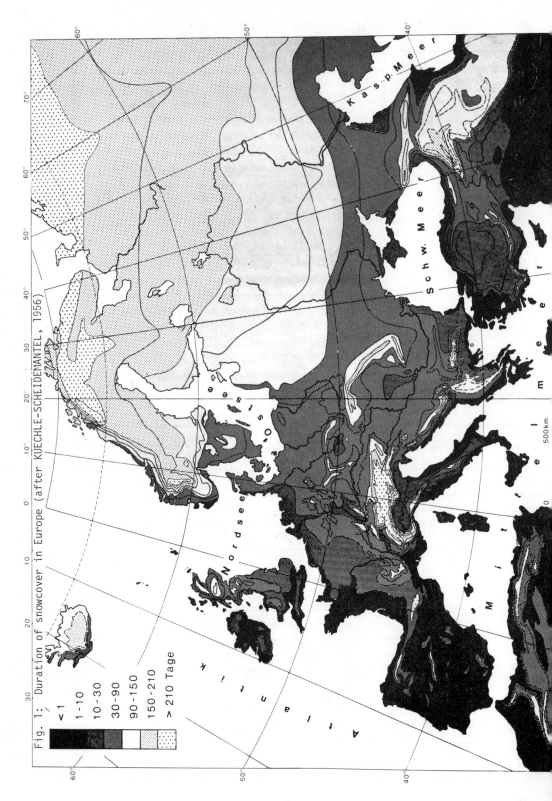

Fig. 1: Duration of snowcover in Europe (after KUECHLE-SCHEIDEMANTEL, 1956)

< 1
1 - 10
10 - 30
30 - 90
90 - 150
150 - 210
> 210 Tage

Atlantik

Nordsee

Ostsee

Kasp. Meer

schw. Meer

Mittelmeer

0   500km

0   2   4   6   8   10 km

N

Fig. 3:   Example of high altitude aerial underflight; Davos and vicinity,
          November 24th, 1975 (Photo by courtesy of LA, Dübendorf)

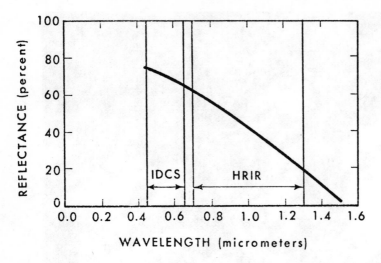

Fig. 4: Spectral reflectance of melting snow (after MANTIS, 1951)

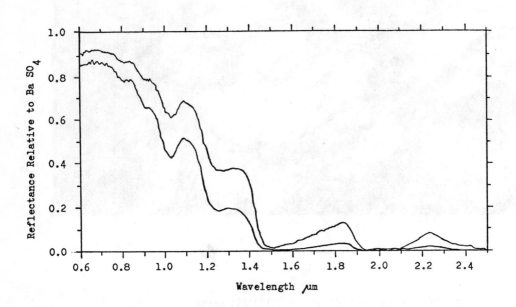

Fig. 5: Changes of spectral reflectance of snow with natural aging:

upper curve: nearly fresh, cold snow
lower curve: two days old, with melting and refreezing effects
(after O'BRIEN & MUNIS, 1973)

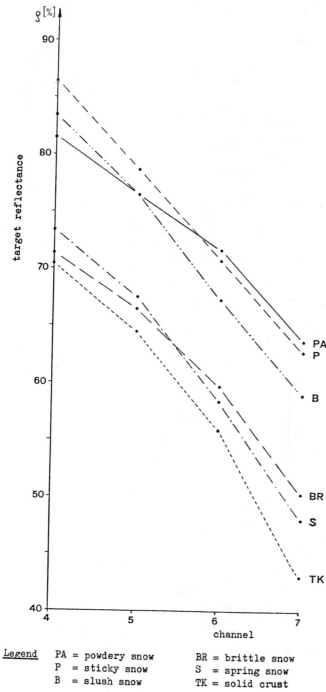

Legend   PA = powdery snow   BR = brittle snow
          P = sticky snow   S = spring snow
          B = slush snow   TK = solid crust

Fig. 6:   Reflectance ϱ (%) of various snowtypes in the four Landsat-MSS-
          channels measured with EXOTECH-100 (after STAENZ, 1976)

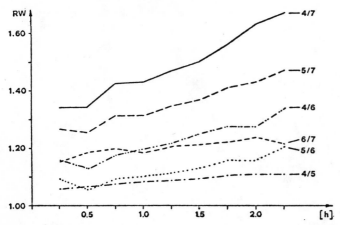

Fig. 7: Ratio values (RW) of target reflectance of EXOTECH-measurements corresponding to Landsat-channels in function of time during melting of snow; near Davos, Grisons, May 20th, 1975 (after STAENZ, 1976)

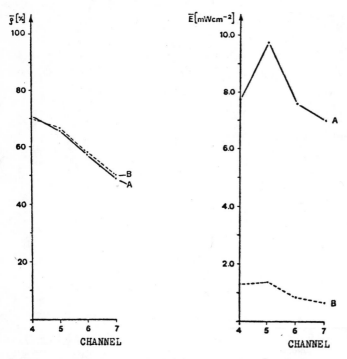

8a) target reflectance ($\bar{\varsigma}$) of brittle snow determined with reference pannel

8b) Radiance ($\bar{E}$)

A = measurement in sun

B = measurement in shadow

Fig. 8: Comparison of reflectance measurements in sun and shadow (after STAENZ, 1976)

354

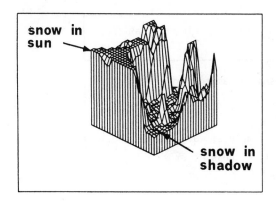

channel 4

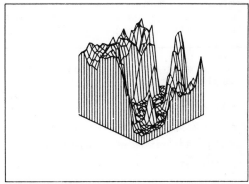

channel 6

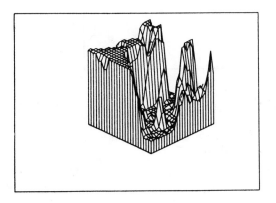

channel 5

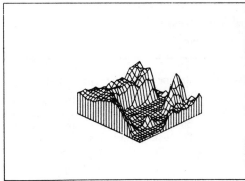

channel 7

Fig. 9: Distribution of snow densities in sun and in shadow in the four
Landsat-channels

Presentation of 25 x 25 pixels by Interactive Graphical System. The
saturation in band 4 and 5 is clearly visible.(after GFELLER, 1975)

Observations of position of transidient snowline (visible observations, photographs, mapping)

Measurements and observations of various snow parameters at selected sampling points:

- snowtype

- characterization of snow surface (form, cleanness, wetness, age etc.)

- characterization of snowpack (granularity, wetness etc.)

- snow temperature

- snowdepth

- snowmass

- air temperature

- air moisture

- slope angle and exposure

- cloudiness

- daytime

Fig. 10: Ground observations and measurements for snowmapping

The following parameters were surveyed at all days of Landsat-2 overflights in 1976 with good weather conditions at various observation points along a route Davos - Tiefencastel - Thusis - Domleschg (Grisons). (after STIRNEMANN, 1976)

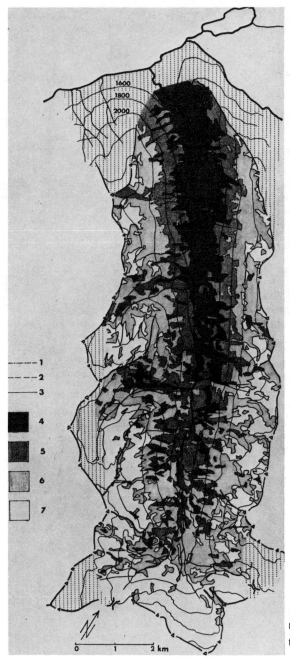

Fig. 11:
Generalized map of snowmelting in
the DISCHMA-valley in spring 1960

Legend: (1) Snowline (separating 87 % snowfree from 2 % snowcovered area) of May 14th, 1960
(2) Snowline of May 31st, 1960   (3) Snowline of June 22nd, 1960
(4) Area snowfree before May 14th, 1960
(5) Area becoming snowfree between May 14th and May 31st, 1960
(6) Area becoming snowfree between May 31st and June 22nd, 1960
(7) Area still snowcovered after June 22nd, 1960

(Interpretation by U. WALDER, M.S., after HAEFNER, 1974)

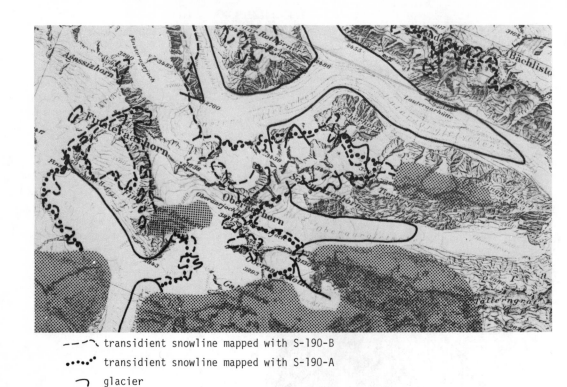

- - - ⌐ transidient snowline mapped with S-190-B

•••••• transidient snowline mapped with S-190-A

⌐ glacier

▓▓▓ clouds

0 1 2 3 4 KM

N

Fig. 12: Comparison of course of transidient snowline mapped with SKYLAB-
S-190-A and S-190-B (after HAEFNER & GEISER, 1975)

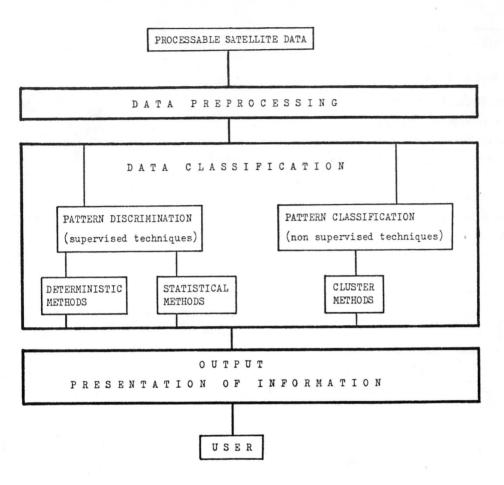

Fig. 13:   Diagram of digital data processing

```
┌─────────────────────────────────────────────────────────────────────┐
│  PATTERN DISCRIMINATION                                               │
│                                                                      │
│  - TRAINING SAMPLES (SUPERVISED LEARNING)                            │
│                                                                      │
│  - DETERMINISTIC METHODS                                            │
│    no probability concepts                                          │
│    discriminant functions                                          │
│    decision: sample in class with largest discriminant score        │
│                                                                      │
│  - STATISTICAL TECHNIQUES                                           │
│    probability distribution                                         │
│    parametric                          non parametric               │
│      distribution is given               unknown distribution       │
│      analytically                                                   │
│      example:                            example:                   │
│        Gaussian distribution               potential functions      │
│                                            histogram method         │
│                                                                      │
│    decision:  Bayes' decision rule                                 │
│               maximum likelihood                                    │
│                                                                      │
├─────────────────────────────────────────────────────────────────────┤
│  PATTERN CLASSIFICATION                                             │
│                                                                      │
│  - NO TRAINING SAMPLES (UNSUPERVISED LEARNING)                      │
│                                                                      │
│  - NATURAL GROUPINGS IN FEATURE SPACE (CLUSTERS)                    │
│                                                                      │
│  - DETERMINATION OF SELECTED CLASSES: afterwards by ground          │
│                                        sampling information         │
└─────────────────────────────────────────────────────────────────────┘
```

Fig. 14: Decision making techniques in digital processing of multivariable data
          (after STEINER, 1972)

MMM3   – TRANSLATION 9-TRACK TO 7-TRACK

DASH2   – COMPUTATION OF SCANLINEFACTORS AFTER GRAMENOPOULOS (1973)
          (DUE TO DIFFERENT SENSITIVITIES OF THE 6 SENSORS IN THE RECORDING SYSTEM)

TOTRAF – AFTER NORMALISATION BY SCANLINEFACTORS STRUCTURING INTO BLOCKS (128x128 PIXELS) WHICH CAN BE
          CALLED BY AN INDEX (RANDOM ACCESS)
          RAFHIS – STATISTICS OF BLOCKS
          RAFPRI – OUTPUT PRESENTATION OF BLOCKS BY CHARACTER-OVERPRINT (STANDARD LINE PRINTER)
          RAFSYM – OUTPUT PRESENTATION OF BLOCKS FOR SELECTED LEVELS BY SELECTED CHARACTERS (LINE PRINTER)
          RAFFL – OUTPUT PRESENTATION OF BLOCKS WITH STANDARD FILMPLOT DEVICE
          RAFUNT – STATISTICS OF ANY SUBSET OF A BLOCK
          BBRUED – 3-DIMENSIONAL PRESENTATION AT THE INTERACTIVE DIGIGRAPHIC DEVICE (BELLBOX)

TOTRAH – AFTER NORMALISATION BY SCANLINEFACTORS STRUCTURING LINEWISE FOR SUBSEQUENT SUBSETTING (BLOCKED
          BINARY)
          ERTUNT – CREATES ANY SUBSET OF A TOTRAH-FILE FOR FURTHER ANALYSIS OR OUTPUT-PROCEDURES
          OPTER2 – CREATES OUTPUT FOR THE PHOTOWRITE SYSTEM (PHOTOMATION) WITH HISTOGRAM FOR RANGE CONTROL
                  OPTIONS: ORIGINAL CHANNELS OR ANY RATIO, SCALE AND SKEW
          UNTSTA – STATISTICS OF ANY SUBSET OR TRAINING AREA
          UNTDIS – SELECTION AND PREPARATION OF THE TRAINING AREAS FOR STEPWISE DISCRIMINANT ANALYSIS
          UNTHIS – STATISTICS OF COMBINED SUBSETS WITH HISTOGRAM
          B7MM   – STEPWISE DISCRIMINANT ANALYSIS OF DATA PREPARED BY UNTDIS (TRAINER PROGRAM)
          CLA7MM – CLASSIFICATION OF A PICTURE ARRAY WITH THE DISCRIMINANT FUNCTION GENERATED BY B7MM
                  OUTPUT FOR PHOTOWRITE SYSTEM PHOTOMATION (MAPPING CONTROL)

Fig. 15:   Computer programs for digital processing of Landsat-MSS-data
          (available at the Swiss Federal Institute of Technology, Zurich)
          (after SEIDEL, 1976)

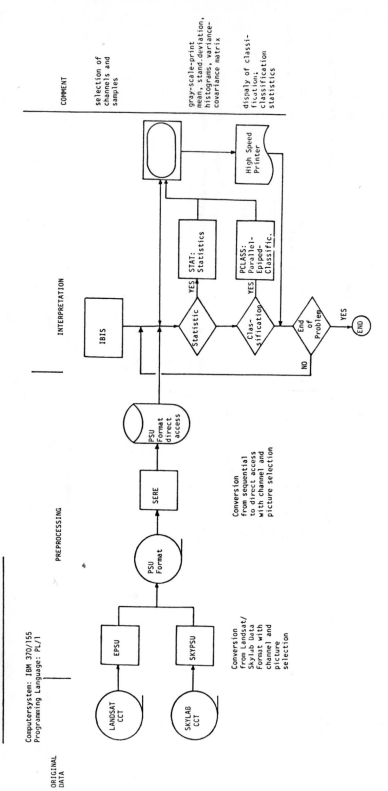

Fig. 16: IBIS interactive interpretation system for IBM 370/155 (available at the Department of Geography, University of Zurich) (after FASLER, 1976)

Cover type	LARSYS Ver. 3	STANSORT-2	GE Image-100
Dry snow	31.8	30.8	30.9
Metamorphic snow	22.1	22.5	21.1
Forest with snow	27.2	27.8	27.1
Interzone	9.4	11.2	4.6
Bare forest/Veg.	6.2	6.0	10.4
Shadow + water	0.4	0.2	0.2
Total snow-covered area	85.8	86.7	81.4
Total area bare of snow	11.3	11.8	12.9
Unclassified area	2.9	1.5	5.7
Accuracy/ Testfields	92 (calc.)	90 (est.)	87 (est.)

Numbers represent areas in percent

Fig. 17: Comparison of results received from three different image
interpretation systems (after ITTEN, 1975)

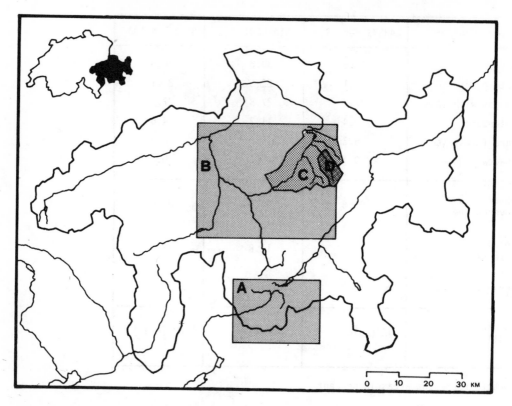

Fig. 18: Location of test areas in Grisons, Switzerland

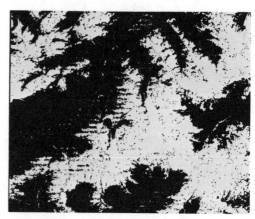

black: snowfree

white: snow and ice

left: classification with set for "normal", e.g. dry snow conditions

right: classification with set for thawing and/or refrozen snow (valid
           for above conditions of October 7th, 1972)

Fig. 19:   Digital snow classification of test site "Bergell" with two different
           classification sets (after GFELLER, 1975)

	X̄	δ		X̄	δ		X̄	δ
①	127.00	00.00	⑧	50.22	7.97	⑭	24.86	1.55
	127.00	00.00		46.21	8.45		25.18	2.21
959	127.00	00.00	131	36.93	7.06	135	36.44	2.73
	62.59	1.18		7.59	2.35		15.68	1.76
②	127.00	00.00	⑨	116.59	10.68	⑮	20.72	1.58
	127.00	00.00		124.17	7.25		18.89	1.70
50	127.00	00.00	29	113.45	11.84	356	28.82	3.17
	51.72	4.42		36.62	5.38		12.46	1.62
③	105.73	9.84	⑩	60.32	17.46	⑯	22.50	4.19
	115.78	9.60		59.75	19.44		19.77	3.96
102	105.13	11.88	44	49.57	19.85	766	23.34	4.63
	31.32	4.42		12.34	6.80		8.25	2.62
④	99.22	10.07	⑪	55.75	6.43	⑰	28.01	2.23
	109.36	10.15		63.33	8.09		29.74	2.97
55	99.86	12.09	48	60.96	8.17	120	34.73	3.36
	30.36	5.25		19.73	2.72		13.05	2.09
⑤	85.69	12.81	⑫	60.74	7.48	⑱	27.88	1.89
	94.93	12.88		71.39	10.73		29.25	4.06
42	85.79	14.07	46	68.11	9.85	8	37.75	7.01
	25.86	4.54		21.96	3.27		15.50	4.96
⑥	69.25	10.60	⑬	72.47	18.61	⑲	30.10	2.89
	74.11	13.65		86.92	22.17		33.61	5.02
36	65.14	12.50	157	87.58	19.96	478	55.80	6.24
	18.42	4.49		29.84	5.90		27.02	4.02
⑦	120.15	8.25	◯	No of training groups				
	125.21	4.50	☐	No of pixels/group				
157	121.50	8.55	X̄,δ	for channels 4–7				
	37.75	4.31						

Legend

no	category		no	category
1	snow SO		14/15	needle-leaf forest SO
2–7	metamorphic snow		16	needle-leaf forest SA
8	snow SA		17	built-up areas snowfree
9	snow intermixed with rocks SO		18	water
10	snow intermixed with rocks SA		19	grass
11/12	built-up areas with snow			
13	transition zone			

SO = sun-exposed   SA = in shadow

Fig. 20: Statistics of training groups (mean values and standard deviation) for test site "Central Grisons" (after STIRNEMANN, 1977)

GR NR	No of pixels	1	2	3	4	5	6	7	8	9	10	11	12	13	14	15	16	17	18	19
1	949	949	10																	
2	45	1	40					4												
3	102			45	32	6		14		5										
4	55		2	9	33	6	2	2		1										
5	42				7	22	9	2		1			1							
6	36				2	1	23		1	9										
7	157		5	22	5			102		23										
8	131					1			81			25						(1)	(23)		
9	29		3	4	3	1		12			6									
10	44				2	5		2	18			15							(2)	
11	48											35	13							
12	46				1							19	22	4						
13	157			4	13	3		3		2		21	21	82		5				3
14	158														116	14	21	4	1	2
15	356														58	292	5	1		
16	766											(1)			1	194	531	15	24	
17	120														10	1	31	78		
18	35						(1)	(1)		(2)		(1)							30	
19	478														39		5			434

| | | | | | | | | | | | SNOW COVER | | | | | TZ | | SNOWFREE AREA | | |

GRNR = No of training groups as in Fig. 20
TZ = transition zone (snow line)

[3] = correct classification
3 = error in same category
(3) = misclassification

Fig. 21: Classification of training groups for snowmapping in test site "Central Grisons", April 22nd, 1975 (after STIRNEMANN, 1977)

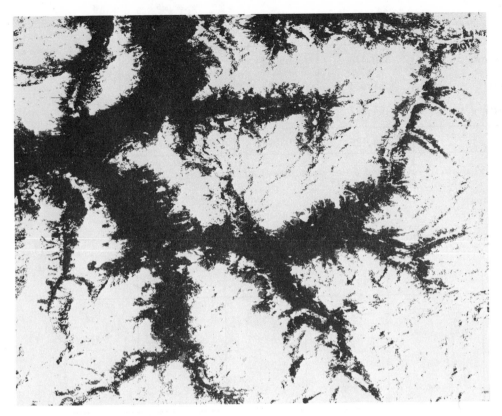

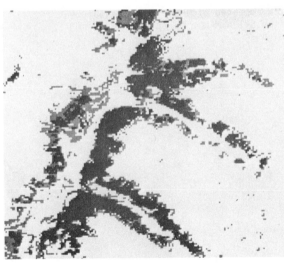

white: totally snowcovered
gray: transition zone
black: totally snowfree

Fig. 22 a&b: Classification of test site "Central Grisons" for April 22nd, 1975
(after STIRNEMANN, 1977)

a) snowmap of total test site

b) enlarged section showing extent and distribution of transition zone

class	No of pixels	%	area in squ.km
snowcover	390,478	65.28	1,770.570
transition zone	53,423	8.93	242.206
snowfree areas	154,227	25.79	600.494
total	598,128	100.00	2,712.270

Fig. 23: Classification and areal measurement of total test site
(after STIRNEMANN, 1977)

size of test area		snowcover		background	
no	squ.km	squ.km	%	squ.km	%
1	25.436	+ 0.700	+ 2.75	- 0.701	- 2.75
2	78.452	+ 6.071	+ 7.75	- 6.071	- 3.87
3	22.597	- 1.049	+ 5.33	+ 6.716	+ 5.18
4	32.046	+ 2.854	+ 6.74	- 2.050	- 7.22

Fig. 24: Accuracy of Landsat classification compared with air photo measurement
for four smaller test areas (after STIRNEMANN, 1977)

Category			Main Category		
Name	No	Symbol for computer prints	Name	No	Symbol for computer prints
snow wet	1	'	snow	1-5	.
snow wet 2	2	'	in		
snow dry	3	.	sun		
glacier	4	,			
snow and rocks	5	"			
snow dry in shadow	6	&	snow	6-9	'
snow dry i.sh. II	7	&	in		
snow and rocks i.sh.	8	%	shadow		
snow and rocks i.sh. II	9	%			
clouds (centers)	10	-	clouds	10-11	=
clouds (edge)	11	=			
water	12	*	snow-	12-17	$
forest	13	/	free	23	
grass	14	?	areas	25	
rocks	15	(	in		
bare soil/grass	16	)	sun		
settlements	17	I			
sediments (detritus)	23	L			
border of reservoir	25	X			
forest in shadow	18	W	snow-	18-22	M
grass i.sh.	19	B	free	24	
rocks i.sh.	20	G	areas		
bare soil/grass i.sh.	21	M	in		
rocks i.sh. II	22	7	shadow		
ricks i.sh. III	24	V			
sediments & rocks i.sh.	26	Z			
bares soil/grass i.sh.	27	S			
grass i.sh. II	28	H			
rocks i.sh. IV	29	A			
others not classified	30	blank			blank

Fig. 25: Categories and main categories of digital snow classification for test site "Zermatt" (after MURI, 1976)

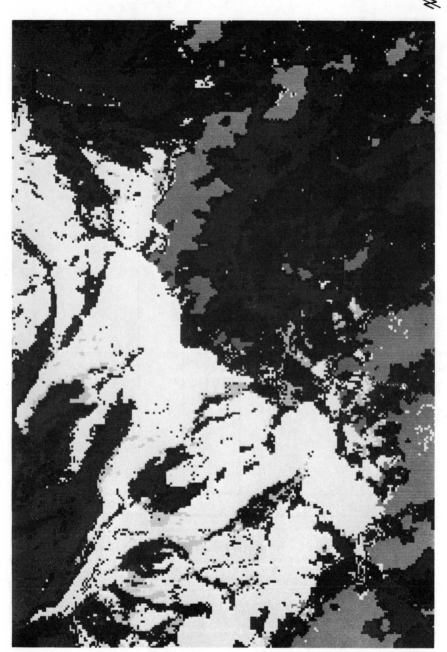

white: snow in sun (and ice)   dark gray: snowfree area in sun
light gray: snow in shadow (and ice)   black: snowfree area in shadow
medium gray: clouds

Fig. 26: Digital mapping of snow and clouds in five main categories for test site
          "Zermatt" (after MURI, 1976)

Fig. 27: Test site "Landwasser" subdivided into three altitudinal zones, as
superimposed on Landsat-data (Fig. 28) for automated areal measurement
(after URFER, 1977)

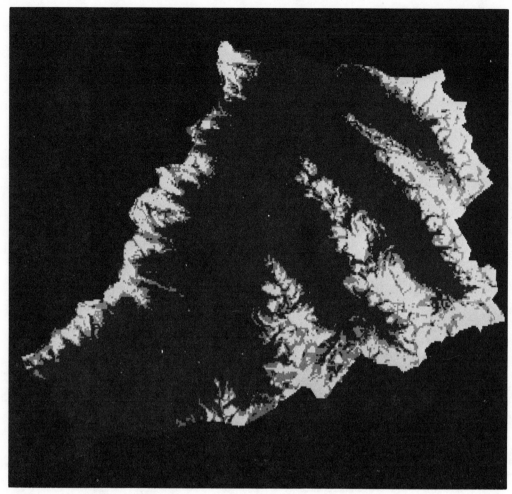

white: totally snowcovered
light gray: transition zone
dark gray: totally snowfree

Fig. 28: Snow classification for test site "Landwasser" (after URFER, 1977)

Computer-aided analysis of satellite and aircraft MSS data for mapping snow-cover and water resources

1 INTRODUCTION

Effective management of our water resources requires current and accurate information about these resources. In some cases, the information needed concerns the areal extent, depth, and condition of the snow-pack so that estimates of the quantity and regimen of the runoff can be obtained. In other instances, the resource manager is more interested in the amount and condition of the water after it has reached reservoirs, lakes and streams.

The developments in remote sensing technology during the past 15 years offer tremendous potentials for the hydrologist to obtain various types of information concerning the water resources--information that previously could not be obtained. Use of thermal infrared scanner systems from aircraft altitudes, for example, has proven the feasibility for accurately obtaining temperature measurements of the surface of water bodies over large areas. With the advent of Landsat satellites, the potential for measuring the areal extent of water bodies at frequent intervals and over very large geographic areas has been proven. Perhaps of even more importance, satellite data offers--for the first time--the opportunity to measure the areal extent of snow-cover in mountainous watersheds in an accurate and cost effective manner. Multi-spectral scanners (MSS) operated from satellite altitudes have also allowed the condition of water bodies to be assessed. For example, some studies have shown that different concentrations of non-organic suspended matter can be delineated on satellite data with a relatively high degree of accuracy.

From these comments, we see that there are several aspects of remote sensing

technology which have application for mapping the extent and condition of snow-cover and surface water resources. This paper will consider the following topics:

• Mapping snow-cover using satellite MSS data and computer-aided analysis techniques.
  - Mapping snow-cover using Landsat data.
  - Snow/cloud differentiation using Landsat data.
  - Snow/cloud differentiation using Skylab data.
  - Mapping snow-cover using Skylab plus topographic data.
• Mapping water resources using satellite MSS data.
  - Areal extent of water bodies.
  - Defining water condition with Landsat data.
  - Water temperature mapping with Skylab data.
• Water temperature mapping using aircraft MSS data.

2 SNOW COVER MAPPING

For over thirty years, snow hydrologists have made numerous attempts to correlate the areal extent of snow-cover with the subsequent runoff. Parshall (1941) and Potts (1944) estimated the areal extent of snow-fields for runoff forecasting using ground photography. Later studies have involved use of aerial photography to measure the areal extent of snowcover (Parsons and Castle, 1959; Finnegan, 1962; Leaf, 1969). It was not until the early 1960's, however, that one could attain a synoptic view of large geographical areas through earth orbiting satellites. Today, a wide variety of environmental satellites are collecting an astronomic amount of data that potentially could be utilized to

map the areal extent of snow-cover in a
repetitive mode. Wiesnet (1974) has
stated that "our capacity for collecting
data on snow and ice far exceeds our
ability to analyze the data". Therefore,
in order to keep up the pace with the
existing and highly advanced data collection
technology, computer-aided analysis techniques (CAAT) have been developed and
evaluated, and needs for further developments and refinements have been defined.

The Laboratory for Applications of Remote Sensing (LARS) was organized at Purdue
University in 1966 with an overall goal
of applying modern computer technology and
pattern recognition theory to the quantitative analysis of multispectral earth
resources data. Several analysis techniques have been developed since that time
and applied to a variety of disciplines.
However, it was not until Landsat-1 MSS
data became available that any of these
computer-aided analysis techniques were
applied to snow/hydrology studies.

2.1 Mapping snow cover using Landsat data

A study involving the use of computer-
aided analysis techniques for accurately
and efficiently mapping snow cover was
conducted in the San Juan Mountains of
southwestern Colorado (Luther et al, 1975).
Cloud-free Landsat data tapes were obtained
for six dates during the 1972-1973 winter.
The boundary of the Animas River Watershed
was delineated on topographic maps and then
transferred to the Landsat data in digital
format as a series of X-Y coordinates.
Such a procedure allows the total area
within the watershed (145 square kilometers
in this case) to be rapidly tabulated by
the computer (see Figure 1).

Preliminary analysis of the data indicated that spectral responses for snow and
"non-snow" areas were significantly different in all four wavelength bands of Landsat data. However, it was also found that
the detectors on the Landsat scanner system
were usually saturated for all resolution
elements containing snow-cover. (This
fact became particularly important later
when we attempted to differentiate the
snow from cloud-cover.) Classification of
the six data sets that were cloud-free
resulted in an accurate identification of
the snow covered areas, so that the total
area covered by snow could be tabulated by
the computer, and the percent snow-cover
within the entire watershed calculated.
The results showed the percent snow-cover
ranged from 62.6% to 87.5%, depending on
the date when the Landsat data was obtained.

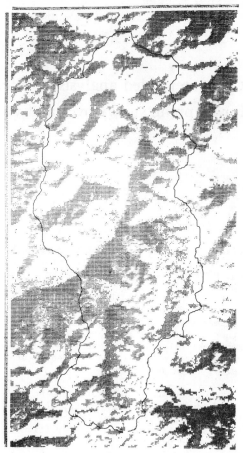

Figure 1. Digital Line-Printer Map of
Landsat Data, With the Boundary of the
Animas Watershed Outlined. (From Luther
et al., 1975.)

Next a study was conducted to show the
location of the changes in snow-cover from
one date to the next. The general technique for determining changes in the condition or type of ground cover between two
different data sets is known as "change
detection". There are at least four change
detection methods that can be used in computer-aided analysis of this type of data:
Delta Transformation, Spectral/Temporal
Concurrent Classification; Spectral/Temporal
Layered Classification; and Post-Classification Comparison.

In this study the Post-Classification
Comparison method was used because of its
simplicity and cost-effectiveness. Each
of the six data sets were digitally over-
layed, resulting in a data tape containing
24 channels of data (i.e., four wavelength
bands for each six different dates). Each

of the six data sets were classified into snow and non-snow categories. Then, the Post-Classification Comparison between any two dates enabled one of four combinations to be defined: (1) snow-cover on both dates; (2) non-snow on both dates; (3) snow-cover on Date 1 changing to non-snow on Date 2; and (4) non-snow on Date 1 changing to snow-cover on Date 2. The results obtained in this phase of the study clearly indicated that if the digital overlay of the data is accurate, simple classifications such as snow and non-snow can be compared so that maps and tables showing the changes in snow-cover from one date to the next can be quickly and easily compiled.

In these initial phases of this study, each resolution element of the Landsat data was being classified. From this preliminary work, it was clear that multispectral classification of enormous quantities of data gathered by earth orbiting satellites can require a moderately large amount of computer CPU (Central Processing Unit) time, even for such relatively simple tasks as determining the areal extent of the snow-cover. Therefore, a third phase of this study was conducted to compare the accuracy of the multispectral classification of snow-cover using different data sampling rates in order to reduce the amount of computer processing time. To do this, an area including approximately 60% of a Landsat scene (therefore involving over 4.3 million data points or resolution elements) was classified into four spectral classes (one "snow" class and three "non-snow" classes) using one visible and one infrared wavelength band. The same set of statistics were utilized to classify the area five times, each classification involving a different sampling rate. After this, the areal extent of snow-cover was computed for each one of five classifications. The results are shown in Table 1.

These results indicate that the percent of the area in snow-cover is not significantly different when every sixteenth column of data is classified as compared to classifying every resolution element in the data, since the difference between the area of snow for the 16 x 16 and the 1 x 1 sampling rates was only 0.4 of one percent. Of particular importance is the fact that the computer classification time was reduced from over 54 minutes to less than one minute for the classification. This would indicate that reasonably accurate estimates of the area of snow-cover can be obtained using a sample of Landsat data rather than using every resolution element present on the data tape. Such a procedure would allow the cost of computer processing to be reduced significantly, while the percent of the area covered by snow could still be determined with a relatively high degree of accuracy. Figure 2 graphically shows the effect of the sampling rate on the computer time used.

2.2 Snow/cloud differentiation using Landsat data

In the early 1960's, the problem of differentiating snow-covered areas from cloud formations was first identified by researchers working with some of the early meteorological satellite imagery. At that time, however, the primary emphasis was on study of cloud types and patterns. For example, while studying cloud patterns as seen on TIROS satellite data, Conover (1964) stated that "clouds are easily confused by the interpreters with snow-cover". Early work with the Landsat data also indicated that snow-cover could be easily confused with clouds. Articles by Meier (1973) and Barnes and Bowley (1973) both pointed out the difficulty in separating clouds from snow on the Landsat data, based upon differences in reflectance. However, by using a manual photointerpretation approach, other characteristics in the data can often be used to advantage to separate cloud-cover from snow. As Barnes and Bowley (1973) pointed out, "although snow and clouds have similar reflectances, mountain snow-cover can be differentiated from clouds primarily because the configuration of the snow patterns is very different from cloud fields and can be instantly recognized. The snow boundaries are also sharper than typical cloud edges, and snow fields usually appear with a more uniform reflectance than do clouds, which have considerable variation in texture. Furthermore, cloud shadows are usually visible, especially with cumuliform clouds, and various terrestrial features can be recognized in cloud-free areas". However, Meier (1973) pointed out that it is often difficult for even an experienced interpreter to distinguish some types of clouds and fog from snow.

In our study, the reflectance characteristics for areas which could be positively identified as snow-cover and as clouds were summarized for data obtained on three different dates. These results are shown on Table 2 and graphically displayed in Figure 3.

375

Table 1. Determination of Areal Extent of Snow-Cover from Landsat MSS Data Using Different Sampling Intervals.

Sample Interval	Number Data Points	Number of Points Classified as Snow	% of Area in Snow	Classification CPU* Time (minutes)
1 x 1	4,330,561	1,385,126	31.99	54.34
2 x 2	1,083,681	345,417	31.87	13.20
4 x 4	271,441	86,433	31.84	4.21
8 x 8	68,121	21,646	31.78	1.52
16 x 16	17,161	5,422	31.60	0.65

*IBM 360 Model 67

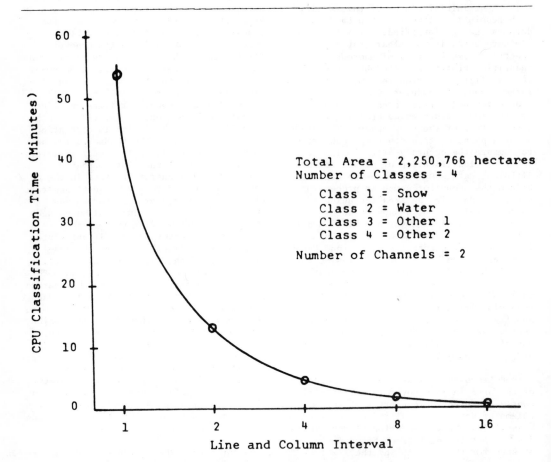

Figure 2. Computer CPU Time Required for Classification of Snow-Cover Using Different Sampling Intervals. (From Luther, et al., 1975.)

A relative response level of 127 indicates the saturation level for wavelength Bands 4, 5 and 6 of the Landsat data, and a relative response level of 63 in Band 7 indicates detector saturation. As can be seen from Figure 3, both snow and clouds saturated all four detectors on two of the dates examined and approached the saturation level for the third date. These results clearly indicate that computer-aided

Table 2. Comparison of Spectral Response of Clouds and Snow Using Landsat-1 Data.

Wavelength Bands

	4	5	6	7
	(0.5-0.6μm)	(0.6-0.7μm)	(0.7-0.8μm)	(0.8-1.1μm)
Clouds	126.6 ± 2.3[1]	126.2 ± 2.8	118.2 ± 6.8	55.6 ± 6.7
Snow	125.4 ± 5.2	125.0 ± 5.6	116.2 ± 10.2	51.2 ± 9.0

[1]Numbers indicate mean relative response ± 1 standard deviation using a combination of approximately 3000 data resolution elements, representing several areas of clouds and snow on each of three dates (1 Nov. '72, 6 Dec. '72, and 18 May '73).

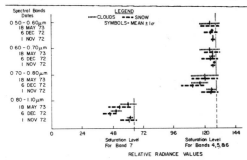

Figure B.15. Statistical analysis of the spectral response of snow and clouds.

Figure 3. Spectral Comparison of Clouds and Snow Using Landsat-1 Data From Three Different Dates. Saturation level was reached in nearly all data sets and the similarity of response indicates lack of spectral separability between those two cover types.

analysis techniques based upon spectral response cannot be reliably utilized to determine the areal extent of snow-cover using Landsat data, because of the limited dynamic range and limited spectral range of the Landsat scanner systems.

2.3 Snow-cloud differentiation using Skylab data

In 1973, the S-192 multispectral scanner was flown on the Skylab space station. This scanner contained 13 wavelength bands, operating in the atmospheric windows in the 0.41 to 12.5μm portion of the electromagnetic spectrum. Analysis of these Skylab data indicated that the limitations of the Landsat data in differentiating snow and clouds could be overcome. The extended spectral range of the Skylab S-192 Multispectral Scanner (0.41-12.5μm) proved to have significant advantages over the more limited spectral range of the Landsat MSS system (0.5-1.1μm). Examination of the Skylab imagery indicated that in the visible (0.4-0.7μm) portion of the spectrum

snow and clouds had similar high reflectance values and appeared white on the imagery. In the thermal infrared band, both clouds and snow had a relatively low spectral response and appeared black on the imagery. This is because of the relatively low temperature for snow as compared to the other cover types, and the fact that in this data set, the clouds had a very similar low temperature. In the near infrared portion of the spectrum (0.7-1.3μm), the snow was white, but the snow-pack appeared to decrease in size with increasing wavelength. However, the clouds had a high reflectance throughout the near infrared portion of the spectrum and did not change in size. In the middle infrared wavelength bands (1.55-1.75μm and 2.10-2.35μm) a very striking difference in spectral response was found between clouds and snow. In this portion of the spectrum, the clouds have a high reflectance and appear white, whereas the snow has a very low reflectance and appears black on the imagery. The reason for the very low response of the snow in the middle infrared portion of the spectrum as well as for the decreased reflectance with increased wavelength in the near infrared portion of the spectrum is indicated by the typical spectral reflectance curve of snow shown in Figure 4. The quantitative spectral responses (means and standard deviations) for snow-cover and clouds in the 13 Skylab S-192 wavelength bands are shown in Table 3. (A spectral response of value of 255 indicates saturation of the detector.)

In order to accurately measure the degree of spectral separability between clouds and snow, a transformed divergence algorithm (Swain, et al., 1971; Swain and Staff, 1972) was utilized. Figure 5 is a bar graph showing the separability (based upon the transformed divergence values) between clouds and snow in the 13 wavelength bands of Skylab data. Transformed divergence

377

Table 3. Mean Spectral Response and Standard Deviation for Snow and Clouds, SL-2 S-192 Data, 5 June 1973.

Wavelength Region	Wavelength Band (m)	Snow		Clouds	
		Mean	S.D.	Mean	S.D.
Visible	0.41-0.46	255	0	254	4
Visible	0.46-0.51	250	16	248	19
Visible	0.52-0.56	229	38	229	37
Visible	0.56-0.61	255	0	254	1
Visible	0.62-0.67	254	6	254	8
Near Infrared	0.68-0.76	246	22	246	22
Near Infrared	0.78-0.88	230	38	230	35
Near Infrared	0.98-1.03	181	44	222	41
Near Infrared	1.09-1.19	165	33	228	32
Near Infrared	1.20-1.30	106	22	210	43
Middle Infrared	1.55-1.75	33	16	163	33
Middle Infrared	2.10-2.35	39	15	160	31
Thermal Infrared	10.2-12.5	67	18	61	14

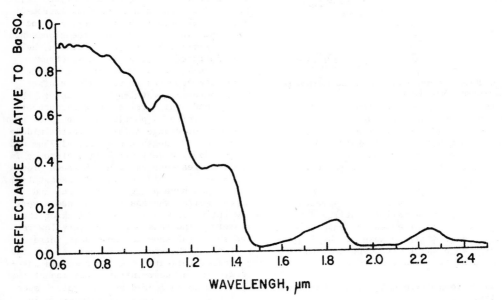

Figure 4. Typical spectral reflectance curve for snow. (After O'Brien and Munis, 1975.)

values of greater than 1750 indicate reliable separability between the spectral classes involved (Swain and King, 1973). The most relevant aspect of Figure 5 is that clouds and snow can be spectrally separated only in the two middle infrared wavelength bands (1.55-1.75µm and 2.10-2.35µm). These results clearly show the advantage for obtaining multispectral scanner data in the middle infrared (1.3-3.0µm) portion of the spectrum when the application involves the mapping of snow cover. These results, as well as others involving computer-aided analysis of vegetative cover types, indicate the need for scanner systems carried by future satellites to contain at least one wavelength band in the middle infrared portion of the spectrum.

2.4 Mapping snow-cover using Skylab plus topographic data

The computer-assisted analysis of the Skylab MSS data for mapping snow-cover involved several processing steps. The first consisted of the definition of the maximum number of spectrally separable classes or categories of snow-cover present in the scene. A clustering or "non-supervised"

378

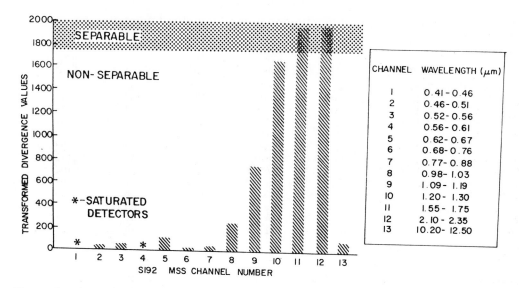

Figure 5. Spectral separability of snow and clouds in the 13 Skylab-2 S-192 wavelength bands. (From Bartolucci, 1975.)

approach was utilized for this phase of the analysis, and five distinct spectral classes of snow-cover were defined. The training statistics thus defined were then used in a "supervised" classification of the entire data set using the maximum likelihood algorithm.

As indicated in Table 4, these five spectral classes of snow-cover were related to differences in reflectance in the individual wavelength bands in the near infrared portion of the spectrum, particularly in this 1.09-1.19μm and 1.2-1.3μm bands. Comparison of the snow-cover classification results with aerial photos taken from two different altitudes one day after the Skylab data was obtained indicated that the five different spectral classes of snow-cover were closely related to different proportions of snow and forest cover present within the individual resolution elements (pixels) of the Skylab MSS data. The S-192 scanner integrates the reflectance from the entire area on the ground within a resolution element (approximately 0.46 hectares). Therefore, a relatively high proportion of coniferous forest cover and a relatively low proportion of snow within a single resolution element will result in a relatively low reflectance as compared to a resolution element containing fewer trees and therefore having a larger proportion of snow-cover.

The spatial distribution of the five spectral classes of snow-cover mapped by the computer was found to be highly correlated to the topography of the area. The "snow-1" spectral class was found only at higher elevations in the areas of alpine tundra. The other four spectral classes were found in lower elevation ranges which generally have an increasing density of coniferous forest cover with decreasing elevation.

To establish a more quantitative correspondence between the spectral classes of forest cover and the topography of the area, digital topographic data were overlaid onto the multispectral scanner data. Tapes of elevation data obtained from 1:250,000 scale USGS topographic maps were reformatted to match the scale the Skylab MSS Data. Then, an interpolation procedure was developed and applied to the digital elevation data in order to obtain data on slope and aspect for each resolution element. This produced a digital data tape containing 13 channels of Skylab data and three channels of topographic data (elevation, slope, and aspect).

To show the value of the topographic overlay data, the elevation data were combined with results of the snow-cover classification, and the area of each of the five classes of snow-cover were determined as a function of elevation, using 100 meter elevation increments. This type of calculation can be accomplished rapidly and effectively using such multiple data sets which have been overlaid in a

Table 4. Mean Spectral Response of Five Snow-Cover Classes and a Forest Class in Skylab S-192 Data.

Band (μm)	Snow 1	Snow 2	Spectral Classes Snow 3	Snow 4	Snow 5	Forest
(0.41-0.46)	255*	255*	255*	255*	255*	205
(0.46-0.51)	255*	255*	255*	197	162	110
(0.51-0.56)	254*	252*	219	120	93	56
(0.56-0.61)	255*	255*	255*	253	251	131
(0.62-0.67)	255*	255*	255*	255*	237	72
(0.68-0.76)	255*	255*	255*	240	166	89
(0.78-0.88)	255*	255*	255*	193	148	113
(0.98-1.08)	255*	255*	194	138	108	102
(1.09-1.19)	251	196	137	104	89	92
(1.20-1.30)	185	148	98	76	68	83
(1.55-1.75)	64	61	56	54	59	72
(2.10-2.35)	19	17	14	11	13	21
(10.2-12.5)	99	100	101	105	110	124

* - Denotes detector saturation

format suitable for computer-aided analysis. The results of this analysis sequence are shown in Table 5. The results in Table 5 indicate the potential for rapidly summarizing the area of the snowpack as a function of elevation and as a function of the spectral class of snow. Such information could be of tremendous help to hydrologists in predicting the amount and timing of runoff from the snowpack in mountainous regions throughout the world.

3 MAPPING WATER RESOURCES USING SATELLITE MSS DATA

In water management, as well as many other areas of resource management, effective planning is dependent upon the accuracy and the reliability of the data which is available for the various planning and management activities. An accurate knowledge of the quantity of water stored in natural and artificial lakes is required to reach sound decisions regarding the conservation and allocation of water for different applications such as recreation, municipal and industrial water supplies, flood control, hydroelectric power generation, and agricultural irrigation.

Because the reflectance characteristics for water are distinctly different from those of other earth's surface features in the near infrared portion of the spectrum, use of remote sensing data in the near infrared region has long been recognized as a very valuable method for locating, mapping, and monitoring water features. Figure 6 shows the distinctly low reflectance of water as compared to soil or green vegetation in the near infrared

portion of the spectrum. Figure 7 shows that the reflectance characteristics of clear water are even lower than for turbid water, particularly in the near infrared and red visible wavelength regions.

3.1 Areal Extent of Water Bodies

The availability and frequency of coverage by Landsat therefore leads to many questions and investigations involving the use of this data for mapping water bodies over large geographic areas of interest. In one study at LARS, it was shown that water bodies of more than three hectares in size could be identified with 100% reliability on the Landsat data through the use of computer-aided analysis techniques (Bartolucci, 1976).

Work and Gilmer (1976) used two different analysis techniques to map open water as an indicator of waterfowl habitat. At present, waterfowl management decisions are based in part on an assessment of the number, distribution, and quality of ponds and lakes in the primary breeding range, which covers a very large geographic area of North Central United States and South Central Canada. At the present time, surveys are conducted annually from low flying light aircraft to monitor the water conditions and provide key information for the necessary management decisions. The results of this study showed that areas of more than 1.6 hectares were nearly always recognized, but ponds smaller than that size were sometimes recognized and sometimes not, depending upon the location of the water body in relation to the resolution element of the Landsat scanner system. Thus, it is problematic as to whether

Table 5. Snowpack Area (in hectares) Within 100 Meter Elevation Increments.

Elevation (meters)	Spectral Class of Snowcover					Total Area (hectares)
	1	2	3	4	5	
Above 3700	1179	2464	308	108	7	4086
3600–3700	400	1914	694	135	37	3180
3500–3600	129	1868	1858	517	61	4433
3400–3500	45	904	1858	1266	280	4353
3300–3400	13	378	1305	1417	812	3925
3200–3300	7	94	922	1258	1298	3579
3100–3200	6	22	529	793	1540	2890
3000–3100		9	213	433	1041	1893
2900–3000		1	38	188	535	752
2800–2900			4	54	289	347
2700–2800			1	13	147	181
2600–2700				1	95	96
Below 2600					79	79
Totals	1779	7651	7730	6183	6221	29564

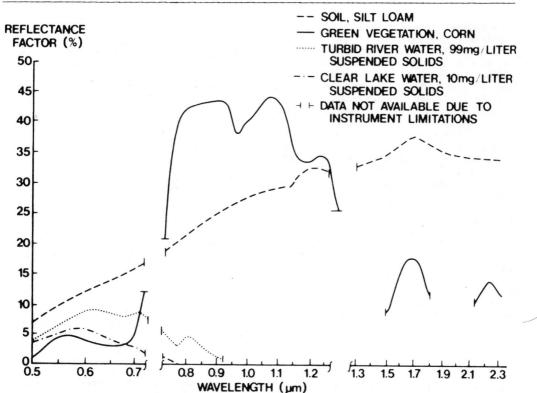

Figure 6. Spectral characteristics of turbid river water, soils, and vegetation.
(After Bartolucci, et al., 1977.)

very small ponds will or will not be recognized on Landsat data. Other studies
have shown some similar results. Future
satellite systems, such as the Thematic
Mapper on Landsat D, or the French SPOT
scanner system will have much better reso-

spatial resolution, and therefore may provide a capability to map smaller ponds
(e.g., < 1 hectare) with a reasonable
degree of reliability.

One difficulty in the use of satellite

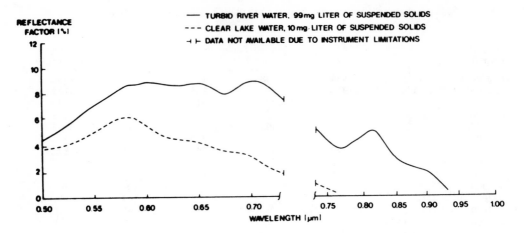

Figure 7. Spectral characteristics of turbid river water and clear lake water. (After Bartolucci, et al., 1977.)

data to identify and map water bodies that should not be overlooked is that in areas of mountainous terrain, topographic shadows have a very low spectral response and can be confused with the low spectral response found in bodies of clear water. Even in flat-land terrain areas, cloud shadows are sometimes confused with water bodies because of the low spectral response. Figure 8 shows the spectral response obtained from Landsat data on four different dates, and shows that the differences between cloud shadows and water are very small, particularly in the near infrared portion of the spectrum. In two visible wavelengths, cloud shadows tend to be slightly lower in spectral response than the water bodies and could be discriminated with reasonable reliability. It is important to recognize, however, that computer-aided analysis of Landsat data for mapping water bodies must take into account the effect of shadow areas present in the data so that these are not mapped as surface water features.

3.2 Defining Water Condition With Satellite Data

In the natural world, water bodies are usually not clear but contain a variety of organic and inorganic materials, some of which are in suspension. These materials cause scattering and absorption of the incident energy, resulting in significant differences in both the transmission and reflectance of incident energy from water bodies. Turbidity caused by suspended sediments is one of the major factors affecting the spectral response of water

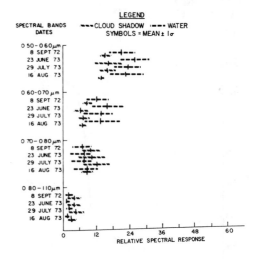

Figure 8. Spectral response of shadows and water bodies. (After Hoffer, et al., 1975.)

features. Previously, in Figure 7 we saw the spectral response for turbid and clear waters under natural conditions. As was seen in this figure, turbid water has a significantly higher reflectance than clear water, and it is also apparent that the peak reflectance for turbid water is at a longer wavelength than the peak reflectance for clear water. In the study by Bartolucci et al, (1977), they found that the reflectance characteristics of the river bottom did not influence spectral

382

response of the turbid water when the water was mere than 30 centimeters in depth. This is a particularly significant finding, in that it indicates that whenever one is interpreting Landsat data for areas in which water is over 30 centimeters in depth, differences in reflectance measured by the Landsat scanner system are due to differences in the water quality rather than reflectance characteristics of the bottom.

A study by Weisblatt (1973) indicated that increased levels of turbidity produce almost a linear relationship with the radiance values measured by the Landsat scanner system in the two visible wavelengths, and a curvilinear relationship in the near infrared wavelength bands. The best relationship between spectral response and water turbidity was found in band 5 (0.6-0.7μm) of the Landsat data. Figures 9a and b show these relationships between Landsat data and turbidity, and for measurements obtained with a reference data source (Exotech 100 C) and water turbidity.

3.3 Water Temperature Mapping With Skylab Data

During the Skylab satellite mission, thermal infrared scanner data was obtained in the 10.2-12.5μm portion of the spectrum for Vallecito Reservoir in the San Juan Mountains of Colorado. The Skylab S-192 Multispectral Scanner System contained a set of internal calibration sources, consisting of two "black bodies" which were maintained at constant and known temperatures, so that while MSS data were being gathered, the rotating scanner mirror viewed these two black bodies during each rotation. Thus, in every scan line of data, a set of two radiance values are recorded which correspond to the energy emitted by the black bodies as a function of their temperatures. Both linear and non-linear calibration procedures were applied to the thermal band of Skylab data. At the time the Skylab satellite passed overhead, a team of researchers were in a boat on the surface of the reservoir taking temperature measurements in a variety of locations. These temperature measurements ranged from 12.0 to 13.1 degrees Centigrade. The linear calibration procedure applied to the Skylab data indicated that the temperature of the surface of Vallecito Reservoir was approximately 8.6 degrees Centigrade, whereas the non-linear calibration procedure indicated a surface temperature of 12.7 degrees Centigrade, which was a better indication of

the true surface temperature of the reservoir.

Although limited in scope, this study indicated the value of the non-linear calibration procedure for use with thermal infrared scanner data. It also indicated that for mountainous areas at high elevations, the need to apply correction factors to account for atmospheric attenuation is minimized, thereby allowing reasonable and accurate surface temperature measurements to be obtained directly from the satellite scanner data.

4 WATER TEMPERATURE MAPPING USING AIRCRAFT MSS DATA

There is a great deal of controversy concerning many of the federal and state guidelines involving quality standards. Often this is because the legislative bodies and the industry both lack accurate and comprehensive factual information about the streams, reservoirs, and lakes with which they are concerned. This is particularly true in situations involving the setting of safe temperature standards for streams affected by the discharge of large quantities of waste heat from power plants. The discharge of heat into the water of the streams and the resulting thermal alteration is one aspect of power generation that is particularly amenable to study by thermal infrared scanner systems operated from aircraft altitudes. Thermal infrared scanner measurements are especially suitable for flowing water systems where no thermal stratification exists and where distinct thermal input may occur at numerous points along the length of the river.

In a study along the Wabash River, two different sets of data were collected and analyzed to determine the accuracy of calibrated remote sensor measurements to determine surface water temperature. One set of the data were collected in 4.5-5.5μm and 8.0-13.5μm portions of the spectrum, and from both 3000 meter and 600 meters altitudes. A second set of data were collected using only 9.3-11.7μm wavelength band from an altitude of 1,500 meters. At the time the scanner data were collected, measurements were also made from a boat on the surface of the river, so that a comparison could be obtained between the temperatures calculated from the scanner data (radiant temperatures) and the temperatures actually measured at the water surface (kinetic temperatures).

Table 6 shows the means for the radiant

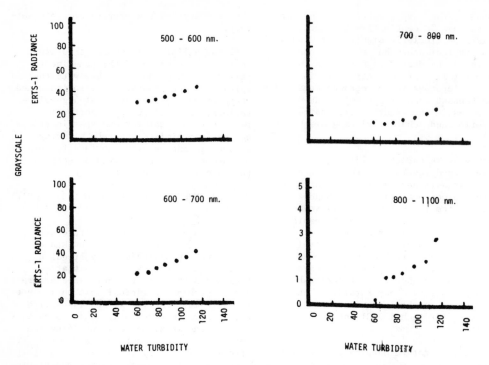

Figure 9a.   Landsat Radiance as a Function of Water Turbidity. (From Weisblatt, 1973.)

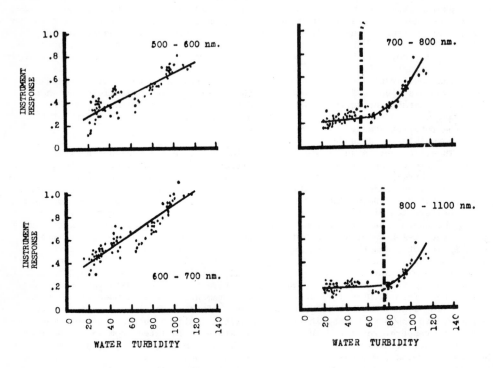

Figure 9b.   Reference Data Showing Response as a Function of Water Turbidity of Bay
Waters.   (From Weisblatt, 1973.)

Table 6. Altitude Effects on Radiant Temperatures.

| Measurement Site | Radiant Temperatures[1] (°C) | | | | Kinetic Temp.[2] (°C) (Contact Thermometer) |
| | 3000 M. Alt. | | 600 M. Alt. | | |
	4.5-5.5µm	8.0-13.5µm	4.5-5.5µm	9.0-13.5µm	
1st River Bend	19.3	24.6	25.8	26.2	26.5
Above Intake	19.2	24.4	25.9	26.2	26.5
Intake	19.3	24.6	25.9	26.1	26.2
Outlet	19.2	24.4	25.8	26.1	26.0
1/2 Mile Below	19.3	24.4	25.6	26.1	26.2

[1]The standard deviation for the radiant temperatures is ± 0.2°C.

[2]All the surface measurements of temperatures listed above were obtained within half an hour from the time the aircraft passed overhead.

temperatures obtained from the thermal scanner data at both the 3000 and 600 meter altitudes, and also for the two wavelength bands (4.5-5.5 and 8.5-13.5µm) that were available during the first data collection mission. The kinetic temperatures measured at the water surface are also shown. From these results, it is quite evident that the aircraft altitude, or more correctly, the atmospheric path length between the target and sensor, plays an extremely important role in the degree of accuracy with which radiant temperatures can be determined from airborne scanner data. For the 600 meter altitude data, water surface temperatures could be determined in the 8.0-13.5µm wavelength band to an accuracy of approximately 0.2 degrees Centigrade. However, at the 3000 meter altitude, the radiant temperatures for the same wavelength band appear to be approximately 2° C. lower than they should be. For the 4.5-5.5µm band, the altitude effects are so pronounced that for the 3000 meter data, the radiant temperatures measured from the scanner data are approximately 7° C. lower than the actual surface temperatures.

The impact of the transmission characteristics of the atmosphere on the data is shown in Figure 10. As can be seen, the so-called 4.5-5.5µm "atmospheric window" has a moderate amount of attenuation. Sensitive detectors, such as the Mercury Cadmium Telluride (Hg: Cd: Te), can obtain thermal infrared scanner data effectively in very narrow wavelength bands and thereby take advantage of the very good transmission characteristics in the 9.3-11.7µm portion of the spectrum. Such very clear transmission characteristics of the 9.3-11.7µm region indicate why accurate measurements of water temperatures could be

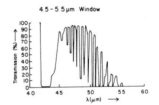

4.5-5.5µm Window

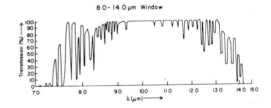

8.0-14.0µm Window

Figure 10. Transmission characteristics of the atmosphere at sea level for a 300 meter pathlength in the two thermal infrared atmospheric windows (4.5-5.5µm and 8.0-14.0µm). (From Bartolucci, et al., 1973.)

obtained with the Skylab S-192 satellite data.

An evaluation of the use of aircraft scanner data for measuring the effluent from a hydroelectric power plant was conducted in conjunction with the Cayuga Power Plant on the Wabash River. This is a fossil fuel plant which utilizes water from the Wabash River for cooling purposes at an approximate rate of 2,200 cubic meters per minute. The temperature of the cooling water is increased by approximately 8° C. as it passes through the power generating units and is discharged back into the Wabash River. Use of the 9.3-11.7μm thermal infrared scanner data from 1,500 meters altitude showed that after the water was discharged into the Wabash River, the river temperature was approximately 4° C. higher than the temperature of the river above the discharge point. A distinct thermal plume could be mapped downstream from the power plant for a distance of approximately 6 kilometers. Use of the 9.3-11.7μm data allowed the water temperatures in the river to be mapped from aircraft altitudes with an accuracy of 0.2° C., as indicated in Table 7.

Table 7. Comparison of Kinetic and Radiant Temperatures at Six Different Locations.

Site & Date	Kinetic Temp. (°C)	Radiant Temp.[1] (°C)
	22.1	21.9
Cayuga 8/9/72	22.5	22.4
	25.7	25.6
	19.8	19.8
Junction 8/10/72	21.5	21.3
	20.4	20.3

[1]From an altitude of 1,500 meters using the 9.3-11.7μm band.

Work with fisheries biologists also indicated that there were distinct changes in the population of various fish species normally found in the Wabash River. Comparisons with data prior to the time the power plant went into operation indicated that large increases in the number of catfish were found in the heated portions of the river where the impact of thermal effluent was the greatest, while the populations of sauger and redhorse decreased in the heated areas.

It is clear from these results that remote sensing techniques involving the use of computer-aided analysis of thermal infrared aircraft scanner data allow an effective method determining the extent and magnitude of thermal effluents from hydroelectric power plants on river ecosystems.

5 SUMMARY AND CONCLUSIONS

Research results such as reported in this and other papers clearly indicate that remote sensing technology can be effectively utilized in mapping and monitoring of snow-cover and water resources. Computer-aided analysis techniques are becoming more cost-effective each year as new hardware and data processing capabilities are developed. Such computer-aided analysis techniques are particularly useful in conjunction with multispectral scanner data collected from satellite altitudes.

Studies such as those reported in this paper involving Landsat and Skylab data have indicated the potential for mapping snow-cover using satellite data and computer-aided analysis techniques. However, the limited spectral range and spatial resolution of the current Landsat scanner systems make them of limited value for mapping snow-cover and small bodies of surface water, but both of these limitations will be overcome by satellites to be launched in the early 1980's. The use of computer data bases for digitally combining topographic data with the satellite scanner data offers additional promise for evaluating and monitoring snow-pack conditions in mountainous areas.

The ability to accurately map surface temperatures of lakes and rivers using calibrated thermal infrared scanners from aircraft altitudes has been developed. The use of this technology will become more common as the availability of calibrated thermal infrared scanners increases, and as the data processing using standardized hardware systems and analysis techniques becomes more cost-effective.

From these comments, one sees that the development and the utilization of remote sensing (and other technologies) must follow a sequence in which there is a proper balance among (1) the research involved in defining the energy/matter interactions involved, (2) availability of appropriate instrumentation to collect and analyze the data, and (3) the refinement of the data processing techniques necessary for providing a cost-effective capability for operational utilization of the technology. It is also apparent that the demand for certain types of information and the time requirements involved will determine the ultimate utilization of remote sensing technology. In the case of the utilization of remote sensing technology for the mapping and assessment of snow-cover and water resources, it is clear that in some areas the technology are ready for operational application and utilization, whereas other aspects of remote sensing technology (particularly the utilization of satellite data) will be considerably improved with the next generation of scanner systems. Analysis techniques have been reasonably well developed, but work remains on the development of cost-effective hardware to be utilized with the data collection systems which will come into existence in the early 1980's. In summary, it seems clear that there are many aspects of remote sensing technology that can and will be utilized in the assessment and monitoring of our water resources.

6 REFERENCES

Barnes, J.C. and C.J. Bowley 1973, Use of ERTS Data for Mapping Snow Cover in the Western United States, Proceedings of the Symposium on Significant Results Obtained from ERTS-1, NASA Report SP-327, p. 855-862. Washington, D.C.

Bartolucci, L.A., R.M. Hoffer, and J.R. Gammon 1973, Effects of Altitude and Wavelength Band Selection on Remote Measurements of Water Temperature, Proceedings of the First Pan-American Symposium on Remote Sensing, p. 147-160. Panama City, Panama.

Bartolucci, L.A. 1975, Hydrologic Features Survey. In Computer-Aided Analysis of Skylab Multispectral Scanner Data in Mountainous Terrain for Land Use, Forestry, Water Resource, and Geologic Applications, by R.M. Hoffer and Staff, LARS Information Note 121275, Laboratory for Applications of Remote Sensing, Purdue University, West Lafayette, Indiana.

Bartolucci, L.A., R.M. Hoffer and S.G. Luther 1975, Snow Cover Mapping by Machine Processing of Skylab and Landsat MSS Data, Proceedings of the Workshop of Operational Applications of Satellite Snowcover Observations, NASA Report SP-391, p. 295-311. Washington, D.C.

Bartolucci, L.A., B.F. Robinson, and L.F. Silva 1977, Field Measurements of the Spectral Response of Natural Waters, Photogrammetric Engineering and Remote Sensing XLIII (5):595-598.

Conover, J.H. 1964, The Identification and Significance of Orographically Induced Clouds Observed by TIROS Satellites, Journal of Applied Meterology 3:226-234.

Finnegan, W.J. 1962, Snow Surveying with Aerial Photographs, Photogrammetric Engineering 28 (5):782-790.

Hoffer, R.M. and LARS Staff 1973, Techniques for Computer-Aided Analysis of ERTS-1 Data, Useful in Geologic, Forest and Water Resource Surveys, Porceedings of the Third ERTS-1 Symposium, 1(A):1687-1708. Goddard Space Flight Center,

Washington, D.C. Also LARS Information Note 121073.

Hoffer, R.M. and Staff 1975, Natural Resource Mapping in Mountainous Terrain by Computer Analysis of ERTS-1 Satellite Data, Agricultural Experiment Station Research Bulletin 919 and LARS Technical Report 061575, Purdue University, West Lafayette, Indiana.

Leaf, C.F. 1969, Aerial Photographs for Operational Streamflow Forecasting in the Colorado Rockies, Proceedings of the 37th Western Snow Conference, p. 19-28.

Luther, S.G., L.A. Bartolucci, and R.M. Hoffer 1975, Snow Cover Monitoring by Machine Processing of Multitemporal Landsat MSS Data, Proceedings of the Workshop on Operational Applications of Satellite Snowcover Observations, NASA Report SP-391, p. 279-294. Washington, D.C.

Meier, M.F. 1973, Evaluation of ERTS Imagery for Mapping and Detection of Changes of Snowcover on Land and on Glaciers, Symposium on Significant Results Obtained from ERTS-1, NASA SP-327, 1(A):863-875.

O'Brien, H.W. and R.H. Munis 1975, Red and Near Infrared Spectral Reflectance of Snow, Proceedings of the Workshop on Operational Applications of Satellite Snowcover Operations, NASA Report SP-391, p. 345-360. Washington, D.C.

Parshall, R.L. 1941, Correlation of Streamflow and Snow Cover in Colorado, Transactions of the American Geophysical Union, Vol. 22, Part 1.

Parsons, W.J. and G.H. Castle 1959, Aerial Reconnaissance of Mountain Snow Fields for Maintaining Up-To-Date Forecasts of Snowmelt Runoff During the Melt Period, Proceedings of the 27th Western Snow Conference, p. 49-56.

Potts, H.L. 1944, A Photographic Snow Survey Method of Forecasting Runoff, Transactions of the American Geophysical Union, 25:149-153.

Swain, P.H., T.V. Robertson, and A.G. Wacker 1971, Comparison of the Divergence and B-Distance in Feature Selection, LARS Information Note 020871, LARS/Purdue University, West Lafayette, Indiana.

Swain, P.H. and Staff 1972, Data Processing I: Advancements in Machine Analysis of Multispectral Data, LARS Information Note 012472, LARS/Purdue University, West Lafayette, Indiana.

Swain, P.H. and A.C. King 1973, Two Effective Feature Selection Criteria for Multispectral Remote Sensing, LARS Information Note 042673, LARS/Purdue University, West Lafayette, Indiana.

Weisblatt, E.A., J.B. Zaitzeff, and C.A. Reeves 1973, Classification of Turbidity Levels in the Texas Marine Coastal Zone, Proc. Symposium on Machine Processing of

Remotely Sensed Data, IEEE Catalog No. 73 CHO 834-2GE, p. 3A-42 to 3A-59, Purdue University, West Lafayette, Indiana.

Weisnet, D.R. 1974, The Role of Satellites in Snow and Ice Measurements, Advanced Concepts and Techniques in the Study of Snow and Ice Resources, p. 447-456. National Academy of Sciences, Washington, D.C.

Work, E.A., and D.S. Gilmer,1976, Utilization of Satellite Data for Inventorying Prairie Ponds and Lakes, Photogrammetric Engineering and Remote Sensing XLII (5): 685-694.

Electromagnetic studies of ice and snow

1 *Radiometry of ice and snow*

1 INTRODUCTION

Microwave radiometry has been used by radio astronomers for many years for detection of radiation from stars and galaxies. In recent years it has also been used for observation of natural thermal radiation from earth surfaces as a special tool in remote sensing.

Being of thermal origin the radiation has a noise-like characteristic and measurement of the emitted energy is effected by integration in time of the signal received by the so-called radiometer. In fact, there is no difference in principle between microwave radiometry and the infrared radiometry except that the energy is several orders of magnitude lower at microwaves. The two techniques may be characterized as passive methods in contrast to radar which uses a controlled microwave source for illumination of the ground and therefore is characterized as an active method.

The remote sensing interest in exploiting microwaves stems from the fact that the influence of the atmosphere on the measurements in general is very small so that an essentially all-weather system can be implemented in contrast to visual and infrared techniques which are influenced by cloud cover, for instance. Also, microwaves give a larger penetration into the surfaces observed than infrared so that sub-surface features may be studied. However, having a smaller resolution capability microwave radiometry may be considered a complement to infrared radiometry for some applications.

From measurements carried out so far it is clear that microwave radiometry has great potential applications in oceanography. Also, there are strong indications that the technique will be very useful for monitoring ice and snow, and, related dynamic processes.

Following tutorial sections on microwave radiometry and techniques, this chapter describes a number of measurements from aircraft and satellite as well as fundamental ground-based measurements. Finally, some numerical models describing various phenomena observed are reviewed.

2 BLACK-BODY RADIATION

Radiometry as applied in remote sensing is basically measurement of the natural radiation from the earth surface. Perfect radiation is obtained from a so-called black body in thermal equilibrium being described by Planck's law:

$$W(f) = \frac{h(f)}{e^{hf/kT}-1} \qquad (W/Hz) \qquad (1)$$

where h is Planck's constant ($6.626 \; 10^{-34}$ J/sec), k is Boltzmann's constant ($1.38 \; 10^{-23}$ J/K), T is the thermodynamic temperature in Kelvin, and f is the frequency of the radiation. In the temperature and frequency ranges where Rayleigh-Jeans approximation (hf << kt) is valid the Planck's law is approximated by

$$W(f) = W_o = kT \qquad (2)$$

It is noted that the inequality hf << kT is valid for f < 100 GHz and T > 20 K i.e. in the microwave region and at normal temperatures encountered in nature.

3 RADIATION FROM NATURAL OBJECTS, EMISSIVITY

Natural objects and surfaces are less

efficient radiators than the black body. For non-ideal radiators the radiated energy can be expressed by

$$W_o = ekT \qquad (3)$$

where the emissivity e is defined as the ratio between the energy emitted by the object and that radiated by a perfect radiator at the same temperature.

Since practical measurements of energy cannot distinguish between the emissivity and the temperature of the object a term: brightness temperature, T_B, is introduced

$$T_B = eT \qquad (4)$$

so that

$$W_o = kT_B \qquad (5)$$

Through this equation it becomes reasonable to describe the radiated energy by the brightness temperature alone.

In the remote sensing situation the emissivity is related to the dielectric properties of the object considered. In the simple case where the object surface forms a plane interface between the object and the air a reflection of energy takes place at the surface so that the transmission coefficient or the emissivity becomes

$$e = 1 - \rho\rho^* \qquad (6)$$

where ρ is the voltage reflection coefficient of the surface.

4 RADIATION FROM LAYERED STRUCTURES

When radiation energy passes from one medium with a certain permittivity, ε_1, through a layer with another permittivity, ε_2, it is subject to reflection from the interface between the two media and from attenuation in the layer. If the layer forms an overburden the radiation will be influenced by another reflection at the surface of the layer.

According to Kirchoff's law, a body in thermal equilibrium at a certain temperature, T_2, will radiate the same amount of energy as it absorps, Thus, in measuring the thermal radiation from a position above the layer surface the energy received can be described by $\qquad (7)$

$$T_A = |\rho_1|^2|\rho_2|^2\tau_2T_1 + (1 - \tau_2)|\rho_2|^2T_2$$

where $\rho_{1,2}$ are the voltage reflection coefficients at the interfaces, τ_2 is the transmission coefficient of the layer

$$\tau_2 = 1 - a_2 \qquad (8)$$

where a_2 is the power absorption coefficient of the layer, and $T_{1,2}$ are the physical temperatures of the media Nos. 1 and 2.

5 INFLUENCE BY THE ATMOSPHERE

When observed from an aircraft or a satellite the radiation from the surface will propagate through the atmosphere. At high frequencies i.e. above 10 GHz the atmosphere attenuates electromagnetic energy by molecular resonance absorption. This attenuation must be taken into account in the same way as above.

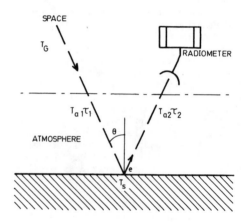

Fig. 1. Measurement situation including influence of the atmosphere and background radiation.

Referring to Fig. 1 and neglecting reflections except at the surface, the total radiative transfer including an isotropic atmosphere may be described by

$$T_A = e\tau_2T_S + (1 - e)\tau_1\tau_2T_G + \qquad (9)$$
$$(1 - e)(1 - \tau_1)\tau_2T_{a1} + (1 - \tau_2)T_{a2}$$

where T_G is the galactic background radiation and we have separated the paths denoted by indices 1 and 2. It is seen that the third term, for instance, describes the radiation from the atmosphere along path No. 1, $(1 - \tau_1) T_{a1}$, reflected at the

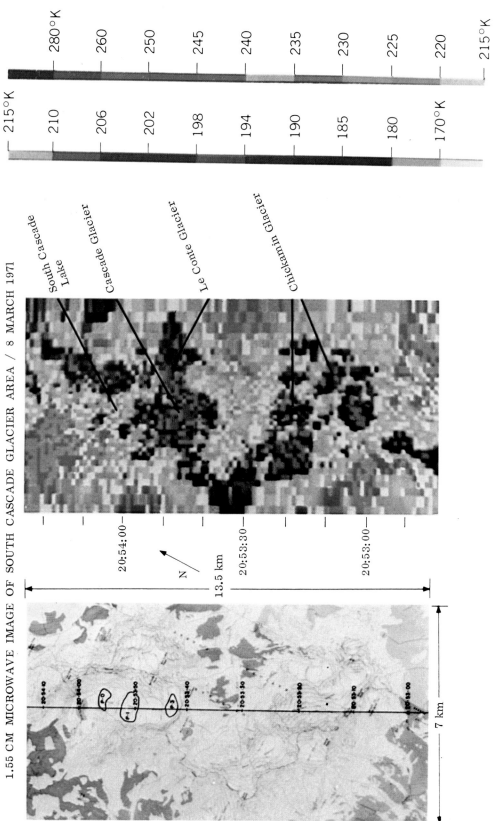

Fig. 6. Microwave radiometer image obtained with an airborne 19.35 GHz scanning radiometer at South Cascade Glacier, March 8, 1971.

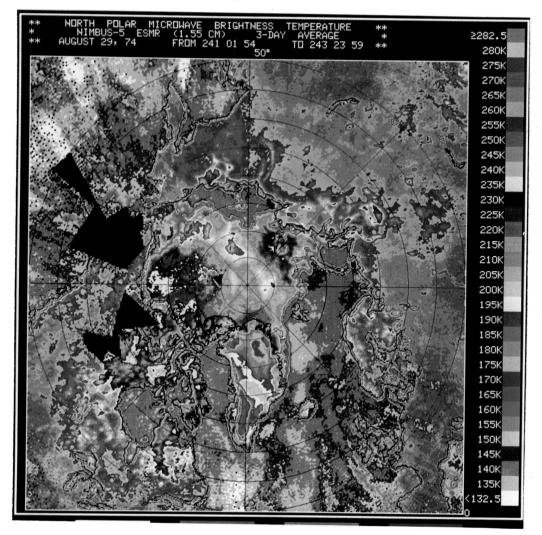

Fig. 7 Microwave radiometer image obtained with the electrically scanning microwave radiometer, ESMR, at 19.35 GHz on board NIMBUS-5. Brightness temperatures are averages of measurements in a three-day period beginning 29 August 1974.

surface with the reflection coefficient of
(1 - e), and attenuated by the atmosphere
along path No. 2 with a tranmsission coefficient of τ_2.

(9) shows that the brightness temperature
of a surface (eT_s) as detected by the radiometer (T_A) is modified by the atmosphere,
clouds etc. due to attenuation and addition
of unwanted energy. Moreover, numerical
examples will show that the atmosphere reduces the contrast between surfaces with
different brightness temperatures. Finally,
it will be found that the modification due
to the atmosphere and reflection of extra-
terrestrial energy in the surface is larger
over water than over land since water has
a comparatively larger reflection coefficient and a lower brightness temperature.

Other attenuation phenomena originate
from scattering from droplets in clouds
and rain. This attenuation may be described
by Rayleigh and Mie scattering from dielectric spheres depending on the size of
the droplets. Rayleigh scattering assumes
that the droplets are much smaller than
the wavelength – as is often the case in
clouds, while the Mie scattering applies
to the case where the droplets are of the
same order as the wavelength which is the
case with rain at frequencies above 10 GHz.

6 POLARISATION

Thermal radiation is a multi-polarisation
phenomenon so that, in principle, there is
no preferred polarisation. However, when
the radiated energy is measured by a microwave receiver an antenna with a well-defined polarisation is involved. Since the
surface reflection coefficient and thereby the emissivity is dependent on the polarisation of the energy, the measured e-
nergy at a certain angle of incidence is
also polarisation dependent. Thus, the
radiation is described as vertically or
horizontally polarized radiation although
the phenomena are related to the antenna
characteristic rather than the radiation.

7 RADIOMETER EQUIPMENT

The basic radiometer is the so-called
Dicke radiometer as shown in the block
schematic Fig. 2. The radiation picked up
by the antenna and presented at the output port of the antenna and the energy
from a reference load at a known temperature are sampled alternatively by the
input switch. Demodulation of the video
signal at the output of the amplifier is
carried out by a synchronous detector with
the switching signal as a reference. The
effect of the integrator is to reduce the
fluctuations of the output signal so that
the true difference between the antenna
energy and that from the reference load
can be measured. The following fundamental
relation is valid for the RMS fluctuations
after the integrator.

$$\sigma = \frac{C(T_A + T_N)}{\sqrt{B \cdot \tau}} \qquad \text{(Kelvin)} \qquad (10)$$

where C is a radiometer constant dependent
on the type of radiometer with C = 2 for
the above-mentioned Dicke radiometer. T_A
is the temperature at the antenna output
port and represents the available power at
this port , cf. (2), which gives the power
when multiplied by the bandwidth, B. Similarly, T_N is the receiver noise temperature referred to the input of the receiver
(T_N is related to the noise figure of the
receiver, F, by the equation $T_N = (F - 1)T_o$
where T_o = 290 K is the reference temperature). B is the receiver bandwidth, and τ
is the integration time.

σ is a measure of the uncertainty of the
measurement, and it is seen that a good
sensitivity, i.e. a small σ, requires low
receiver noise temperature, a large bandwidth and a long integration time. In practice, the magnitude of the bandwidth is
determined by the possible interference
from other services picked-up by the antenna. The maximum integration time is determined by the time needed to sample the
earth surface. Considering the velocity of
the spacecraft and the scan rate of the
imaging antenna (see below) integration
times of the order of 50 milliseconds are
possible. With this integration time the
sensitivity becomes 1 to 2 K dependent
upon the frequency and the type of radiometer.

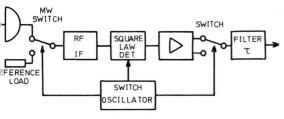

Fig. 2. Dicke radiometer where the antenna
output power is compared with that from
a reference load at known temperature to
reduce the influence of microwave receiver
gain instabilities.

More sophisticated radiometers have been designed which give a smaller value of C(C = 1 ideally). However, the Dicke radiometer is normally used due to its simplicity.

8 RADIOMETER ANTENNA AND RESOLUTION

The antenna temperature, T_A, is a function of the brightness temperature of the surfaces "seen" by the antenna weighted by the antenna radiation pattern:

$$T_A = \frac{1}{4\pi} \int_0^{2\pi} \int_0^{\pi} T_B (\theta, \phi) \, G(\theta, \phi) \, \sin\theta d\theta d\phi \quad (11)$$

where the function $T_B(\theta, \phi)$ is the angular distribution of brightness temperatures seen from the antenna, and $G(\theta, \phi)$ is the radiation pattern gain function of the radiometer antenna.

A simplified version of (11) is obtained when only a single angular variable is considered

$$T_A(\theta_o) = \int_0^{2\pi} T_B(\theta) \, G(\theta - \theta_o) d\theta \quad (12)$$

where $G(\theta - \theta_o)$ is the normalized antenna gain and θ_o is the direction of the main lobe of the antenna. It is seen that T_A may be considered as the convolution of the brightness temperature function with θ and the antenna pattern. In principle, the brigthness temperature distribution, $T_B(\theta)$, may be obtained by Fourier inversion of (12), for instance.

(11) and (12) show the influence of the radiation pattern on the measured brightness temperatures. Obviously, the main beam of the radiation pattern in the direction θ_o should be narrow in order to obtain a good resolution of the T_B-distribution. But it is also important that the side-lobes are small in order that high values of $T_B(\theta)$ off the θ_o direction give only a small contribution to T_A.

Traditionally, the resolution of a radiometer antenna is described by the 3 dB beamwidth of the main lobe of the antenna.

$$\theta_b = 1.22 \frac{\lambda}{D} \quad (13)$$

which is valid for a paraboloid reflector antenna and where λ is the radio wavelength and D is the diameter of the reflector. The resolution problem in radiometry may

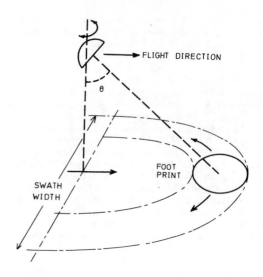

Fig. 3. Conically scanning measurement by imaging radiometer system with moving antenna reflector. In a satellite system the swath width could be of the order of 1 000 km.

be illustrated by a numerical example pertaining to the NIMBUS-5 satellite radiometer where λ = 1.55 cm and D = 70 cm whereby θ_b = 0.027 radians or 1.6 degrees which at an orbit altitude of 1 112 km yields a 3 dB resolution of 30 km at nadir. The same radiometer will give an 800 meter resolution when used in an aircraft at 3 000 meter above the terrain. However, a numerical study of the resolution problem on the basis of (12) has shown that for a paraboloid reflector antenna the resolution is slightly coarser than above, the 6-dB beamwidth being a more realistic estimate.

From the discussion of the polarisation effect above it is understood that the polarisation purity of the antenna is of importance. Normally, an antenna has a high polarisation purity, but with scanning antennas there is a degradation with increasing scan angle. This applies in particular to the case of phased arrays and reflector antennas where the feed horn or the reflector is moved separately. However, with a knowledge of the polarisation properties of a scanning antenna some correction procedure may be accomplished in the data processing.

Fig. 3 shows a mechanically scanning antenna which performs a conical scan so that the angle of incidence is constant over the scan range. By proper data processing such an antenna system will produce an image of brightness temperatures over the swath determined by the maximum scan angles. Such system is used on the NIMBUS-7 satellite. Another imaging antenna is the electrically scanning antenna (phased array) on NIMBUS-5, which scans in the plane at right angle to the flight direction. A conical scan may also be arranged with such an antenna system, however, as exemplified on NIMBUS-6.

9 RADIOMETER MEASUREMENTS

Radiometry has some potentialities in observation of snow and ice and eventually for monitoring seasonal variations and melt processes. A number of experimental and theoretical studies have been carried out and some are being undertaken. Some results of experimental activities will be discussed below.

The first examples are data obtained from airborne measurements in the western mountains of U.S.A. with six radiometers (Schmugge et al. 1973). Relevant parameters for these measurements are given in Table 1 where H and V stands for horizontal and vertical polarisation, respectively. Different altitudes over the terrain were in fact employed so the altitude of 3 500 m should be considered an average altitude only. The 19.35 GHz radiometer being the prototype for NIMBUS-5 scans at right angle to the flight direction the 3 dB foot print diameter stated is valid only for the antenna pointing to nadir.

Figs. 4 and 5 show the emissivity versus measurement frequency. Data at all frequencies except 37 GHz are from nadir pointing radiometers and may therefore be compared directly. The 37 GHz data are obtained with an angle of incidence of 45 degrees with vertical and horizontal polarisation. However, from theory it is known that the brightness temperature and the emissivity increases with the angle relative to nadir (angle of incidence) upto about 60 degrees whereas a drop in brightness temperature is experienced with horizontal polarisation. The magnitude of deviation from the nadir brightness temperature depends on the structure of the snow cover investigated, but an average of the vertical and the horizontal brightness temperatures is a good approximation to the nadir temperature.

The emissivity has been calculated from the measured brightness temperature and the measured infrared temperature i.e. the surface temperature. It should therefore be considered an approximation since the physical temperature of the snow may be very different from that of the surface. In general, there will be a variation of the physical temperature through the snow cover. However, the emissivity is rather insensitive to errors in the reference temperature (physical temperature). Thus, a 10 degree error gives only a 4 % error in emissivity, for instance.

Fig. 4 shows a high emissivity at the mid-frequencies and a rather low emissivity at high frequencies. With reference to the discussion above about the 37 GHz emissivity there seems to be a minimum at about 20 GHz which is ascribed to scattering

Table 1

Freq. GHz	Wavelength cm	Pointing relative to nadir	3 dB beam width	RMS temp. sens	Altitude above ground m	3 dB foot print diameter m
1.42	21	0^0	15^0	5.0K	3500	920
2.69	11	0^0	27^0	0.5K	-	1680
4.99	6.0	0^0	5^0	15.0K	-	300
10.69	2.8	0^0	7^0	1.5K	-	430
19.35H	1.55	scanner	2.8^0	1.5K	-	170
37.00V	0.81	45^0	5^0	3.5K	-	300x950
37.00H	0.81	45^0	5^0	3.5K	-	300x950
infrared	10μm	14^0	$<1^0$	<1.0K	-	∿ 35

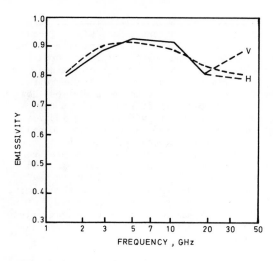

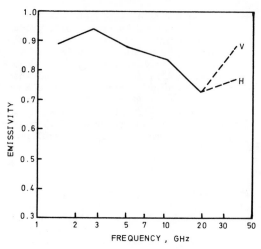

Fig. 4. Emissivity versus frequency at Steamboat Springs, Colorado, with a snow cover of 0.75 - 0.90 m over unfrozen soil, the snow-density being 260-330 kg/m³. The infrared temperature was -7⁰ C. The dotted curve is calculated on the basis of a model of Kong 1974.

Fig. 5 Emissivity versus frequency at South Cascade Glacier, Washington, with a snow cover of 8.4 m and a snow density of 200 kg/m³. The infrared temperature was -5⁰ C. There was no liquid water in the top meter of the snow at any point on the glacier.

from the ice particles in the snow. At 1.43 GHz a low brightness temperature of 212 K was observed (emissivity 0.798). This is attributed to the transparency of the snow at that frequency so that the low emissivity of the unfrozen gound with a soil moisture of about 35 % by weight, essentially determines the composite emission.

Fig. 5 also shows a minimum emissivity at 20 GHz and is in fact very similar to other measurement with dry snow over glacial ice. In comparing these measurements of a thick layer of dry snow with those above it may be concluded that at the high frequencies the penetration into the snow is less than 0.8 m due to scattering.

By means of the 19.35 GHz scanning radiometer mentioned in Table 1 an image of the South Cascade Glacier area was obtained shown in Fig. 6. The data in Fig. 5 were obtained from the galcier marked P3 on the map in the figure. In the centre of the image the resolution is about 130 meter while at the edge it is 130 m times 450 m.

In spite of the fact that a comparatively fine resolution is achieved from the low-flying aircraft the image appears rather coarse. However, valuable information of gross features may be obtained which might

prove useful in hydrological applications. In particular, this is true if repetitive coverage of the same area could be undertaken so that temporal variations could be observed and melt processes be studied.

An image obtained with the electrically scanning microwave radiometer (ESMR) on board the NIMBUS-5 satellite is shown in Fig. 7 which is a polar projection of the northern hemisphere north of 55⁰ N latitude composed by data received in the three-day period beginning 29 August 1974 obtained from about 30 orbits (Zwally 1975). This pseudo-colour map has a scale of 32 colours covering the brightness temperature range of 132.5 K to 282.5 K in 5 K increments. The resolution or the pixel size is about 30 km x 30 km. An interesting brightness temperature pattern is observed in Greenland. The brightness temperature of the inland ice varies between 165 K and 240 K, a temperature interval much larger than the span of physical temperatures of the surface. The relative high temperatures in central Greenland coincide with the summit of the northern dome of the inland ice where the physical temperature is the lowest.

It should be mentioned that the winter image shows largely the same pattern as the summer image but presents lower brightness temperatures. The seasonal changes in brightness temperature - in central Greenland from 205 K to 190 K, for instance - points to a scattering mechanism in the firn where the physical temperature varies with season.

Of particular interest are the high brightness temperatures - 240 K to 245 K - on the western and the southeastern side of the inland ice indicating areas with summer melting with high liquid water content in the firn i.e. a layer of high absorption. Such "pink patterns" (of much larger extent in July, for instance), have been used to delineate areas of summer melt in Greenland. (Gloersen & Salomonson 1975).

(Note the influence of clouds and/or precipitation over the Atlantic Ocean causing an increase in apparent brightness temperature from about 132 K (in the cloud free case) to about 165 K as described by eq. 9).

Other measurements at 1.5 GHz - airborne as well as ground based - showed the brightness temperature approaching the annual mean temperature (10-meter temperature). This is true in north Greenland where the snow particle scattering is negligible at that frequency. In contrast, low brightness temperatures were observed in the southern part of the inland ice due to scattering from ice lenses in the firn, i.e. ice layers of limited horizontal extent formed by refrozen melt water (Voldby 1975). This is the same scattering mechanism as observed in the radio echo soundings in Greenland.

Ground-based measurements are being carried out in the Swiss Alps to improve the knowledge of the microwave radiative properties of snow for optimization of airborne and satellite snow sensors and for interpretation of large-scale snow images obtained by such sensors. Microwave brightness temperatures are measured at five frequencies 4.9 GHz, 10.5 GHz, 21 GHz, 36 GHz and 94 GHz at horizontal and vertical polarisation and at angle of incidence between 0 degrees (nadir) and 65 degrees (Hofer & Schanda 1978).

An example of the measurements is shown in Fig. 8 where the brightness temperature is shown for horizontal and vertical polarisation during morning hours (upper part of the figure) and during afternoon hours on the same day and with the same snow.

During the morning measurements the air temperature was so low that there were no liquid water in the snow. A surface melt during the previous day resulted in a refrozen hard layer of 4 centimeter with a mean grain size of 2.3 millimeter. Two thin ice sheets were observed 5 and 10 centimeter below the surface the grain size of the intervening snow being 0.25 to 1 millimeter.

The figure shows (morning) that the 94 GHz signal experiences very small variations with look angle and polarisation indicating that the surface acted almost like a diffuse scatterer of the sky radiation. With decreasing frequency penetration into the snow increases and ice particle scattering (volume scattering) becomes more and more pronounced causing an increase in the emissivity and a corresponding decrease of the diffuse reflection of the sky radiation. This behaviour is clear for vertical polarisation in particular. The crossing of the 21 GHz and 36 GHz curves indicates that the transition between diffuse surface scattering and volume scattering takes place in this frequency range. The wavy behaviour of the curves for horizontal polarisation is attributed to interference of the radiation in the stratified snow.

During the afternoon measurements the uppermost snow layers were very humid and almost at melt. Comparing these measurements with the morning measurements it is found that there is a complete spectral reversal in brightness temperature. At 4.9 GHz and 10.5 GHz the wet snow behaves as a homogeneous dielectric with a Brewster angle behaviour for vertical polarisation and monotonically decrease with look angle for horizontal polarisation. The very high brightness temperatures at 36 Ghz are related to the surface layer of melting snow of a few millimeter thickness and so is the wavy pattern at 94 GHz. However, a contribution of diffusely scattered solar radiance is also a possibility.

The interpretation of these ground-based measurements has been described in some details first of all because they show the general feature of a typical spring situation with a multi-frequency microwave signature reversal between the frozen morning state and a wet snow afternoon situation. They do also show the complexity of the interpretation caused by the composite structure of the snow and the influence of external radiation. In addition, it should be mentioned, that measurements on fresh

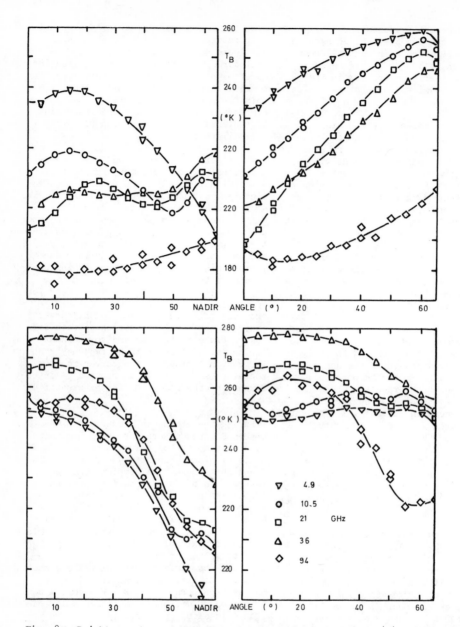

Fig. 8. Brightness temperature versus look angle off nadir. (a) horizontal polarisation, (b) vertical polarisation; (a) and (b) morning measurements (0910-1020 h). (c) horizontal polarisation, (d) vertical polarisation; (c) and (d) afternoon measurements (1405-1515 h). Note the difference in scale of brightness temperature between morning and afternoon measurements.

light powder snow in a winter situation showed no variations between the morning and the afternoon measurements. The brightness temperatures were largely independent of frequency, so that volume scattering is assumed to be the dominant mechanism.

10 REVIEW OF RADIOMETRY MODELS

In remote sensing models of the earth one often deals with a medium that is relatively uniform in a direction parallel to the surface but has a permittivity and a temperature that varies with depth. In the case of such (stratified) media with low losses near to the surface the well-known Fresnels reflection coefficients are not useful for calculation of the reflection and emission characteristics of the medium, and more complicated models are developed.

It is not the intention to give a full account of such models but this section reviews in a qualitative way some models related to stratified media of ice and snow developed in an attempt to describe some of the observed radiation phenomena which must be understood before interpretation of radiometry data can be made safely. Some characteristic results are given.

A simple layered model consisting for example of a layer of snow on a semi-infinite soil will first be considered (Voldby 1975). It is assumed that the layer is homogeneous and at a constant temperature and so is the underlying soil. A relatively simple solution to the radiative problem shows that the emissivity of the surface is a function of the angle of incidence (look angle), the permittivity and the temperature of the media and the electrical depth of the layer. For a given frequency and with a thin layer the emissivity will be an oscillatory function of depth with amplitudes which decrease with increasing depth due to absorption in the layer. The emissivity approaches that of a semi-infinite medium with the same permittivity as the layer. This effect is often described as an interference effect. A similar oscillatory behaviour is obtained with a constant depth but varying look angle as observed in one of the measurements reported previously. However, since a radiometer must have a certain - large - bandwidth (to obtain a reasonable accuracy) the oscillatory function of emissivity is smoothed due to bandwidth integration.

A more complicated model is a vertically structured medium where the temperature and the permittivity may be arbitrary functions of depth (Voldby 1975). This model has been employed for the Greenland ice taking into account (a) the permittivity variations in the ice derived from measured density-depth profiles, (b) the temperature variations in the same region, and (c) a range of microwave frequencies. It was found that for a medium with a slowly increasing density and thereby slowly increasing loss from the light snow surface and downwards, it should be possible to measure the physical temperature deep in the ice sheet. The measurement would have

an absolute inaccuracy of less than 0.5 K (in contrast to the radiometer sensitivity) if a frequency below 1.5 GHz is chosen. Measurements in north Greenland at this frequency confirm this approximately while in south Greenland the brightness temperature was 10-50 K lower than calculated (depending on angle of incidence) probably due to scattering from ice lenses.

In a far more advanced model the emission from a random inhomogeneous layer over a homogeneous medium is described by Tsang & Kong 1977. The permittivity of the random medium is given by

$$\varepsilon_1(z) = \varepsilon_{1m}(z) + \varepsilon_{1f}(z) \qquad (14)$$

where $\varepsilon_{1m}(z)$ is the non-random part of the permittivity as a function of depth, z, and $\varepsilon_{1f}(2)$ characterizes its random part. The random part is described by its variance, $\delta(z)$, and its correlation length, $\ell(z)$, and the analysis is carried out with small losses.

Numerical examples of emission in direction of nadir derived from this model are shown in Fig. 9 which gives the brightness temperature as a function of layer thickness and of frequency for a laminar structure of snow over sand soil. In the first instance the frequency is fixed at 10 GHz and in the other the snow thickness is constant at 5.1 meter. The two curves denoted (a) and (b) refer to different values of permittivity variance. We see that the brightness temperature decreases with layer thickness because the emission from snow is lower than that from the soil owing to scattering. The curves (a) have higher brightness temperatures because there is less scattering.

The model is extended also to cover the case where there are random variations horizontally and numerical examples show that this effect may have a marked influence on the brightness temperature versus look angle relationship. With a layer that is so thick that it may be considered a half-space random media, Fig. 10 shows (Tsang & Kong 1976) the brightness temperature as a function of the observation angle for horizontal polarisation (left) and vertical polarisation (right). It is seen how the horizontal correlation length, l_ρ, modifies the radiation both in magnitude and in range of temperatures.

This example is calculated for a certain temperature profile (slightly increa-

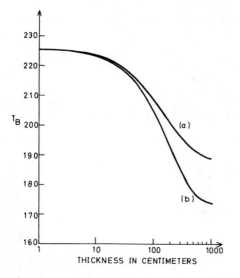
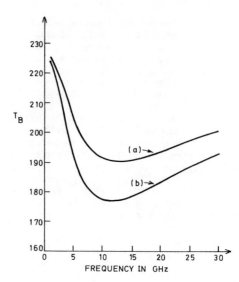

Fig. 9. Brightness temperature versus layer thickness at 10 GHz (left) and versus frequency at a layer thickness of 5.1 meter (right). Parameters: $\varepsilon_{1m} = 1.8\varepsilon_o$, loss tangent 0.0005, $l(z) = 2$ mm, layer temperature constant = 225 K, sub-layer temperature 225 K and permittivity $\varepsilon_2 = 4(1 + i \quad 0.1)\varepsilon_o$. Curve (a): variance $\delta(z) = 0.002(1 + \exp(-0.012))$ and (b): $\delta(z) = 0.004$.

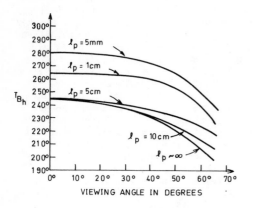
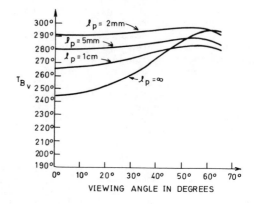

Fig. 10. Brightness temperature versus observation angle for horizontal polarisation (left) and vertical polarisation (right) with the horizontal correlation length as a parameter. Other parameters are: $\varepsilon = 1.8(1 + i \ 0.0005)\varepsilon_o$, $f = 10$ GHz, $l(z) = 2$ mm, $T(2) = 300$ K, and $\delta(2) = 0.002$.

sing temperature with depth). Examples with different temperature profiles as observed in nature show that the shape of temperature profile is of minor consequence when considering brightness temperature versus frequency (Tsang & Kong 1975).

The examples derived from this model show a great variability in brightness temperature for variations of a number of

snow parameters in case of small losses i.e. in cold non-melting snow. It is concluded that more numerical and experimental work is needed in order to ascertain the number of frequencies, for instance, needed for a safe estimate of snow layer thickness under various snow conditions. So far, none of the models are applicable to wet snow.

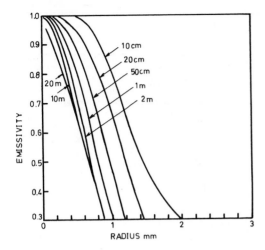

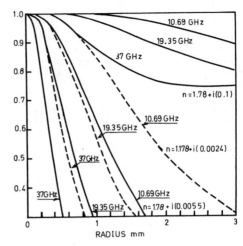

Fig. 11. Emissivity at nadir versus ice particle size for snow layers of various thickness: The emissivities are calculated for 19.35 GHz and a refractive index of the ice particles of n = 1.78 + i 0.00055. The layers have a constant temperature of 250 K while the sub-layer medium has a temperature of 270 K. The latter temperature is chosen as reference temperature for calculation of the emissivity from the calculated brightness temperature. Negligible emissivity variations occur for layers thicker than the asymptotic thickness of about 5 meters.

Fig. 12. Emissivity at nadir versus ice particle size for a snow layer equal to the asymptotic thickness at the three frequencies 10.69 GHz, 19.35 GHz and 37 GHz. Three cases are given for different absorption represented by the imaginary part of the refractive index, n. The dotted curves are for n = 1.78 + i(0.0024).
n" = 0.00055 pure ice at
        -20°C, loss tangent = 0.00062,
n" = 0.0024 pure ice at
        0°C loss tangent = 0.0027,
n" = 0.= first year sea ice
        0°C, loss tangent = 0.112.

It has been suggested that scattering from the ice particles in the snow could account for the low brightness temperatures observed in Greenland, for instance. Thus, a model considers the firn to consist of a collection of ice particles (crystals) in the shape of spheres which scatter the incoming energy but do not interact coherently (Chang et al. 1976). Under the assumption that the particle size is comparable to the wavelength in the medium the theory of Mie scattering may be employed.

For a simple case with a single layer of a certain thickness at a temperature of 250 K on a bed at 270 K the brightness temperature has been calculated as a function of the ice particle radius. Fig. 11 shows the corresponding emissivity for different snow thickness using 270 K as a reference temperature. The figure is derived for a frequency of 19.35 GHz (λ = 1.55 cm) and a refractive index of the ice particles of n = 1.78 + i 0.00055, i.e. for very small losses (loss tangent = 0.00062). It is seen that the scattering

has a very marked influence on the emissivity and that the emissivity variations observed in Greenland (0.65 - 0.85) may be accounted for by this theory. Also, it is seen that there exists an asymptotic thickness of about 5 meters at this frequency defining an effective penetration depth. (If no scattering was present the usual penetration depth would be about 10 meters for the snow density of 450 kg/m^3 assumed for these calculations. For 37 GHz the asymptotic thickness is about 2 meter, while it is about 10 meter at 10.69 GHz for the same loss tangent).

Fig. 12 shows the emissivity as a function of the ice particle radius at the frequency of 10.69 GHz (2.81 cm), 19.35 GHz (1.55 cm), and 37 GHz (0.81 cm) and for three values of the imaginary part of the refractive index. From this figure the following observations may be made:

(a) The emissivity is frequency dependent the variation of the emissivity being more pronounced with small losses.
(b) For a given frequency and ice par-

ticle radius the emissivity is dependent
on the dielectric losses.

(c) It seems possible to determine the
ice particle radius from brightness temperature - and thereby n" is known. Thus,
the measurements at station P3 on the
South Cascade Glacier reveal radii of o.5
mm in reasonable agreement with ground data. Similarly, the emissivities observed
at 19.35 GHz of the Greenland inland ice
reveal particle radius of 0.5 mm in north-
west Greenland, and 0.9 mm in southeast
Greenland with the emissivity of 0.65
which is plausible judged from observational data.

The model assumes that there is no interaction between the ice particles. Considering the composition of snow this assumption may not ever be fulfilled. Also, it
assumes a constant temperature in the snow
layer while the temperature variation in
the top of a continental ice sheet may be
as much as 20 K. Finally, it seems difficult to explain the emissivity minima observed at about 20 GHz (Figs. 4 and 5).
Consequently, it may be concluded that
this model of Mie scattering requires
further controlled experiments to ascertain its validity.

11 CONCLUSIONS

The preceding sections describe some examples of microwave radiometry on ice and
snow selected from the literature and some
theoretical/numerical models proposed in
support of the interpretation/inversion
problem.

So far the information available is in-
conclusive since it represents few special
cases of the great many possible combinations present in nature. However, some
of the latest controlled measurements reveal features of the dynamic snow processes involved which indicate the great
potentialities of radiometry for snow monitoring. More measurements on a continuous basis are needed but coupled to numerical results from promising models,
they may lead to a firm basis for interpretation of space data.

From the introductory sections and some
of the examples it is clear that radiometry has a resolution problem when applied
in regions like the Alps and the Pyrenees.
However, it seems to have great potentialities in areas like northern Scandinavia
and Greenland for studies of large scale
features. Future satellite systems may
carry very large antennas so that resolutions of about 5 km may be obtained.

12 REFERENCES

Chang, T.C., P. Gloersen, T. Schmugge,
T.T. Wilheit, H.J. Zwally 1976,
Microwave emission from snow and glacier
ice, J. Glaciology 16 : 23-29.
Gloersen, P., V.V. Salomonsen 1975,
Satellites - new global observing tech-
niques for ice and snow, J. Glaciology
15 : 373-389.
Hofer, R., E. Schanda 1978, Signatures of
snow in the 5 to 94 GHz range, Radio
Science 13,: 365-369.
Schmugge T., T.T. Wilheit, P. Gloersen,
M.F. Meier, D. Frank, I. Dirmhirn 1973,
Microwave signature of snow and fresh
water ice, NASA Goddard Space Flight
Center, Report X-652-73-335.
Tsang, L., J.A. Kong 1975, The brightness
temperature of a half-space random medi-
um with nonuniform temperature profile,
Radio Science, 10 : 1025-1033.
Tsang, L., J.A. Kong 1976, Thermal micro-
wave emission from half space random
media, Radio Science, 11 : 599-609.
Tsang, L., J.A. Kong 1977, Thermal micro-
wave emission from a random inhomogeneous
layer over a homogeneous medium using the
method of invariant imbedding, Radio
Science, 12 : 185-194.
Voldby, J. 1975, On thermal radiation
from layered media, Licentiate Thesis
LD 25, Electromagnetics Institute, Technical University of Denmark.
Zwally, H.J. 1975, Private communication
on NIMBUS-5 imagery, NASA Goddard Space
Flight Center, Greenbelt, Maryland.

P.E.GUDMANDSEN
Electromagnetics Institute, Technical University of Denmark, Lyngby, Denmark

Electromagnetic studies of ice and snow

2 *Radio echo sounding*

1 INTRODUCTION

Radio echo sounding of today is a well developed technique for glaciological studies of polar ice caps, i.e. large masses of ice with an average temperature well below the pressure melting point. Examples are the inland ice in Greenland and Antarctica to mention the largest. Radio echo sounding of temperate glaciers, i.e. glaciers with an average temperature near to or at the melting point is a more difficult task and has only been succesful very recently. Whereas polar ice caps may be investigated by means of airborne soundings with continuous recording of ice profiles along the flight tracks, the temperate glaciers are sounded from the surface with a technique allowing measurement only point by point. In contrast sounding of snow cover is in a development phase with a view to measure snow thickness and the water content of a snow cover using airborne techniques.

This chapter describes the technique of radio echo sounding with emphasis on the physical aspects and system considerations rather than on the electronics and the detailed field techniques.

2 RADIO WAVE PROPAGATION IN ICE

The radio echo sounding technique is influenced by the dielectric properties of the ice i.e. the permittivity

$$\varepsilon = \varepsilon_o \varepsilon'(1 - j\frac{\varepsilon''}{\varepsilon'})\qquad(1)$$

where $\varepsilon_o = 10^{-9}/36\pi$ Farads/meter is the permittivity of free space and ε' is the dielectric constant, and

$$\tan\delta = \frac{\varepsilon''}{\varepsilon'}\qquad(2)$$

is the loss tangent which determines the dielectric absorption of the radio wave (Gudmandsen 1971).

The propagation constant for a plane radio wave in a homogeneous medium is

$$\gamma = \alpha + j\beta = j\omega\sqrt{\mu\varepsilon}\qquad(3)$$

with the angular frequency $\omega = 2\pi f$ where f is the radio frequency, and $\mu = \mu_o = 4\pi \cdot 10^{-7}$ Henry/meter is the permeability of the medium which in the non-magnetic case is equal to that of free space, μ_o, as stated.

For low-loss dielectrics like ice, i.e. when $\tan\delta < 0.1$ the attenuation constant α may be approximated by

$$\alpha = \frac{\omega}{2c} \sqrt{\varepsilon'}\tan\delta$$

$$= 9.1 \cdot 10^{-2}f\sqrt{\varepsilon'} \tan\delta\qquad(4)$$

where f is the radio frequency in MHz and c is the velocity of light in free space $c = 2.988 \cdot 10^8$ m/sec. The phase velocity may in this case be approximated by

$$v = \frac{c}{\sqrt{\varepsilon'}}\qquad(5)$$

For a dielectric of thickness d the total delay time in the dielectric in the radar case becomes

$$t = \frac{2d}{v} = \frac{2d\sqrt{\varepsilon'}}{c}\qquad(6)$$

The reflection from a plane interface between two media with different permittivities, ε_1 and ε_2, at normal incidence is described by the voltage reflection coefficient

$$\rho = \frac{K - 1}{K + 1} \qquad (7)$$

where $K = \sqrt{\varepsilon_1/\varepsilon_2}$ assuming wave propagation from medium 1 with ε_1 towards medium 2 with ε_2.

The power reflection coefficient becomes

$$|\rho|^2 = \left|\frac{K - 1}{K + 1}\right|^2 = \left|\frac{\sqrt{\varepsilon_1/\varepsilon_2} - 1}{\sqrt{\varepsilon_1/\varepsilon_2} + 1}\right| \qquad (8)$$

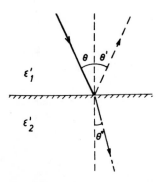

Fig. 1. Reflection and refraction at a plane interface.

At a plane interface between the two media at any angle to the normal, Fig. 1, it is required that the tangential phase velocities are the same in the two media,

$$\frac{v_1}{\sin \theta} = \frac{v_1}{\sin \theta'} = \frac{v_2}{\sin \theta''} \qquad (9)$$

where θ is the angle of incidence, θ' is the angle of reflection (both in medium 1) and θ'' is the angle of the transmitted wave in medium 2 (angle of refraction).

Combining (9) and (5) one obtains Snell's law

$$\frac{\sin \theta''}{\sin \theta} = \frac{\sqrt{\varepsilon_1}}{\sqrt{\varepsilon_2}} = \frac{n_1}{n_2} \qquad (10)$$

where $n_{1,2}$ designate the refractive indices of the media. It is seen that the angle of refraction is smaller than the angle of incidence if $\varepsilon_2 > \varepsilon_1$ i.e. the wave is refracted towards the normal in the medium having the larger permittivity.

The expression for the reflection coefficient (7) is also valid in case of angle of incidence different from normal (Fig. 1) if K is redefined. There are two cases: (a) a wave with the electric vector in the plane of incidence (the plane of Fig. 1), (b) with the electric vector normal to the plane of incidence. These two cases are often referred to as vertical and horizontal polarisation, respectively.

In the first case - polarisation in the plane of incidence - K assumes the value

$$K = \frac{\sqrt{\varepsilon_1}}{\sqrt{\varepsilon_1}} \frac{\sqrt{1 - \left(\frac{v_2}{v_1}\right)^2 \sin^2\theta}}{\cos \theta} \qquad (11)$$

and inserted in (7) it is seen that total reflection, i.e. $|\rho| = 1$, occurs for K = 0, or for

$$\sin\theta = \frac{v_1}{v_2} = \frac{\sqrt{\varepsilon_2}}{\sqrt{\varepsilon_1}} \qquad (12)$$

A real solution for θ exists only if $\varepsilon_2 < \varepsilon_1$, however. In the radar case where the transmitter and the receiver are on the same place with a common antenna, reflections are obtained from facets at right angle to the wave normal. Consequently, features with a slope larger than the "critical" angle given by (12) may not be observed when embedded in a dielectric such as ice, for instance.

Total reflection also occurs for K = ∞, i.e. for $\theta = 90^0$, but this is a trivial case, since the wave propagates parallel to the interface.

In the case of polarisation normal to the plane of incidence

$$K = \frac{\sqrt{\varepsilon_1}}{\sqrt{\varepsilon_1}} \frac{\cos \theta}{\sqrt{1 - \left(\frac{v_2}{v_1}\right)^2 \sin^2\theta}} \qquad (13)$$

and total reflection is obtained for K = ∞, i.e. for the same condition as in (12) and for K = 0 which similarly is a trivial case.

It is interesting to note that the refraction effect gives rise to a focusing of radio waves when propagation takes place from the medium with low refractive index towards that with higher index. For small angles of incidence, i.e. $\theta < 10^0$, the focus gain may be approximated by

$$q = 20 \log \left[\frac{\frac{h}{d} + 1}{\frac{h}{d} + \frac{n_1}{n_2}} \right] \text{ dB} \qquad (14)$$

where h is the height of the radar above the interface (air/ice), d is the thickness of the dielectric, and $n_{1,2}$ are the refractive indices on the two sides of the interface. With $n_2 > n_1$ an appreciable gain is obtained for small values of h/d.

3 DIELECTRIC PROPERTIES OF ICE

Polar ice in the state found in Greenland and Antarctica is a low-loss dielectric, so that the velocity of the radio waves is solely determined by the refractive index of the ice, $n = \sqrt{\epsilon'}$. Laboratory measurements on solid ice have revealed $n = \sqrt{3.2}$ = 1.78 so that the velocity of the radio waves in ice is 169 m/µsec (Evans 1965). This value is for all practical purposes independent of the radio frequency, the ice temperature (below the pressure melting point), and the pressure. However, the refractive index is a linear function of the ice density which is of consequence for the velocity in the upper part of an ice sheet, the firn, where the density increases exponentially with depth. In contrast, the loss tangent is dependent upon the frequency and the temperature as well as the ice density. As to the frequency dependence the absorption (expressed in dB/m, for instance) is constant in the frequency range 10-400 MHz, while at higher frequencies the absorption increases drastically. The absorption is a highly non-linear function of temperature. At 150 MHz the absorption is 0.01 dB/m at -30^0C, while it is 0.035 dB/m at -5^0C, for instance.

The frequency and temperature dependence of polar ice is shown in Fig. 2 by plotting f · tanδ (logarithmic scale) versus the frequency with temperature as a parameter assuming $n = \sqrt{\epsilon'}$ = 1.78, (Evans 1965), cf. eq. (4). It is seen that minimum losses occur at frequencies below 400 MHz and that the frequency variations of the loss are largest at low temperatures.

The dielectric properties and their effects on the radio wave propagation as described above have the following main con-

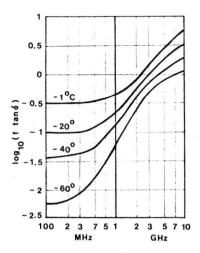

Fig. 2. Influence of frequency and temperature on the dielectric loss of solid ice.

sequences for the radio echo sounding techniques, (Bogorodsky 1968, Robin et al. 1969, Gudmandsen 1971):

(a) Due to the increase in absorption which takes place at high frequencies the radar frequency is generally chosen to be below 400 MHz. Normally a low frequency is chosen since scattering from a rough bedrock may reduce the reflection coefficient when the roughness is of the order of the radio wavelength. Also, a low frequency will be chosen to overcome scattering from ice irregularities such as ice lenses and water inclusions present in temperate glaciers in particular. For airborne sounding of the large polar ice caps frequencies above 30 MHz have to be used in order to avoid interference from other services via long distance ionospheric propagation.

(b) In a polar ice sheet the snow density increases gradually from the surface into solid ice at a depth of about 100 meters. Therefore, the radio wave velocity decreases gradually from about 220 m/µsec to that in solid ice at 169 m/µsec. If radar time measurement are converted into thickness using the velocity in solid ice a figure of 8-16 meters should be added in order to account for the higher velocity in the upper layers of the ice, the actual correction being dependent upon the density/depth profile. With a velocity of 169 m/µsec the resolution of a pulsed radar becomes 85 meter when operated with a 1 µsec pulse.

403

(c) In order to obtain the full benefit of the refraction gain expressed by (14) it is necessary to fly low over the ice surface. (A maximum refraction gain of 4.8 dB may be obtained). This is also of importance in order to limit the reflections from the ice surface and near-surface irregularities.

(d) With an undulating surface which is found in areas with a mountainous bedrock structure the stratification in the density-varying upper part of the ice will show curvatures. Consequently, due to refraction the radio waves will undergo a focusing or a defocusing effect which is of importance for echoes from the stratification in the ice in particular. (Harrison 1973)

(e) With a refractive index of 1.78 the critical angle is 34 degrees. In the radar case where the transmitter and the receiver are at the same position this means that slopes of the bedrock larger than this angle are not observable.

(f) With a solid ice surface with the refractive index of 1.78 the power reflection coefficient becomes -11 dB and the transmission coefficient -0.4 dB. With a snow covered surface the reflection coefficient of the surface becomes less, -20 dB, for instance. At the ice-rock interface the reflection coefficient becomes of the order of -15 to -20 dB depending on the rock permittivity.

(g) The total absorption in the ice is dependent on the temperature and the ice thickness. In cold polar ice the absorption is relatively small since the temperature is low through most of the ice. However, near the bottom the temperature increases due to the geothermal heat and internal friction. The temperature profile depends upon the surface temperature, the accumulation rate, and the velocity of the ice movement (Gudmandsen 1971). At places where the accumulation rate is small the bedrock temperature may reach the pressure melting point and a very large absorption is obtained. Calculations for various points in Greenland showed the two-way absorption to be between 30 dB and 90 dB. Also it was found that in southwest Greenland melting at the bedrock is likely to occur, for instance.

4 RADIO ECHO SOUNDING TECHNIQUES

The principle of a pulse radar system for airborne radio echo sounding is shown in Fig. 3. It is seen that the same antenna

is used for transmitting and receiving by means of the transmit/receive switch, which is common for the transmitter and the receiver. The received signals are presented on oscilloscopes in two ways - in the so-called A-scope and the Z-scope representations. The A-scope gives the actual received signals (after detection) i.e. the transmitted pulse, the pulse reflected from the surface, and that reflected from the bedrock as shown schematically in Fig. 3.

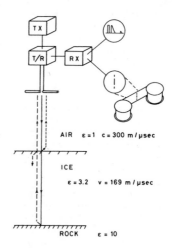

Fig. 3. Principles of radio echo sounding.

In the basic Z-scope representation the intensity of the received signals modulate the cathode ray tube so that lines of echoes are obtained on a continuously moving film, the intensity being a function of the signal strength. In the examples shown later, the intensity modulation is performed after differentiation of the received signals, so the intensity is proportional to the time variation of the received signals. This technique enables observation of weak signals superimposed on strong signals and increases the contrast.

The performance of a radar system may be described by the system sensitivity, S, which is the ratio between the transmitter power and the minimum detectable signal measured with the transmitter and the receiver back-to-back. Systems with sensitivities in the range of 128 dB to 189 dB have been designed. It should be noted that the system sensitivity is dependent upon the receiver bandwidth which again is chosen to match the pulse. A system with a short pulse length and therefore a wide bandwidth will have a comparatively low system sensitivity.

The parameters for the latest version of the 60 MHz system used in Greenland and Antarctica are given here as an example (Gudmandsen et al. 1975). Peak power, 10 kilowatts; mean power with 1-microsecond pulse, 125 to 250 watts for pulse repetition frequencies of 12.5 to 25 kHz; pulse length 60, 125, 250, 500, 1 000 nanoseconds; receiver bandwidths, 1, 4, 14 MHz; receiver noise figure, 2.5 dB. The antenna system for this radar is a linear array of four dipoles (with a total length of 12 m) suspended a quarter of a wavelength (1.25 m) under the wing of the aeroplane using the wing surface as a reflector. The antenna radiates the power in a beam of 22 degrees transverse to the flight direction and about 110^0 in the fore-and-aft direction. Including this antenna system the total system sensitivity is about 205 dB with a receiver bandwidth of 1 MHz.

The figures of sensitivity quoted above are measurable figures but a more practical way of describing the radar performance as a radio echo sounder is shown in Fig. 4. It describes for a given receiver bandwidth (and pulse length) the total amount of losses which may be overcome by the system in addition to the normal loss caused by the spread (diffraction) of energy (distance squared attenuation). It is based on the usual radar equation which in the radio echo sounding case takes the form

$$\frac{1}{S(B)} = \frac{P_{rmin}}{P_t} = \left(\frac{G^2 \lambda^2 \cdot q}{64\pi^2 (h+d)^2 L_s^2}\right) \cdot \left(\frac{\rho_r}{L_i^2}\right)$$

$$(15)$$

where S is the measured system sensitivity, B is the receiver bandwidth, P_{rmin} is the minimum acceptable received signal power, which is dependent of the noise figure of the receiver and the pulse integration performed, P_t is the transmitter power, G is the gain of the antenna, λ is the radio wavelength, q is the refraction gain, h is the height above the surface, d is the ice thickness, L_i is the one-way dielectric loss in the ice which is distance dependent, ρ_r is the power reflection coefficient at the rock-ice interface, and L_s is the transmission loss through the surface. Note that the first paranthesis of (15) contains known quantities or quantities (q and L_s^2) which are known to a good approximation. The second paranthesis pertains to the unknown medium under the ice surface.

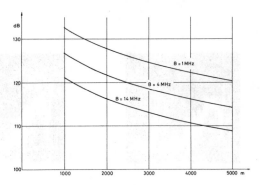

Fig. 4. Permissible extra loss as a function of ice thickness, with receiver bandwidth as a parameter.

For the 60 MHz system mentioned above Fig. 4 gives curves for the permissible extra loss i.e. the ratio of reflection coefficient, ρ_r, and the two-way dielectric absorption in the ice, L_i^2, (the second paranthesis in (15)) as a function of the ice thickness when the aeroplane is 300 meters (=h) above the ice surface. The parameter, B, refers to the receiver bandwidth employed. Assuming a reflection coefficient of −20 dB it is seen that an echo is detectable if the total ice absorption in a 3 000 meter ice column is smaller than 105 dB when a 1 MHz bandwidth is employed, for instance.

5 EXAMPLES OF RADIO ECHO SOUNDING

The following examples of radio echo sounding are obtained in Greenland and Antarctica with equipment at 60 MHz and 300 MHz designed by the Technical University of Denmark (TUD). The soundings in Greenland were carried out by TUD as part of the international Greenland Ice Sheet Program (GISP) while those in Antarctica were made by Scott Polar Research Institute (SPRI), Cambridge, England, as part of a cooperation between National Science Foundation (NSF), Washington D.C., SPRI and TUD.

Fig. 5 shows 60 MHz recordings with a 250 nanosecond pulse of relatively shallow ice in east Greenland ($69^041'N$, $35^053'W$). The two small photos in the upper part of the figure show the received signals with a time interval of 15 seconds as seen on the A-scope. The first narrow pulse is the

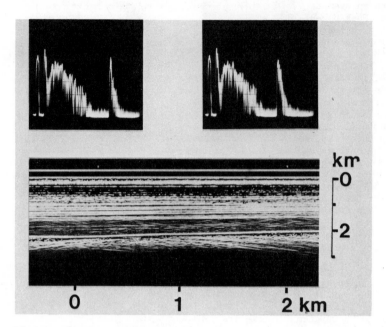

Fig. 5.   60 MHz sounding with a continuous Z-scope recording and simultaneous intermittant A-scope recording at intervals of 15 second.

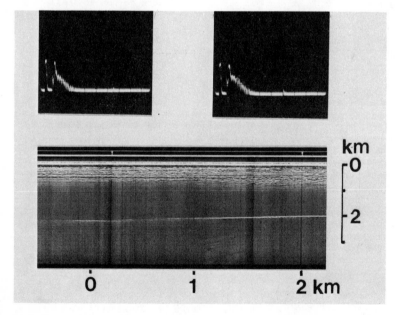

Fig. 6.   300 MHz sounding carried out simultaneously with the 60 MHz sounding shown in Fig. 5.

transmitted pulse, the next wide pulse is the signal from the surface (modified in the beginning by a time dependent suppression to avoid receiver overload superimposed by a number of small pulses, and the last pulse is the signal reflected from the ice-rock interface. The lower photo is a reproduction of the 35 mm film of the Z-scope operated simultaneously. Time marks are seen on both films and coincidence is ensured by proper time indications on both films (not shown in the reproduction) corresponding to those on the magnetic recording of the relevant position coordinates determined by an inertial navigation system. The upper horizontal line is the transmitter pulse and the next line is the ice surface about 2 800 meters above sea level, the aircraft flying 450 meters above the ice. The skew heavy line is the bedrock which is covered by about 2 100 meters of ice. The thin lines appearing in the ice are due to stratification; about 20 lines may be observed down to 1 580 meters and four additional lines in the range 1 840 to 2 010 meters. The corresponding echo signals are clearly seen on the A-scope film, the first series of echoes being superimposed on the surface pulse while the last series emerges from the receiver noise. The superimposed echoes are only observable because a signal differention has been made before recording. In most cases the echoes from the layers are weak compared to those from the surface and the bedrock. The echoes observed below the bedrock echo are due to the rough surface of the bedrock seen at an angle mostly in the fore-and-aft direction due to the very wide radiation pattern often more than six kilometers from the aircraft, i.e. seen at an angle of 85 degrees from nadir. In fact these lines are the asymptotes of the observation hyperbola of a point reflector passed by the radar, (Harrison 1973).

Fig. 6 is a recording of the 300 MHz signal made simultaneously with that in Fig. 5. In this case the bedrock echoes are limited almost to the nadir point partly due to the confined radiation and partly due to the smaller sensitivity of this system. The extra losses which may be overcome by this system are 28 dB smaller than those at 60 MHz due to the smaller wavelength, cf. eq. (15), the larger receiver bandwidth employed (14 MHz) and the smaller transmitter available (1.6 kW). For the same reason stratification echoes are only observed to a depth of 985 meters but due to the better range resolution obtained with a pulse duration of 60 nanoseconds, more layers are observed in this

range interval. It is also noticeable that echoes due to ice surface irregularities are lacking. (The heavy horizontal line observed just below the surface echo is an equipment feature created at the end of the signal suppression period).

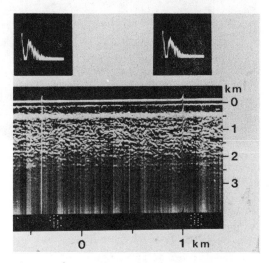

Fig. 7. 60 MHz recording in southwest Greenland with excessive scattering from ice lenses.

The surface irregularities are a disturbing effect at 60 MHz all over Greenland which limits the altitude of the aircraft to a few hundred meters above the ice surface, particularly when the stratification is being studied. In southwest Greenland very serious scattering occurs which prevents even the bedrock echo from being observed. Fig. 7 is an example of heavy scattering through which a very weak bedrock echo is recorded at about 2 200 meters at $67^0 48'N$, $44^0 50'W$. The scattering is caused by ice lenses in the firn, i.e. refrozen melt water which form horizontal plates of solid ice in the snow near the surface. These lenses create a large echo and are therefore seen by the radar at long distances. In the same area the ice absorption is large because of the high temperature of the ice at this low latitude, approaching the pressure melting point at the bedrock, so that the bedrock echoes have a magnitude comparable to those from the surface solid ice inclusions. The only way to overcome this problem is to fly low over the surface or to increase the spatial resolution. In fact the 300 MHz radar was designed with this in mind since, a better antenna radiation pattern could be obtained at this frequency.

407

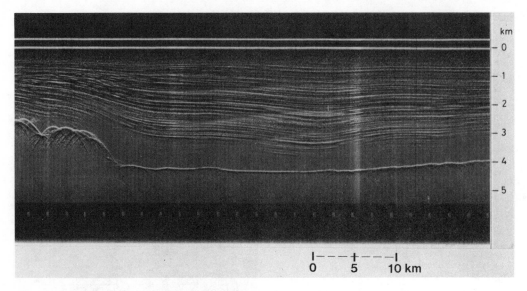

Fig. 8. Compressed and enlarged 60 MHz recording from central east Antarctica with submerged mountains and a subglacial lake.

A special way of presentation of a recording is shown in Fig. 8. It is obtained from the original film by enlargement of the film at the same time as a compression is carried out lengthwise. The recording was made in Antarctica with the 60 MHz system and a pulse length of 250 nanoseconds and shows interesting features. The mountains to the left covered by about 2 500 meters of ice are represented by the hyperbolas referred to above. In principle they represent only the summit of the mountain but often modifications by the mountain structure result and a method of reconstruction of the relief has been worked out (Harrison 1970). The maximum ice thickness observed is about 4 300 meter in the neighbourhood of the Dome C (76^{0}S, 125^{0}E (Gudmandsen et al. 1975). At that place the bottom echo shows a smooth pattern in contrast to the echoes from the ice-rock interface at both sides which partly show the usual hyperbolic shape. This feature extends over about 4 kilometers and is attributed to refelction from a water surface - a subglacial lake.

The figure also shows stratification to great depth. These layers which extend over very great distances establish time horizons in the ice. Studies of similar recordings from the Greenland inland ice compared with data from ice core drilling have shown that it is possible to identify specific layers such as the layer deposited just after the last ice age, for instance (Gudmandsen 1975). A discussion of the accuracy of the measurement of position of the individual layers by the radar method is given in (Gudmandsen 1977). Also, it has been found that the stratification as observed by radar in Greenland may be related to large vulcanic eruptions. In one case, the radar layer has been referred to a large volcanic event in Iceland in 1783. Probably the eruptions create an acidic precipitation resulting in a lossy layer in the ice. It is interesting to note that the stratification also may be utilized for ice dynamics studies.

6 ICE PROFILES AND CONTOURING

One of the aims of the radio echo sounding is to produce ice profiles along the flight tracks and from these profiles to work out contour maps of the surface and the subglacial relief of ice caps. Examples from Greenland will be shown briefly.

Fig. 9 is a compressed image of the recording of soundings at 60 MHz on a flight line along the ice divide from Thule over Camp Century, CC, (77^{0}11'N, 61^{0}09'W), North Site, NS (76^{0}46'N, 42^{0}27'W) to Crête, CR (71^{0}01'N, 37^{0}19'W) in the centre of the ice cap. The variations observed in the recording are due to changes of the equipment operation during the flight. Thus, differences in resolution of the stratifica-

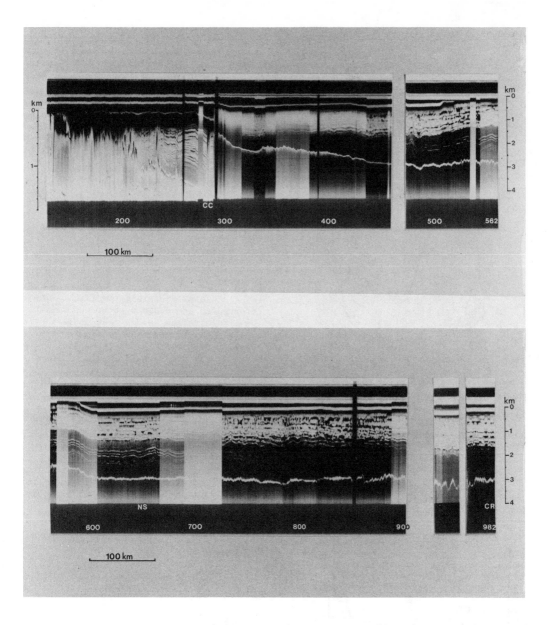

Fig. 9. Compressed and enlarged 60 MHz recording made on a flight line along the central ice divide in northern Greenland. The variation in appearance is due to variations in the operation parameters of the radar - change of pulse width and receiver bandwidth, for instance. The numbers are reference numbers for all recordings made: A-scope, Z-scope, altimeter, and coordinates. This recording is the basis for the profile shown in Fig. 11.

tion are due to changes of the pulse length and the receiver bandwidth, for instance. The flight line is shown as a heavy line on the map, Fig. 10 where a great number of other flight lines also are shown.

The flight lines in Fig. 10 have been drawn by a computer based on a digital recording of the position measurement by the inertial navigation system of the aircraft. The recording is made every 15 second corresponding to a distance interval of about 2 km and an average accuracy of the position indication will be of the order of one to two nautical miles (2 to 4 km). At the same time the static pressure is measured so that the altitude of the aircraft is determined when related to a calibration of the pressure gauge.

Based on this information and digitization of the radar recording giving the altitude of the aircraft over the ice surface and the ice thickness, an ice profile may be constructed by computer as shown in Fig. 11. In this figure the step curve shows the altitude of the aircraft and the two other curves show the surface and the ice-rock interface along the flight line referred to above. On part of the flight, the aircraft flew so low that it is impossible to distinguish the transmitted pulse from that of the surface. Similar profiles may be drawn for the individual layers of the radar stratification if these were digitized in the same way as the ice-rock interface.

Contour maps may be constructed on the basis of the same data used for the profiles: coordinates of position and corresponding altitudes of the surface and the ice-rock interface. An example of a contour map of the ice-rock interface is shown in Fig. 12 for the central part of Greenland of 600 km by 776 km.

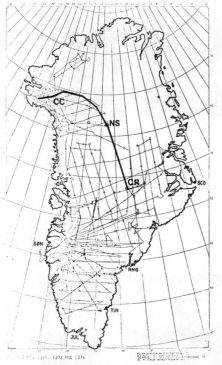

Fig. 10. Computer-drawn flight lines in Greenland based on digital recording of position coordinates.

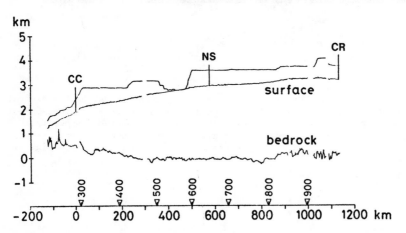

Fig. 11. Computer-drawn profile along the central ice divide in northern Greenland. Heights are given relative to mean sea surface level.

410

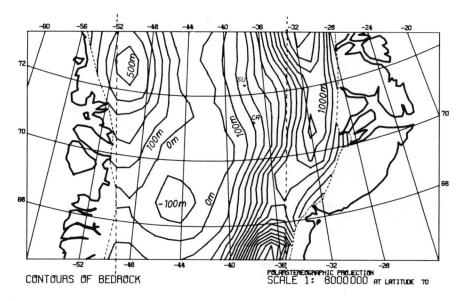

CONTOURS OF BEDROCK

POLARSTEREOGRAPHIC PROJECTION
SCALE 1: 8000000 AT LATITUDE 70

Fig. 12. Contour map of the ice-bedrock interface in central Greenland.

Here, a number of intersecting flight lines at an interval of about 70 km have been completed as may be seen in Fig. 10. A computer programme divides the area into 700 smaller squares of 30 km by 30 km and averages the elevations determined in each rectangle as a basis for the construction of the contours. Data for squares with no crossing flight lines are obtained by area interpolation.

From Fig. 12 it is noted that in a great part of central Greenland the subglacial relief is below sea level. Also, it is seen that this depressed area extends towards the large and very productive glacier on the west coast, Jakobshavn Isbræ (69°11'N, 49°57'W).

Based on surface and sub-glacial contour maps a computer graphic of the same area of the inland ice has been constructed, Fig. 13. The dramatic effects of this drawing stem from the fact that the elevation scale is exaggerated by a factor of about 100 in relation to the horizontal distances. From the slope of the surface it is noted that the Jakobshavn Isbræ is draining a very large area of the central part of the ice cap - up to 550 km from the glacier. Similarly, the drainage basin

of glaciers on the east coast of Greenland may be delineated from this figure.

The accuracy of the contouring and therefore of the graphic of Fig. 13 is dependent on the accuracy of the measurement of the altitude of the aircraft. An evaluation has shown that this accuracy is no better than 50 meter although fix points on the ice surface determined by means of satellite position system have been used for up-dating the flight altitude. However, satellite altimetry employing a microwave radar altimeter and a very accurate tracking system for determination of the satellite orbit has potential capabilities for profiling of large ice caps as exemplified by data from the satellite GEOS-3.

7 SOUNDING OF SHALLOW ICE

Shallow polar ice presents a problem when sounded from an aircraft. Due to the relative high reflection coefficient and the proximity of the surface the echo pulse from the surface becomes very large, see Figs. 5 and 6. Of equal importance is the scattering from the rough surface illuminated by the antenna. Therefore, although the receiver input circuit is designed with a very short time constant it is

411

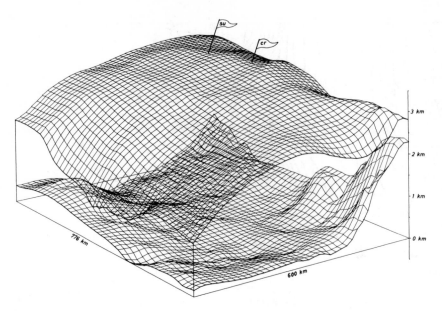

Fig. 13. Computer graphic of the inland ice in central Greenland with the vertical
scale exaggerated relative to the horizontal one by a factor of about 100.

often difficult to detect a small bottom
echo received shortly after that of the
surface. This is in contrast to the resolution at great depths which in solid ice may
attain 5 meter with a pulse length of 60
nanoseconds. With the existing 300 MHz system the minimum measurable thickness is
in the order of 30 to 50 meter, to a great
extent determined by the surface roughness
(Gudmandsen et al. 1974). However, when the
equipment is used from the ice surface the
minimum measurable thickness may be 10 to
20 meter.

This problem may be overcome by means of
a short pulse radar, i.e. a system with a
pulse width of the order of a few nanoseconds. Since airborne low-frequency antennas - and other electronic circuits -
with a sufficiently large bandwidth to
handle the short pulse are very difficult
to implement, a relative high carrier frequency should be used. A frequency upto
2 GHz may prove useful when considering
that the thickness is small and consequently the dielectric losses are moderate. Also, at that frequency a narrow beam may be
obtained with an antenna of reasonable
size, so that scattering from surface
irregularities and ice inclusions may be
overcome to a great extent. Work in this
direction is in progress.

Whatever frequency chosen airborne sounding of an area with crevasses is almost impossible due to excessive scattering. However, when carried out at a low frequency,
5 MHz for instance, and from the surface,
echoes from the ice/rock interface may be
detectable, if the crevasses do not extend
to the bottom (Vickers and Bollen 1974).

The measurement of shallow ice is often
connected with some engineering application such as mining or the planning of
hydro-electric power plants (Gudmandsen
1977). In this context the determination of
the position of the measurements have to
be more accurate than obtainable by the
inertial navigation equipment of an aircraft. Soundings will therefore have to be
supported by an automatic position system
with the master on board the aircraft and
two radio transponders (slave stations)
placed at well-defined high points prior
to the measurements. An accuracy of about
5 meter in position recording may thereby
be obtained.

8 SOUNDING OF TEMPERATE GLACIERS

Temperate glaciers are defined as glaciers
with an average temperature throughout the

glacier at the melting point. All glaciers in Europe - including Iceland, are of that type and even some of the local glaciers in the Greenland area may be characterized as such. Due to the high temperature the glacier contains a great deal of water in the form of water pockets and streams at almost all depths. Since water has a large refractive index, n = 9 in contrast to the value for ice n = 1.78, the water inclusions give a very large reflection so that an image obtained by the usual radio echo sounding technique will be very similar to Fig. 7 as reported from measurements on Hardangerjøkull, Norway, using a 480 MHz sounder (Ewen Smith and Evans 1972).

In order to overcome the scattering of the water inclusions it is necessary to employ a radio wavelength much longer than the dimension of the reflecting bodies. Successful sounding of temperate glaciers have been made by means of a special type of radar working as a sort of a spark generator which could be operated at a number of frequencies between 1.4 MHz and 100 MHz (Vickers and Bollen 1974). The transmitter consists of a series of transistors operating in the avalanche mode giving a balanced output connected directly to the antenna. The peak output power was approximately 5 kW, with a pulse length of 600 nanosecond at 5 MHz or 1.5 microsecond at 1.4 MHz. The waveform was approximately two cycles of the carrier frequency.

The antennas used were resistive loaded dipoles with lumped resistances given an approximately exponential loading of the wave propagating along the antenna. By this non-resonant design ringing is almost avoided but energy is dissipated in the antenna. The transmitter and the receiver use separate antennas.

The receiver is an oscilloscope coupled to the antenna through a balancing network and the data were recorded on poloroid film. Each trace starts with a pulse propagating through the air from one antenna to the other, which is then followed by subsurface returns.

By means of this technique temperate glaciers like the South Cascade Glacier, Washington, (maximum thickness measured at 1.4 MHz: 1000 meter) have been investigated. Measurements were carried out point by point and were only slightly disturbed by crevasses - in fact the antennas were placed on the surface bridging water filled crevasses in some cases. An example of an

Fig. 14. Example of oscilloscope recording of sounding signal obtained with a 1.4 MHz single-pulse system on a temperate glacier (Columbia Glacier, Washington).

oscilloscope recording is shown in Fig. 14 obtained at 1.4 MHz on the Columbia Glacier showing an echo from approximately 500 meter.

A similar surface technique has been used during the summer 1977 by a team from Cambridge University to sound the largest ice cap in Iceland, Vatnajøkull, where a maximum thickness of about 900 meter was measured.

9 SOUNDING OF SNOW

Sounding of snow is faced with two problems. One is related to the dielectric properties and the stratification of the snow, and the other is created by the relative small thickness to be measured.

The dielectric problem is summarized in the subsequent paragraphs:

(a) The refractive index of snow is dependent upon the density according to the formula

$$n = 1 + 0.85 \, \rho \qquad (16)$$

where n is the refractive index and ρ is the density in g/cm^3 (Evans 1965). Since the density varies through a snow cover the radio wave velocity will also vary, cf. eq. (5), so that the conversion of radar data into snow thickness becomes inaccurate.

(b) In general, a snow cover is a stratified medium with a random density profile due to layers of light and heavy snow depending upon the characteristics of the precipitation (snow crystal size) and the temperature. In between the layers there may be solar ice crusts and high density

413

ice refrozen from melt water. With a sounding system of high resolution these high density layers may be observed due to interface reflections but their actual depth in the snow become uncertain with no ground information available.

(c) The considerations above refer to dry ice. An additional problem arises from the presence of free water in the pack, due to the high permittivity of water (n = 9) as opposed to dry snow (n = 1.5). The effect is a decrease of the pulse travel time which may be quite appreciable. Thus, a free water content of 10 % increases the refractive index to n = 1.87 so that the radio wave velocity decreases from 200 m/µsec in dry snow to 160 m/µsec in wet snow.

(d) Similarly, the dielectric absorption in dry snow is in general so small that it may be neglected considering the thickness to be measured. This is true for frequencies upto about 3 GHz, for instance. However, when free water is present very large absorption takes place so that rather powerful systems may be needed. Thus, the loss tangent increases by a factor of 10 when the water content increases from 0 % to 1.6 % by weight in snow with a density of 0.38 g/cm^3 at 0^0 C as found by measurement at 9.4 GHz (Evans 1965). This means that the attenuation constant increases by the same factor with this rather moderate water content, cf. eq. (4).

At a first glance sounding of a snow pack with some accuracy appears almost hopeless. However, experiments have shown (Vickers and Rose 1972), that the snow depth may be determined from soundings with an accuracy of better than 10 % when compared with gamma ray measurements taken as a standard.

Fig. 15 shows an example of simultaneous measurements with a 2.7 GHz sounder with 1 nanosecond pulses and with a gamma ray density profiler (Vickers and Rose 1972). It will be noted that solar ice crusts shown by the increased density at A, B, and C, are well-defined in the radar signature. The disparity in relative placement of these features observed, is ascribed to differences in the snow structure at the sites of the sounder and that of the gamma ray instrument.

In remote sensing, interpretation may often be facilitated by means of recurrent measurements recording the variation with time of the phenomenon in question,

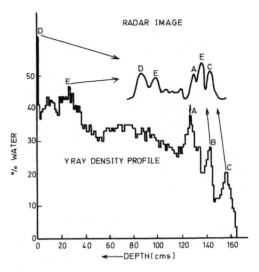

Fig. 15. Comparison between snow data obtained by means of a 2.7 GHz sounder (oscilloscope recording) and a density profile measured by means of a gamma ray profilometer.

eventually related to other observations. This is certainly the case with snow monitoring where a reference could be established at a number of points by placing chicken nets, for instance, on the ground before snow fall. They will form strong echoes that may be detected even if the snow absorption becomes large, and are used for recurrent updating of the measurements. Recording of meteorological parameters at a few points in the area being studied, such as temperature and solar flux, may assist in the interpretation of stratification features like those shown in Fig. 15, and thereby in the estimation of the density profile. If such measures are taken surveying of large snow fields may be carried out with great accuracy in spite of the variability of the medium considered.

The other problem is the one of obtaining sufficient resolution to be able to carry out measurement of small thicknesses. This is a technical problem which may be solved by using a narrow pulse, for instance. With a pulse length of τ, the resolution in range is given by

$$s_r = \frac{\tau}{2} \cdot v \tag{17}$$

where v is the radio wave velocity in the medium, cf. eq. (5). Thus, a resolution

414

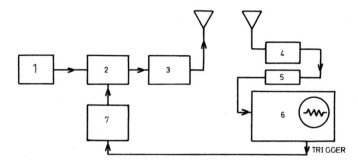

Fig. 16. Block schematic of a 2.7 GHz short-pulse sounder. 1 = 2.7 GHz microwave generator, 2 = fast switch, 3 = microwave amplifier (travelling wave tube), 4 = microwave amplifier (travelling wave tube), 5 = envelope detector, 6 = sampling oscilloscope, 7 = 1 nanosecond pulse generator.

of 10 cm in snow with a refractive index of 1.5 requires a pulse length of 1 nanosecond. Such system can be implemented, but since reproduction of a pulse with a length of τ requires a bandwidth of

$$B = \frac{1}{\tau} \tag{18}$$

a very large bandwidth of 1 GHz is needed in the transmitter and the receiver including the antennas. A high carrier frequency is therefore necessary.

An example of a short-pulse system is the one used for the measurements of Fig. 15, shown in Fig. 16 (Vickers and Rose 1972). The carrier frequency is 2.7 GHz pulse-modulated by a 1 nanosecond pulse generator. The microwave amplifiers are travelling wave tubes. The system may be used from helicopter, and from an aircraft if the sampling oscilloscope is replaced with other faster recording device. A similar system employing only one antenna common to the transmitter and the receiver has been developed at about 10 GHz (Page and Ramseier 1975).

The pulse systems described above may be characterized as time-domain devices. Another approach is application of the frequency-domain technique in which a spectral analysis of the return information-bearing signal is conducted. An example is the FM-CW radar (Chudobiak et al. 1974, Page and Ramseier 1975, Christoffersen and Gudmandsen 1970). In this type of radar, the output frequency of the system varies linearly with time, Fig. 17.
The signals reflected from the air-snow interface and the snow-rock (ice) interface are mixed in the receiver with the instantaneous transmitter frequency to pro-

duce two audio frequency beat signals whose separation in the frequency domain is a measure of the snow thickness. The beat signals may be displayed on a spectrum analyzer for a real-time visual readout, Fig. 18, or they may be tape recorded in a digital format for subsequent computer processing (filtering).

A frequency sweep between 8.2 GHz and 12.4 GHz, for instance, corresponds to a pulse length of about 0.25 nanosecond in the time-domain, and gives a range resolu-

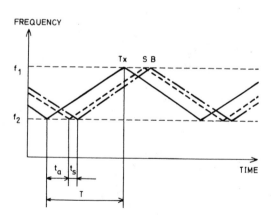

Fig. 17. Frequency-time diagram for CW-FM radar with a linear triangular swept frequency between f_1 and f_2. The solid line, marked Tx is the transmitted signal, the dotted line (S) is the surface reflected signal, and the dashed line (B) is the signal from a subsurface interface. The time differences t_a and t_s are related to the distances involved; thus $t_s = 2d_s/v$, where d_s is the distance in the medium and v the corresponding wave velocity.

415

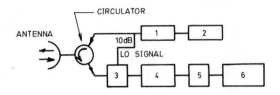

Fig. 18. Block schematic of an FM-CW
sounder. 1 = microwave oscillator, 2 =
triangular voltage generator, 3 = first
mixer (microwave mixer), 4 = intermediate-
frequency amplifier, 5 = second mixer and
local oscillator, 6 = low-frequency
(0.50 kHz) spectrum analyzer.

tion of about 2.5 cm in a rather dense snow
pack with a refractive index of 1.5. It is
clear that application of these high frequencies limits the capability of the system of measuring snow pack with a high
water content.

10 REFERENCES

Bogorodsky, V.V. 1968, Physical methods
  for investigation of ice (in Russian),
  Leningrad, Hydrometeorological Publishing Co.
Christoffersen, P.D., & P. Gudmandsen 1970,
  Experiments with electromagnetic probing of sea ice, Report R 83, Laboratory
  of Electromagnetic Theory, Lyngby, Denmark.
Chudobiak, W.J., R.B. Gray, R.O. Ramseier,
  V. Makios, M. Vant, & J.L. Davis 1974,
  Radar remote sensors for ice thickness
  and soil moisture measurements, Proc.
  Second Canadian Symp. on Remote Sensing,
  2 : 417-424.
Evans, S. 1965, Dielectric properties of
  ice and snow - a review, J. Glaciology,
  5 : 773.792.
Ewan Smith, B.M., & S. Evans 1972, Radio
  echo sounding: Absorption and scattering
  by water inclusion and ice lenses, J.
  Glaciology, 11 : 133-146.
Harrison, C.H., 1970, Reconstruction of
  subglacial relief from radio echo sounding records, Geophysics, 35 : 1099-1115.
Harrison, C.H., 1973, Radio echo sounding
  of horizontal layers in ice, J. Glaciology, 12 : 383-397.
Gudmandsen, P., 1971, Electromagnetic pro-
  ping of ice, in Electromagnetic probing
  in geophysics (ed. J.R. Wait), pp. 321-
  343, Boulder, Colorado, Golem Press.
Gudmandsen, P., Layer echoes in polar ice
  sheets, J. Glaciology, 15 : 95-101.
Gudmandsen, P., 1977, Studies of ice by
  means of radio echo sounding, in Colston

Papers No. 28, Remote sensing of terrest-
  rial environment (eds. Peel, Curtis,
  Barrett), pp. 198-211, Bristol, Colston
  Papers.
Gudmandsen, P., 1977, Remote sensing of
  snow and ice with radio waves, Vannet
  i Norden, 10, No. 4 : 39-44.
Gudmandsen, P., E. Nilsson, M. Pallisgaard,
  N. Skou, F. Søndergaard 1975, New equip-
  ment for radio echo sounding, Antarctic
  Journal of the United States, 10 : 234-
  236.
Gudmandsen, P., N. Skou, F. Søndergaard
  1974, Radioglaciology - soundings near
  Isua, southwest Greenland, Report R 224,
  Electromagnetics Institute, Lyngby, Den-
  mark.
Page, D.F., & R.O. Ramseier 1975, Applica-
  tion of radar techniques to ice and snow
  studies, J. Glaciology, 15 : 171-191.
Robin, G de Q., S.Evans, J.T. Bailey 1969,
  Interpretation of radio echo sounding in
  polar ice sheets, Phil. Trans. Royal Soc.
  London, Ser. A 265, No. 1166. 437-505.
Vickers, R.S., & G.C. Rose 1972, High re-
  solution measurements of snowpack stra-
  tigraphy using a short pulse radar, Proc.
  Eight Intl. Symp. on Remote Sensing of
  Environment, 1 : 261-277.
Vickers, R.S., & R. Bollen 1974, An experi-
  ment in the radio echo sounding of tem-
  perate glaciers, Final Report, Stanford
  Research Institute, Project 3606.

ERIC C. BARRETT
Department of Geography, University of Bristol, UK

26

The use of radar and satellites in rainfall monitoring

1 THE NEED FOR REMOTE SENSING INPUTS INTO RAINFALL MONITORING PROGRAMMES

Water is one of the most abundant, yet also one of the most intensively exploited, resources of the world. Shortages of fresh water for domestic consumption, as well as suitable water for agriculture and industry have long been chronic in some parts of the world. However, additional acute shortages in widely separated regions - some low-consumption areas like the Sahel, but others, more significantly, high consumption areas like north-western Europe and parts of North America - have recently brought the problem into sharper focus. As a consequence few eyebrows were raised when the following pronouncement was made by a spokesman for the U.S. Department of Commerce at the World Water Conference in 1976 (Reuter News Report):

"If, today, we are suffering from problems associated with the provision of a suitable supply of oil, at a price we can afford, so, by 2000 AD we shall be suffering from comparable problems associated with the provision of a supply of water."

Although this suggestion was accepted without challenge at the time it has become clear since then that a proper appreciation of the immediacy of this problem has yet to be impressed upon the general public, politicians, planners - even upon some in the water business itself. However, signs that even traditionally abundant water supplies are under increasing strain are appearing in the most unlikely places. Who would have predicted only a decade ago that hotel tables in the USA would be adorned in 1977 with cards advising the patrons that iced water was available only on request - the shortage of water being cited as the principal reason?

Looked at differently, as I have said elsewhere (Barrett, in press), water supply problems already restrict significantly the lives of men and the economies of entire nations through by far the greater part of the world.

Of course, water resource evaluation for many purposes depends more on assessments of water on or below, rather than above, the ground. But a good knowledge of rainfall is essential for a broad spectrum of dependent studies, ranging from research questions, such as those which arise in theoretical modelling of hydrological cycles at different scales, to intensely immediate questions such as river management and flood prediction and control.

Rainfall monitoring may be divided into two categories on the basis of the timely availability of the resulting data:
1. Real-time monitoring, in which rainfall data are available to the user community immediately after observations have been made. Such data are a very small fraction of all rainfall observations.
2. Non-real-time monitoring, in which rainfall data are not transmitted so rapidly from the observing stations, and may not become available for analysis by the user community until days, weeks, or even many months afterwards. Such monitoring provides the bulk of the rainfall data throughout the world.

In this essay we will be concerned directly with the first type of rainfall monitoring alone, for remote sensing has most to offer in contexts in which immediacy and comprehensiveness of view are paramount: it is in the shorter time-scale that conventional rainfall data are most seriously deficient. It should be stressed, however, that remote sensing is at present – and will continue, perhaps indefinitely to be – a supplement to, rather than a replacement for, more traditional methods of rainfall assessment. The measurement of rainfall by raingauges is fraught with many problems, but those relatively simple instruments will long continue to provide the data against which rainfall assessments by other means must be adjudged.

Living as we do in an era of enhanced consciousness of "the environment" it is easy to assume that man is monitoring his surroundings more closely than he has done hitherto. Work in progress in the University of Bristol has revealed that, with rainfall, as with many other environmental parameters, this is manifestly not so. For example:
1. There are fewer operational rainfall stations today of both types specified by (1) and (2) above than in the past. In many tropical regions the peak station density was as long ago as the 1930's, and the ratio of present to peak station densities is, in some cases, lower than 1:3.
2. The Global Telecommunication System (GTS), whereby weather reports from principal observing stations are circulated rapidly to central weather bureaux around the world is suffering from both practical and political difficulties. Demonstrably data from some regions are highly spasmodic, usually because of organisational deficiencies at one level or another, so that 50% of the anticipated data for any synoptic reporting hour is a good haul from some regional blocks. Meanwhile, data from certain regions are being denied access to the data stream on account of poor international relations.
3. There is clear evidence that the quality of both real-time and non-real-time reports has declined in many regions of the world.

Bearing such considerations in mind, and adding to them the realisations that we need more, not less, information on rainfall – not only in relatively data-rich regions, but also in traditionally data-sparse regions including the world's oceans - it is obvious that the need for assistance by remote sensing is both large and growing. Indeed, it is possible to foresee a time when the remote sensing

approach may cease to be subordinate to the in situ approach, despite the fact that remote sensing observations of rainfall are generally less direct, and are subject to calibration against raingauge measurements. In many respects raingauge data are, and will continue to be, preferred over remote sensing data: it is the failure of the conventional approach to meet our needs which leads us to remote sensing of this - and, likewise many other – aspects of our environment.

The following text reviews past and present uses of radar and satellites in monitoring rainfall where the need is for real-time information. It concludes with reference to future possibilities, and the proposal of an ultimate fully-integrated scheme involving both remote sensing and in situ observations.

2 THE USE OF RADAR IN RAINFALL MONITORING

2.1 Introduction

Radar is a family of active remote sensing systems, most of which operate in the microwave(1mm - 1m) region of the electromagnetic spectrum, though recent additions have exploited acoustic principles (sodar) or other regions of the electromagnetic spectrum (lidar). Some radars have been designed for operation on the ground, and others on aircraft and even satellites. In rainfall monitoring ground-based microwave radars have been strongly predominant. Hence, we will focus our attention on these systems.

A basic distinction within microwave radars separates coherent from non-coherent systems. Non-coherent systems lack stable transmitter frequency, and are used principally to observe the location and pattern of echoes, to measure the intensity of the back-scattered signals, and, perhaps, to detect pulse-to-pulse changes in signal strength so as to provide estimates of the relative motion of the targets. Coherent systems use very stable trans-mitter frequencies, permitting these 'Doppler' systems to measure very precisely the shift in microwave frequency caused by moving targets. Unfortunately these coherent radars are beset by two difficulties which have effectively limited their applications in hydrometeorology to specialised research studies, e.g. in the mapping of 3D fields of motion in severe convective storms (e.g. Ray et. al. 1975). These difficulties are:

1. There is a restriction on the range of Doppler radars resulting from range and velocity ambiguities where a high pulse repetition frequency is used. Many Dopplers operate out to only several tens of km.
2. Doppler radars only measure the line-of-sight velocity component of the targets; hence, relatively sophisticated equipment and procedures are required to yield 2D or 3D patterns.

The uses of non-coherent radar in rainfall monitoring include the following (Browning, 1978):
1. Qualitative determinations of the dynamical structures of clouds and precipitation structures, e.g. assessment of echo shape, and patterns of precipitation associated with different weather systems such as mid-latitude depressions and severe storms.
2. Short-period forecasting of rainfall, e.g. involving simple extrapolation, or the expected development and/or decay of rain systems, taking account of the particular topography of the forecast area.
3. Quantitative measurement of precipitation. It is this use with which we shall be most concerned here.

2.2 Physical principles involved in the quantitative measurement of precipitation by radar

Radar systems have become quite widely used in rainfall monitoring especially at the mesoscale, operating out to ranges of the order of 100 km, although even here research and development has outweighed fully-operational application.

Three different operating modes are possible when the aim is to scan and evaluate rainfall rates over an area centred on a radar facility. Of these, two are impracticable operationally because they depend on relationships between rainfall rate and attenuation of the radar signal, either separately, or in conjunction with signal reflection from the areas of rain: they require large numbers of remote targets (receivers) placed around the edge of the area, adding to problems of system cost and security. Hence, the mode which employs a combined receiver/transmitter for signal propagation and evaluation of energy reflected back to it from rain in the target area is of more general use.

Given appropriate knowledge of the technical specification and performance of a radar installation the radar equation basic to most radar studies of rainfall may be represented as:

$$\overline{P}_r = \frac{cZ}{r^2}$$

where $\overline{P}_r$ is the mean received power (the "signal") from the target, c is a constant (the speed of light), Z a reflectivity factor specific to the type of precipitation in the target areas, and r the range (or distance) between the radar and the reflecting shower of rain. Of special significance is the reflectivity factor (Z), which represents a set of physical conditions prone to considerable variation both from time to time and place to place. Unfortunately, it is usually necessary to evaluate this factor empirically on a region-by-region basis, over lengthy periods of time. Inevitably, its use leads to operational errors in the evaluation of rainfall, especially on short-term bases. Much better results may be obtained by relating the average reflectivity observed over a period in the environment of a well-exposed raingauge to the rain recorded by that gauge, the recorded amounts being telemetred to the radar and the relationship calculated at, say, hourly intervals (e.g. Grinsted, 1974). Unfortunately, such refinements are not always possible. In any case, the key assumption is that, for different latitudinal zones, and different types of rain, the radar signal is proportional to the rate of rainfall divided by the square of the range.

2.3 Practical problems in the quantitative measurement of precipitation by radar

Radar has a unique capability to observe the areal distribution of precipitation at frequent intervals, giving information from quite large areas to a single centre with a minimum of telemetry requirements. However, radar techniques have yet to become the standard method of supplying the user community (meteorologists, hydrologists, agriculturalists, communications engineers etc.) with the mesoscale rainfall data they desire. A plethora of practical problems explains most, if not all, of the remaining consumer resistance. These problems include and involve the following:
1. Signal fluctuations from rain. The radar signal from rain is the sum of the signals from all the raindrops distributed, and moving at random, in each pulse volume. Not surprisingly the result is a

strongly fluctuating signal, whose average most be established and interpreted with care.

2. Variations of reflectivity in the vertical. Since the surface of the Earth is curved, and radar beams travel in relatively straight lines (even allowing for their curvature by atmospheric refraction), they sample the atmosphere at progressively greater heights as they move away from the transmitter (Battan, 1973). For a horizontally-operating radar the bottom of a beam rises to c.1km at a range of 100km. Since precipitation layers are usually quite shallow this effect, more than any other, restricts the radar range, and limits the accuracy of estimates of rainfall intensity at some distance from the radar facility. In middle/high latitudes a further related problem is the low height of the melting layer, in which the reflectivity factor may be enhanced ten times by changes in dielectric constant and fall speed, and aggregation of wet snowflakes. No adequate method has been devised yet to correct for the resulting "bright band" echoes.

3. Effects of screening and ground clutter. Radar beams are reflected not only by cloud drops and /or raindrops, but also by upstanding features on the ground, e.g. tall buildings and relief. The usual solution for this problem is to map the strong, permanent, echoes caused by such features, and substract them from the operational echo pattern, and interpolate isohyets across them. This problem, often in combination with the one which preceded it, places significant restrictions on the utility of radar in hilly or mountainous terrain.

4. Calibration of radar echoes in terms of rainfall intensity. Although estimates of rainfall intensity can be made in the absence of calibrating gauges, such gauges should be operated whenever possible. Experience in field tests have further shown that there is a marked improvement in accuracy if the method of calibrating the radar against a value of rainfall is obtained from a gauge within the actual rain area itself. Gauge-based estimates of hourly rainfall over sub-catchments in the Dee Weather Radar Project in Wales had a probable error of some 5-10%, neglecting any errors in the gauge readings themselves. In the same study agreements between gauge and radar estimates for sub-catchments under optimum conditions were within c.20% for hourly periods, improving to 8% for three hourly periods over the whole study area.

These results were better than most previous results elsewhere - perhaps because the areas quantified extended out to only 50km from the radar.

5. Staffing needs. Adequate staff training and supervision is vital if a radar facility or network is to be operated efficiently. Conditions for this are less than optimal in many countries, especially in lower latitudes where the need for improved rainfall intelligence is greatest. The associated problems often became particularly acute when foreign technical assistance ceases - an important fact of scientific life which, perhaps, only an independent witness is able to stress.

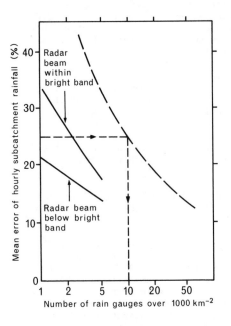

Figure 1. Radar mean error versus the number density of calibrating raingauge sites (solid lines), and the mean error of hourly subcatchment totals for mean rainfall events in middle latitudes in the absence of radar (broken line). (After Browning, 1978).

6. System costs and cost-effectiveness. It is in this key area of evaluating the utility of radar that generalisations are most difficult to make. However, Fig. 1 is suggestive in its comparison between radar rainfall mean error curves established for networks of 1-5 calibrating gauges in temperate mid-latitudes, and a raingauge mean error curve for "average" rainfall events.   This indicates that, for a mean error of, for example, 25%,

420

10 telemetring gauges would be required over 1000km^2 to yield similar results to those obtainable from a single radar. Estimates by the U.K. Water Resources Board have further suggested that, if such an accuracy is required, the cost-effectiveness of a calibrated radar system exceeds that of a telemetring raingauge network once measurements are required over areas exceeding 3000km^2.

2.4 The future of radar in rainfall monitoring

Especially on grounds of cost-effectiveness there would appear to be a strong case for widespread employment of radar for rainfall monitoring - especially since more qualitative information such as system growth cycles and movement can be obtained at the same time in support of studies of rainfall behaviour. However, there are very few localities in the world in which continuous-recording raingauge networks are as dense as the 1 gauge per 10km x 10km cell which the Water Resources Board figures imply. Indeed, densities of gauges of all types are overwhelmingly much lower. Langbein (1960) proposed that a reasonable - and realistic - minimum density for a precipitation network would be as low as 2 gauges per 1000km^2. Britain, with one of the densest networks in the world, can boast of about 1 gauge per every 40km^2. However, these figures include, and are strongly dominated, by, daily and twice-daily, accumulating gauges whose data therefore have a much lower temporal resolution than more expensive observation systems would provide.

In conclusion, then, the situation is this: radar can indeed provide valuable rainfall data, especially when integrated with a network of calibration gauges, but almost all present users of rainfall data have to cope at present with small to tiny fractions of the data radar systems would yield at break-even points, established in comparison with conventional gauge networks. Thus, it is generally difficult to convince funding agencies of the need for the much greater volume of data radar could provide, especially in view of the substantial increase in costs that would be incurred over and above actual present expenditure on rainfall monitoring.

Some consequences of the foregoing arguments are that meteorological radar is in widespread use only in North America and Western Europe, but even there its present applications to rainfall monitoring are mostly qualitative. Its key use in North America is in short-term forecasting of tornadoes and other severe storms; in Western Europe it is used mainly as an auxiliary tool for weather analysis and forecasting. Plans have been developed, however, for a radar-augmented quantitative rainfall mapping programme covering much of the British Isles (Browning et al. 1977). Here computers will be used to up-date continually the reflectivity factor (Z) through reference to calibration gauges, and to print out rainfall maps at half-hour intervals for grid squares with a highest resolution of 1km. It is proposed that the resulting information be made available immediately to agencies with an established need-to-know, and also to the general public via the developing Prestel (or Ceefax) directories on television. However, it is not likely that ambitious schemes of this kind will be installed in most countries of the world in the foreseeable future; nor even is it likely that radar will be utilised widely for rainfall monitoring except for research and development.

The need for improved rainfall data, in the short- to medium-terms at least, must be fulfilled by other remote sensing means. Satellites offer more hope in these time frames.

3 THE USE OF SATELLITES IN RAINFALL MONITORING

3.1 Introduction

Although the first weather satellite (Tiros I) was launched in 1960, it was not until early 1966 that the first operational weather satellite system was inaugurated using two polar-orbiting satellites of a "cartwheel" variety (Essa 1 and 2). Not surprisingly, attention in the early years of satellite meteorology was focussed on those atmospheric phenomena which could be observed relatively directly in the visible and infrared wavebands (e.g. cloud types and systems) rather than others which could be assessed relatively indirectly or inferentially (e.g. rainfall rates and distributions). Thus the earliest paper which proposed a method for the systematic evaluation of rainfall from the cloud contents of weather satellite images appeared as late as the tenth year of satellite operations (Barrett 1970). In this, and its immediate successors (e.g. Follansbee 1973; Barrett 1973, Follansbee and Oliver 1975) it was implicitly assumed that satellites were, in effect, in competition with conventional systems in the monitoring of precipitation. However, it has been recognised more recently that it is better to consider the data from such widely different sources as complementary. The prevalent view today is that satellite data may be employed to fullest advantage in the extension of reliable mapping of rainfall (Barrett 1975) and other meteorological parameters, e.g. temperature (Giddings 1976) from areas rich in conventional observations into those more sparsely provided with ground observing stations.

3.2 The use of satellite visible and infrared data in the monitoring of rainfall

A wide variety of approaches have been followed by workers seeking to extract the most useful information on rainfall from satellite imagery in the visible and/or infrared wavebands. The range includes the following:
1. Cloud brightness techniques. These depend on the assumption that precipitating clouds are often brighter than others, so that automatic brightness thresholding and even brightness contouring techniques can be used to map rainfall from satellite visible imagery, and that calibration of rainfall amounts can be achieved using raingauge and/or radar observations. (See, e.g. Woodley et al, 1972; Griffith et al. 1976). Unfortunately such approaches, which have worked well in some convective situations, seem less likely to yield useful data where stratiform cloud types are dominant (Barrett & Grant 1977) because of the greater range of the rainfall intensity: cloud brightness relationship observed in this case.
2. Infrared techniques. Essentially, these are similar to (1), but are based on cloud top temperatures rather than cloud top reflectivities. Layered stratiform, and thick middle to upper level cloud again pose problems, but simple cloud development models have been invoked to improve rainfall estimates, especially from mesoscale convective cloud systems (see, e.g. Follansbee & Oliver 1975; Schofield & Oliver 1977). A quasi-global operational scheme involving such an approach has been envisaged recently by Griffith et al. (1978).
3. Parameterization techniques. Simple models have been devised to describe the physics of convective cloud growth in the tropics, incorporating satellite evidence of the areal extent of cloud. Verification of the results has not been easy, for tests have been mainly oceanic (See, e.g. Kuo 1965).
4. Basin hydrology techniques. Geostationary satellite data have been used to indicate total cloud amount over selected catchments, and the time changes of the cloud. Basin-wide calibration has been undertaken by reference to steam-gauge (basin discharge) records. (See, e.g., Grosh et al. 1973; Amorocho 1975). This type of approach is attractive where geostationary satellite data are available, and the mass of rain falling over an area is more significant than its spatial distribution.
5. Cloud indexing techniques. These were pioneered by the present author (Barrett 1970, 1971), and have been developed by him (e.g. Barrett 1975, 1977a) and others. (See e.g. Follansbee 1973, Merritt & Sabitini 1977). Satellite cloud images are ascribed indices relating to cloud cover, and the probability and intensity of associated rain. Different methods have been used to calibrate the indices to give rainfall estimates. In its most developed form the cloud indexing approach has become a virtual extension of classical synoptic meteorology (Barrett 1977b).
6. Rainfall climatology techniques. In middle latitudes, where rain is associated mostly with large, well-organised cyclonic systems, synoptic-scale cloud models have been used in conjunction with regional mean annual rainfall distributions to yield appropriate 12-hourly patterns of rainfall, which are calibrated through assessments of the annual frequencies of such patterns and the proportions of the mean annual rainfall they may be expected to bring. (Follansbee 1976).

It is the cloud indexing types which have, in one form or another, shown most flexibility and which have yielded the first results in support of operational rainfall monitoring programmes. For example, they have been used by the present author in support of irrigation design in Indonesia (Barrett 1975), water resource evaluation and management in Oman (Barrett 1977c),and desert locust control in North-west Africa (Barrett 1977b). Results from the African study, compiled on a 12-hourly basis in conjection with GTS weather reports, were later seen to be superior to rainfall intelligence upon which the FAO periodic publication "Foodcrops and shortages" depended, illustrating the further value of satellite-assisted rainfall mapping methods in agroclimatology. Cloud indexing method have been used also by the EarthSat Corporation of the USA in support of crop prediction in Iran, and, using some supporting hardware to permit a degree of man-machine interaction, in conjunction with the prediction of spring wheat yields in North America (Merritt & Sabatini 1977). Experience shows that rainfall maps based on both conventional and satellite data are more realistic in spatial detail than maps based on gauge data alone, and therefore reveal more accurately areas of average, above average and below average precipitation (See also Figure 3). Given a suitable arrangement for the acquisition of input data of different kinds cloud indexing methods can, and have been, tailor-made to suit local needs, and to provide information cheaply on a near real-time basis.

The object of such methods is to maximise the utility of contemporaneous conventional and satellite data through the recognition that, whilst raingauge data give acceptable measures of rain accumulated through time at separated point locations, satellite data give valuable indications of the distributions of rain spatially at separated points in time. Figure 2 illustrates the steps entailed by such an approach. In the compilation of the regressions relating rain to cloud through an historical period the following type of relationship is invoked:

$$R = f (Ca, Ct, Sw)$$

where R is rainfall at a rainfall station, Ca is cloud amount in a grid square of selected standard size, Ct is satellite-observed cloud type, and Sw is the synoptic weather situation. Ranks and scales for Ct and Sw are established for each individual area of study, and for subordinate monphoclimatic regions where necessary. The resulting regression diagrams are "contoured" parallel to the computed best-fit regression line to facilitate the floating upwards or downwards of estimates each day according to the mean observed relationship between rain and cloud at the surface stations affected by given weather structures. Available ground observations of rainfall are than combined with satellite estimates of rainfall in grid squares lacking raingauges to provide rainfall maps demonstrably more detailed and accurate than they would have been if based on gauge data alone. The required range of hydrological statements dependent on rainfall information can then be prepared with increased confidence.

3.3 The use of satellite microwave data in the monitoring of rainfall

Passive microwave radiometers on Nimbus 5 and 6 (the Electrically Scanning Microwave Radiometers (ESMR))have yielded measures of naturally-emitted microwave radiation from the surface of the Earth, and the water content of the atmosphere. ESMR data have been processed to give, amongst other outputs, maps of instantaneous precipitation intensities. Passive microwave approaches are being developed further through experimentation with systems on Nimbus-G and Seasat-A, both launched in 1978. But there is no plan as yet to place a system on an operational spacecraft series such as the new Tiros-N.

Allison et al. (1975) have shown that ESMR data is very useful in delineating areas of rainfall over the oceans and gives qualitatively good comparisons with rainfall data. Wilheit et al. (1977) used a theoretical model which included scattering effects and concluded that, despite the difficulties in interpreting rain rate using land-based radars, T_B can be related to rain rate within a factor of two over the range from 2-20mm.hr^{-1}.

Unfortunately, results overland with the Nimbus 5 ESMR (19.35 GHz) have shown that, at this wavelength, the T_B field in general can neither unambiguously locate active rain areas, nor quantize their rain-fall rates (Meneely 1975), though processing of data from the Nimbus 6 ESMR (37 GHz) holds out more hope in these respects. Unfortunately, the best spatial resolution here is rather low (c.25km at nadir). A global climatology of rainfall over the world's oceans has been prepared for 1974, and this is undoubtedly superior to any similar statement based on observations gleaned mostly from ship reports (Rao et al. 1976). Further experience over sea surfaces should soon make this type of method at least as dependable as the cloud indexing approach over land. Although calibration over the sea surfaces is clearly more difficult than over land, maritime precipitation is less noisy in space and time than continental precipitation.

3.4 Opportunities for global rainfall monitoring

Drawing together the key points in the earlier discussion, we may sketch a precipitation monitoring system for which we already have the basic technology and analytical expertise for a successful operation in the near future. This might involve the following (after Barrett, 1977d);(see Figure 4):
1. The identification of those existing rainfall stations from which dependable data are, or could reasonably be, expected.
2. The augmentation of the network in (1) by as many additional stations as circumstances might permit, with standardisation of observational method and quality of the resulting records on the critical criteria.
3. The mapping of areas which could not be represented adequately by the data from

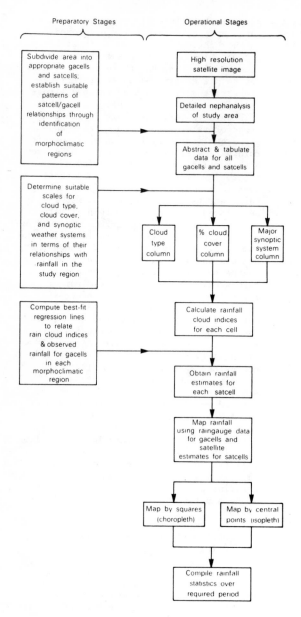

Figure 2. A flow-diagram summarising the procedures typically followed in a rainfall mapping programme based on cloud indexing principles, and integrating data from both conventional and satellite sources. (From Barrett 1977d).

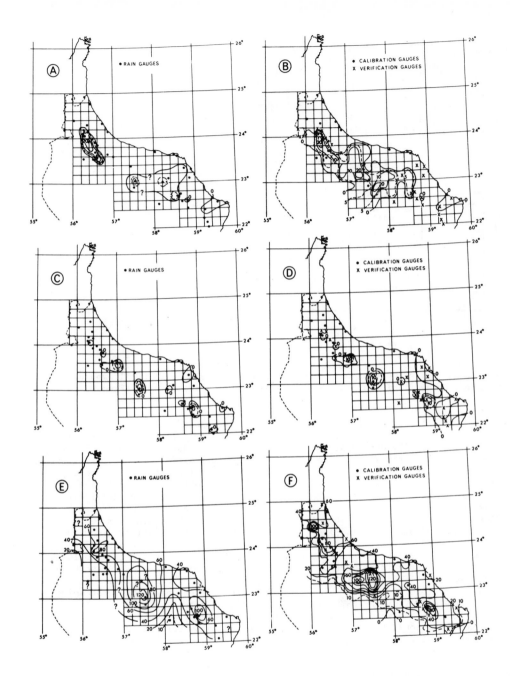

Figure 3. Rainfall maps prepared from gauge data only (A, C, E) and gauge plus satellite data using a cloud indexing method (B, D, E) for June 1974 (A, B), July 1974 (C, D), and the year 1974 (E, F) over eastern Oman, in the south-eastern corner of the Arabian peninsula. In some periods the satellite-assisted maps (e.g. B, F) are significantly different from the conventional maps (A, E), whilst in others the differences are slight (D compared with C). Over the year as a whole a depth/area estimate c.30% below the conventional estimate was obtained from gauge plus satellite data. This is very significant for water resource evaluation and exploitation. (From Barrett, 1977c).

425

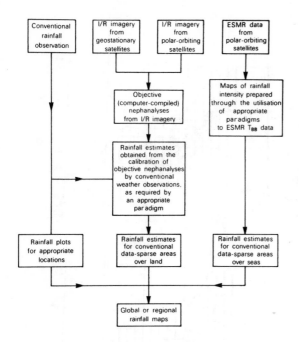

Figure 4. The key components of a proposed scheme for global rainfall mapping using data from presently-available raingauges and satellite sensing systems. (From Barrett, 1977d).

the augmented network; these are the areas for which satellite estimates of rainfall would be most usefully obtained.
4. The selection of the appropriate satellite technique for rainfall estimation in each geographical region. In the immediate future the following set of choices would seem to be most appropriate:
(a) Infrared data from the Visible-Infrared Spin Scan Radiometers (VISSR) of geostationary satellites contributing to the World Weather Watch. From such data estimates could be made of daily rainfall over land areas in the tropics.
(b) Microwave data from the ESMR systems of polar orbiting satellites, hopefully of the operational Tiros-N series: at present there is no firm plan to fly ESMR-type systems on any of these spacecraft. It is to be hoped that this need will be met, so that precipitation may be mapped over the world's oceans in low and middle latitudes, and, perhaps, over frozen land and sea surfaces in high latitudes also.
(c) Infrared data from Scanning Radiometers on operational satellites of the Tiros-N series. From these, rainfall may be assessed over land areas in middle latitudes.

Clearly such a scheme would not be simple, particularly on account of the large volumes of data to be handled daily, and the difficulties of interfacing data of several different types. However, there is little doubt that, if relatively homogeneous and regular rainfall data are required from mesoscale areas, or larger, these could be provided more readily, and probably more cheaply - given the prior existence of suitable satellite platforms and sensing systems - than would be possible through augmented gauge networks and/or radar/raingauge networks.

The biggest problems to be solved would seem to be:
1. The integration of physical models of cloud behaviour into cloud brightness/ temperature algorithms.
2. The organisation of data handling and analysis at an appropriate central facility, employing the highest possible level of automation.

Operational research should concentrate on these issues if satellites are to contribute to rainfall monitoring on a uniform global basis instead of the nonuniform and fragmented bases we see today.

4 FUTURE POSSIBILITIES

4.1 Active microwave systems

Considerable problems attend the assessment of rainfall by remote sensing means, whether surface radar, or satellite visible, infrared or passive microwave systems are employed to provide the basic data for processing as required. Surface radar has the advantage of being physically direct, but has limited range and is subject to a multiplicity of often unhelpful physical influences. Satellite visible and infrared systems provide globally comprehensive data, but rainfall distributions and amounts can only be established inferentially from their data. Satellite passive microwave systems function promisingly over sea surfaces but disappointingly over land.

Recent discussions with scientists in NASA's Goddard Space Flight Center, Greenbelt, Md., have revealed that the capability already exists to fly an active microwave (radar) system on an orbiting satellite to improve upon the results obtainable from passive microwave systems of the ESMR family. (Eckerman & Wolff 1975). It is likely that a downward-looking, scanning pulse multibeam radar with a very large dynamic range will be flown as a Space Shuttle experiment in the early 1980's. Design considerations indicate that a successful rainfall radar system of a similar kind could be flown on a satellite operating at Noaa/Tiros N altitudes. In the system currently envisaged, roughly semicircular patterns of pulse transmissions would be made ahead of the polar orbiting satellite platform to facilitate the differentiation of rain echoes from surface echoes (Eckerman, personal communication). Even with present technology it is thought that a ground resolution as good as 1km would be attainable if required, accepting that the unit areas investigated by the system should be contiguous on the ground.

Clearly such a system, were it proved to be capable of revealing rainfall areas and intensities both finely and accurately over both land and sea, would have great advantages over current systems being used for rainfall monitoring from satellites. Undoubtedly there would be the familiar range of problems to solve in the extraction of dependable rainfall intensities and amounts from the recorded echoes, but there is no reason to suppose that these problems would be more difficult than those associated with the interpretation of ground-based radar, whilst it is certain they would be less intransigent than some associated with the evaluation of satellite visible, infrared, and passive microwave data. Although the technical problems of flying active microwave systems for rainfall evaluation on geostationary satellites are such as to rule this out for, perhaps, the rest of this century, one polar orbiter, suitably equipped, could provide twice-daily rainfall data round the whole globe, whilst two polar orbiters would yield six-hourly patterns of instantaneous rainfall intensities. Since the hydrometeorologist is generally more interested in total rainfall catches than instantaneous rainfall intensities, algorithms would have to be developed for the estimation of periodic rainfall totals from the samples of instantaneous rainfall rates.

4.2 An integrated global rainfall monitoring system

If it is accepted that spaceborne meteorological radar is feasible, it is not too soon to design a system which would maximise the use of the new data anticipated therefrom, in conjunction with existing methods of rainfall assessment. As with the scheme outlined for immediate operation in Section 3.4 it would seem that a complex system of inter-related sensors of several different types would yield the best results, and provide maximum flexibility. Three components would appear to be vital in a scheme designed to meet the needs of most - operational users of rainfall data; of these, the first two would provide spatial continuity, and the third continuity through time:
1. A satellite-borne active microwave system to provide routine evaluation of the field of instantaneous global rainfall rates.
2. An adequate network of recording raingauges to provide the means for constantly checking the calibration and interpretation of the instantaneous rainfall echo patterns.
3. An adequate network of accumulating-raingauges to provide data for use in computer programmes designed to print out realistic maps of rainfall totalled through standard periods of, say 6, 12 and 24 hours, and multiples of the same.

Table 1 summarises the range of resolutions across the print-outs which might be prepared for different applications. Clearly the type of operation envisaged here is one that could only be undertaken successfully by an international centre

427

Table 1 Suggested resolutions of rainfall outputs from an integrated global system.

Spatial Resolution of output	Temporal Resolution of output	Principal Application
5° grid squares	5 days	Monitoring global rainfall, and modelling the global hydrological cycle.
$2\frac{1}{2}^{\circ}$ grid squares	Twice daily	Assessing latent heat releases from rain clouds to aid numerical modelling of meteorological processes.
1° grid squares	2-4 times daily	Assessing broad regional rainfall patterns for numerous applications in pure and applied hydrology.
$\frac{1}{2}^{\circ}$ grid squares	4-6 times daily	Assessing rainfall inputs to large basin hydrologies; flood forecasting and river management; global crop predictions.
$\frac{1}{2}^{\circ}$ grid squares, down to the maximum resolution of the system, say 2.5km with an accuracy of $\pm$ 1km.	4-6 times daily, or greater if system permits	Detailed hydrological applications in local regions of high spatial variation in rainfall.

with ready access to both in situ and remote sensing data. Of the various choices that have been considered the most appropriate would seem to be a facility of the World Meteorological Organization. This body is sufficiently international, and influential in the spheres of decision-making for global meteorology, to be able to initiate a scheme which, in many ways, might be seen to be a natural extension of its present operations. The aim should be to design, implement, and operate a rainfall programme intended to meet most needs for rainfall information beyond purely local needs for especially detailed data related to the requirements of the research community.

Of course, no such scheme could become a reality without:
1. The recognition, in the right quarters, of the need for a global rainfall monitoring system incorporating satellite-bourne radar.
2. The establishment of the organisational machinery which would be required for the smooth and effective operation of a global rainfall-monitoring system.

It is unlikely that either will come about without the protestation of the case by an adequate number of present and potential users of rainfall intelligence. In the past the remote sensing user

community has rarely received the data it would like most because it has failed to agree on its needs and to express them both cogently and forcefully in the corridors of power. In the future, opportunities to monitor, and exploit, our environment in the best possible way will continue to be lost if politicians and administrators are not better convinced of the urgency of particular needs for the right types of integrated on-going remote sensing systems to compensate for deficiencies of data obtained in more traditional ways.

It is hard to overstress the need for better rainfall data; the technology is already in hand to place global rainfall monitoring on a much firmer and more satisfactory basis as early as the mid-1980's: the critical question now is 'How may this be achieved?'

5 REFERENCES

Allison, L.J. et al. 1975, Tropical cyclone rainfall as measured by the Numbus 5 ESMR. Bull. Amer. Soc., 55: 1074-1089.
Amorocho, J. 1975, An application of satellite imagery to hydrologic modeling the Upper Sinu River Basin, Colombia. In MMH Symposium Report, Bratislava.
Barrett, E.C. 1970, The estimation of

monthly rainfall from satellite data, Mon.Wea.Rev. 98: 322-327.

Barrett, E.C. 1971, The tropical Far East: ESSA satellite evaluations of high season climatic patterns, Geog.J., 137: 535-555.

Barrett, E.C. 1973, Forecasting daily rainfall from satellite data, Mon.Wea.Rev. 101: 215-222.

Barrett, E.C. 1975, Rainfall in northern Sumatra: analyses of conventional and satellite data for the planning and implementation of the Krueng Jreue/Krueng Baro irrigation schemes, Final Report to Binnie & Partners (Consulting Engineers), London, 50pp.

Barrett, E.C. 1977a, Applications of satellite data in mapping rainfall for the solution of associated problems in regions of sparse conventional observations. In R.F. Peel, L.F. Curtis & E.C. Barrett (eds.), Remote Sensing of the Terrestrial Environment, p.126-142, London, Butterworths.

Barrett, E.C. 1977b, Rainfall monitoring in the region of the N.W. African Desert Locust Commission in 1976-77 integrating data from synoptic weather stations and meteorological satellites, Consultants Report to Food and Agriculture Organisation, W/K8207, Rome, FAO, 43pp.

Barrett, E.C. 1977c, The assessment of rainfall in north-eastern Oman through the integration of observations from rain-gauges and meteorological satellites, Consultants Report to Food and Agriculture Organisation, W/K7629, Rome, FAO, 55pp.

Barrett, E.C. 1977d, Monitoring precipitation: a global strategy for the 1980's, Monitoring Environmental Change by Remote Sensing, Remote Sensing Society, U.K.: 53-58.

Barrett, E.C. & Grant, C.K. 1977, Studies of cloud and rainfall based on Landsat 2 imagery of the British Isles, Final Report, ERTS Follow-on Program Study No.2962A, Greenbelt, Md., NASA, 122pp.

Battan, L.J. 1973, Radar observation of the atmosphere, Chicago, Chicago University Press.

Browning, K. 1978, Meteorological aspects of radar, Reports on Progress in Physics, 41: 761-806.

Browning, K.A., Bussell, R.B. & Cole, J.A. 1977, Radar for rain forecasting and river management, Water Power and Dam Construction, 24: 37-42.

Earthsat Corporation 1976, Earthsat Spring Wheat yield system test, 1975, Final Report, Contract NAS 9-14655, Chevy Chase, Washington, D.C., Earthsat Corporation.

Eckerman & Wolff 1975, Spaceborne meteorological radar measurement requirements meeting, NASA X-900-75-198, Greenbelt, Md., NASA, 73pp.

Follansbee, W.A. 1973, Estimation of average daily rainfall from satellite cloud photographs, NOAA Technical Memorandum, NESS 44, Washington, D.C., U.S. Dept. Commerce, 39pp.

Follansbee, W.A. 1976, Estimation of daily precipitation over China and the USSR, using satellite imagery, NOAA Technical Memorandum, NESS 81, Washington, D.C., U.S. Dept. Commerce, 30pp.

Follansbee, W.A., & Oliver, V.J. 1975, A comparison of infrared imagery and video pictures in the estimation of daily rainfall from satellite data. NOAA Technical Memorandum, NESS 62, Washington, D.C., U.S. Dept. Commerce, 14pp.

Giddings, L.E. 1976, Extension of surface data by use of meteorological satellites, Technical Memorandum, LEC-8377, Houston, Texas, Lockheed Electronics Company, Inc., 33pp.

Griffith, C.G. et al. 1976, Rainfall estimation from geosynchronous satellite imagery during daylight hours, NOAA Technical Report, ERL 356 - WMPO 7, Boulder, Colo., U.S. Dept. Commerce, 106pp.

Griffith, C.G. et al. 1978, Rain estimation from geosynchronous satellite imagery: visible and infrared studies, Mon.Wea. Rev. 106(8): 625-643.

Grinsted, W.A. 1974, The measurement of areal rainfall by the use of radar. In E.C. Barrett & L.F. Curtis (eds.) Environmental Remote Sensing: Applications and Achievements, p.267-284, London, Arnold.

Grosh, R.C. et al. 1973, Cloud photographs from satellites as a hydrological tool in remote equatorial regions, J. Hydrol. 20: 147-161.

Langbein, W.B. 1960, Hydrologic data networks and methods of extrapolating or extending available hydrologic data, Hydrologic Networks and Methods, U.N. Flood Control Series No.15, Bangkok: 13-41.

Meneely, J.M. 1975, Application of the Nimbus 5 ESMR to rainfall detection over land surface, NASA Contractors Report, NAS 5-20878, ES-1008.

Merritt, E.S. (ed.) 1976, Earthsat Spring Wheat Yield System Test 1975, Final Report, Contract NAS 9-14655, Houston, Texas: NASA.

Rao, M.S.V., Abbott, W.V., & Theon, J.S. 1976, Satellite-derived global rainfall atlas, NASA 5P-410, Washington, D.C., NASA, 31pp.

Ray, P.S. et al. 1975, Dual-Doppler observation of a tornadic storm, J.Appl. Met., 14: 1521-1530.

Schofield, R.A. & Oliver, V.J. 1977, A scheme for estimating convective rainfall from satellite imagery, NOAA Technical Memorandum, NESS 86, Washington, D.C., U.S. Dept. Commerce, 47pp.

Wilheit, T.T., et al. 1977, A satellite technique for quantitatively mapping rainfall rates over the oceans, J. App. Met., 16: 551-560.

Woodley, W.L. et al. 1972, Rainfall estimation from satellite cloud photographs, NOAA Technical Memorandum, ERL OD-11, Boulder, Colo., U.S. Dept. Commerce, 43pp.

E.M.MORRIS, K.BLYTH & R.T.CLARKE
Institute of Hydrology, Wallingford, UK

27

Watershed and river channel characteristics, and their use in a mathematical model to predict flood hydrographs

1 THE HYDROLOGICAL CYCLE

The hydrological cycle is the recirculatory process whereby water is evaporated from the sea surface, is advected over land masses where it falls as precipitation, and is eventually returned to the ocean as flow, either in river channels or from aquifers around the sea's perimeter. The land phase of the hydrological cycle is the set of physical processes by which precipitation is transformed into streamflow; studies of this land phase of the cycle, and of how it is affected by human activity, constitute the research programme of the Institute of Hydrology, a component body of the United Kingdom Natural Environment Research Council.

Of particular interest to the Institute is the effect of alternative land uses on the quantity and distribution in time of streamflow from river basins. The Plynlimon study, described below, was established in 1969 for the hydrological comparison of two adjacent river basins, one of which is under hill pasture of relatively low productivity and the other is planted with coniferous forest; other studies assess the hydrological effects, on volume and distribution of streamflow, of establishing new towns in river basins that were previously agricultural.

2 THE PLYNLIMON STUDY

The Institute of Hydrology's study of the headwater catchments of the Wye and Severn, like much other recent work on the losses of water from coniferous forest, has its origins in the meticulously executed study of Law (1956, 1957) on the Hodder Catchment in the Yorkshire Pennines.

Concerned about the absence of research on the water balance of woodland, and about possible lower water yields from forested catchments than from catchments planted with short herbaceous vegetation, Law made a careful study of the water balance of a small (0.045 hectare) natural lysimeter in a plantation of Sitka spruce (*Picea sitchesis*) set in a slightly larger block of woodland (area 0.24 hectares). Based on measurements collected over the period 4 July 1955 to 8 July 1956, Law found that the precipitation above the forest canopy was 984 mm; of this, 630 mm reached the forest floor, from which 273 mm appeared as runoff. The total water loss was therefore 711 mm; over the remainder of the Hodder Catchment, however, the total water loss was 421 mm. Law concluded that the loss of water from the forested plantation was the greater by 290 mm.

Law's results attracted the attention of many research workers and the scepticism of not a few: scepticism based upon the small size of the plantation used in his study which, it was argued, would have led to the introduction of edge effects in both radiative and aerodynamic aspects of vegetation. Rutter (1964), summarizing evidence from his own study of water relations of *Pinus sylvestris* under plantation conditions, from work reported by Deij (1956) on the Castricum lysimeters in the Netherlands, and from East African work by Pereira, Dagg and Hosegood (1962), observed that evaporation from forests had generally been found to lie between 0.8 and 1.0 times the Penman estimate of evaporation from an open-water surface (although, as Rutter states, Penman himself drew no such firm conclusion in his monograph 'Vegetation and Hydrology' (1963)). Rutter's own evidence suggested that actual evaporation from the plantation exceeded open-water evaporation by 10-20%, a high figure by comparison with most other results except those of Law. The main causes suggested for the high values of actual evaporation from forests were the smaller albedo of the vegetation, expecially of conifers, giving greater absorption of energy; and greater aerodynamic roughness.

Law's result, if valid, was clearly of the greatest import at a time when water supply undertakings were being encouraged to afforest their water catchments; several industrial conurbations, such as those of the Midlands, draw water supplies from the wetter West and North of the United Kingdom where the climate is such that land use is largely restricted to a choice between softwood production on the one hand and upland pasture of relatively low productivity on the other. Law's result suggested that extensive afforestation would result in reduced water yield in a period during which domestic consumption may be expected to rise from its present level of 168 litres (37 gallons) per head daily to 273 litres (60 gallons) by the year 2000, and during which industrial water demand for cooling and processing is likely to increase significantly. Millis (1975) estimated that 44000 gallons, 30000 gallons and 44 gallons of water are required to produce one ton of steel, 1 ton of aluminium and 1 pint of beer respectively.

To attempt verification (or possibly, refutation) of Law's result, the Institute of Hydrology began a programme of research during the 1960s which consisted initially of two catchment studies. In the first, two adjacent catchments on the slopes of Plynlimon, in Central Wales, were intensively instrumented for the measurement of precipitation, river discharge and soil moisture change; one of the catchments is the headwaters of the Wye (area 1055 hectares, almost entirely upland pasture) whilst the other is the headwaters of the Severn (area 870 hectares, of which slightly more than two-thirds is coniferous forest, principally Sitka spruce and Norway spruce, but with an admixture of Japanese larch). This study has two objectives, best formulated as questions: (i) is the mean annual loss (precipitation minus streamflow) greater for the forested Severn than for the hi pasture of the Wye, and if so, how far is the difference explicable in terms of different land-use? (ii) does the rapidity and magnitude of response to unit depth of precipitation differ for the two catchments, and if so, how far are the

differences explicable in terms of different land-use?

Results leading to the fulfilment of the first objective have been reported elsewhere (Institute of Hydrology, 1976); it is the purpose of this paper (a) to describe the modelling methods by which the second objective is being achieved, and how they may be extended to areas of different topography, soil type and underlying geological structure; (b) to describe methods by which the extrapolation may make use of remote sensing methods.

It must be emphasized at the outset that the modelling studies at present in progress do not make use of remotely-sensed data: the application of remote-sensing methods is unnecessary in research catchments that are both well-mapped and readily accessible by field staff. Where predictions are required of the effects of changing land-use on flood flows on areas that are either poorly mapped or remote, it appears possible that remote-sensing methods will find considerable application.

3 THE PROCESSES TO BE MODELLED

The physical processes requiring mathematical description and which affect the transformation of storm precipitation to streamflow in the Plynlimon catchments are (a) interception of precipitation by vegetative canopies; (b) infiltration of water falling through the canopy, and down plant stems, to the soil surface; (c) runoff from areas of the catchment where the net precipitation (throughfall plus stemflow) exceeds infiltration; (d) vertical and lateral movement of soil moisture through the soil, regarded as a porous medium; (e) movement of water into stream channels and along their length. Where flood flows are to be predicted for a heavy storm of given intensity, however, certain of these processes are of minor importance; the vegetative canopy, for example, can be regarded as a reservoir which must be filled before net precipitation can begin to infiltrate soil, but since the depth of this reservoir is no more than a few millimetres even for a forest stand, the small abstraction from precipitation which it causes can be safely neglected. (Similarly, transpiration losses, which are rarely more than a few millimetres per day at Plynlimon, even in summer, can be safely neglected for the few hours of a heavy storm's duration and have therefore not been listed as a process to be modelled when the objective is to predict storm flows). For the purposes of this paper, we therefore concentrate on three processes affecting the magnitude and distribution in time of streamflow following a heavy storm: (a) infiltration into the soil surface and through the soil profile; (b) surface runoff; (c) channel flow.

4 THE MODEL

The catchment is divided into areas represented in the model by 'slope elements' or 'channel elements'. A slope element represents part of a hillside and consists of a layer of soil with constant thickness resting on an inclined impermeable stratum. The area and inclination of the rectangular slope element are arranged to be equal to the area and average inclination of the hill slope it represents. Channel elements represent sections of the major streams, and are taken to be straight with rectangular cross-sections. The length of a channel element is the shortest distance between the upstream and downstream boundaries of the corresponding section of stream and its width is the average width of the stream (Figure 1).

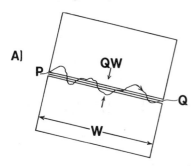

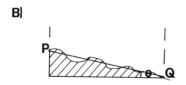

Figure 1   Calculation of length and width of a channel element

The purpose of this division is to simplify the application of equations of flow used in the model. These equations are solved for each element separately, and within each element the roughness and slope of the surface and the depth of the soil are assumed constant. Soil properties are allowed to vary within an element if this is necessary. There are many ways of

dividing up a catchment into elements; the geometric divisions between slope elements are taken to lie along the main streams, along ridges, or at interfluvial plateaux. Figure 2 shows a map of slope type and curvature in the Wye catchment (from Newson, 1976) and Figure 3 shows the slope and channel elements obtained as a result of geometric division.

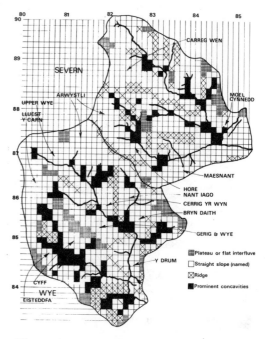

Figure 2   Slope type and curvature, Plynlimon experimental catchments (from 1:25,000 map)

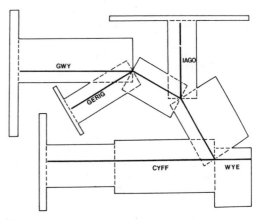

Figure 3   Slope and channel elements for the Wye catchments

433

One tributary, the Cyff, has been divided
into two channel elements and its valley
sides into four slope elements; this was
because there is a marked change in average
slope between upper and lower reaches.
Soils, which could form another criterion
for division of the catchment into slope
elements, vary over the Plynlimon catchments as shown in Figure 5; elements could
also be defined according to vegetation
(Newson, 1976, Figure 4). Clearly if all
three criteria (slope, soil type,
vegetation) were used together, the number
of elements would become very large; slope
elements have therefore been so delineated
that no great change in vegetation or soil
type occurs within them.

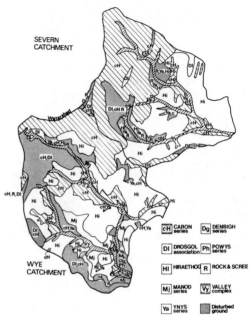

Figure 5 Soils of Plynlimon experimental
catchments (prepared by
C C Rodeforth, Soil Survey of
England & Wales)

5 INFILTRATION

The equation of flow of water in a non-
swelling porous medium is:

$$C\frac{\delta\psi}{\delta t} = \nabla.(K \nabla \psi) + \frac{\delta K}{\delta z} \qquad (5.1)$$

Here, ψ is the capillary pressure potential,
K the hydraulic conductivity and C the
specific moisture content, $\delta\theta/\delta\psi$. C and K
are functions of ψ. For saturated flow,
with $\psi \geqslant 0$, K is a constant and C = 0.
Thus, for saturated soil,

$$\delta\nabla^2\psi = 0 \qquad \psi \geqslant 0 \qquad (5.2)$$

Equation (5.1) can be written in a non-
dimensional form

$$C^* \frac{\delta\psi^*}{\delta t^*} = \nabla.(K^*\nabla\psi^*) + \frac{\delta K^*}{\delta z^*} \qquad (5.3)$$

where

$$K^* = \frac{K}{K_{sat}} \quad \psi^* = \frac{\psi}{\psi_o} \quad \theta^* = \frac{\theta}{\theta_{sat}} \qquad (5.4)$$

$$x^* = \frac{-x}{\psi_o} \quad y^* = \frac{-y}{\psi_o} \quad z^* = \frac{+z}{z_o}$$

$$t^* = \frac{+K_{sat} \, t}{\psi_o \theta_{sat}}$$

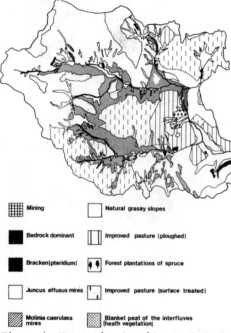

Mining

Bedrock dominant

Bracken(pteridium)

Juncus effusus mires

Molinia caerulaea mires

Natural grassy slopes

Improved pasture (ploughed)

Forest plantations of spruce

Improved pasture (surface treated)

Blanket peat of the interfluves (heath vegetation)

Figure 4 Vegetational guides to hydrology
of Wye catchments (prepared by
M D Newson)

Water input to a slope element may flow
through the soil or over the surface.
Flow on the surface of a slope or in a
channel may be described by the St Venant
equations for shallow water flow over a
plane surface. Flow through the soil as
infiltration and throughflow, is described
in the model by Richard's equation for
flow in a porous medium.

are dimensionless variables; x, y, z are space co-ordinates, t is time and θ is the volumetric water content. The normalising factors are the hydraulic conductivity and moisture content at saturation, K_{sat} and θ_{sat}, and the air entry capillary potential, ψ_o.

Several authors have suggested relationships of the form

$$C^* = \alpha(\psi^*)^{-\beta} \qquad \psi^* > 1$$
$$= 0 \qquad\qquad \psi^* \leqslant 1 \qquad (5.5)$$

$$K^* = (\psi^*)^{-\gamma} \qquad \psi^* > 1$$
$$= 1 \qquad\qquad \psi^* \leqslant 1 \qquad (5.6)$$

for the dimensionless soil characteristics and give empirically or theoretically derived functions connecting the constants α, β and γ. Ghosh (1977), for example, gives the following empirical equations derived from measurements on a number of contrasting soils:-

$$\beta = \alpha - 1 \qquad (5.7)$$
$$\gamma = 3 - 6a$$

Figures 6 and 7 show the experimentally determined characteristic curves for Yolo light clay and the corresponding theoretical curves with $\alpha = -0.1976$.

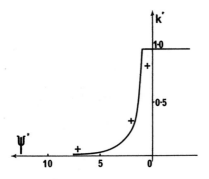

Figure 6

If we suppose that this sort of description holds reasonably well for the range of soils to be modelled, then the form of the solutions for unsaturated flow can be discussed in terms of one parameter only.

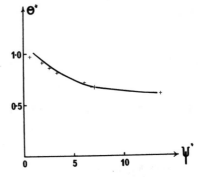

Figure 7

Equation (5.3) can be simplified using the Kirchoff transformation

$$v^* = \frac{1}{V} \int_{\psi^*_{min}}^{\psi^*} K^*(\psi^*)d\psi^*$$

$$V = \int_{\psi^*_{min}}^{\psi^*_{max}} K^*(\psi^*)d\psi^* \qquad (5.8)$$

The transformed equation is

$$F^*(v^*) \frac{\delta v^*}{\delta t^*} = \nabla^2 v^* + G^*(v^*) \frac{\delta v^*}{\delta x^*} \quad (5.9)$$

where $F^*(v^*) = \alpha(\psi^*)^{0.5-2\alpha}$

$$G^*(v^*) = (3\alpha - 1.5)(\psi^*)^{-1}$$

It becomes evident that the relative importance of the diffusion and conduction terms depends on $|\alpha|$. For example, if $\alpha = -71$ (Botany sand), then

$$\frac{\delta v^*}{\delta z^*} G^*(v^*) \simeq 0(10^2)$$

and

$$\nabla^2 v^* \simeq 0(1) \text{ for } \psi^* \simeq 0(1).$$

Diffusion is much more important than conduction. Most soils have much lower values of $|\alpha|$ however, so that the two terms are of much the same importance. Figure 8 shows the normalized capillary pressure at a distance $z^* = 1$ below the surface of a soil layer as a function of normalized time t^*. The values of α range from -0.1 to -0.7 covering for example, Ida silt loam (-0.1), Yolo light

435

clay (-0.2) and Adelanto sandy loam (-0.3). The soil is initially unsaturated with $\psi^* = 1.5$ at all points, and a rainfall input greater than the infiltration demand is applied to the surface. The normalized rate of increase in soil moisture is greatest for the lowest value of $|\alpha|$.

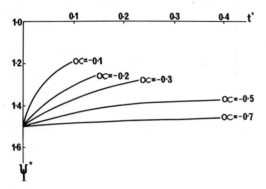

Figure 8

Soil characteristics are, of course, derived from measurements on fairly small homogeneous samples. Within a slope element of $\simeq O(10^5 - 10^6 \text{ m}^2)$ there will be considerable variation in soil type and structure; one of the most noticeable features of the structure of the peat at Plynlimon is the system of pipes which run in a general downslope direction, with a spacing $\sim O(1-10\text{m})$. On a scale larger than this spacing, the 'effective' hydraulic conductivity will be greater than that of a small peat sample. The form of the effective functions $C(\psi)$ and $K(\psi)$ to be used for a slope element in the model must be determined on the large slope element scale. However, once this has been done for a particular type of land use on a given soil, it seems reasonable to suppose that the results can be used for similar soils in other catchments.

6 OVERLAND FLOW

The St Venant equations for shallow water flow over a sloping plane are

$$\frac{\delta h}{\delta t} + u\frac{\delta h}{\delta i} + h\frac{\delta u}{\delta i} = q \qquad (6.1)$$

which is the continuity equation; and

$$\frac{\delta u}{\delta t} + u\frac{\delta u}{\delta i} + g\cos\theta\,\frac{\delta h}{\delta i} = g(\sin\theta - S_f) - q\frac{u}{h} \qquad (6.2)$$

which is the equation for conservation of momentum. The quantities h and u are the depth and velocity of the water layer, i is the downslope coordinate, θ the angle of the slope, q the rainfall less the infiltration (or plus seepage) and S_f is the friction slope. S_f is defined by the Chézy equation

$$S_f = \frac{u^2}{C^2 h} \qquad (6.3)$$

C, the Chézy roughness, is assumed to be a constant for a given slope element. The input q is a function of i and t.

The equations may be written in a dimensionless form

$$\frac{\delta h^*}{\delta t^*} + u^*\frac{\delta h^*}{\delta i^*} + h^*\frac{\delta u^*}{\delta i^*} = 1 \qquad (6.4)$$

$$\frac{\delta u^*}{\delta t^*} + u^*\frac{\delta u^*}{\delta i^*} + \frac{1}{F_o^{\,2}}\,\frac{\delta h^*}{\delta i^*} = k\{1 - \frac{u^{*2}}{h^*}\} - \frac{u^*}{h^*} \qquad (6.5)$$

where u^*, h^*, i^* and t^* are dimensionless variables.

It is now apparent that there are only two independent parameters to be specified when the equations are solved; these are the Froude number

$$F_o = C\{\frac{\tan\theta}{g}\}^{\frac{1}{2}} \qquad (6.6)$$

and

$$k = \{\frac{g^3 L\sin\theta}{C^4 q^2}\}^{-1/3} \qquad (6.7)$$

L is the distance down the plane at which the velocity u and depth h of the water layer are assumed to be related by the equation $u^2 = C^2 h$, and at which the dimensionless variables u^*, h^* both equal unity.

Figures 9 to 15 show normalised rising and recession hydrographs for steady rainfall and various values for k and F_o. The dotted line shows the hydrograph which would be obtained using the kinematic wave approximation, in which all the terms of the momentum equation except the body force $g\sin\theta$ and the retarding fraction gS_f are assumed to be negligible. The momentum equation (6.2) simplifies to

$$u^2 = C^2 \sin\theta h \qquad (6.8)$$

The shape of the hydrograph can vary considerably with F_o when k is low, but for the hydrographs approach the kinematic wave hydrograph when k is large.

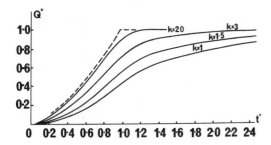

Figure 9

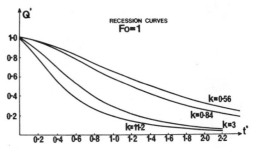

Figure 13

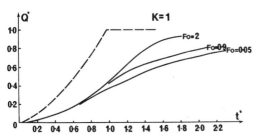

Figure 10

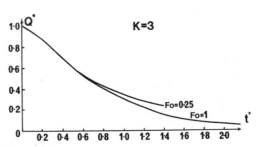

Figure 14

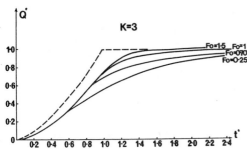

Figure 11

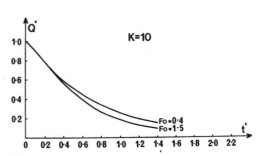

Figure 15

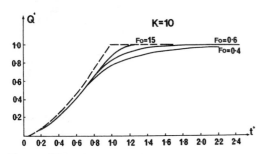

Figure 12

In the kinematic wave approximation, the normalised rising hydrograph is given by

$$Q^* = \{t^*\}^{3/2} \qquad (6.9)$$

where $Q^* = \dfrac{Q}{qL} \qquad (6.10)$

and $t^* = \{\dfrac{\sin\theta C^2 q}{L^2}\}^{1/3} t \qquad (6.11)$

For most conditions in the Plynlimon catchment, the kinematic wave approximation will give a reasonable estimate of the depth and velocity of overland flow down to the 10 m scale. The Chézy roughness of grass is $C \simeq 3$ m $^{\frac{1}{2}}$s^{-1} and a normal

437

storm rainfall is $q \simeq 10^{-5} ms^{-1}$; so for a slope with $\sin\theta = 0.1$ $k = 4885$ on the 10 m scale. A larger storm (for example the August 1977 storm at Plynlimon with $q = 2.6 \ 10^{-5} ms^{-1}$) on a shallow bare slope ($\sin\theta \simeq 0.01$ $C \simeq 30 m^{\frac{1}{2}} s^{-1}$) would lead to a value of $k \simeq 56$ on the same length scale.

Since the slope elements will contain small streams, concavities, sub-surface cracks and pipes, all of which may provide faster pathways for the rain than overland flow on the grass, the effective value of the Chézy roughness coefficient for a whole element may be greater than the value for the vegetation itself. White and Jaywardena (1975) used an optimised value of $C \simeq 30 m^{\frac{1}{2}} s^{-1}$ for the grassy slope of the Wye. C must be determined from field experiments on the slope element scale in terms of the vegetation type and the mesoscale geometry of the element surface. Table 1 shows the value of C for various surfaces (from Woolhiser, 1974).

Table 1. The value of C for various surfaces

Surface	Chezy $C/m^{\frac{1}{2}}s^{-1}$
Concrete or asphalt	40 - 21
Bare sand	36 - 18
Gravelled Surface	21 - 10
Bare Clay-Loam Soil (eroded)	20 - 8.9
Sparse Vegetation	6.1 - 2.8
Short grass prairie	3.6 - 2.0
Blue grass sod	2.3 - 1.0

The variation in C is about 30% within each class but errors in C will be larger than this on the slope element scale.

If the input rainfall is known the errors in the estimation of F_o and k are

$$\frac{\Delta F_o}{F_o} \sim \frac{\Delta C}{C} + \frac{1}{2} \ cosec \ \theta \ sec \ \theta \ \Delta\theta \quad (6.12)$$

$$\frac{\Delta k}{k} \sim \frac{1}{3} \frac{\Delta C}{C} + \frac{1}{3} cot\theta \ \Delta\theta - \frac{4}{3} \frac{\Delta C}{C} \quad (6.13)$$

As $\theta \to 0$ the slope error term becomes more and more important; this problem shows up in the kinematic approximation also. For the rising hydrograph, the error in the output is

$$\frac{\Delta Q}{Q} \sim \frac{\Delta C}{C} + cot\theta \ \Delta\theta \quad (6.14)$$

Again, as $\theta \to 0$ any error in the slope is more important than the error in estimating the roughness.

7 CHANNEL FLOW

The equation for channel flow are again the St Venant equations (6.1) and (6.2), this time with q representing the lateral input per unit width of the stream. The channel element is assumed to have a constant rectangular cross-section and a constant slope. These "effective" values for the channel element have to be derived from point values for the real stream and the Chezy roughness can be estimated from visual inspection of the river. (See, for example, the U.S. Geological Survey's photographic catalogue of rivers classified according to their roughness).

The design study for the Plynlimon flumes used a value of $C \sim 20 m^{\frac{1}{2}} s^{-1}$. Given a storm of $10^{-5} \ ms^{-1}$ on two lateral hill slopes of length ~ 500 m leading to a stream of width 1 m, the lateral inflow per unit width of stream is $\sim 10^{-3} ms^{-1}$. On the length scale $L \sim 30$m and with slope $\sin\theta \sim 0.03$, we have $F_o = 1.11$ (supercritical flow) and $k = 2.37$; if the solution is to be fully distributed, the full St Venant equations must be used since k is so small. Even if a valid solution is required only on the channel element scale, $L \sim 1000$ m, k is still rather small (7.63) for the kinematic wave solution to be accurate.

8 APPLICATIONS FOR REMOTE SENSING METHODS IN WATERSHED MODELLING

Whether surface runoff and channel flow are described by the full St Venant equations or by kinematic wave approximations, the following procedure is adopted when formulating a distributed model of the surface hydrological behaviour of a basin:-

1. The physical extent of the basin is determined;
2. The topography of the basin is idealized, for example by representing each valley slope as a rectangular 'slope element'. The delineation of the elements is achieved by eye; in the case of the Institute of Hydrology's model of the Plynlimon catchments, each slope element is regarded as a layer of soil of constant thickness resting on an inclined stratum of bedrock with negligible permeability. The area

and inclination of a slope element is arranged to be equal to the area and average inclination of the area of hillside that it represents;

3 The geometry of the stream channel is idealized, for example by representing it as a series of line segments, each of which has uniform rectangular cross-section;

4 Estimates are obtained for parameters describing the hydraulic roughness of the slope elements and stream channels.

5 Impermeable areas within the basin are delineated, on which all rainfall becomes flow over the surface.

To model the response of those areas of the basin that are not impervious requires the separation of rainfall into a part which infiltrates and the remainder which flows over the surface. The greater the extent of saturated soil over the basin's surface, the greater will be the surface runoff component and the smaller will be the infiltration; the state in which the entire basin is saturated provides an upper bound to that component of precipitation input that leaves the basin as rapid runoff. As a storm proceeds over a basin that is not entirely saturated, the extent of the saturated area will increase; provided that the soil moisture characteristic (relating volumetric water content θ to capillary pressure potential ψ) and the unsaturated hydraulic conductivity characteristic (relating $K(\psi)$ to ψ) are known over the basin area and throughout the soil depth, the increase in saturated area can be predicted.

Application for remote sensing methods in formulating such a distributed model of basin hydrological behaviour therefore include the following:-

delineation of basin boundaries;
division of basin area into slope elements, and estimation of the area of each;
estimation of mean slope within each slope element;
estimation of the roughness parameter appropriate to each slope element;
delineation of the stream channel and its cross-sectional geometry;
delineation of impervious areas within the basin;
delineation of areas of saturated soil within the basin.

We consider, below, the application of remote sensing to these aspects of watershed modelling.

9 APPLICATIONS OF REMOTE SENSING METHODS TO WATERSHED MODELLING

Although many remote sensing methods of providing river catchment physical data are economically best suited to regional scale studies, they may in some instances still remain cost effective for small catchment area surveys, especially if detailed information is required. No single method of remote sensing can be recommended as being suitable for all situations and therefore it is essential that modelling requirements be clearly defined in terms of measurement accuracy in order that the most suitable sensing technique can be applied.

9.1 Delineation of Watershed Boundaries

For the Plynlimon flood prediction study, catchment modelling is carried out on a relatively small scale which requires accurate topographic information. This requirement was satisfied by the commission of a 1:5000 scale aerial photogrammetric survey of the 1925 ha catchments. A topographic map was produced with contours at 2.5 m intervals over a vertical range of about 400 metres. This high quality base map has subsequently been used for the delineation of catchment and sub-catchment watersheds, for the delineation of slope elements and measurement of slope element detail, and for extraction of stream channel information. When accuracy is a prime requirement, therefore, photogrammetric surveying using stereo aerial photography should be given first consideration regardless of the physical characteristics of the area of interest.

Where lower standards of accuracy can be tolerated, cheaper forms of sensing should first be considered. In areas of medium or high relief, watersheds can often be adequately defined from conventional non-stereo aerial photography; however, as an aid to the enhancement of surface texture and relief, photographs should be taken at a time of low sun angle. Ideally the sun should be normal to the longest side of the river catchment area so that the majority of the watershed is defined by the highlight/shadow interface. Exposure and development of the film should be controlled so as to retain image detail within the shadow areas, and in this respect, colour film may prove very suitable (Higuchi and Iozawa, 1971). In poorly mapped areas where satellites can supply low cost imagery of adequate detail (eg Landsat), imagery acquired in autumn or spring should be chosen when sun illumination angles are low. Using Landsat imagery, maps can be constructed at scales of up to 1:100,000 whilst conforming to acceptable cartographic accuracies. (Mott 1973).

Aerial imaging in visible and near visible wavelengths, especially when required for photogrammetric purposes, requires very clear and cloud free atmospheric conditions. For regions where weather conditions are persistently poor, all-weather active microwave (radar) sensing offers a valuable alternative for the delineation of watershed boundaries. Unlike short wavelength sensing where the electromagnetic return signal is highly dependent upon the chemical composition of the body, the return signal from an active microwave sensor is less affected by body composition, but is highly dependent upon its surface physical state. This is because the ground surface is illuminated from a single source (the antenna) and the intensity of the reflected return signal is largely dependent on the angle of incidence of the body surface to this source. Thus hill slopes facing the antenna will appear brighter than similar slopes falling away from the antenna (see Fig 16) with the result that interfluvial boundaries are strongly accentuated in a way similar to low sun angle photography. Although algorithms can be constructed to allow the calculation of ground scale from radar images, inaccurate image positioning can still occur as a result of layover relief effects or variable errors caused by uncorrected platform movements or navigational drifts. It is desirable therefore to have accurate base maps onto which radar data can be projected, rather than relying on it as a basic mapping system. As a result of the high cost of radar imaging from aircraft, its use for catchment delineation purposes would normally be restricted to large area surveys, probably of a regional scale, where the advantage of large numbers of catchment samples would tend to outweigh the loss of performance in planimetric accuracy. For areas of low relief, low antenna slant angles would be used in order to enhance even slight variations in slope (Deane, 1973). Catchment boundaries should normally be easily identified on radar imagery, but care must

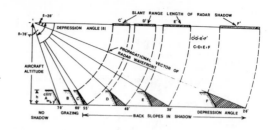

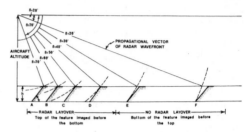

Figure 16 Layover and shadowing effects on S.L.A.R. imagery

be taken as changes in density are not solely dependent on topographic effects.

9.2 Division of Catchment into Slope Elements

Although in many instances catchment slope elements can be defined from single frame aerial imagery, the application of stereoscopic viewing enables positive identification of slope changes to be made. By visual interpretation of stereo-paired aerial photographs, adequate grouping of slope elements should be possible without resorting to the costly height contouring process; 5 or 6 categories of surface steepness should be achievable. As with the delineation of catchment boundaries, a low sun angle will cause shadows to be cast where changes of slope occur, but some loss of detail may result in areas of deep shadow. In order to achieve similar classification accuracies on all slope aspects, it would be advisable to acquire at least morning and evening imagery so that all points of interest are suitably illuminated. By utilising the on-board illumination source of a side-look radar system, for example, a series of different low illumination angles can be achieved in a single sortie, thereby reducing time in the air. Major slope elements should appear as areas of fairly homogeneous image density, but because of the complexities of radar interpretation

440

it is advisable that some form of photographic record also be taken as an added
check. One of the advantages of side-look
radar over high resolution photography
is that unwanted image clutter, such as
small scale surface texture effects, need
not be recorded if longer wavelengths are
chosen, and thus subtle slope variations
may become more evident. These effects
can be further eliminated by the use of
multifrequency radars so that only major
image variations remain.

Computer controlled classification of
slope elements is quite feasible using
either multifrequency radar data or
automatic contouring systems based on
photogrammetric quality stereo photographs,
but unless vast quantities of data are
to be processed, it is unlikely that costs
could be justified for what is essentially
a simple image interpretation process.

Before the length and area of slope
elements can be determined from remotely-
sensed data, the mean slope of each element
must first be measured.

9.3 Determination of Mean Slope of Slope Elements

Whilst classification of slope steepness
can often be achieved on an empirical
basis, an empirical measure of slope angle
would generally introduce too great an
error into the model. It may be possible
to overcome this by sufficient ground
sampling of mean slope in typical slope
element groups, but in the majority of
cases a direct measure of mean slope angle
will be required from the remotely sensed
data. Accurate aerial measurement of slope
angle can only be achieved over large areas
by stereoscopic imaging techniques. The
most accurate form is, as previously
mentioned, photogrammetry. Where standard
stereo-plotting machines are used, height
contour maps must first be produced from
which mean slope angles can be calculated.
Alternatively, slope angles between sample
spot heights taken within each slope
element can be averaged if a contour map
is not required. The measurement of slope
angle can also be achieved by radar
stereoscopy, after suitable data processing.
When an object is imaged twice at two
different look angles from the same
altitude, or at two different altitudes
from the same look direction, then radar
stereoscopy can be obtained, (Matthews,
1974) where dual beam radar is available,
single flight stereosocopy may be possible
(Figure 17).

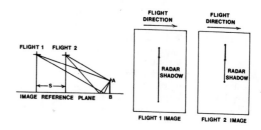

a) TWO FLIGHT STEREO RADAR TECHNIQUE

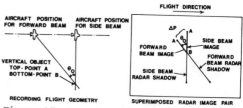

b) SINGLE FLIGHT STEREO RADAR TECHNIQUE

Figure 17   Geometry of stereo radar

Once again, multifrequency or continuously
variable frequency radars are most
suitable, as surface roughness and dielectric constant variation effects can be
reduced. However, in areas with steep
mean slope angles, radar showing may
result in a significant loss of data
which may be sufficient to negate the use
of stereoscopy. In areas of shallow
slope, this technique is worthy of consideration as the determination of θ
becomes more critical as 0° is approached.
Slope measurement may therefore be quicker
by radar, especially where forest or
other tall vegetation would cause problems
with photogrammetric contouring. Rectification of radar data is often carried
out by digital processing, so it would be
a small step to achieve direct machine
calculation of mean slope angle of given
slope elements.

If photographic imagery is handled
through a stereoplotter fitted with a
co-ordinate digitising facility,
programmes could be developed to calculate
mean slope angle directly rather than by
reference to the produced contour map.
(Proceedings of Commonwealth Survey
Officers Conference, London 1975.

9.4 Determination of Roughness Parameters for Each Slope Element

Only active microwave sensors offer a
direct method of estimating surface
roughness, due largely to the fact that

441

their wavelengths are of a similar scale to surface variations. Although it has been shown possible to classify surface roughness values using single frequency radars (MacDonald & Waite, 1973), variations due to changes in surface soil moisture, vegetation, etc cannot be adequately accounted for, and therefore multi-frequency dual polarised systems are favoured. There remains also the problem of relating remotely-sensed measures of surface roughness to the Chezy coefficient C.

When a radar pulse is incident on completely smooth surface, the proportion of the signal which is not absorbed by the surface is reflected in one direction only - the specular angle (See Fig 18).

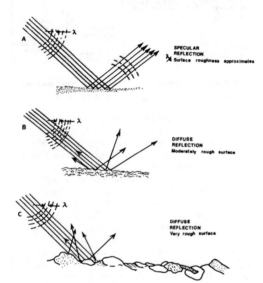

Figure 18   Reflection of electromagnetic radiation from surfaces of different roughness

As the surface becomes rougher in relation to the incident wavelength, the proportion of the signal reflected through the specular angle is reduced and signal scattering in all directions occurs. A surface is considered rough if its undulations are comparable in size to the wavelengths of the microwave sensor. By transmitting radar signals which are either vertically or horizontally polarised and subsequently measuring the degree of scattering of the return signal away from that polarised plane, an indication of whether the surface is rough or smooth in relation to the transmitted wavelength is obtained. If a series of different wavelengths pulses is applied to a given surface, their return signals can be analysed to assess the actual surface roughness of

the reflecting surface. For very rough and complex natural surfaces, such as trees, the varying wavelength may not have a great deal of effect on return signal scatter, but for fairly uniform surfaces such as grass, crops, or bare soil, a good measure of surface roughness is possible.

If radar sensors are not available to provide a measure of surface roughness, indirect sampling methods must be used. Land use types can be identified using remote sensing techniques as described in section 9.5, and these can be further sub-divided if necessary to account for local surface variations. Once the main surface elements have been delineated, sample areas which are representative of each element can be chosen from the aerial data. Ground measurements of surface roughness can then be made at these selected points in order to establish relations between the Chezy coefficients C shown in Woolhiser's table, and the remotely-sensed indices of surface roughness.

9.5   Land Use Classification for Delineation of:-

(a)   Soil Characteristics and Sub-surface Conductivity

For both cultivated and relatively unworked land such as the upland grassland areas of Plynlimon, variations in subsurface soil types and their soil moisture regimes, which affect the infiltration and throughflow component, are frequently reflected in variations in surface vegetation. Such relationships have long been recognised, and much work has been carried out in relating species types and their condition to the effect of near surface soil state (Nefedov & Popova   , Chikishev, 1965) and similar criteria were used by Newson in his soil classification of Plynlimon. Verification of postulated reasons for vegetation distribution is necessary in the form of ground sampling of soil capillary conductivity and specific moisture, but in areas of simple lithology such as Plynlimon the relationships are so well defined that extensive sampling is not necessary for accuracies acceptable for this type of catchment model. In the Wye grassland catchment of Plynlimon, this vegetation/subsurface relationship is defined well enough for mapping to be carried out from the vantage point of hill tops. However, vertical aerial photography would enable better cartographic accuracy of vegetation plots to be achieved, and at this site the use of natural colour photography would be ideal for the purpose. In addition, colour infrared film may be used to advantage in that, within a given plant species, changes in foliage area, density and condition can often be detected as variations in the magenta near infrared sensitive layer of the film. Such changes in plant condition may be indicative of hydrologically related factors such as available soil moisture, soil drainage conditions etc. For example, on the hill slopes where bedrock is near the surface, the main routes of subsurface water flow are strongly delineated by vegetation flushes.

For large area surveys, semi-automatic pattern or texture recognition techniques can be applied to single band imagery in order to identify areas of similar surface type. Where colour or multispectral images are available, texture effects can be combined with spectral signatures to allow more positive identification of ground types (Herzog 1975). Spectral techniques alone are of course capable of providing very accurate land use classifications (Hoffer *et al* 1971) but with the addition of texture grouping, regions of different lithology may be more easily pinpointed. Visual interpretation of imagery will generally provide adequate information to enable slope element groupings to be made, but where more specific information regarding individual species distributions are required, computer analysis of spectral data may be necessary. Care must be taken that too many data are not gathered, causing the number of elements in the model to become unworkable. In this respect, multispectral analysis of satellite data could provide satisfactory land use inputs for this type of model.

(b) Impervious Surfaces

Impervious areas such as exposed bedrock or man-made structures like roads and buildings, can be readily identified on most types of remotely-sensed data. Man-made structures are normally characterised by their smooth boundaries, geometric shapes, and frequently even tone. In thermal wavelengths, impervious surfaces are normally at contrast to water or vegetated surfaces, due to their lower thermal inertia and lack of evaporative cooling. Because of these marked contrasts, automatic recognition of impervious areas is generally straightforward,

using either edge finding, spectral, texture or thermal classification techniques (Haralick and Shanmugan, 1973). After classification of impervious surfaces, ground investigations should be carried out to establish how they react to intensive rainfall events, especially with regard to routing of storm water.

(c) Saturated Soil

As storm water throughflow rates are much higher in saturated soil conditions, it is desirable to delineate both areas which are in a permanently saturated state, and those which will achieve saturation very quickly. For bare soil conditions, wet zones exhibit a markedly lower reflectance in visible wavelengths. This relationship is not always meaningful, however, as soil surface effects as recorded by the majority of sensors give little indication of sub-surface conditions. For example, at times of high evaporation a thin crust of dried soil can be formed even over very wet subsoil. Hoffer and Johannsen (1969) noted that this could occur within a few hours of heavy rainfall. Thermal sensing has been widely used for the delineation of saturated areas, the most successful technique being through the measurement of soil thermal inertia. In theory, high diurnal temperature ranges should result from dry soils and low diurnal temperature ranges from wet soils. In practice, however, ground temperature is also dependent on air temperature, wind velocity, surface roughness, ground exposure, and most importantly, the nature of any vegetation cover. Wet areas will only be easily identifiable when they are at extreme contrast to their surroundings or when atmospheric and surface conditions are very stable.

For direct delineation of saturated zones, only long microwave wavelengths appear capable of providing other than surface information, and they have the potential to integrate sub-surface moisture values from a depth which largely dependent on the degree of soil saturation. Tests by Schmugge *et al*, (1974) and others suggest that wavelengths in excess of 20 cm are most suitable for soil moisture evaluation in the top few centimetres of soil and that vegetation up to 15 - 20 cm high does not significantly effect the basically linear response relationship of signal return to soil moisture. With regard to choice of sensor, a multi-frequency dual; polarised system would be most successful in separating soil dielectric constant effects from those of surface roughness, slope angle etc.

443

If one considers the present operational cost of this type of sensor and the difficulty of obtaining such equipment for non-military purposes, one realises that for applications such as catchment modelling, more empirical measures will have to suffice, unless regional scale studies are envisaged. For small catchment studies, it is likely that vegetationally derived measures of soil moisture will be more appropriate. Plant root systems react to soil conditions throughout the root zone so that indications of soil moisture in the top 20 - 30 cm of soil at least can be expected.

9.6 Determination of Stream Parameters

Again, when faced with the problem of remotely measuring slope angle, photogrammetric techniques must first be considered, as they are undoubtedly the most comprehensive information on stream bed configuration. In order to obtain all of this information, however, quite large-scale imagery may be required and it may be prohibitive to provide this for the whole of a river network. The most practical approach, therefore, would be to carry out an aerial survey of medium accuracy for the whole river network, and then select representative sections where high-accuracy measurements can be carried out. For this to be achieved satisfactorily at least two overflights would be required, the higher accuracy work being undertaken after analysis of the general survey. In remote regions, this may be impracticable, and so 'representative' stream sections must be chosen *a priori* from whatever information is available. Where base maps are unreliable or non-existent, satellite imagery may prove to be invaluable for this purpose, and even for well mapped areas, their initial use can often provide a better understanding of the main topographic influences which should be included in the catchment model. Sensing in the near-infrared region at times of high-water level will provide the best land/water definition from which channel widths and patterns can be determined. Photogrammetric measurements can then be taken along the land/water boundary and mean stream slopes determined. Marked variations in slope will become apparent during this process. The use of colour infrared film is recommended for most situations, and its advantages over normal colour film become apparent when direct stream visibility is impaired by overhanging vegetation etc.

Stream segment boundaries are normally drawn at stream junctions which can be readily identified, and also at points where obvious changes in slope, cross-sectional dimension or channel roughness occur. If imagery is taken at times of low water level, considerable information on channel state can be obtained and, for example, photogrammetric cross sections can then be taken at given intervals along the channel length. The mean cross sectional area for each channel segment can thus be simply derived. Where river water levels do not fall appreciably in dry conditions, sensing of underwater features may be attempted. True colour film has been shown to have good water penetration capabilities (Lockwood *et al*, 1974) but narrow band sensing may be more successful in revealing stream bed conditions as film sensitivity, through suitable filtration, can be 'tuned' to suit the prevailing water conditions. In clear water, maximum water penetration occurs in the blue/green region around 0.5 microns whilst for more heavily sedimented water a colour shift generally towards the yellow will yield best results, depending on the sediment colouration. Underwater photogrammetry has been used extensively in coastal waters, but little work appears to have been done in small river situations. This technique cannot be considered where sediment concentrations are very high or where high turbulence reduces the water clarity. Photogrammetry is also easier when the river bed possesses a reasonable amount of texture due to the presence of rocks or vegetation; smooth muddy bottoms presenting the greatest problems.

In rivers unsuitable for photogrammetric determination of cross sectional area, laser profiling could be considered. With this technique a pulsed coherent light source is aimed vertically from any sensor platform so as to strike the stream surface at right angles (Figure 19). A strong reflected signal is received from the water surface followed closely by a weaker reflected signal from the river bed. The calculation of water depth is dependent on the time lapse between the strong and weak reflected return signals. Laser profilers allow the most accurate measurement of water depth, but only at a point or along a flight line. Profiles are therefore best used in conjunction with imaging depth estimation techniques where they provide an accurate depth reference along the centre of the image.

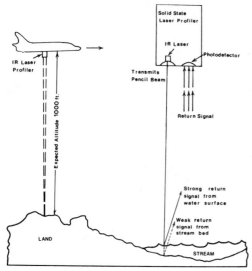

Figure 19

10 CONCLUSION

Remote sensing methods undoubtedly have applications in catchment modelling, particularly for the derivation of surface characteristics of the basins to be modelled; their use will become more extensive where their cost effectiveness can also be demonstrated.

11 REFERENCES

Chikishev, A. G. (ed.) 1965, Plant indicators of soils, rocks and subsurface waters. New York, Consultants Bureau.

Deane, R. A. 1973, Side-looking radar systems and their potential application to earth-resources surveys, basic physics and technology, ESRO Contractor Report No. 136.

Deij, L. J. L. 1956, Contribution to a discussion on evaporation, Neth. J. agric. Sci. 4: 92

Ghosh, R. K. 1977, Determination of unsaturated hydraulic conductivity from moisture retention function, Soil Science 124: 122

Haralick, R. M. & K. S. Shanmugan 1973, Combined spectral and spatial processing of ERTS imagery data, Paper I-16. Symposium on significant results obtained from the Earth Resources Technology Satellite-1, Maryland, 1973.

Herzog, J. H. 1975, Textural features and their role in classification and change detection in remote sensing. Ispra, Italy, Commission of the European Committees Joint Research Council.

Higuchi, K. & T. Iozawa 1971, Atlas of perennial snow patches in central Japan. Japan, Water Research Laboratory, Faculty of Science, Nagoya University.

Hoffer, R. M. & C. J. Johannsen 1969, Ecological potentials in spectral signature analysis, Remote sensing in ecology, University of Georgia.

Hoffer, R. M. et al 1971, Application of ADP techniques to multiband and multi-emulsion digitized photograph, LARS information note 091071. Indiana, Purdue University.

of Hydrology 1976, Water balance of the headwater catchments of the Wye and the Severn 1970-1975, Report No. 33

Law, F. 1956, The effect of afforestation upon the yield of water catchment areas, J. Brit. Waterworks Ass. 38: 489-494

Law, F. 1957, Measurement of rainfall, interception and evaporation losses in a plantation of Sitka Spruce trees, IUGG/IASH Gen. Assembly, Toronto II: 397-411.

Lockwood, H. E. et al 1974, Water depth penetration film test, Photogrammetric Engineering.

MacDonald, H. C. & W. P. Waite 1973, Imaging radars provide terrain texture and roughness parameters in semi-arid environments, Mod. Geol. 4.

Mathews, E. (ed.) 1974, Active microwave workshop report. Lyndon B. Johnson Space Centre.

Millis, L. W. F. 1975, Water and water management: the developing philosophy, IHD/ICE/IH Conference, Engineering Hydrology Today, 3-9.

Mott, P. G. 1973, Mapping from ERTS and imagery, British Interplanetary Society Symposium on earth observation satellites.

Nefedov & Popova, Deciphering of Ground-water from aerial photographs.

Newson, M. D. 1976, The physiography, deposits and vegetation of the Plynlimon catchments, Institute of Hydrology, Report No. 30.

Pereira H. C., M. Dagg & P. H. Hosegood
    1962, The development of tea estates
    in tall rain forest: the water balance
    of both treated and control valleys,
    E Afr. agric. For. J. 27: 36-40.

Penman, H. L. 1963, Vegetation and
    hydrology. Tech. Comm. 53, Common-
    wealth Bureau of Soils: Commonwealth
    Agricultural Bureaux, Farnham Royal,
    Bucks.

Rutter, A. J. 1964, Studies on the water
    relations of Pinus sylvestris in
    plantation conditions. II The annual
    cycle of soil moisture change and
    derived estimates of evaporation, J.
    appl. Ecol. 1: 29-44.

Schmugge, T. *et al* 1974, Remote Sensing
    of soil moisture with microwave radio-
    meters, J. Geophys. Res. 79(2).

White, J. K. & A. W. Jayawardena 1975,
    A finite element approach to watershed
    runoff – a discussion, J. Hydrol. 27(314):
    357-358.

Woolhiser, D. W. 1977, in Mathematical
    models for surface water hydrology.
    Ciriani, T. A., U. Malone & J. R. Wallis
    (eds.) Proceedings of the Workshop
    held at the IBM Scientific Center, Pisa,
    Italy 1974, London, John Wiley.

12 ACKNOWLEDGEMENTS
This paper appears by permission of the
Director, Insitute of Hydrology.

J.MARTINEC
Federal Institute for Snow and Avalanche Research
Weissfluhjoch/Davos, Switzerland

28

Hydrologic basin models

1 INTRODUCTION

One of the main tasks of hydrology consists in developing methods for describing and forecasting the hydrological regime of rivers. This is a basic information for hydraulic engineering projects, such as the water power generation,water supply,flood control,and generally for the planning of the water management. In view of the rapidly increasing needs a rational use of water resources becomes imperative.

The hydrologic cycle is easily understood qualitatively but a quantitative assessment depends on measurements of all components and their continuous interactions. A model of a hydrological process has a better chance to approach the reality if the variables can be measured as accurately and efficiently as possible. This is where the remote sensing can significantly contribute to the progress of the hydrological science. For their part, the hydrologists are glad to prove the economical benefit of these data by applying them to solve practical problems.

Thank to the remote sensing, more extensive informations are becoming available for example on the surface temperature, runoff conditions (soil moisture, sheet flow, flooding) and in particular on the areal extent of the snow cover. The significance of this last item for runoff models will be dealt with in more detail.

2 TYPES OF RUNOFF MODELS

2.1 Definitions

Hydrologic models are mathematical formulations to simulate natural hydrological phenomena which are considered as processes or as systems (Chow 1964). This definition applies evidently to mathematical models.

Physical models, in other words laboratory models, are small-scale representations of existing or fictitious basins on which hydrographs resulting from simulated rainfall can be studied.

Representative basins can be considered as physical models in the real scale where parameters necessary for the mathematical models can be studied in undisturbed natural conditions.

2.2 Runoff components

Fig.1 shows the runoff components resulting from precipitation in a basin.

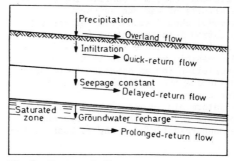

Fig.1 Types of flow resulting from precipitation (drawn after Jamieson & Amerman 1969)

Already the separation of these components, that is to say an assessment of their respective proportions in the runoff, is a difficult task. Environmental isotopes emerged as a new tool in this field and the remote sensing might also contribute to the solution. This is illustrated by Fig.2 which shows the continuously changing conditions after the start of rain.

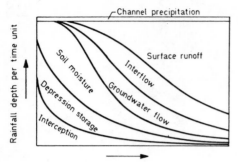

Fig.2 Temporal variations of runoff
components after the start of a rainfall
(drawn after Linsley,Kohler,Paulhus 1958)

The proportions of the surface runoff,
interflow, groundwater flow etc. are indicated only by way of an example. It is at
present not possible to measure these components directly and continuously.Even so,
the outline in Fig.3 (Freeze & Harlan 1969)
gives an idea of how the channel flow ori-
ginates. It is an example of a runoff
model.

2.3 Characteristics of models

It does not seem necessary to contribute
any further to the existing classification
of mathematical models (for ex.Chow 1964,
Overton 1977), which has become very elaborate and sometimes conflicting due to
the efforts of various authors.

A model should transform the input (precipitation into the output (discharge) in
order to approximate the reality in the
best possible way. There is no general
prescription for an approach to be used
considering the available data in the
given system.

A deterministic model should have a
structure based on physical laws and on
the measurement of the input. Such system

transforms the input into the output for
the given initial conditions without an
element of chance. Due to the complexity
of the runoff process, simplifications and
approximations are unavoidable even in
this approach. However, coefficients and
parameters of the model should be measured
or at least indirectly determined by a verification of the model results in typical
basins.

In a parametric model the values of
coefficients and parameters are determined
by a statistical procedure on a computer.
They are optimized in order to obtain the
best approximation of output from a given
input, using the available set of data.
For example, if the optimization is based
on 10 years records, it is assumed that
the performance of the model with values
of coefficients thus determined will be
good in other years, even in other basins.
At the same time, little is known about
the physical meaning of the optimized
values.

In a probabilistic model the information
on the significance and sequence of vari-
ates involved in the process is lacking
altogether and so it is ignored.The chance
of their occurrence is assumed to follow
a definite probability distribution (Chow
1964). An example of a probabilistic model
is a flow-duration curve.

2.4 Merits of the deterministic and parametric approach

A deterministic model developed and verified in a representative basin has a good
chance to perform well in another basin,
in which at least basic hydrological measurements are available. Since the magnitude of various coefficients is physically
or at least logically explained, the model
can be readily adjusted to the other basin
with the new conditions (size, altitude,
slope etc.) in mind.

However, the transfer of knowledge from

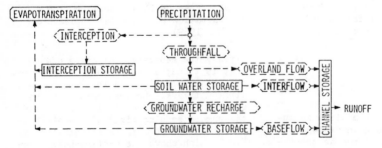

Fig.3 Graphical outline of a runoff model (drawn after Freeze & Harlan 1969)

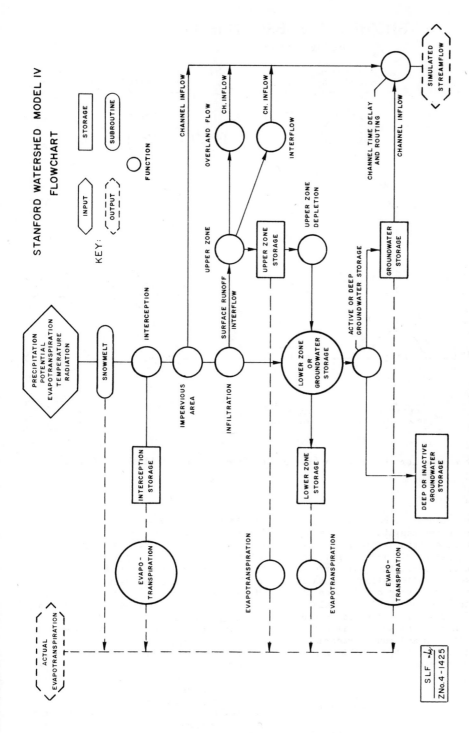

Fig. 4 The Stanford IV model (drawn after Crawford & Linsley 1966)

SNOWMELT SUBROUTINE IV
FLOWCHART

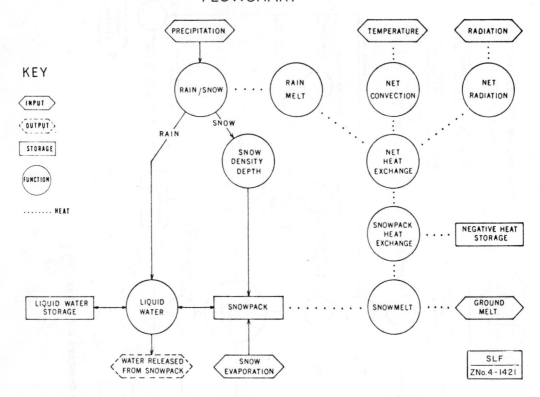

Fig.5 A subroutine for the snowmelt (reproduced from Crawford & Linsley 1966)

one basin to another one has its limits. Coefficients to be substituted in the new basin are in some cases difficult to determine and a parametric model might come to the rescue. However, the good performance achieved by optimization in a basin might result from a lucky combination of false coefficients and this is revealed by a failure if the model is applied in another basin. This can be avoided if sufficient precipitation and runoff records are available in the new basin, enabling a new optimization to be made.

2.5 Example of a general simulation model

Fig.4 shows the flowchart of the Stanford-IV model which is intended to represent mathematically the whole hydrologic cycle (Crawford & Linsley 1966). If a certain process is of a special importance in a given basin, it is dealt with in more detail by a subroutine. A flowchart of a subroutine for the snowmelt is shown in Fig.5. Certainly a universal model for the entire range of conditions to be met in all climates would be the ultimate solution. But the available data and the present knowledge of hydrologic and hydraulic pro-.cesses involved are hardly sufficient. Consequently, the increasing complexity of a mathematical model brings about difficulties of obtaining a solution. On the other hand, an exaggerated simplicity increases the risk of not representing the system (Overton 1977). This dilemma is illustrated by Fig.6.

An alternative solution is to develop a

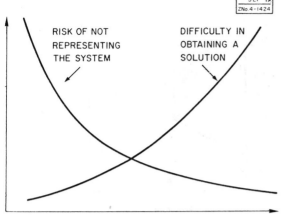

RISK OF NOT   DIFFICULTY IN
REPRESENTING   OBTAINING A
THE SYSTEM   SOLUTION

COMPLEXITY OF MATHEMATICAL MODEL

Fig.6 Effect of the complexity of
a mathematical model on the solution
(drawn after Overton 1977)

model with certain characteristic conditions in mind, for example snowmelt,which
is a dominant factor in the mountains.
A universal model needs for verification
data from all parts of the world,including
developing countries and remote areas.
This is difficult to achieve, even if conventional hydrometeorological networks are
combined with measurements by the remote
sensing. By concentrating on one important
item it is possible to formulate the needs
for improved measurements in a more con-
crete way and to obtain results quicker.

The snowmelt runoff model which will be
dealt with needs periodical evaluations of
the areal extent of the snow cover. This
can be achieved from aeroplanes or, in a
more efficient way, from satellites which
have an adequate resolution.

3 MONITORING THE SEASONAL SNOW COVER

3.1 Role of snow and ice in the hydrologic
cycle

From the distribution of fresh water on
earth in different forms of presence, it
would seem that the polar ice and glaciers
are the most important item, representing
according to estimates 24,8 million km^3 of
water or 77 % of the total. The volume of
water in lakes and rivers amounts only to
0,132 million km^3 or 0,44 % of the total
(Martinec 1976).

However, for the practical utilization
of water resources, the renewal rates of
the different modes of occurrence of water
are more important than the stationary
volumes. The renewal rate is obtained from
the equation

$$t_r = \frac{V}{v_e} \qquad (1)$$

where t_r is the average residence time
in years

V is the volume of water in the given
medium in km^3
v_e is the velocity of exchange by
input and output in km^3 per year

If the total annual discharge of glaciers
to the seas is estimated at 2500 - 3000
km^3 of water (Kotlyakov 1970),a residence
time of the order of 10 000 years results
from Eq.(1).

The average total flow of the world's
rivers is about 36 . 10^3 km^3 per year,
which gives an average residence time of
12 days for river channels only and less
than four years if the lakes are included.

It is the seasonal snow cover, formed
every year in winter and disappearing in
summer, which influences the river flow
in many industrialized countries, that is
to say in the areas where a rational water
management is especially important.

3.2 The changing areal extent of the snow
cover

Fig.7 shows characteristic depletion curves of the snow coverage in an alpine basin,which can be described by the equation

$$S = \frac{100}{1 + e^{-b(t_{50}-t)}} \qquad (2)$$

where S is the snow-covered area in %
t_{50} is time at which S = 50
b is a coefficient
e is the base of natural logarithms

The gradual decrease of the snow coverage
has two reasons:
1. The irregular deposition of snow
which results, even on a plain, in variable
snow depths. The subsequent melting of snow

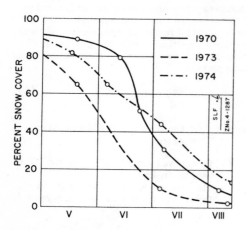

Fig.7 Depletion curves of the snow coverage in the Dischma basin in the Swiss Alps, in various years

Fig.9 Area-elevation curve of the Dischma basin with elevation zones

layers leads to the disappearance of snow in a gradually increasing part of the total area as is illustrated in Fig.8.

2. Due to the temperature lapse rate, the snowmelt is progressing from the lower parts of the basin to the upper parts. The characteristic form of the area-elevation curve of a basin shown in Fig.9 is reflected in the S-shaped depletion curve of the snow coverage.

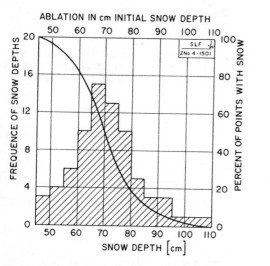

Fig.8 Frequency distribution of snow depths and the resulting gradual diminution of the snow-covered area

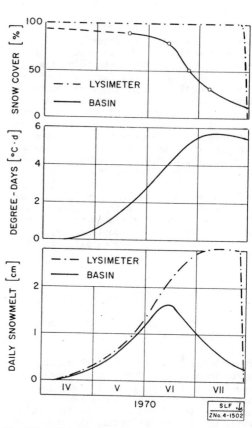

Fig.10 Meltwater production in a basin and from a small snow-covered plot (snow lysimeter)

452

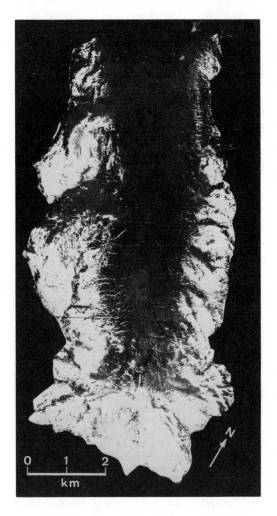

Fig.11 Snow cover in the Dischma basin on 8 June 1976, evaluated from Landsat 2 data by H.P.Urfer.Reproduced by courtesy of the Dept.of Geography, University of Zurich

Fig.12 Orthophoto of the snow cover in the Dischma basin on 8 June 1976

Fig.10 demonstrates the practical importance of this phenomenon for the river flow: In contrast to a small snow-covered area, the meltwater production from a basin is already on the decline while the temperature is still rising,so that the risk of a flood from snowmelt only is reduced.

The need of the snow mapping for hydrological purposes has been recognized long before the advent of the remote sensing. For example the areal extent of the snow cover in an experimental basin in the Rocky Mountains (Garstka,Love,Goodell,Bertle 1958) was estimated by dividing the total area of 93 km^2 into 16 topographic compart-

ments in which the snow coverage was observed and evaluated. This laborious procedure had to be naturally repeated in not too long intervals in order to draw a depletion curve.

4 EVALUATION OF REMOTE SENSING DATA

From the various satellites monitoring the earth, Landsat has a sufficient resolution of about 80 m to be used for the snow mapping in mountain basins. Its four channels in the range from 0,5 to 1,1 μm make possible a refined evaluation of data (Haefner

& Seidel 1974, Rango & Itten 1976). The information stored on magnetic tapes can be converted into an image as is illustrated by Fig.11. A simultaneous orthophoto is shown in Fig.12.

For hydrological purposes it is essential to be able to process the data not only in terms of rectangular sections,but also for catchment areas or basins of irregular shapes. The areal extent of the snow cover must be evaluated even for partial areas of a basin separately if the elevation range is great. If the Landsat data are to be used for operational discharge forecasts results should be delivered in few days after the overflight of the satellite.

Another requirement is a periodical repetition of these evaluations during the snowmelt period since a dynamic process is being monitored in contrast for example to a geological survey. Consequently, there must be a reasonable chance of a good visibility in the monitored area. The 1,75 μm channel envisaged for Landsat D will help to identify the clouds but will not remove them. Other methods of data processing,for example a completion of partial images of the snow cover by the pattern recognition method, are being developed along with new methods of remote sensing in order to overcome this difficulty.

So far, the exploitation of Landsat data in the Alps for snow hydrology is restricted by this problem while it is at once possible in other areas.For example in the Wind River mountain range in Wyoming,U.S.A. the depletion curves of the snow cover have been evaluated in four consecutive years practically without cloud problems. As an example, Fig.13 shows depletion curves of 1976 already in a form suitable for the runoff model, that is to say separately for the elevation zones given in Tab.1.

Table 1. Elevation zones of the Dinwoody Creek basin, Wyoming.

Zone	Elevation		Area
	ft a.s.l.	m a.s.l.	km^2
A	6500 - 8000	1981 - 2438	13
B	8000 - 9500	2438 - 2896	28
C	9500 -11000	2896 - 3353	95
D	11000 -13785	3353 - 4202	92

A depletion curve should refer to the snow cover accumulated in winter and gradually melted in the following months. A snow storm during the snowmelt period can cover the basin for a few days or hours and such

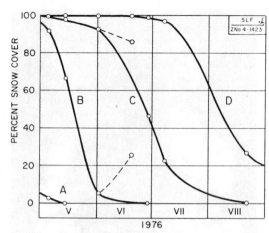

Fig.13 Depletion curves of the snow coverage in the Dinwoody Creek basin, Wyoming, U.S.A.,evaluated from Landsat 1 and Landsat 2 data by NASA/Goddard Space Flight Center in Greenbelt, U.S.A.

situation can be accidentally monitored by the satellite. However,the depletion curve is interpolated from periodical readings and cannot include short-term deviations.

On 19 June 1976, as shown in Fig.13, a snow storm just preceding the satellite overflight increased optically the snow cover to 86 % in the zone C and to 26 % in the zone B. It can be assumed that this new snow disappeared in a short time and that the normal course of the depletion curve was restored.Therefore it is advisable to eliminate such deviations and to consider the snowfall as precipitation to be added to the snowmelt which is computed from the area covered by the old snow. To this effect, satellite images of a short-lived new snow cover must be identified and omitted. In the given example,the distortion of the depletion curve on 19 June 1976 was evident in the zone B but might have escaped attention, without a proper analysis, in the zone C.

5 COMPUTATION OF THE SNOWMELT RUNOFF

Depletion curves of the snow cover are directly used by the snowmelt runoff model developed in the representative basin Dischma in the Swiss Alps (Martinec 1975). Each day the meltwater production is computed as a product of the snowmelt and of the snow-covered area:

$$V_m = M \cdot A \qquad (3)$$

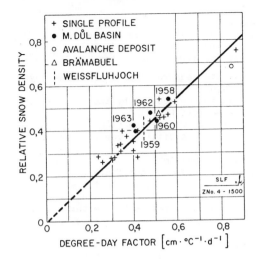

Fig.14 Relation between the density of snow and the degree-day factor

Values of the degree-day factor which summarizes the complicated energy balance are determined from the empirical equation

$$a = 1,1 \ \frac{\rho_S}{\rho_W} \qquad (5)$$

where ρ_S , ρ_W is the density of snow and water, respectively

Experimental measurements from which this relation has been derived are plotted in Fig.14.

A hypothetical example in Table 2 illustrates the serious errors which result if the temperature and the areal extent of the snow cover are taken only as an average for the whole basin or if a not well defined snowline is used.

If the snowline is set at 50 % snow coverage, the area below this line is interpreted as snow-free (I) and the area above this line as totally snow covered (II). This gives a lower meltwater volume than the total of the partial meltwater volumes obtained from the three elevation bands. If an average value of the snow coverage is applied and the meltwater production is calculated for the whole basin by the average number of degree-days, the obtained volume appears to be too high.

This example was computed for an assumed difference of 500 m between the average elevations of the zones A,B,C and for a temperature lapse rate of 0,6 °C per 100 m.

In order to avoid such errors, the Dischma basin was divided into three elevation bands as illustrated in Fig.15.

where V_m is the meltwater production
$m^3.d^{-1}$
M is the snowmelt $m.d^{-1}$
A is the snow-covered area m^2

The snowmelt is computed from the air temperature and from the variable degree-day factor:

$$M = a \ . \ T \qquad (4)$$

where M is the snowmelt $cm.d^{-1}$
a is the degree-day factor
$cm.^{\circ}C^{-1}.d^{-1}$
T is the number of degree-days per day $^{\circ}C$

Table 2. Meltwater production from different assessments of the snow coverage

Zone	Area m^2	Snow coverage	Degree-days $^{\circ}C.d$	Degree-day factor $cm.^{\circ}C^{-1}.d^{-1}$	Meltwater production m^3
A	10.10^6	0,25	7,5	0,5	93 750
B	20.10^6	0,5	4,5	0,5	225 000
C	10.10^6	0,75	1,5	0,5	56 250
Total	40.10^6				375 000
I	20.10^6	0	6,75	0,5	0
II	20.10^6	1,0	2,25	0,5	225 000
Total	40.10^6	0,5	4,5	0,5	225 000
Entire basin	40.10^6	0,5	4,5	0,5	450 000

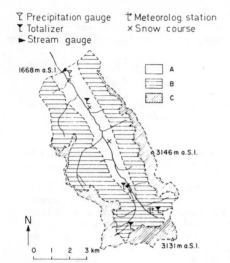

Ⴤ Precipitation gauge   ⴑ Meteorolog. station
Ⴤ Totalizer   × Snow course
► Stream gauge

1668 m a.S.l.

☐ A
☐ B
☐ C

3146 m a.S.l.

N

0 1 2 3 km

3131 m a.S.l.

Fig.15 Situation of the representative
basin Dischma

The sequence of daily flows during a snowmelt season is then computed as follows:

$$Q_n = c_n \left\{ [a_n(T_n + \Delta T_A) \cdot S_{An} + P_{An}] \frac{A_A \cdot 10^{-2}}{86400} + \right.$$

$$+ [a_n(T_n + \Delta T_B) \cdot S_{Bn} + P_{Bn}] \frac{A_B \cdot 10^{-2}}{86400} +$$

$$+ \left. [a_n(T_n + \Delta T_C) \cdot S_{Cn} + P_{Cn}] \frac{A_C \cdot 10^{-2}}{86400} \right\}(1 - k) +$$

$$+ Q_{n-1} \cdot k \qquad\qquad\qquad (6)$$

where
Q is the average daily discharge $[m^3 s^{-1}]$
c_n is the runoff coefficient
a_n is the degree-day factor $[cm \cdot {}^\circ C^{-1} \cdot d^{-1}]$
T_n is the measured number of degree-days
    $[{}^\circ C \cdot d]$
ΔT is the correction by the temperature
    lapse rate $[{}^\circ C \cdot d]$
S is the snow coverage (100 % = 1,0)
P is the precipitation contributing to
    runoff $[cm]$
A is the area $[m^2]$
k is the recession coefficient
n is an index referring to the sequence
    of days
A,B,C as index refers to the three eleva-
    tion bands

An important feature of this model is
the variable recession coefficient k which
is always related to the current intensity
of runoff. Fig.16 illustrates experimental
results from two basins. In Dischma, the
following equation has been derived:

$$k_n = 0,85 \, Q_{n-1}^{-0,086} \qquad (7)$$

The general trend of the variables S, T,
a, k during a snowmelt season is shown
in Fig.17.
The transformation of the meltwater
production in the snowmelt season 1973
into the discharge from the basin by the
model is illustrated in Fig.18.

Depletion curves of the areal extent of
the snow cover, used to calculate the
snowmelt in the respective elevation zones
(Fig.19) are based on orthophotos.

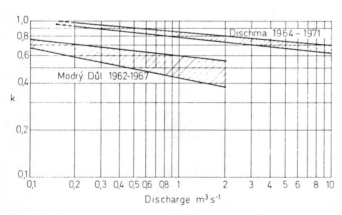

Fig.16 Relation between the recession coefficient k and
the current runoff in two representative basins

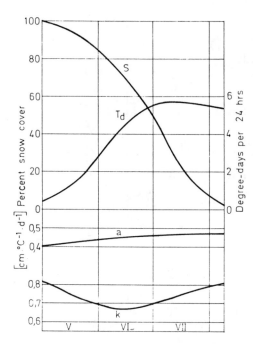

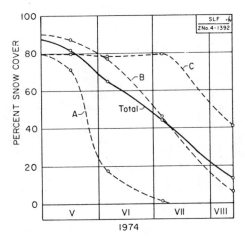

Fig.19 Depletion curves of the snow coverage in the Dischma basin evaluated separately for the respective elevation zones

Fig.17 Hypothetical seasonal changes of variables in a snowmelt runoff model

A comparison of the discharge thus computed with the directly measured outflow from the Dischma basin is shown in Fig.20 and for the next season 1974 in Fig.21.

This example shows that a deterministic approach in snowmelt-runoff modelling depends on a periodical mapping of the seasonal snow cover. The monitoring by the Landsat is already achieving this task in areas with favourable conditions. There are good prospects that the remaining obstacles will be overcome by further developments of the remote sensing and data processing.

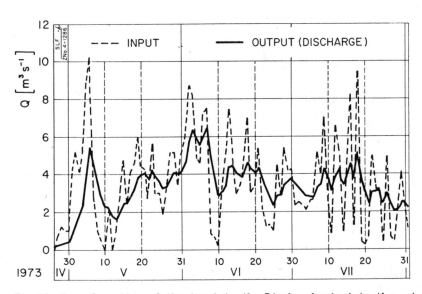

Fig.18 Transformation of the input to the Dischma basin into the output

457

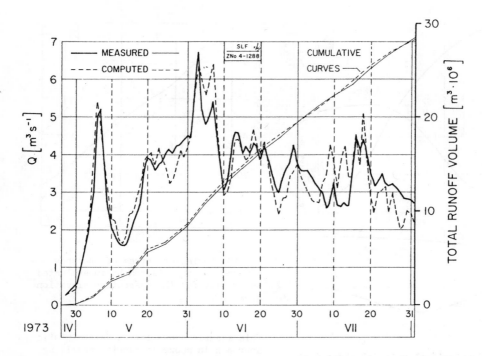

Fig.20 Computed and measured discharge from the Dischma basin, cumulative curves of the runoff volumes, snowmelt season 1973

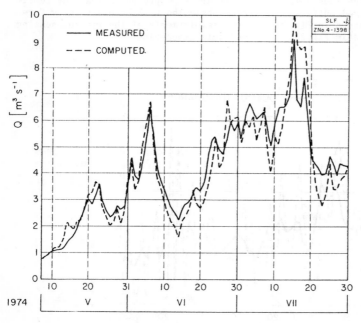

Fig.21 Computed and measured discharge from the Dischma basin, snowmelt season 1974

6 REFERENCES

Chow, V.T. 1964, Statistical and probability analysis of hydrologic data. Section 8, p.8-9. In V.T.Chow (ed.), Handbook of applied hydrology. McGraw-Hill, New York.

Crawford, N.H. & R.K.Linsley 1966, Digital simulation in hydrology: Stanford watershed model IV. Stanford University, Dept. of Civil Engineering, Tech.report No.39.

Freeze, R.A. & R.L.Harlan 1969, Blueprint for a physically-based, digitally simulated hydrologic response model, Journal of Hydrology, November 1969, Vol.9, No.3: 237-258.

Garstka, W.U., L.D.Love, B.C.Goodell, F.A. Bertle 1958, Factors affecting Snowmelt and Streamflow, Frazer Experimental Forest, p.51. U.S. Government Printing Office.

Haefner, H. & K.Seidel 1974, Methodological aspects and regional examples of snow-cover mapping from ERTS-1 and EREP imagery of the Swiss Alps, Proceedings Symposium of Frascati 1974, ESRO SP-100: 155-165.

Jamieson, D.G. & C.R. Amerman 1969, Quick return subsurface flow, Journal of Hydrology, June 1969, Vol.VIII, No.2: 122-136.

Kotlyakov, V.M. 1970, Land glaciation part in the earth's water balance, IAHS-UNESCO Symposium on World Water Balance, Reading, Vol.I, IAHS Publ.No.92: 54-57.

Linsley, R.K., M.A.Kohler, J.L.H.Paulhus 1958, Hydrology for Engineers. McGraw-Hill, New York.

Martinec, J. 1975, New methods in snowmelt runoff studies in representative basins, IAHS Symposium on the Hydrological Characteristics of River Basins, Tokyo, IAHS Publ.No.117: 99-107.

Martinec, J. 1976, Snow and Ice. In J.C. Rodda (ed.), Facets of Hydrology, p.85-118. John Wiley & Sons Ltd., London-New York-Sydney-Toronto.

Overton, D.E. 1977, Catchment hydrology, Reporter's comments, Third International Hydrology Symposium, Fort Collins, Colorado (in press).

Rango, A. & K.I.Itten 1976, Satellite Potentials in Snowcover Monitoring and Runoff Prediction, Nordic Hydrology, 7: 209-230.

ACKNOWLEDGEMENT
The author thanks Mr.J.von Niederhäusern and Mr.E.Wengi for the graphical layout.

Satellite data collection systems – Hydrologic application

For the sections 1 to 4 see Chapter 18: Satellite data collection systems — Agricultural applications

4 HYDROLOGIC DATA COLLECTION

4.1 Hydrologic data

The water data collection programs are designed to provide -generally national- water data bases to guide the development
and conservation of a critical natural resource. Hydrological data collection can be
divided in two groupings :
- that required for archiving for future
use (historical data)
- that required for real time decision
making (real time data)

Historical water data are used for design of
dams, reservoirs, water supply sources, waste treatment facilities, flood prevention
structures, irrigation projects, flood-plain
management plans, navigation facilities.
Or in other words historical hydrologic data are being collected and archived for :
planning, designing, constructing, operating water resources systems and for forecasting of hydrologic conditions ranging
from droughts to floods.
A small but growing part of the water data
collection programs presently is structured
to provide real-time hydrologic data.
Real time data are being collected for real
time decision making and provided to computer models that attempt to predict river
conditions and optimize the operation of
water resource systems.
These models involve flood control operations,
flood warnings and forecasts, and river
forecasts for the managment of electric
power generation, navigation, water supply,
irrigation and water quality.

Hydrologic data also are collected to support interdisciplinary research, in such as:
  Agriculture (rain, snow, hail)

Biology : monitoring of water systems:
water flow and temperature for study of mi-
gration routes of animals and nutrient budget, rates of water release from impound-
ments influence both warm and cold water
fisheries, competing uses of water for man
and animals have to be evaluated (Ecosystem cost/ use analysis)
Ecology : monitoring water quality parameters : water pollution arising from the
activities of man by industrial effluents.
Meteorology : precipitation, snow accumu-
lations...

4.2 Major hydrologic data collection sites existing in USA (1975)

Table 4.2 summarizes the major hydrologic
data collection sites existing in the US
in the year 1975 by major agency and type
of measurement.
This table shows the dominant role of USGS
hydrologic network which is the largest operated by a single agency.
Not all of the 47,000 sites are potential
candidates for telemetry or satellite relay (at some sites , the current state of
the art of sensors, and/or budgetary constraints mandate that the data be collected
physically. At other sites the data load is
so small or is required so unfrequently as
to negate investment for telemetry or satellite relay).

For example :
for two major users

Total of existing sites :   42,000   4,000

Table 4.2.

Agency	Water stage	Water quality	Ground water	Total
National Oceanic and Atmospheric Administration (NOAA)	1,000			1,000
US Geological Survey (USGS)	18,000	6,000	18,000	42,000
Environmental Protection Agency (EPA)		600		600
Corps of Engineers (COE)	2,000	2,000		4,000
BUREC	200	300	5,000	5,500
STATE	1,500	1,000		2,500
	≈23,000	≈10,000	≈23,000	

total : 47,000

Total of potentially amenable to data telemetry (or satellite relay) :   24,000

Total addressable market :   11,000   4,000
(sites already equipped with data storage capability*)

* the addition of telemetry for sites, already equipped with data storage capability represents a relatively manageable expense.

4.3 Types of hydrologic parameters and estimate of the total number of stations of each type to be needed in North America in 1975.

The hydrologic parameters listed in table 4.3 are essential to measure the phenomena described in paragraph 4.1. The numerator under the heading "Total number of sites" is the estimate of the total number of stations of that parameter type, to be needed in North America in 1975.
The denominator is the estimate of the number of stations of that parameter type where real time data is required.
The table 4.3 gives also :
- For measurement :
  The sampling rate : number of measurements taken from a single sensor per unit time
  Frequency of transmission : number of measurement (or collection) transmissions per unit time.
When real time is not needed this frequency is not critical and can be once a day or less often.
- For sensor (developped in 1975)
  Life : period of operation of the sensor with repair or attention (cleaning...)

Repair interval : period of time without repair or attention.

4.4 Data collection hydrologic stations

4.4.1 Definition of the number of bits per sample or measurement

The number of binary digits which express the number of possible levels of the measure range.
Example :
- range   0-100 deg. C
- resolution 0.1 deg. C

- Number of levels   : $\dfrac{\text{Range}}{\text{Resolution}} = 1,000$

Express number of levels in binary bits :

Since 512 can be represented by 9 binary bits (2^9) and 1024 can be represented by 10 bits (2^{10}).
Bits per sample : 10

4.4.2 Data collection hydrologic stations
      Number of bits per sample

Many of the preceeding types of parameters can be colocated at a common observation point or data collection hydrologic stations. For example, a water level station with two sensors :
- water level sensor
- precipitation sensor
The number of bits per sample is calculated in detail (according to the characteristics of the sensors).

TABLE 4.3.

TYPE	USA - CANADA 1975 Total number of sites to be needed	MEASUREMENT		SENSOR		REMARKS
		Sampling rate	Frequency of transmission	LIFE	REPAIR INTERVAL	
Precipitation	20,000 / 5,000	15 minutes	3 hourly 5%-10% on demand	10 years	1 month	5% - 10% required on demand
Water level	6,000 / 1,200	15 minutes	3 hours	<10 years	6 months	
	3,000 / 400	30 minutes	3 hours	"	"	
	3,000 / 400	1 hour	3 hours	"	"	
Ground water Depth to water table	10,000 / <100	1 hour	Daily	10 years	6 months	
Snow depth	300 / <10	Daily	Daily	10 years	1 year	
Snow moisture equivalent	2,000 / 200	Daily	Daily	5 years	1 year	
Evaporation	1,000 / 10	Daily	Daily	5 years	Weekly	
Water temperature	900 / 100	1 hour	3 hours	1 year	1 week	
Wind direction	500 / 300	5 minutes	1 hour	5 years	6 months	
Wind velocity	500 / 300	5 minutes	1 hour	5 years	6 months	
Soil moisture	500 / <10	12 hours	12 hours	5 years	1 month	
Specific conductance	400 / 75	1 hour	3 hours	1 year	1 month	
Ice thickness	300 / <10	Daily	Daily	5 years	1 year	
Air temperature	300 / 300	1 hour	1 hour	10 years	1 year	
Dew point-frost point	300 / 300	1 hour	1 hour	10 years	6 months	
Dissolved oxygen	200 / 50	1 hour	3 hours	6 months	1 week	
PH	150 / 50	1 hour	3 hours	1 year	1 week	
Water velocity	100 / <10	6 minutes	3 hours	5 years	1 month	USA only
Incoming solar radiation	50 / <10	Daily	Daily	1 year	1 month	
Net radiation	25 / 10	Daily	Daily	6 months	1 month	
River discharge	10 / 5	3 minutes	3 hours	5 years	6 months	
Ice presence	5	Daily	Daily	1 year	1 year	

Water level station
Water level sensor
- Range   : 0 to 99 feet
- Resolution : 0.01 foot

Number of levels : $\frac{99}{0.01} \simeq 10,000$

2^{13} = 8,192 levels

2^{14} = 16,284 levels ——— 14 bits per sample

Precipitation sensor
- Range   : 0 to 99 inches/hr
- Resolution : 0.1 inches/hr

Number of levels : $\frac{99}{0.1}$ = 1,000
10 bits per sample.

Total number of bits per sample, each sample including water level and precipitation measurements : 24 bits.

The same evaluation has been done for other types of stations such as :

Snow station
	Number of bits per sample(see definition)
-Snow moisture equivalent	10
-Snow depth	12
	22

Atmospheric station
	Number of bits per sample (see definition
- Air temperature	12
- DEW/FROST point	11
- Wind direction	9
- Wind velocity	8
- Incoming solar radiation	7
- Net radiation	7
	54

Water quality station
- Water temperature	10
- Specific conductance	6
- Dissolved oxygen	6
- PH	7
- Water level	14
	43

Water level station
- Water level	14
- Precipitation	10
	24

Ice presence
- Ice presence	1
- Ice thickness	8
	9

These numbers of bits must be compared to the characteristics of satellite DCP.

Report message

Number of bits for sensor data (see 3.2.) (per message)

- 2000 to 5200 bits for synchronous system
- 256 bits for ARGOS System

There is no data limitation problem with the two systems if each message sent by the DCP is really received by the satellite. That is always the case for geosynchronous systems : the satellite remains in view of the DCP.
For low altitude systems using drifting satellites, a satellite is not generally in view of the DCP when the measurements are made so that DCP must have storage capability.

4.4.3 Use of a low orbit system
Characteristics of TIROS N-ARGOS system

A) Collection transmission frequency

The table 4.4.3.(1) shows the visibility performance of the TIROS/ARGOS System depending on the hydrologic site latitude. The number of passes per day is summarized in table 4.4.3(2), (the average duration of a pass is ten minutes).

The collection transmission frequencies per 24 hours (according to the latitude) are roughly, the numbers of passes.

At the poles, the collection period is 50 minutes (in average). At the Equator, the time between each of the 7 passes is not constant (variation from 100 minutes to 300 minutes).

Table 4.4.3. (2)

Latitude	Number of passes per 24 h	
± 0°	7	At the Equator transmitted data is collected 7 times per day
± 15°	8	
± 30°	9	
± 45°	11	
± 55°	16	At the poles transmitted data is collected 28 times a day or every 50 minutes
± 65°	22	
± 75°	28	
± 90°	28	

B) Number of samples per collection interval

Definition : total number of measurements of the parameters taken during the collection interval (or period).
(Number of bits per sample) x (Number of samples per collection interval*)= Number of bits to be transmitted.

* assuming that DCP has a storage capability.

The ARGOS message capacity for sensor data is 256 bits.

Example : Data from "water quality station" (see 4.4.2.) 43 bits per measurement
$\frac{256}{43}$ ≈ 5 successive measurements can be stored without exceeding the ARGOS message.
The needed sampling rate for water quality parameters(see table 4.3) is 1 hour.
The maximum collection interval can be 5 hours (without exceeding the 256 bits Argos message capacity).
Argos low orbit satellite system meets the data requirements (for the "water quality station") at any latitude for the number of bits to be collected and transmitted (maximum time interval between two passes 5 hours).
Argos low orbit satellite system does not meet the data frequency transmission requirements of 3 hours (see table 4.3).
This example shows the capability of this system in terms of data collection (in the worst case i.e. at the Equator).
It shows also the limitations of this system in terms of measurement-to-collection delays.

4.5 Users of data collection in Hydrology

4.5.1 U.S.G.S. (USA)

The US Geological Survey operates a network of hydrologic sensors that automatically record the data on sites. These data routinely :
- are manually retrieved at intervals of 4-6 weeks,
- are manually preprocessed,
- entered into the geological survey's national telecomputing network.
(computer center in Reston, Virginia with two 370/155 computers and a network of over 180 remote computer terminals accross the USA).

This network is used to perform most of the Survey's basic hydrologic data processing and hydrologic analysis with the Survey's WATSTORE System (water data storage and retrieval system). WATSTORE is a collection of computer programs and files that are used by the Survey to process virtually all the water-resources data that the survey collects.

One of the tasks being undertaken is the experimental use of the WATSTORE system for processing and filing the satellite data relayed from three different satellite data collection systems (see figure 4.5.1)
All DCS data relayed through the Landsat system are routinely sent in real time to the Landsat Operation Control Center in the National Aeronautics and Space Administration (NASA) Goddard Space Flight Center in Greenbelt, Maryland. Under a NASA-US Geological Survey agreement, these data also are sent in real time via a dedicated line to Reston where they are recorded on a 9-track magnetic tape recorder. Periodically these data are transferred to an on line disk file in the Reston computer center, where programs are available to retrieve, process and disseminate the data over the remote terminal network.
All DCS data relayed through the SMS/GOES system are routinely received by a National Oceanic and Atmospheric Administration/National Environmental Satellite Service (NOAA/NESS) tracking station at Wallops Island, Virginia, and sent in real time to the NOAA World Weather Building in Suitland Maryland. These data are filed in a NOAA/NESS mini-computer. Under a NESS-Geological Survey agreement this mini-computer, one or more times a day, will sign on to the Reston computer and will enter a computer job containing all of the Survey's DGS data that have been accumulated since the last time of data entry. These data then are processed and placed in an on-line file in the computer for retrieval.

The DCS data to be relayed through the commercial COMSAT General system will be accumulated at the COMSAT General earth station in Southbury, Connecticut. The DCS data will be periodically and automatically transferred from a Comsat General Computer to the Reston computer where they will be filed on an on-line disk file. In addition to the data being made available to the Reston computer, it also will be possible for the two US Geological Survey district offices in Harrisburg, Pennsylvania, and Fortland, Oregon, to use their Reston computer - compatible terminals to occasionnally connect to the COMSAT computer to directly retrieve unprocessed real-time data.

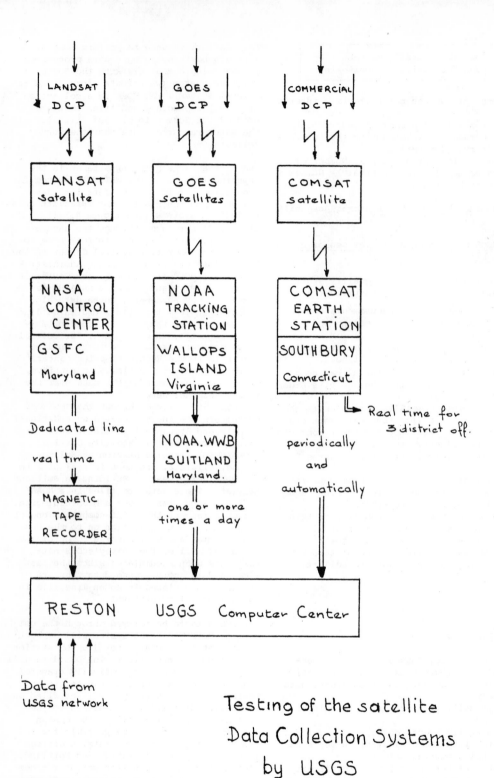

FIG. 4.5.1.

Testing of the satellite
Data Collection Systems
by USGS

The automatic collection of data in real time offers two benefits :

- the first is that the real time processed data can be used to monitor the performance of the instrument network. (stations are visited only when instruments fail or conditions warrant the collection of supplementary data - when water quality or discharge is outside the normal range).
- the second is that a real time data service can be offered to the water resources management community (growing pressure of municipal, industrial use and environmental protection).

The increasing cost of manpower, the decreasing cost of electronics, and cost effectiveness of satellite telemetry probably will result in an eventual automation of the collection of hydrologic data.

Note 1 : USGS made in 1975 an experimental integration of the LANDSAT data collection system with the geological survey's network. The objective of this experiment was to simulate an operational system for collecting, relaying, processing and dissemination hydrologic data.
About 90 DCP in 24 USGS districts, were used for the experiment.

Note 2 : The water resources division of the USGS in installing (1977) over 120 self-timed DCPs to operate with GOES for the collection of hydrologic data.

4.5.2 Canadian User

Applied Hydrology Division,
Department of the Environment
Principal Investigator : R. Halliday

The role of Water Survey of Canada (similar to USGS in USA) is that of monitoring river flows and lake levels at about 2400 gauging stations under agreement with the Provincial governments and making the data available to users. The water survey operates about 100 water telemetry systems using telephones lines. For the vast majority of the gauging stations, however it is prohibitively expensive to install either telephone or radio telemetry systems.
In response to a demand for near real time data from additional sites, it was decided to implement a network of about 30 sites on a quasi operational basis using LANDSAT 1. Some typical examples of data uses are as follows :
- 6 DCPs were used in the Mackensie River basin to provide data for the preparation

of daily water level forecasts during the short navigation season.
- 3 DCPs were installed in the Ottawa river basin and one on the Saint John River to provide data for input into streamflow synthesis and reservoir regulation models of the watersheds.
- 1 DCP was installed on the Severn River. This gauging station on the Trent-Severn Waterway is below the confluence of several small streams, all of which are regulated. The river is also regulated downstream from the DCP site. Water level and water velocity data are used to compute river discharge. Water temperature data are also transmitted to assist in winter flow computations.

The sensors used with the DCP are summarised as follows :

Parameter	Sensor
Water level	Float pressure
Water velocity	Electromagnetic acoustic
Ice condition	Electro-mechanical
Precipitation	Weighing type
Air temperature	Platinium resistance bulb Thermister
Water temperature	Thermister
Snow water content	Snow pillow

Data handling and processing

All messages relayed by LANDSAT are received in Alaska, California and Maryland then sent over Nascom lines to the GSFC Maryland.

- The Canadian messages are sent by teletype to the Canada Centre for Remote Sensing in Ottawa, usually within 15 to 20 minutes after each LANDSAT pass. A software data retrieval system sorts the user data platforms, reformats the data into engineering units and stores individual user files on disk. A user may then access the file usually daily using a teletype or telex remote terminal.

- Data are also sent by punch cards and uncalibrated computer listings which arrive about two weeks aftet transmittal by the DCP. The data received on a near real-time basis are usually discarded a short time after use but the data received in card form are retained for archival purposes and to develop statistics on DCP performances.

The location of the sites are illustrated in figure 4.5.2. (1).

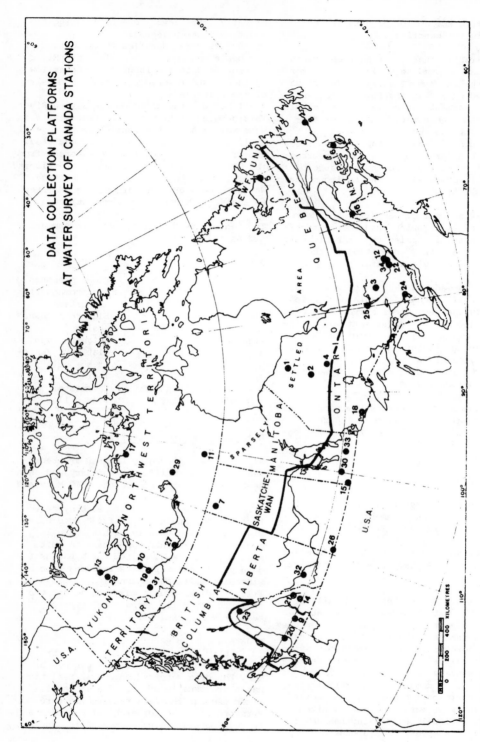

FIG. 4.5.2. (1)

468

Significant results

The LANDSAT program has demonstrated that the polar orbiting satellites can be used to relay hydrologic data from any part of Canada to a user without difficulty and at low cost. These data can be used for many operational purposes such as :
- hydroloelectric power plant operation
- water supply for municipalities, industries and irrigation
- navigation
- flood forecasting
- operation of flood control structures and systems
- recreation

Benefits : There are several ways in which real time data acquisition can aid hydrometric field operations : these are :
- planning of field trips : If the real time data indicates that all sensors at a site are operating normally and if flow conditions are such that a discharge measurement is not required, then a visit to the station can be omitted.
- planning of sensor maintenance :If real time data indicate a sensor malfunction it is usually possible to diagnose the problem by examining the incoming data. A decision can be made whether immediate repairs are warranted, or whether maintenance should be included on the next scheduled trip into an area. In either case the repair is completed in one trip where otherwise two may be needed (one to discover the problem, the second to do the repairs).
- Filling in missing record : when field recorders have stopped but sensors continue to function, the real time data can be used to fill in data that would otherwise be missing.
- Primary collection of data : If the cost of acquiring real time data and the reliability of the system proved better than using in situ recorders, then satellite telemetry could be used as a primary means of data collection.

The current capital cost of satellite telemetry of water level data in Canada is about $5,000 for a DCP plus $1,000 for a water level encoder.
Operating costs of a DCP are small and about $100 a year (repairs). These costs does not include processing and dissemination charges.

Future : the deployment of DCPs in Canada proceeded at a relatively slow pace. The principal reason has been the lack of an operational satellite system. The service provided by LANDSAT is excellent but the system is experimental and therefore cannot

be used as the basis for any long term project.
The operational GOES system often cannot be used in mountainous or high latitude areas because of antenna aiming problems. It now seems likely that the Water Survey of Canada will select GOES self-timed DCPs at all sites where it is technically feasible and will use the TIROS N-NOAA-ARGOS system where GOES is not feasible (this assumes of course, that the ARGOS System meets published specifications). A contract for development of an ARGOS-GOES DCP has recently been awarded to a Canadian company.

5 REFERENCES

- Use of Earth Satellites for Automation of Hydrologic Data Collection (USGS), Richard Paulson, July 1976, NASA Report.

- Performance of the Landsat Data Collection System in a total system context (USGS), Richard W. Paulson and Charles F. Merk, November 1975, International Seminar on organization and operation of hydrologic services, Ottawa, Canada.

- Potential impact of Satellite Data Relay Systems of the Operation of Hydrologic Data Programs, D.H. Moody and D.M. Preble (USGS), December 1975, 2nd World Congress on Water Resources of the International Water Resources Association, New Delhi, India.

- Retransmission of Hydrometric Data in Canada, Notes for an oral report to NASA, November 1976, Applied Hydrology Division, Department of the Environment, Ottawa, Canada, R.A. Halliday, Principal Investigator.

- Satellite Data Collection User Requirements Workshop, Final Report, June 1975, Edward A. Wolff , Charles E. Cote, J. Earle Painter.

- Satellite Data Collection Newsletter, edited by Dr Enrico P. Mercanti, code 952, NASA-GSFC, Greenbelt Md. USA.

- GOES DCP User's Guide, NOAA NESS, November 1976.

_ METEOSAT DCP User's Guide ESA - MPO March 1977.

- ARGOS DCP User's Guide, Service Argos, CNES, 18 av. E. Belin, Toulouse, France.

- IInd Argos User's Meeting, November 2-3, 1977.

- New Ventures in satellite Communications
Comsat General Corp. 950 L'Enfant Plaza
S.W. Washington, DC 20024, USA.

- Use of ERTS and Data Collection System ima-
gery in reservoir management and operation,
Saul Cooper, US Army Corps of Engineers,
Waltham, Massachussetts.

- Water quality parameter study in Warrior
River and Mobile Bay using DCP's & ERTS imagery, Harrold Henry, Alabama Geological
Survey, Univ. of Alabama, Tuscaloosa, AL
35486, R.C. "Red" Bamberg, Director, Alabama

Development Office, State Office Building,
Montgomery, AL 36104.

- DCS platforms were used to provide opera-
tional data to control water flow during
flood stages of Salt River in March 1973.
DCS System proved more reliable and useful
than microwave system now in use.(State
Department of Water Resources USGS Watershed
Division, Salt River Project,William Warskow,
Lead Watershed Specialist, Watershed Div.
Salt River Project, Po Box 1950, Phoenix,
Arizona 85001,(602)273-5680, Herb Schuman,
USGS,Phoenix, Arizona, 8(602) 262-318 b.

- Florida Fish & Wildlife Office USGS,
Corps of Engineers US Forest Service.
A DCP network (20 platforms) and ERTS imagery are being used to monitor and make environmental assessments and manage the water resources of 3500 sq. kilometers in the
Everglades. Proved particularly useful during the 1973-74 winter spring draught where
damage was minimized by planning based upon
accurate knowledge of the water stored in
the various lakes, canals and conservation
areas. Soil moisture sensors on the DCP's
aᵢe also being used to warm of potential
fire hazards.
David Cox (305) 724-1571, Edward Vosatka,
Florida Game and Fresh Water Fish Commission, 7630 Coral Drive, West Melbourne,
Florida 32901.

- Retransmission of Hydrometric Data in
Canada, Final Report, April 1978, Principal
Investigator R.A. Halliday, Applied Hydro-
logy Division, Department of Fisheries and
the Environment, Ottawa, Ontario, Canada
KIA OE7.

B. STURM

Electronics Division

Joint Research Centre, Ispra Establishment, Ispra (Varese), Italy

Optical properties of water-applications of remote sensing to water quality determination

1 INTRODUCTION

The study of the interaction of sunlight with terrestrial water bodies forms a main concern of so-called Optical Oceanography which is a branch of Oceanography. In analogy to the term Optical Oceanography it may be allowed nowadays to speak of Optical Limnology, since many studies of light propagation in water have been done also in fresh-water lakes. The interest in the light propagation in water bodies is, however, not limited to Oceanography only. Also biology is concerned with the optical properties of ocean water since more than 70% of the earth's surface is covered by ocean and the major part of biological productivity takes place in the upper part of it. Meteorology is interested in the absorption of sunlight by the ocean since this determines the evaporation of water into the atmosphere and the heat energy stored by the ocean.

Recently the study of optical properties of water has received great interest, also from environmental science and resource management, due to the rapidly evolving techniques of remote sensing, both from airplanes and satellites. In this case it is mainly the spectral composition of the upwelling light, the water colour, which contains the desired information.

In the following we will concentrate on this last mentioned aspect of optical oceanography, namely the question: How does downwelling light from sun and sky interact with natural water bodies and how can the upwelling light be used for obtaining detailed information about the quality and the quantity of organic and inorganic matter, suspended in the water.

For the moment we will not consider the effect of the atmosphere, that is: neither the alteration of spectral composition and intensity of the upwelling light during its way to the sensor, nor the light scattered into the sensor by dust particles and air molecules.

Our problem can therefore be characterized by the following (Fig. 1):

Direct sun irradiance is falling with a zenith angle, θ_0, on a water surface, S, a part of it - roughly 2-3% for $\theta_0 \leqslant 45°$ - is reflected without being detected by the sensor looking vertically down. The part which is transmitted into the water is either absorbed (A) or scattered (S) by suspended matter or water molecules. The scattered light going in the direction of the sensor can again either be absorbed or scattered or leaves the water surface being detected by the sensor.

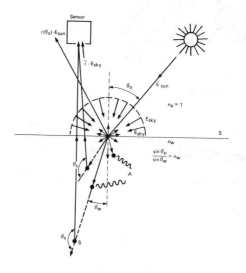

Fig. 1

The diffuse irradiance falling on the water surface is E_{sky}; a part of it is again reflected (6.6%) and the part transmitted into the water can again be absorbed or scattered, coming finally into the sensor.

2 DEFINITION OF RADIOMETRIC QUANTITIES

In the following we will use some radiometric quantities (see Figs. 2 – 6):

- the radiant flux, F: energy per unit time flowing across a given surface

$$F = \frac{Q}{t} \quad (\text{Watt, J.sec}^{-1}, \text{erg.sec}^{-1})$$

- the radiant intensity, I: radiant flux emitted into an infinitesimal solid angle, $d\omega$, divided by $d\omega$

$$I = \frac{dF}{d\omega} \quad (\text{Wsr}^{-1})$$

- the radiance; L: radiant flux per unit solid angle per unit projected area of a surface (Fig. 3a,b)

$$L = \frac{d^2 F}{dA\cos\epsilon d\omega} \quad (\text{Wm}^{-2}\text{sr}^{-1})$$

L has two important properties:

. the invariance along the propagation path in the vacuum; (Fig. 4a)

. discontinuity at the boundary of two materials with different indices of refraction. (Fig. 4b)

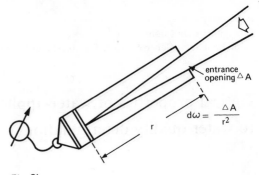

Fig. 3b

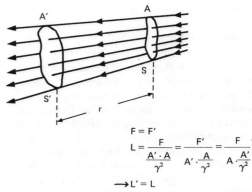

$$F = F'$$
$$L = \frac{F}{\dfrac{A' \cdot A}{\gamma^2}} = \frac{F'}{A' \cdot \dfrac{A}{\gamma^2}} = \frac{F}{A \cdot \dfrac{A'}{\gamma^2}}$$

$$\longrightarrow L' = L$$

Fig. 4a Invariance of radiance along rays

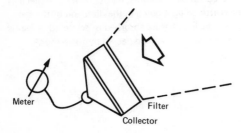

Meter

Filter

Collector

Fig. 2a

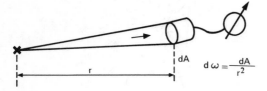

Fig. 2b

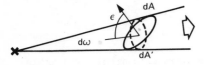

Fig. 3a

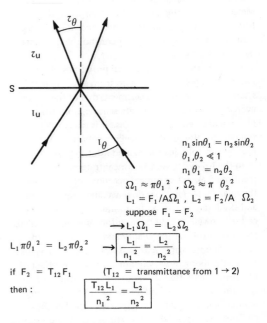

$$n_1 \sin\theta_1 = n_2 \sin\theta_2$$
$$\theta_1, \theta_2 \ll 1$$
$$n_1 \theta_1 = n_2 \theta_2$$
$$\Omega_1 \approx \pi\theta_1{}^2, \quad \Omega_2 \approx \pi\,\theta_2{}^2$$
$$L_1 = F_1/A\Omega_1, \quad L_2 = F_2/A\,\Omega_2$$
suppose $F_1 = F_2$
$$\longrightarrow L_1\Omega_1 = L_2\Omega_2$$

$$L_1\pi\theta_1{}^2 = L_2\pi\theta_2{}^2 \quad \longrightarrow \boxed{\frac{L_1}{n_1{}^2} = \frac{L_2}{n_2{}^2}}$$

if $F_2 = T_{12}F_1$ $\quad (T_{12} = \text{transmittance from } 1 \to 2)$

then : $\quad \boxed{\dfrac{T_{12}L_1}{n_1{}^2} = \dfrac{L_2}{n_2{}^2}}$

Fig. 4b n^2-Law for radiance

472

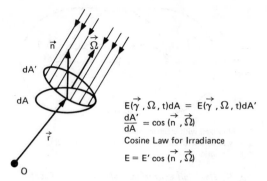

$$E(\vec{\gamma}, \Omega, t)dA = E(\vec{\gamma}, \Omega, t)dA'$$

$$\frac{dA'}{dA} = \cos(\vec{n}, \vec{\Omega})$$

Cosine Law for Irradiance

$$E = E' \cos(\vec{n}, \vec{\Omega})$$

Fig. 5a

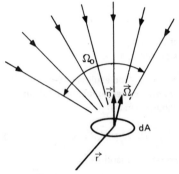

$$\frac{dF}{dA} = \int_{\Omega_0} L(\vec{\gamma}, \Omega, t) \cos(\vec{n}, \vec{\Omega})d\Omega$$

Fig. 5b

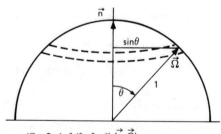

$$d\Omega = 2\pi \sin\theta \, d\theta \quad \theta = \measuredangle(\vec{n}, \vec{\Omega})$$

$$E = \frac{dF}{dA} = \int_{2\pi} L(\vec{r}, \vec{\Omega})\cos\theta \, d\Omega = L \int_0^{\pi/2} 2\pi \cos\theta \sin\theta \, d\theta$$

$\longrightarrow$ $\boxed{E = \pi \cdot L}$ Scalar irradiance E_0: $E_0 = \int_{4\pi} L(\theta, \varphi)\sin\theta \, d\theta \, d\varphi$

Fig. 6 Constant radiance from a hemisphere

the irradiance, E: radiant flux on an infinitesimal element of surface, divided by the area of that element

$$E = \frac{dF}{dA} \quad (Wm^{-2})$$

E depends generally on the orientation of the normal vector of dA to the direction of the (collimated) light source. (Fig. 5a)

For radiation coming from distributed light sources E defined by an integral Fig. (5b)

For constant radiance from a hemisphere (Fig. 6)

$$E = \pi \cdot L$$

3 DEFINITION OF MAIN OPTICAL PROPERTIES OF WATER

Optical properties of water are divided into two classes (Preissendörfer)

a) inherent properties, that are independent from radiance distribution;

b) apparent properties, that depend from radiance distribution.

In the first group we have (see Figs. 7 and 8):

- absorption coefficient a, absorptance A;
- scattering coefficient b, scatterance B;
- attenuation coefficient c, attenuance C;
- volume scattering function $\beta(\theta)$.

In the second group we find most of the optical properties which interest us here in this context. For the moment we mention two of them:

- radiance attenuation coefficient, k;
- irradiance attenuation coefficient, K.

4 SCATTERING AND ABSORPTION IN PURE WATER

Scattering in pure water is attributed to molecular

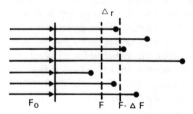

$$A = \frac{F_a}{F_0} \qquad a = -\frac{\Delta A}{\Delta r} = -\frac{\Delta F_a}{F \Delta r} \; (m^{-1})$$

$$F(r) = F_0 e^{-\int_0^r a(r)dr}, \text{ for homogenous medium } F(r) = F_0 e^{-ar}$$

Attenuance C, attenuation coefficient c

$$C = -\frac{F_c}{F_0} = -\frac{F_a + F_b}{F_0}$$

$$c = -\frac{\Delta F_c}{F \cdot \Delta r} = -\frac{\Delta F_a + \Delta F_b}{F \cdot \Delta r} = a + b \; (m^{-1})$$

Optical length (depth)

$$\tau = \int_0^r c(r)dr \text{ for homogenous medium } \tau = c \cdot r$$

Fig. 7 Absorptance A, absorption coefficient a

473

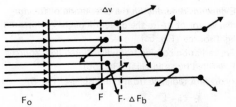

Definition: Scatterance B, scattering coefficient b

$$B = \frac{F_b}{F_0} \qquad b = -\frac{\Delta B}{\Delta r} = -\frac{\Delta F_b}{F \Delta r} = \frac{4\pi \int I_b d\Omega}{E \Delta A \Delta r} = \frac{4\pi \int I_b d\Omega}{E \Delta v} \quad (m^{-1})$$

Volume scattering function β

$$\beta(\theta) = \frac{dI(\theta)}{E \, dv}$$

$$b = \int_{4\pi} \beta(\theta) d\Omega = 2\pi \int_0^\pi \beta(\theta) \sin\theta d\theta \quad (m^{-1})$$

$$b_f, b_b : \quad 2\pi \int_0^{\pi/2} \dots, \quad 2\pi \int_{\pi/2}^\pi \dots$$

Fig. 8 *Scattering: Deviation of light from rectilinear propagation*

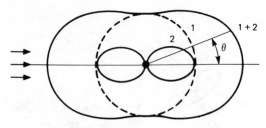

1 = polarized with electric vector $\perp$ plane of drawing
2 = polarized with electric vector in plane of drawing
1 + 2 = total

Fig. 9 *Rayleigh scattering: Polar diagram of scattered intensity if incident radiation is unpolarized*

with n = refractive index,
 r = radius of molecule.

For a unit volume with N molecules per cm^3 (see definition of the volume scattering function $\beta(\theta)$ in Fig. 8), $\beta(90)$ is then given by

$$\beta(90) = \frac{N \cdot I(90) d^2}{I_0} = \frac{1}{2} N k^4 p^2 = N \frac{8\pi^4}{\lambda^4} p^2 \quad (4.3)$$

and the Rayleigh ratio is therefore

$$R = \beta(90) = N \frac{8\pi^4}{\lambda^4} r^6 \left(\frac{n^2-1}{n^2+2}\right)^2 \quad (4.4)$$

If we express the volume function scattering, $\beta(\theta)$, by $\beta(90)$, we can write:

$$\beta(\theta) = \beta(90)(1 + \cos^2\theta) \quad (4.5)$$

and by integrating over 4π we have

$$b = 2\pi \int_0^\pi \beta(\theta) \sin\theta d\theta = \frac{16\pi}{3} \beta(90) \quad (4.6)$$

and

$$b = \frac{16\pi}{3} 8N \frac{\pi^4}{\lambda^4} r^6 \left(\frac{n^2-1}{n^2+2}\right)^2 \quad (4.7)$$

In the case of anisotropic molecules the Rayleigh ratio, R, as defined by eq. (4.4) has to be modified by the so-called Cabannes factor (Cabannes, 1920):

$$f = \frac{6 + 6\delta}{6 - 7\delta} \quad (4.8)$$

where δ is the depolarization ratio at $\theta = 90°$

$$\delta = \frac{i_\varrho(90)}{i_r(90)} \quad (4.9)$$

where i_r is perpendicular component,
 i_ϱ is horizontal component,
  both in the scattering plane.

The Rayleigh ratio becomes then:

---

scattering and can qualitatively be explained by the Rayleigh theory (Rayleigh, 1871): A particle (molecule) placed in an electrical field, $E \sim e^{i\omega t}$, behaves like a dipole of momentum, $P = p \cdot E$, where p is the polarizability. Generally p is a tensor.

According to electromagnetic theory the oscillating dipole emits an electromagnetic field in all directions, the scattered radiation. In the case where p is a scalar, i.e. the polarizability is isotropic, the intensity of this field is given by (Fig. 9)

$$I(\theta) = \frac{I_0 k^4 p^2}{2d^2}(1 + \cos^2\theta) \quad (4.1)$$

where $k = \frac{2\pi}{\lambda}$ is the wave number,
  d is the distance from the scattering molecule, and
  θ is the scattering angle.

If $\beta(\theta)$ is the volume scattering function, we define $\beta(90)$ as the Rayleigh ratio R. For isotropic molecules p is given by the Lorentz-Lorenz formula

$$p = \frac{n^2-1}{n^2+2} \cdot r^3 \quad (4.2)$$

$$R_{tot} = R_{iso} \cdot f = N \frac{8\pi^4}{\lambda^4} r^6 \left(\frac{n^2-1}{n^2+2}\right)^2 \cdot \frac{6+6\delta}{6-7\delta} \quad (4.10)$$

and the scattering coefficient b:

$$b = \frac{8\pi}{3} \beta(90) \cdot \frac{2+\delta}{1+\delta} \quad (4.11)$$

and the volume scattering function:

$$\beta(\theta) = \beta(90) \cdot \left(1 + \frac{1-\delta}{1+\delta} \cos^2\theta\right) \quad (4.12)$$

$\frac{1-\delta}{1+\delta} = \frac{i_\varrho - i_r}{i_\varrho + i_r}$ is the degree of polarization.

Rayleigh's molecular scattering theory, as expressed in formulas (4.4) and (4.7), considers free molecules and does not take into account the interaction between molecules in a liquid. This is the reason why the theory overestimates the scattering intensity by a factor of roughly 100 as experience shows. However, as far as dependence from wavelength and scattering angle and polarization are concerned, the Rayleigh theory is in good agreement with experience. A better approach for the calculation of the isotropic part of the Rayleigh ratio is provided by the theory of fluctuations formulated by Smoluchowski (1908) and Einstein (1910). In this theory scattering is considered to be caused by random density fluctuations in the liquid which cause fluctuations in the dielectric constant, ϵ; the expression for the isotropic part of the Rayleigh ratio, as given by this theory, is:

$$R_{iso} = \frac{\pi^2}{2\lambda_0^4} \Delta V \langle \Delta\epsilon \rangle^2 \quad (4.13)$$

where $\langle \Delta\epsilon \rangle^2$ is the mean square of the fluctuation of the dielectric constant in a small volume, ΔV, λ_0 is the wave length in the vacuum.

The fluctuation in ϵ can be considered as a result of fluctuations of density, ρ, therefore

$$\langle \Delta\epsilon \rangle^2 = \left(\frac{d\epsilon}{d\rho}\right)^2 \cdot \langle \Delta\rho \rangle^2 \quad (1.14)$$

For the product $\Delta V \langle \Delta\rho \rangle^2$ the statistical thermodynamics gives

$$\Delta V \langle \Delta\rho \rangle^2 = KT \cdot \beta_T \cdot \rho^2 \quad (4.15)$$

where β_T is the compressibility, $\frac{1}{p}\left(\frac{\Delta v}{v}\right)_T$ in atm^{-1} or cm$^2 \cdot$ dyn^{-1}, K is the Boltzmann constant (1.38·10^{-16} erg·grd^{-1}), T is the absolute temperature.

R_{iso} can therefore be written as:

$$R_{iso} = \frac{\pi^2}{2\lambda_0^4} KT \cdot \beta_T \rho^2 \left(\frac{d\epsilon}{d\rho}\right)^2 \quad (4.16)$$

and since $\epsilon = n^2$

$$R_{iso} = \frac{2\pi^2}{\lambda_0^4} KT \cdot \beta_T \left(\rho n \frac{dn}{d\rho}\right)^2 \quad (4.17)$$

for the relation between the refractive index, n, and the density, ρ, one can use for example the Lorentz-Lorenz formula

$$\frac{n^2-1}{n^2+2} \cdot \frac{1}{\rho} = \text{const.} \quad (4.18)$$

Differentiating this, we find

$$\rho \frac{d\epsilon}{d\rho} = \frac{(n^2-1)(n^2+2)}{3} \quad (4.19)$$

and introducing this into eq. (4.16) we obtain the final formula for R_{iso}:

$$R_{iso} = \frac{\pi^2}{2\lambda_0^4} K \cdot T\beta_T \frac{(n^2-1)^2(n^2+2)^2}{9} \quad (4.20)$$

and

$$\beta(\theta) = \beta(90)(1 + \cos^2\theta) = R_{iso}(1+\cos^2\theta) =$$

$$= \frac{\pi^2}{2\lambda_0^4} K \cdot T\beta_T \frac{(n^2-1)^2(n^2+2)^2}{9}(1+\cos^2\theta)$$

$$(4.21)$$

and after integration over 4π:

$$b = 2\pi \int_0^\pi \beta(\theta) \sin\theta d\theta = \frac{16\pi}{3}\beta(90) =$$

$$= \frac{16\pi}{3} \cdot R_{iso} =$$

$$= \frac{8\pi^3}{3} K \cdot T\beta_T \frac{(n^2-1)^2(n^2+2)^2}{9} \cdot \frac{1}{\lambda_0^4} \quad (4.22)$$

Introducing the constants, K = 1.3807 · 10^{-16} erg.grad^{-1} and β_T = 46 · 10^{-12} cm^2.dyn^{-1} into eq. (4.22), we find:

$$b = 1.434 \cdot 10^{-4} \cdot \frac{f(\lambda_0)}{\lambda_0^4} \cdot \frac{T}{273} \text{ (m}^{-1}) \quad (4.23)$$

where λ_0 is given in μ and $f(\lambda_0)$ is a slightly dependent function of λ_0, and is obtained from the wavelength dependence of n.

The isotropic part of the Rayleigh ratio, R_{iso}, is then:

$$R_{iso} = \frac{3}{16\pi} \cdot b = 8.556 \cdot 10^{-6} \cdot \frac{T}{273} \cdot \frac{f(\lambda_0)}{\lambda_0^4} \quad (4.24)$$

In Table 4.1 R_{iso} is given for T = 0 and T = 20°C, as function of $\lambda_0(\mu)$; n is obtained from data published by Lauscher (1955).

By fitting the data of Table 4.1 to a curve, $A \cdot \lambda_0^{-b}$ we find:

$$R_{iso}(T=0°C) = 7.437 \cdot 10^{-2} \lambda_0^{-4.21} \text{ (m}^{-1})$$

λ_0 in μ.

Table 4.1 : R_{iso} ($.10^4 \, m^{-1}$) calculated by Smoluchowski - Einstein Theory

$\lambda_0 (\mu)$	n	$f(\lambda_0) = \dfrac{(n^2 - 1)^2 \, (n^2 + 2)^2}{9}$	$T = 0°C$ R_{iso} $(.10^4 \, m^{-1})$	$T = 20°C$ R_{iso} $(.10^4 \, m^{-1})$
.308	1.3569	1.1600	11.03	11.84
.359	1.3480	1.0809	5.57	5.98
.400	1.3433	1.0407	3.48	3.73
.434	1.3403	1.0157	2.45	2.63
.486	1.3371	.9895	1.52	1.63
.589	1.3330	.9566	0.680	0.730
.768	1.3289	.9246	0.227	0.244
1.000	1.3247	.8926	0.076	0.082

Table 4.2: Absorption of water in the IR

$\lambda (\mu)$	1	1.5	2	3	4	6	8
$c = a$ (cm^{-1})	0.42	21	69.2	2700	300	2500	800

Anisotropic scattering can also be explained by the fluctuation theory when fluctuations are considered to prefer certain orientations of anisotropic molecules. Work by Cabannes (1929), Prinz and Prinz (1956), Benoit and Stockmayer (1956) shows that formula (4.10) remains valid.

If an electrolyte is added to pure water, an increase of scattering is observed (Sweitzer (1927), Lochet (1953), Morel (1966), Pethica and Smart (1966). The fluctuation theory explains this increase by fluctuation in the concentrations of the electrolyte. For a 35 $^o/oo$ NaCl solution, which represents very well the sea water of 38 $^o/oo$ salinity, as far as the Cl^- concentration is concerned, an increase of $1.18 - 1.20$ times the scattering of pure water is found. Considering the presence of all the other cations and anions, an estimation (Morel, 1974) gives an increase in scattering of sea water of $35 - 38 \, ^o/oo$ salinity compared to pure water by a factor of 1.30 (Fig. 10).

The total attenuation coefficient of pure water was measured by several authors. The data, as shown in Fig. 11 are not very consistent, due to experimental difficulties, such as:

- radiometrical problems, since incident and transmitted flux are of the same order of magnitude;
- problems of stray light, high collimation - low acceptance angle;
- problems of purification: since c is small, big quantities of pure water are necessary.

Generally, path lengths between 1 m and 4 m are used.

The attenuation of pure water has a minimum at $\lambda = 470$ nm; at this wave length the contribution of

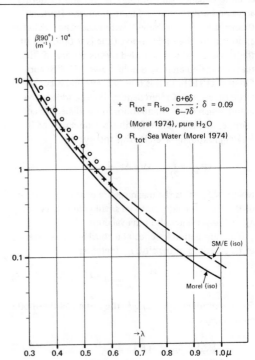

Fig. 10 Scattering of water as a function of wavelength

scattering to the extinction is approximately 20% (b/c = 0.2). For increasing wave length the scattering becomes less important and for $\lambda > 600$ nm more than 99% of the extinction is due to absorption. The absorption in the IR is very high; some values are reported in Table 4.2 (Palmer, Williams, 1974).

476

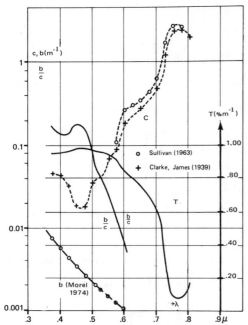

Fig. 11   Attenuation, scattering and transmittance of pure H_2O

Few data exist in the ultraviolet range, because of experimental difficulties (strong absorption by impurities, strong scattering).

Values obtained by Lenoble, Saint Guily (1955 indicate rather high transparency for UV (200 − 300 nm, c $<$ .1 m^{-1}).

The influence of temperature on the attenuation coefficient seems to be small (Collins, 1925).

5 SCATTERING AND ABSORPTION IN NATURAL WATER

Attenuation of light by natural water is strongly dependent from the content of suspended and dissolved matter. Whereas the scattering is mainly influenced by suspended organic and inorganic particles, the absorption is mainly influenced by dissolved matter. The definition of suspended matter is given by the pore diameter of the filter used for filtration, i.e. all the matter retained by a filter (usually 0.45 μ pore diameter) is called suspended matter (SM). The total amount of SM in the open ocean waters is of the order of 0.06 mg/l, the organic fraction being estimated to 20 − 60% with decreasing values for deep water. In coastal zones SM concentrations can reach rather high values (18 mg/l have been found in coastal zones of the Pacific, 30 mg/l in the North Sea).

The most important parameter for the scattering characteristics of natural water is the particle size dis-

tribution of SM. Microscopic examination, photographic techniques and Coulter counting techniques are used for the determination of size distributions. A hyperbolic distribution

$$N \sim d^{-\gamma} \tag{5.1}$$

is assumed now to represent best the particle size distributions measured by the various techniques. N is the number/cm^3 of particles with diameter greater than d, γ is an exponent ranging from 0.7 to 6. Fig. 12 shows some typical examples of cumulative size distribution.

From these results it appears that distributions with mode diameters of the order of 2 − 20 μ are normally found in ocea waters. In theoretical calculations of light scattering by particles it is often necessary to start from a size distribution n (r) instead from the integral of it (cumulative distribution N). In Fig. 13 we have assumed the distribution function

$$n(r) \sim r^6 e^{-2r}, \qquad r_{max} = 3\mu \tag{5.2}$$

which has been used by Kattawar and Plass (1969) and by Kattawar and Humphreys (1974) for Monte

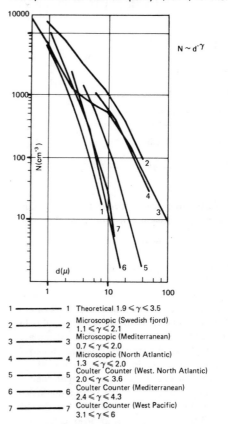

1 ——— 1	Theoretical	$1.9 \leqslant \gamma \leqslant 3.5$
2 ——— 2	Microscopic (Swedish fjord)	$1.1 \leqslant \gamma \leqslant 2.1$
3 ——— 3	Microscopic (Mediterranean)	$0.7 \leqslant \gamma \leqslant 2.0$
4 ——— 4	Microscopic (North Atlantic)	$1.3 \leqslant \gamma \leqslant 2.0$
5 ——— 5	Coulter Counter (West. North Atlantic)	$2.0 \leqslant \gamma \leqslant 3.6$
6 ——— 6	Coulter Counter (Mediterranean)	$2.4 \leqslant \gamma \leqslant 4.3$
7 ——— 7	Coulter Counter (West Pacific)	$3.1 \leqslant \gamma \leqslant 6$

Fig. 12   Cumulative particle size distributions

477

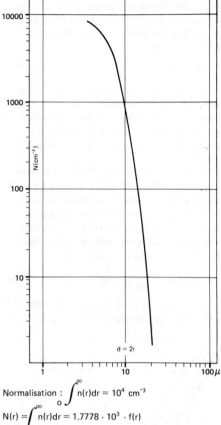

Normalisation : $\int_0^\infty n(r)dr = 10^4 \text{ cm}^{-3}$

$N(r) = \int_r^\infty n(r)dr = 1.7778 \cdot 10^3 \cdot f(r)$

$f(r) = e^{-2r}(5.625(1 + r) + 10.25r^2 +$
$\qquad + 7.5r^3 + 3.75r^4 + 1.5 \cdot r^5 + 0.5 \cdot r^6)$

$N(r) \sim r^{-\gamma}, \ 1.5 \leqslant \gamma \leqslant 6.2$

Fig. 13   Cumulative particle size distribution

Carlo calculations of light distribution in ocean. Calculating the cumulative size distribution

$$N(r) = \int_r^\infty n(r) \, dr \qquad (5.3)$$

and fitting it to Const. $r^{-\gamma}$ we find γ-values in the range of 1.5 to 6.2.

Although the mode radia of the shown distributions are of the order of $1 - 10 \ \mu$, the large particles are nevertheless of importance for the scattering of natural water, as is shown by Figs. 14 and 15. Here the distribution of the geometric cross sections of particles of different diameter is plotted. It appears that though few particles of large size are present, their contribution to the total scattering area is considerable. In any case it is clear that for the theoretical calculation of light scattering by suspended matter in water we have to assume that the particle radius is

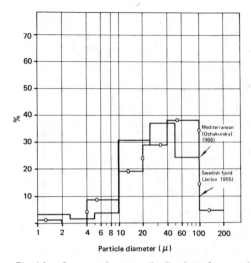

Fig. 14   Cross-section area distribution of suspended matter
(Normalized: total area = 100%)

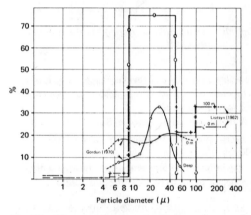

Fig. 15   Cross-section area distribution of suspended matter
(Total area = 100%)

greater than the wavelength of the scattered light and that the calculation has to consider a polydispersed system.

The theory that calculates the scattering of electromagnetic radiation by spherical particles has been developed by Mie (1908). It starts from the rigorous solution of the Maxwell equations for a sphere with complex refractive index $m = \tilde{n} - i\chi$ and radius, a, which is interacting with a plane monochromatic wave described by

$$\vec{E} = \vec{a}_x e^{-ikz+i\omega t}$$

$$\vec{H} = \vec{a}_y e^{-ikz+i\omega t} \qquad (5.4)$$

478

$\vec{E}$ and $\vec{H}$ being the electric and magnetic field vectors respectively, $\vec{a}_k$, $\vec{a}_y$ are unit vectors in the direction of the x and y-axes.

The wave is propagating in positive z direction. Its wavelength is λ, $k = 2\pi/\lambda$ and frequency ν, $\omega = 2\pi\nu$.

As a final result of the theory one obtains the function which describes the intensity of the scattered wave in the direction θ, φ (see Fig. 16), it is given by

$$F(\theta, \varphi) = i_2(\theta) \cos^2\varphi + i_1(\theta) \sin^2\varphi \qquad (5.5)$$

with

$$i_1(\theta) = |S_1(\theta)|^2 = |\sum_{n=1}^{\infty} \frac{2n+1}{n(n+1)} \{a_n \pi_n (\cos\theta) + b_n \tau_n (\cos\theta)\}|^2$$

$$i_2(\theta) = |S_2(\theta)|^2 = |\sum_{n=1}^{\infty} \frac{2n+1}{n(n+1)} \{b_n \pi_n(\cos\theta) + a_n \tau_n (\cos\theta)\}|^2$$

$$(5.6)$$

describing the intensities in perpendicular and parallel direction with respect to the scattering plane, and

$$a_n = \frac{\psi'_n(y)\psi_n(x) - m\psi_n(y)\psi'_n(x)}{\psi'_n(y)\varphi_n(x) - m\psi_n(y)\varphi'_n(x)} \qquad (5.7)$$

$$b_n = \frac{m\psi'_n(y)\psi_n(x) - \psi_n(y)\psi'_n(x)}{m\psi'_n(y)\varphi_n(x) - \psi_n(y)\varphi'_n(x)}$$

where $y = m.ka$ and ψ_n, φ_n are the Riccati-Bessel functions differing from spherical Bessel functions $j_n(z)$ and $h_n^{(2)}(z)$ by the factor z:

$$\psi_n(z) = zj_n(z) \qquad \varphi_n(z) = z.h_n^{(2)}(z) \qquad (5.8)$$

τ_n and $\pi_n(\cos\theta)$ are defined by

$$\pi_n(\cos\theta) = \frac{1}{\sin\theta} P_n^1(\cos\theta)$$

$$\tau_n(\cos\theta) = \frac{d}{d\theta} P_n^1(\cos\theta) \qquad (5.9)$$

$P_n^1(z)$ being associated Legendre polynoms. The intensity of the scattered light at a distance r from the scattering particle, is given by

$$I = I_0 \cdot \frac{F(\theta, \varphi)}{k^2 r^2} \qquad (5.10)$$

where I_0 is the intensity of the incoming light. The scattering cross section, C_{sca}, is then defined from energy conservation principles: the total scattered energy is that incident energy falling on the cross section C_{sca}. From eq. (5.10) we have

$$I_0 \cdot C_{sca} = \int_{4\pi r^2} I \, df = I_0 \frac{\int F(\theta, \varphi) \sin\theta \, d\theta \, d\varphi}{k^2} \qquad (5.11)$$

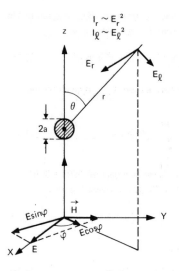

$$\vec{E} = \vec{a}_x e^{-ikz + i\omega t}, \quad \vec{H} = \vec{a}_y e^{-ikz + i\omega t}; \quad k = \frac{2\pi}{\lambda}, \quad \omega = 2\pi\nu$$

Scattering center : sphere radius a, $x = \frac{2\pi a}{\lambda}$

Refr. Index $n = m - ix$

Scattering crossection :

$$C_{sca} = \pi a^2 \cdot Q_{sca}, \quad Q_{sca} = \frac{2}{x^2} \sum_{n=1}^{\infty} (2n + 1) \{|a_n|^2 + |b_n|^2\}$$

Extinction crossection :

$$C_{ext} = \pi a^2 Q_{ext}, \quad Q_{ext} = \frac{2}{x^2} \sum_{n=1}^{\infty} (2n + 1) Re(a_n + b_n)$$

Polydispersed systems (N_i spheres/cm^3 of radius a_i)

Scattering coefficient :

$$b = \pi \sum_{i=1}^{\infty} Q_{sca,i} \cdot N_i \cdot a_i^2$$

Fig. 16   Mie-scattering theory

and

$$C_{sca} = \frac{1}{k^2} \int_0^{2\pi} \int_0^{\pi} F(\theta, \varphi) \sin\theta \, d\theta \, d\varphi \qquad (5.12)$$

and using eq. (5.5):

$$C_{sca} = \frac{\lambda^2}{4\pi} \int_0^{\pi} (i_2(\theta) + i_1(\theta)) \sin\theta \, d\theta \qquad (5.13)$$

The ratio of the scattering cross section, C_{sca}, to the projected surface of the scattering sphere, πa^2, is called the efficiency factor for scattering:

$$Q_{sca} = \frac{C_{sca}}{\pi a^2} \qquad (5.14)$$

From eq. (5.13) and $x = 2\pi a/\lambda$ we have:

$$Q_{sca} = \frac{1}{x^2} \int_0^{\pi} (i_1(\theta) + i_2(\theta)) \sin\theta \, d\theta \qquad (5.15)$$

479

Introducing i_1 and i_2 from eq. (5.6) into eq. (5.15) and integrating over θ, one obtains for Q_{sca}:

$$Q_{sca} = \frac{2}{x^2} \sum_{n=1}^{\infty} (2n+1)\left\{|a_n|^2 + |b_n|^2\right\} \qquad (5.16)$$

The extinction cross section, C_{ext}, is given by the theory as

$$C_{ext} = \frac{4\pi}{k^2} \cdot \text{Re}\left\{S(0)\right\} \qquad (5.17)$$

where $\text{Re}\left\{S(0)\right\}$ is the real part of $S(0) = S_1(0) = S_2(0)$ (see formulas (5.6)). Considering that

$$\pi_n(1) = \tau_n(1) = \frac{1}{2}n(n+1) \qquad (5.18)$$

one finds for C_{ext}

$$C_{ext} = \frac{2\pi}{k^2} \sum_{n=1}^{\infty} (2n+1)\, \text{Re}\,(a_n + b_n) \qquad (5.19)$$

a_n and b_n from eq. (5.7). And, if we define an efficiency factor for extinction, Q_{ext}, it becomes:

$$Q_{ext} = \frac{C_{ext}}{\pi a^2} = \frac{2}{x^2} \sum_{n=1}^{\infty} (2n+1)\, \text{Re}\,(a_n + b_n) \quad (5.20)$$

The absorption cross section, C_{abs}, is obtained by difference

$$C_{abs} = C_{ext} - C_{sca} \qquad (5.21)$$

It is equal to zero for real refractive index $\tilde{n} = m$, $x = 0$.

Before discussing polydispersed systems, we should note the limiting cases of the Mie-theory. The characteristic parameters of the Mie-theory are (see formulas (5.7)):

$$x = \frac{2\pi a}{\lambda} \quad \text{and} \quad y = m.k.a = m.x$$

for x and y $\ll 1$, i.e. a $\ll \lambda$, the Mie theory gives the same result as the Rayleigh theory, i.e. λ^{-4}-dependence of the scattering and efficiency factors proportional to $(m-1)^2 x^4$. For large x and y the Mie theory describes the classical diffraction theory as formulated by Fresnel from the Huygens principle, many years before Mie. It results in an efficiency factor, $Q = 2$, independent of wavelength. A special case being $m \to 1$ and $|m - 1| \ll 1$ (low absorption of the spheres), called anomalous diffraction, is characterized by strong forward scattering.

In a medium with N scattering spheres per cm³ the scattering coefficient, b, is then given by

$$b = N.C_{sca} = N.\pi a^2 \cdot Q_{sca} \qquad (5.22)$$

and for the case of polydisperse system

$$b = \pi \sum_{i=1}^{n} Q_{sca,i} \cdot N_i a_i^2 \qquad (5.23)$$

with N_i the number of particles of radius a_i per cm³. The volume scattering function for a single particle is then

$$q_{sca}(\theta,\varphi) = C_{sca} \cdot \frac{F(\theta,\varphi)}{\int F(\theta,\varphi)\,d\omega}\; ; \quad d\omega = \sin\theta\,d\theta\,d\varphi \quad (5.24)$$

For natural light with random polarisation equation (5.5) can be averaged over φ, and since

$$\frac{1}{2\pi} \int_0^{2\pi} \cos^2 \varphi\,d\varphi = \frac{1}{2\pi} \int_0^{2\pi} \sin^2 \varphi\,d\varphi = \frac{1}{2}$$

the scattered intensity becomes

$$I = I_0 \frac{i_1(\theta) + i_2(\theta)}{2k^2 r^2} \qquad (5.25)$$

and in analogy with the total scattering cross section, we find the angular scattering cross section

$$C_{sca}(\theta) = \frac{\lambda^2}{8\pi^2} (i_1(\theta) + i_2(\theta)) \qquad (5.26)$$

For a polydisperse system we have, in analogy to eq. (5.16), to sum over the contributions from particles of various radii:

$$\beta(\theta) = \sum_{i=1}^{n} N_i C_{sca,i}(\theta) \qquad (5.27)$$

The interest in computation of scattering by the Mie-theory lies in the fact that the various scattering properties (intensity, angular distribution and polarization) depend on particle size distribution and refractive index of the particles. By comparison of experiments with theoretical evaluations, it is possible to derive valuable information about particle size and refractive index.

Experimental data on particle scattering* in ocean water are obtained from in situ and in vitro measurements of the volume scattering function $\beta(\theta)$. In situ measurements are generally more reliable because they are not hampered by the risk of contamination and changes in particle characteristics during the time lapse between sampling and measurement. Fig. 17 gives some examples for both types of measurements that cover a rather wide range from very turbid to very clear waters (Jerlov, 1976). The following features appear as characteristic of the particle scattering volume function of natural water: a very strong forward scattering compared to pure water Rayleigh

---

*   particle scattering = difference between total scattering and pure water scattering.

type scattering, and a pronounced difference in scattering in forward and backward direction, the slope of the forward scattering being greater for turbid waters.

Fig. 17d shows an example of a comparison of measured and calculated volume scattering function (Kullenberg, 1968). For the measurements at $\lambda = 632.8$ nm made in the surface layer of the Sargasso Sea, the best fit to the observations is found for an exponent of the cumulative particle size distribution function, $\gamma = 2$, and a particle refractive index, $m = 1.2$. Kullenberg (1974) has made extensive comparisons of experimental and theoretical results for angular volume scattering functions. For typical ocean waters (Sargasso Sea, Mediterranean) he finds that the scattering is essentially determined by particles of sizes, larger than $1 - 2\,\mu$. The average refractive index (relative to water) is of the order of 1.15 to 1.20. For coastal waters, represented by measurements made in the Baltic Sea, the average refractive index of the particles seems to be lower, $= 1.04$ and the exponent, γ, of the cumulative particle size distribution varies between 1.25 and 1.5. However, the agreement between calculation and experiment is less good as for ocean waters.

The wave length dependence of scattering by natural waters is determined essentially by three effects:

- selectivity of scattering by pure water;

- selectivity of scattering due to the presence of small particles;
- absorption by scattering particles. $m = \tilde{n} - i\chi$, $\chi \neq 0$, $\chi\,(\lambda)$.

The first effect is especially important for the backscattering from clear waters, whereas the forward scattering for practically all natural waters depends only slightly on wave length. Selectivity due to scattering by small particles is considered a minor contribution because of the small total area. The third effect can give rise to a certain inconsistency of experimental data. Whereas in almost all measurements the scattering in the blue predominates over scattering in the green and in the red, there exist also results (Jerlov, 1953) showing higher scattering coefficients in the red than in the blue.

The dissolved optically active compounds of natural waters are known under the collective name "yellow substance"; their formation from carbohydrates is favoured by alcalic reaction, heat and the presence of amino acids (Kalle, 1962, 1966). The absorption coefficient of the yellow substance is studied in vitro and in situ, the first method being more suited for turbid waters, whereas the second is used in clear ocean water (Jerlov, 1974). The in vitro method consists in filtering the water probe from all particulate matter and by determining the absorption coefficient, a_y, at the wave length of 380 nm. For determining a_y in situ, transmittance meters are used and the total attenuation coefficient, $c\,(m^{-1})$, is determined at two wave lengths, 380 nm and 655 nm. Knowing the extinction coefficient, c_w, at these two wave lengths

$$c_w\,(380\ nm) = 0.04\ m^{-1}$$
$$c_w\,(655\ nm) = 0.30\ m^{-1}$$

the differences $(c-c_w)_{380}$ and $(c-c_w)_{655}$ are determined. Typical results are plotted in Fig. 18. (Jerlov, 1974, 1976). The major part of these measurements can be represented by a straight line

$$(c-c_w)_{380} = 1.8\,(c-c_w)_{655} \tag{5.28}$$

Since
$$(c-c_w)_\lambda = b_{p\lambda} + a_{p\lambda} + a_{y\lambda} \tag{5.29}$$

$b_{p\lambda}$ = scattering coefficient for wave length λ,
$a_{p\lambda}$ = particle absorption coefficient for wave length λ,
$a_{y\lambda}$ = yellow substance absorption coefficient for wave length λ,

and since $b_{p\lambda}$ is very slightly depending on wave length, the results indicate that the selectivity of the quantity $c-c_w$ is mainly due to absorption by particles and by yellow substance.

From particle scattering measurements at the two wave lengths 380 nm and 655 nm, Jerlov (1974) has derived linear relations between $c-c_w$ and b_p:

$$\begin{aligned} &\text{for } \lambda = 380\ nm \quad c-c_w = 2.1\,b_p \\ &\text{for } \lambda = 655\ nm \quad c-c_w = 1.43\,b_p \end{aligned} \tag{5.30}$$

If we assume that $a_y\,(655\ nm) = 0$ (see Fig. 19), then, by dividing the two results, we find

$$(c-c_w)_{380} = 1.5\,(c-c_w)_{655} \tag{5.31}$$

which is, considering the accuracy of the measurements given in Fig. 18, well consistent with eq. (5.28). From Jerlovs results (see eq. (5.30)) two useful relations between particle absorption, a_p, yellow substance absorption, a_y, and particle scattering, b_p, can be derived:

$$\begin{aligned} &\text{for } \lambda = 380\ nm \quad a_p + a_y = 1.1\,b_p \\ &\text{for } \lambda = 655\ nm \quad a_p = 0.43\,b_p \end{aligned}$$

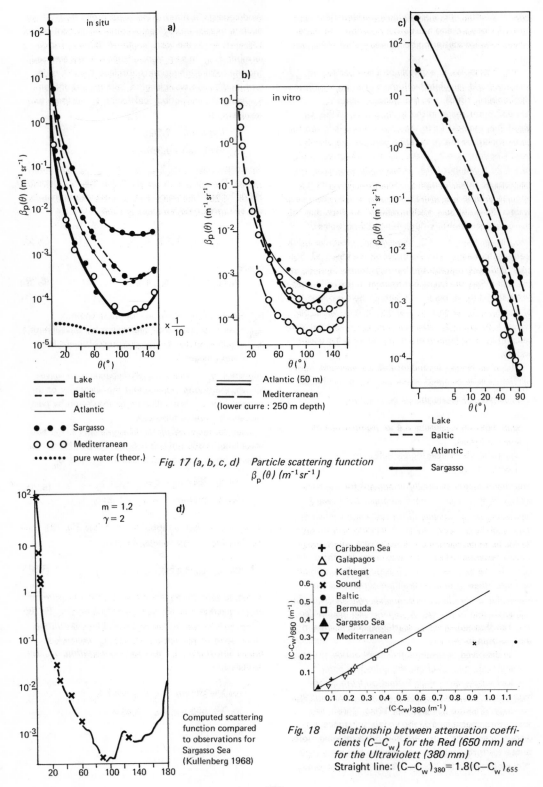

a) in situ

$\beta_p(\theta)$ (m^{-1} sr^{-1})

$\times \frac{1}{10}$

$\theta(°)$

—— Lake
- - - Baltic
—— Atlantic
● ● ● Sargasso
○ ○ ○ Mediterranean
•••••• pure water (theor.)

b) in vitro

$\beta_p(\theta)$ (m^{-1} sr^{-1})

$\theta(°)$

══ Atlantic (50 m)
━━ Mediterranean
(lower curre : 250 m depth)

c)

$\beta_p(\theta)$ (m^{-1} sr^{-1})

$\theta(°)$

—— Lake
- - - Baltic
—— Atlantic
━━ Sargasso

Fig. 17 (a, b, c, d) Particle scattering function
$\beta_p(\theta)$ (m^{-1}sr^{-1})

d)

m = 1.2
γ = 2

Computed scattering
function compared
to observations for
Sargasso Sea
(Kullenberg 1968)

+ Caribbean Sea
△ Galapagos
○ Kattegat
✕ Sound
● Baltic
□ Bermuda
▲ Sargasso Sea
▽ Mediterranean

$(C-C_w)_{650}$ (m^{-1})

$(C-C_w)_{380}$ (m^{-1})

*Fig. 18 Relationship between attenuation coeffi-
cients (C−C$_w$) for the Red (650 mm) and
for the Ultraviolett (380 mm)
Straight line: (C−C$_w$)$_{380}$= 1.8(C−C$_w$)$_{655}$*

482

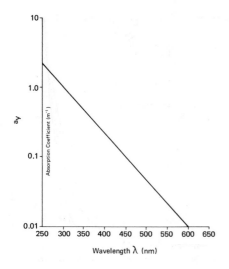

Fig. 19   Absorption curve for yellow substance

6 GLOBAL RADIATION INCIDENT ON THE SURFACE, ITS REFLECTION AND REFRACTION

The global irradiance incident on a natural water surface is composed, for the case of cloudless sky, of direct solar radiation and diffuse sky radiation. The unique source for both components is the extra-terrestrial sun-irradiance falling on the top of the atmosphere, E_0 (mW/cm^2). Its spectral composition has been studied extensively, a compilation of most reliable data has been published by Thekaekara (1972) and Labs, Neckel (1971) (see Fig. 20).

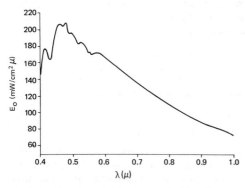

Fig. 20   Extraterrestrial solar spectral irradiance (from Thekaekara (1972))

Integrating E_0 over the wave length, the solar constant S_0, is obtained; its value is 0.136 W/cm^2.

Depending from the Earth-Sun distance, this value has to be multiplied by a factor, $f(t)$, which varies from 0.967 in July to 1.034 in January. If i is the zenith distance of the sun, then the spectral irradiance due to direct solar radiation at the water surface is

$$E_s(\lambda) = E_0(\lambda) \cos i \cdot T(\lambda)^{1/\cos i} \qquad (6.1)$$

where $T(\lambda)$ is the transmittance of the atmosphere for collimated light of wave length λ. The total irradiance is

$$E_{tot}(\lambda) = E_s(\lambda) + E_d(\lambda) \qquad (6.2)$$

with $E_d(\lambda)$ being the diffuse irradiance.
Both $T(\lambda)$ and $E_d(\lambda)$ depend strongly on the scattering and absorption properties of the components of the atmosphere. Scattering is mainly due to air-molecules and dust-particles, whereas absorption is mainly caused by H_2O, CO_2 molecules and ozone (O_3). The scattering by air-molecules is strongly selective (Rayleigh scattering) and proportional to λ^{-4}, while dust-particle scattering is approximately proportional to λ^{-1}. As a consequence, the diffuse spectral irradiance is shifted to the short-end part of the spectrum. It is clear that also the total diffuse irradiation is strongly influenced by the turbidity of the atmosphere. Fig. 21a shows some indicative evaluations of the percentage of diffuse irradiance as function of solar zenith angle, i, for various wavelengths and Fig. 21b shows the global spectral irradiance (sun plus sky) in relative units.

Experience shows that the spectral distribution of global irradiance is only slightly affected by turbidity of the atmosphere and that it is also little affected by the sun zenith angle up to $i < 60°$; only at low sun heights the diffuse sky radiation predominates, shifting the spectrum towards short wave lengths.

The diffuse irradiance, $E_d(\lambda)$, is an integral over the radiance coming from the various directions of the sky:

$$E_d(\lambda) = \int_0^{2\pi} \int_0^{\pi/2} L(\lambda, \theta, \varphi) \cos\theta \sin\theta \, d\theta \, d\varphi \qquad (6.3)$$

where φ is the azimuth angle and θ the zenith angle of the direction to the point of the sky with radiance $L(\theta, \varphi)$. Even for a cloudless sky the function $L(\lambda, \theta, \varphi)$ cannot be given in analytic form because it depends in complicated form on the turbidity of the atmosphere and sun zenith angle. For a completely cloud-overcast sky Moon and Spencer (1942) have proposed

$$L(\lambda, \theta, \varphi) = L(\lambda, \frac{\pi}{2})(1 + 2\cos\theta) \qquad (6.4)$$

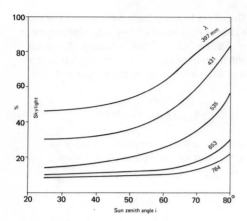

Fig. 21a   Percentage of skylight in the global irradiance

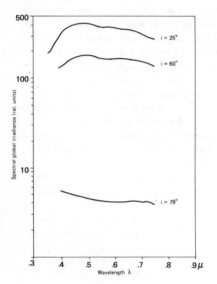

Fig. 21b   Spectral distribution of irradiance at different solar zenith angles

which, when introduced into eq. (6.3) gives

$$E_d(\lambda) = \frac{7}{3}\pi\, L(\lambda, \frac{\pi}{2})$$ (6.5)

where $L(\lambda, \frac{\pi}{2})$ is the spectral radiance from the zenith. If we assume that L does not depend on θ and φ, we find

$$E_d(\lambda) = \pi \cdot L(\lambda)$$ (6.6)

For the treatment of reflection and refraction of light at the air-water interface we consider first the direct

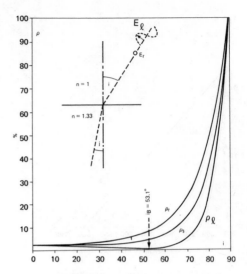

Fig. 22   Reflectance at a calm water surface
ρ_r : perpendicular $\left.\right\}$ to plane of incidence
ρ_ℓ : parallel

radiation. In this case, for a plane water surface, the formulas of Fresnel give the reflectances for polarized light perpendicular and parallel to the plane of incidence

$$\rho_r = \frac{\sin^2(i-j)}{\sin^2(i+j)}$$ (6.7)

$$\rho_\ell = \frac{tg^2(i-j)}{tg^2(i+j)}$$ (6.8)

where i is the angle of incidence and j the angle of refraction (see Fig. 22). For solar radiation, which is unpolarized, the reflection is the mean value of the two quantities

$$\rho_s = \frac{1}{2}(\rho_r + \rho_\ell) = \frac{1}{2}\frac{\sin^2(i-j)}{\sin^2(i+j)} + \frac{1}{2}\frac{tg^2(i-j)}{tg^2(i+j)}$$ (6.9)

From the refraction law of Snellius

$$\frac{\sin i}{\sin j} = \frac{n}{n'}$$ (6.10)

where n is the refractive index of water and n' that of air, which we assume here equal to 1, we can evaluate the refraction angle, j, as function of the incidence angle and calculate ρ_ℓ, ρ_r and ρ_s as function of i. For $n = 1.333$ ($\lambda = 580$ nm) these curves are given in Fig. 22. Since n is only slightly dependent on λ ($n = 1.348$ for $\lambda = 359$ and $n = 1.325$ for $\lambda=1000$ nm), these reflectance values can be used for all the visible part of the spectrum.

484

Table 6.1

m	-1	0	0.1	0.5	1.0	2.0	∞
$\rho_d(\%)$	17.73	6.6	5.45	4.90	4.65	3.0	2.04

Bouguer has given a form of equation (6.9) which is often useful when integration over the incidence angle is required:

$$\rho_s = 0.020 + 0.28\,(1-\cos i)^3 + 0.30\,(1-\cos i)^6 +$$
$$+\ 0.40\,(1-\cos i)^9 \tag{6.11}$$

As seen from eq. (6.8) and Fig. 22, ρ_ℓ becomes zero for $(i+j) = 90°$. Introducing $j = 90-i$ into eq. (6.10) gives the incidence angle, i_B, for which this condition is fulfilled:

$$\text{tg}\ i_B = n,\quad \text{and}\quad i_B = 53.1° \tag{6.12}$$

This equation is called Brewster's equation and the angle i_B is called Brewster's angle.

The reflectivity of a calm water surface for the diffuse sky irradiance can in principle be calculated from the equation

$$\rho_d = \frac{\int_0^{2\pi} d\varphi \int_0^{\pi/2} di\,\rho(i)\,L(i,\varphi)\cos i\,\sin i}{\int_0^{2\pi} \int_0^{\pi/2} L(i,\varphi)\cos i\,\sin i\,di\,d\varphi} \tag{6.13}$$

Using Bouguer's formula for (i) and

$$L(i,\varphi) = L(\pi/2)\cdot\cos^m i \tag{6.14}$$

Lauscher (1955) has calculated ρ_d for various exponents m (see Table 6.1).

Using Bouguer's formula (6.11), we can calculate easily the reflectance of calm water for a completely overcast sky, the radiance of which is given by eq. (6.4). The result is $\rho_d = 5.08\%$.

Now, if we want to evaluate the total reflectance of a calm water surface, we have to use the reflectance values, ρ_s, from eq. (6.9) or Fig. 22, for the direct sun irradiance, eq. (6.1) and ρ_d for the diffuse irradiance. Doing this, we can formulate the total reflectance

$$\rho(i,\lambda) = \frac{\rho_s(i)\cdot E_s(i,\lambda) + \rho_d\,E_d(i,\lambda)}{E_s(i,\lambda) + E_d(i,\lambda)} \tag{6.15}$$

or, if we call n the ratio of diffuse to global irradiance

$$n(i,\lambda) = \frac{E_d(i,\lambda)}{E_s(i,\lambda) + E_d(i,\lambda)} \tag{6.16}$$

we can write for $g(i,\lambda)$:

$$\rho(i,\lambda) = \rho_s(i)\,(1-n(i,\lambda)) + \rho_d\,n(i,\lambda) \tag{6.17}$$

Neumann and Hollman (1961) have used these formulas to calculate the total amount of energy reflected from a calm water surface. Their results are shown in Fig. 23 as function of solar elevation and n. For low levels of diffuse irradiance the reflected energy has a maximum at a rather low solar elevation of 20° and a minimum at about 50°. For turbid atmosphere and higher contribution from diffuse irradiance the maximum reflected energy is found for high solar elevation. All curves intersect at about 29° solar elevation ($i = 61°$) where $\rho_s = \rho_d = 0.066$.

An important implication of the refraction at the water-air boundary is that only a small part of the light directed from the water into the air is transmitted by the water surface, due to total reflection. The maximum angle of inclination for an upwelling light beam at which the light leaves partially a calm water

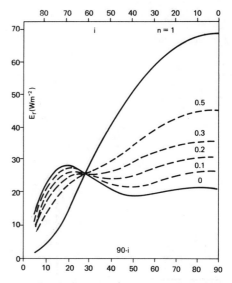

Fig. 23 Reflected energy as a function of solar elevation for different portions of diffuse radiation, n

surface is $j = 48.6°$. It follows from Snellius' low (eq. (6.10)) for $i = 90°$:

$$j = \text{arc sin} \frac{1}{n} = 48.6° \qquad (6.18)$$

In terms of radiometry this has the consequence that radiance leaving the water is reduced by a factor of $1/n^2 \cong 0.56$ due to the fact that the solid angle in air is n^2 times the solid angle in water (see Chapter 2 and Fig. 4b.

Austin (1974) gives data for reflectance and transmittance for both directions air-water for calm and also for a rough water surface. These results are given in Fig. 24a and b. The calculation of reflectance and transmittance for a rough sea is generally made by using statistical models for the distribution of inclination angles for surface elements of water roughened by a given wind speed. Cox and Munk (1955) have derived such a distribution from observations of sunglitter:

$$\gamma(z_x, z_y)\delta z_x \delta z_y = \frac{1}{\pi\sigma^2} \exp\left(\frac{z_x^2 + z_y^2}{\sigma^2}\right)\delta z_x \delta z_y \quad (6.19)$$

where $\gamma(z_x, z_y)$ is the probability to find the inclinations of a surface element in the range $z_x \pm \frac{1}{2}\delta z_x, z_y \pm \frac{1}{2}\delta z_y$. x, y and z are the axes of a cartesian coordinate system (see Fig. 25). The inclinations z_x and z_y can be obtained from the azimuth and zenith angle of the normal of the surface element α and β.

$$z_x = \sin\alpha \cdot tg\beta$$
$$z_y = -\cos\alpha \cdot tg\beta \qquad (6.20)$$

The standard deviation of the distribution is related to the wind velocity, v, by the empirical relation:

$$\sigma^2 = 0.003 + 5.12 \cdot 10^{-3}v \pm 0.004 \qquad (6.21)$$

where v is the wind velocity in m/sec.

The reflected irradiance for a rough surface, defined by the equations above, is now calculated by integrating over all the inclinations $z_x z_y$:

Let $E(i)$ be the irradiance falling on a horizontal plane from above, then a surface element inclined by β with respect to the horizontal plane receives the irradiance

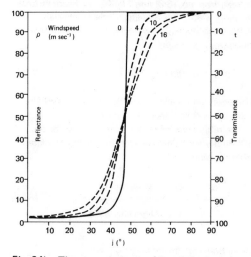

Fig. 24b   Time averaged water/air reflectance and transmittance from below

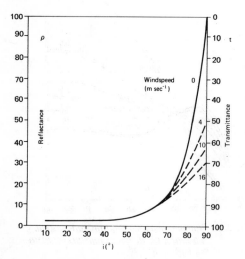

Fig. 24a   Time averaged air/water reflectance and transmittance from above

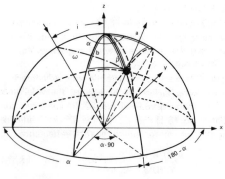

$\cos\omega = \cos i\cos\beta + \sin i\sin\beta\cos\alpha$
$z_x = tgb = tg\beta\cos(\alpha{-}90) = tg\beta\sin\alpha$
$z_y = tga = \cos(180{-}\alpha)\cdot tg\beta = -\cos\alpha\, tg\beta$

Fig. 25   Coordinate system for Cox, Munk equation

486

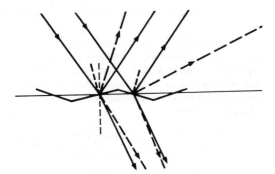

Fig. 26 *Reflection of sunlight at a calm and a rough surface (Sunglitter)*

$$E(i) \cdot \cos \omega \cdot \frac{1}{\cos \beta} \cdot \gamma(z_x, z_y) \delta z_x \delta z_y \qquad (6.22)$$

where ω is the angle between sun direction and the surface normal a (see Fig. 25). ω is given by:

$$\cos \omega = \cos i \cos \beta + \sin i \sin \beta \cos \alpha \qquad (6.23)$$

In order to obtain the reflectivity of a rough water surface, we have now to multiply eq. (6.22) by $\rho(\omega)$ from Fresnel equations, substituting i by ω and to integrate over all surface elements i.e. over all possible inclinations:

$$\rho(i) = \frac{E_r(i)}{E(i)} = \int_{z_x} \int_{z_y} \rho(\omega) \frac{\cos \omega}{\cos \beta} \gamma(z_x, z_y) dz_x dz_y$$

$$(6.24)$$

Transformation from variables z_x, z_y to variables α, β (see eq. (6.20) and Fig. 23) gives (Raschke (1971)):

$$\rho(i) = \int_0^{\pi/2} \int_0^{2\pi} \rho(\omega) \frac{\cos \omega}{\cos \beta} \frac{\mathrm{tg}\beta}{\cos^2 \beta} \cdot \frac{1}{\pi \cdot \sigma^2} e^{-\frac{\mathrm{tg}^2 \beta}{\sigma^2}} d\alpha d\beta$$

$$(6.25)$$

This is in principle the solution which leads to the results shown in Fig. 24a and b. It must be pointed out that this approach does not take into account the shadowing effect at great sun zenith angles ($i > 80°$). Cox and Munk (1956) have evaluated this effect, which results in a decrease of reflectivity above $i = 80°$ and consequently in an increased transmittance of sunlight into the water. Cox and Munk (1956) and Burt (1954) have evaluated the reflectance of a rough water surface for diffuse sky light. The result is that the roughness influences only slightly the diffuse reflectivity. For a uniform sky radiance 6.6% of the light is reflected by a calm surface, whereas for rough

sea the value is 5.7% (Burt, 1954) and 5.0–5.5% (Cox, Munk, 1956). The effect of the roughness of the sea is, as shown schematically in Fig. 26, a widening of the reflected sundisk resulting in a cone of reflected sunlight. The angle of this cone is related to the roughness of the water surface, i.e. the wind speed. The intensity of reflected light decreases towards the outer parts of the cone due to the low probability high slope angles. This phenomenon is known as sunglint.

7 THEORY OF RADIATIVE TRANSFER IN WATER

The simplest approach to calculate the light distribution below the water surface, is to consider the incoming sunlight as the source of radiation and to integrate over the contribution from single scattering events (Lauscher, 1955, Jerlov and Fukuda, 1960). The validity of this method is evidently limited to water of low turbidity and to low depth.

We consider the light intensity at a point Q (see Fig. 27) due to light scattered at P. If E is the sun irradiance just below the water surface, then the irradiance at P is given by

$$\frac{E}{\cos j} e^{-c \frac{x}{\cos j}}$$

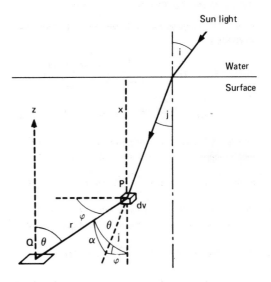

Radiance at Q: $dL = \dfrac{E}{\cos j} e^{-\frac{cx}{\cos j}} \cdot \beta(\alpha) \cdot e^{-cd} dr$

$z = x + r \cos \theta$ $\qquad \cos \alpha = \cos j \cos \theta + \sin j \sin \theta \cos \varphi$

Fig. 27 *Geometry for the evaluation of scattered light*

487

where x is the depth of point P and c the extinction coefficient. The light intensity scattered by the volume element, dv, at P, into the direction Q, is then by definition (see Fig. 8):

$$dI = \frac{E}{\cos j} e^{-\frac{cx}{\cos j}} \beta(\alpha) dv \qquad (7.1)$$

The irradiance at Q, normal to the direction, r, which is due to scattering at P, is then

$$dE_{sc} = \frac{E}{\cos j} e^{-\frac{cx}{\cos j}} \beta(\alpha) dv \cdot \frac{1}{r^2} e^{-cr} \qquad (7.2)$$

and $dv = r^2 \cdot d\omega dr = r^2 \cdot \sin\theta d\theta d\varphi dr$. The radiance $dE/d\omega$ at Q becomes then

$$dL = \frac{E}{\cos j} e^{-\frac{cx}{\cos j}} \beta(\alpha) e^{-cr} dr \qquad (7.3)$$

Introducing $x = z - r\cos\theta$ into eq. (7.2) and integrating over r from 0 to $\frac{r}{\cos\theta}$ (surface) and over φ and θ from 0 to 2π and 0 to $\pi/2$ respectively, we find the downwelling scattered irradiance at Q:

$$E_{sc} = \frac{E}{\cos j} e^{-\frac{cz}{\cos j}} \frac{1}{c} \int_0^{2\pi} \int_0^{\pi/2} \frac{\beta(\alpha)\sin\theta}{\frac{1}{\cos\theta} - \frac{1}{\cos j}} \cdot$$

$$\cdot (1 - e^{-cz(\frac{1}{\cos\theta} - \frac{1}{\cos j})}) d\theta d\varphi \qquad (7.4)$$

Adding to this the downward solar irradiance,

$E \cdot c^{-\frac{cz}{\cos j}}$ we have the total downwelling irradiance at depth z:

$$E_d(z) = E \cdot e^{-\frac{cz}{\cos j}} (1 + \frac{1}{c \cdot \cos j} \int_0^{2\pi} \int_0^{\pi/2} \frac{\beta(\alpha)\sin\theta}{\frac{1}{\cos\theta} - \frac{1}{\cos j}} \cdot$$

$$\cdot (1 - e^{-cz(\frac{1}{\cos\theta} - \frac{1}{\cos j})}) d\theta d\varphi \qquad (7.5)$$

For $j = 0$ (sun in the zenith and $\theta = \alpha$) and small optical depth ($cz \ll 1$) the equation becomes:

$$E_d(z) = E \cdot e^{-cz} (1 + 2\pi z \int_0^{\pi/2} \beta(\theta)\sin\theta d\theta) \qquad (7.6)$$

Due to the fact that backscattering is much smaller than forward scattering, we can write approximately

$$2\pi \int_0^{\pi/2} \beta(\theta)\sin\theta d\theta \approx 2\pi (\int_0^{\pi/2} \beta(\theta)\sin\theta d\theta +$$

$$+ \int_{\pi/2}^{\pi} \beta(\theta)\sin\theta d\theta) = b \qquad (7.7)$$

and

$1 + bz \approx e^{+bz}$ since $bz \ll 1$

and we obtain finally because of $c = a + b$

$$E_d(z) = E \cdot e^{-cz + bz} = E \cdot e^{-az} \qquad (7.8)$$

This indicates that the downwelling irradiance at low optical depth is determined by absorption only.
By integrating eq. (7.3) over r from 0 to ∞ and over θ from $\pi/2$ to π and over φ from 0 to 2π, we can obtain the upwelling irradiance at depth z (Lauscher (1955)):

$$E_u(z) = \frac{E}{\cos j} e^{-\frac{cz}{\cos j}} \frac{1}{c} \int_0^{2\pi} \int_0^{\pi/2} \frac{\beta(\alpha)\sin\theta d\theta d\varphi}{\frac{1}{\cos\theta} + \frac{1}{\cos j}} \qquad (7.9)$$

If we integrate eq. (7.3) from $r = 0$ to $r = z/\cos\theta$ for $\varphi = 0$ (i.e. in the plane to the direction observer-sun), we obtain for the downward scattered radiance

$$L(\theta) = \frac{E}{c} \frac{\beta(\theta - j)}{\cos j - \cos\theta} (e^{-\frac{cz}{\cos j}} - e^{-\frac{cz}{\cos\theta}}) \qquad (7.10)$$

$$0 \leqslant \theta < \pi/2$$

and for the upward scattered radiance

$$L(\theta) = \frac{E}{c} \frac{\beta(\theta - j)}{\cos j - \cos\theta} e^{-\frac{cz}{\cos j}} \qquad (7.11)$$

$$\pi/2 \leqslant \theta < \pi$$

From eq. (7.10) it appears that the downward scattered light is zero at the surface ($z = 0$) and has a maximum at

$$z_m = \frac{1}{c} \frac{\ln\cos j - \ln\cos\theta}{\frac{1}{\cos\theta} - \frac{1}{\cos j}} \qquad (7.12)$$

which, for $\theta = j$ becomes

$$z_m = \frac{1}{c} \cos j \qquad (7.13)$$

The upward directed radiance is highest at the surface and decreases exponentially with depth. If we call

$$\frac{\beta(\theta)}{b} = P(\theta); \quad 2\pi \int_0^{\pi} P(\theta)\sin\theta d\theta = 1$$

the phase function, we can write eq. (7.11) in the form

$$L(\theta) = E \cdot \frac{b}{c} \frac{P(\theta - j)}{\cos j - \cos\theta} e^{-\frac{cz}{\cos j}} \qquad (7.14)$$

which shows that the upwelling radiance just below the water surface is proportional to the ratio b/c and to the scattering phase function in the backward direction $\pi - j$.
Equations (7.4), (7.5), (7.9), (7.10) and (7.11) can in principle be used in order to calculate irradiance and radiance distribution near the surface. When empi-

488

rical values of the scattering function β are used and the contribution from skylight is taken into account, reasonable agreement with measured values is obtained. An indicative result of measurements made by Lundgren (1976), is shown in Fig. 28.

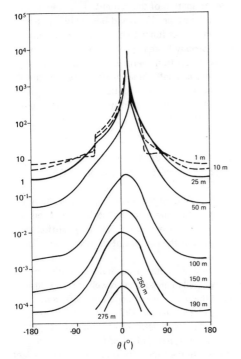

Fig. 28   *Radiance distribution at various depths. Solar zenith angle varies between 14° (1m) and 37° (275m)*

The radiance distribution in the upper layers exhibits high radiance values in the cone -48.6 to $+48.6°$, the angle of total refraction. Beyond this angle the radiance meter "looks" into deep water. For near surface measurements a weak maximum is found between 70 and 90° when near surface water is "seen" through total reflection at the surface. For angles greater than 90° the radiance distribution becomes rather flat with a minimum at $\pm 180°$. For greater depth the drop at 48.6° disappears probably because of the finite resolution of the radiance meter, and the total radiance decreases rapidly.

A more precise solution of the radiative transfer problem, than given by the equations above, has to start from the general radiative transfer equation

$$\frac{dL(z,\theta,\varphi)}{dr} = -cL(z,\theta,\varphi) + L_*(z,\theta,\varphi) \quad (7.15)$$

which equates the net change of the radiance at point z in direction θ,φ along the path dr to the loss due to attenuation at this point (1. term on the righthand side), and the contribution due to scattering at z into the direction θ,φ from all other directions, which is

$$L_*(z,\theta,\varphi) = \int_0^{2\pi}\int_0^\pi \beta(\theta,\varphi;\theta',\varphi')\sin\theta'd\theta'd\varphi' \quad (7.16)$$

L_* is called the path function.

The various methods of solution of equation (7.15) have been discussed extensively by Jerlov (1976) and by Zonneveld (1974). We will not go into further detail here, but discuss some indications regarding the measurement of the extinction coefficient, c, and absorption coefficient, a.

By writing eq. (7.15) in the form

$$c = -\frac{dL}{Ldr} + \frac{L_*}{L} \quad (7.17)$$

it appears that by minimizing L_* c can be determined from simple radiance transmittance measurements. This method is used in beam transmittance meters with fixed distance. Another method is to minimize the first term of eq. (7.17) by looking horizontally $(dL/dr \approx 0)$ against a black screen, so measuring L_* and, when the screen is removed, L; so c is obtained from (Le Grand, 1939):

$$c = \frac{L_*}{L} \quad (7.18)$$

Another important relation can be derived from eq. (7.15) by integrating it over all angles and using the relation $z = r\cos\theta$, then

$$\int_0^{2\pi}\int_0^\pi \frac{dL}{dz}\cos\theta\,\sin\theta\,d\theta\,d\varphi = -c\int_0^{2\pi}\int_0^\pi L(z,\theta,\varphi)\sin\theta\,d\theta\,d\varphi$$

$$+ \int_0^{2\pi}\int_0^\pi\int_0^{2\pi}\int_0^\pi \beta(\theta,\varphi,\theta',\varphi')\,L(z,\theta',\varphi',)\sin\theta',d\theta'd\varphi'\sin\theta\,d\theta\,d\varphi \quad (7.19)$$

If we define the scalar irradiance

$$E_0(z) = \int_0^{2\pi}\int_0^\pi L(z,\theta,\varphi)\sin\theta\,d\theta\,d\varphi \quad (7.20)$$

and consider that

$$\int_0^{2\pi}\int_0^\pi \beta(\theta,\varphi,\theta',\varphi')\sin\theta\,d\theta\,d\varphi = b$$

and

$$\frac{d}{dz}\int_0^{2\pi}\int_0^\pi L\cos\theta\,\sin\theta\,d\theta\,d\varphi = \frac{d(E_d - E_u)}{dz}$$

we find

$$\frac{d(E_d - E_u)}{dz} = -c\cdot E_0 + b\cdot E_0 = -a\cdot E_0 \quad (7.21)$$

489

which allows the measurement of the absorption coefficient, a, from the measurement of up and downwelling irradiance and from measurement of E_0 by a spherical collector.

We can now define some apparent properties:

- The radiance attenuation coefficient:

$$k = -\frac{1}{L}\,\frac{dL}{dz} \tag{7.22}$$

- The downward irradiance attenuation coefficient:

$$K_d = -\frac{1}{E_d}\,\frac{dE_d}{dz} \tag{7.23}$$

- The irradiance attenuation coefficient:

$$K_E = -\frac{1}{E_d - E_u} \cdot \frac{d(E_d - E_u)}{dz} \tag{7.24}$$

From eq. (7.21) it follows that

$$\frac{a}{K_E} = \frac{E_d}{E_0}\left(1 - \frac{E_u}{E_d}\right) \tag{7.25}$$

The ratio E_u/E_d, which is called internal reflectance R_z, can be calculated from equations (7.8) and (7.9). We assume for simplicity $j = 0$ (zenith sun) and constant backscattering coefficient, β_b; then

$$R_z = \frac{E_u}{E_d} = \frac{\dfrac{2\pi\,\beta_b}{c}\displaystyle\int_0^{\pi/2}\dfrac{\sin\theta}{1 + \dfrac{1}{\cos\theta}}\,d\theta}{e^{-az}}\,e^{-cz} \tag{7.26}$$

The integral is equal to $1 - \ln 2$ and we have

$$R_z = \frac{E_u}{E_d} = 2\pi\,(1 - \ln 2)\,\frac{\beta_b}{c}\,e^{-bz} \tag{7.27}$$

A more exact evaluation of the internal reflectance for $z = 0$ by using the Monte Carlo method has been given by Gordon et al. (1975):

$$R_0 = 0.0001 + 0.3244x + 0.1425x^2 + 0.1308x^3$$

$$x = \frac{b_b}{a + b_b} \tag{7.28}$$

This relation can be fairly well approximated by

$$R_0 = C \cdot \frac{b_b}{a + b_b} \tag{7.29}$$

where

$b_b = 2\pi\displaystyle\int_{\pi/2}^{\pi}\beta(\theta)\sin\theta\,d\theta$ (see Fig. 8), with C varying from 0.32 to 0.37.

These results indicate that, since b_b is only slowly dependent on wave length, λ, the absorption coefficient, a, will be the quantity which influences most the spectral behaviour of the internal reflectance, R. As we have seen before, the selectivity of a is mainly determined by absorption by particles and dissolved organic matter (yellow substances). Therefore a measurement of the spectral behaviour of R_0 can be used for the determination of the content of the various materials in the water. Smith and Baker (1977) have evaluated $R_0(\lambda)$ as function of λ for various contents of chlorophyll-like pigment. Fig. 29 shows the results. The absorption coefficient of chlorophyll-like pigments has been determined from in situ measurements of the irradiance attenuation coefficient k_d. Similar results have been recently reported by Morel and Prieur (1977).

8 APPLICATION OF WATER OPTICS TO REMOTE SENSING OF WATER QUALITY

As shown at the end of the preceding chapter, there is an indication that the spectral distribution of the upwelling light from a water body contains quantitative information about the content of suspended and dissolved materials.

There exist many data on spectral distribution of up- and downwelling irradiance as well as on spectral variation of the irradiance attenuation coefficient, K_d, (see eq. (7.23). Jerlov (1976, 1968) has used these data for a classification of oceanic water types and coastal water types. This classification is given in Fig. 30, where the irradiance transmission coefficient for downward irradiance, e^{-K_d} per meter is plotted against wave length.

The significant difference between oceanic and coastal waters is the lower transmittance for shortwave radiation. This is a consequence of the presence of yellow substance. The regional distribution of wa-

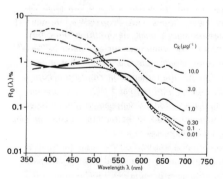

Fig. 29 Internal reflectance $R_0(\lambda)$ as a function of wavelength for various values of chlorophyll-like pigment concentration C_k ($\mu g/l$)

490

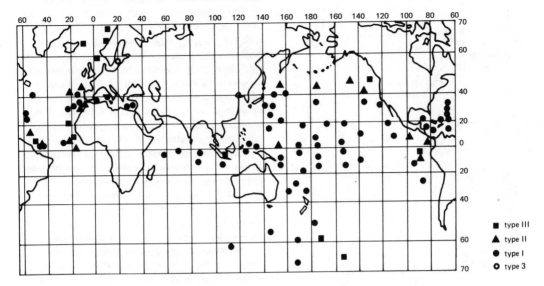

Fig. 30   Optical water types (Jerlov 1976)

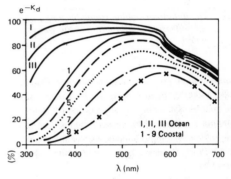

ter types as shown also in Fig. 30 shows that type III is strongly present in upwelling areas and that clear water (types I and II) are present mainly in the eaquatorial region ($\pm 20°$). It is also seen that for certain regions, such as South Pacific and South-Atlantic, few data are available.

From detailed analysis of the existing experimental data Jerlov (1976) suggests that some of the features exhibited by type III water are due to absorption by chlorophyll.

In recent years the possibility of using satellites and high-flying aircrafts for the observation of optical properties (colour) of water has become very attractive due to the development of radiometrically calibrated multispectral imaging systems (multispectral scanners (MSS), Vidicon and very recently "push broom scanners" .

The MSS data from LANDSAT 1 and 2 satellites have been extensively used for the study and mapping of sediment and chlorophyll distributions in near coastal zones and great inland lakes (Wezernak (1975), Rogers (1975), Rogers et al. (1975)) and for remote bathymetry (Polcyn, Lyzenga (1975)).

As a preliminary conclusion of these studies, it can be stated that since they were optimized for land, the four broad LANDSAT channels (0.5 – 0.6, 0.6 – 0.7, 0.7 – 0.8, 0.8 – 1.1 μ) offer a limited usefulness to water quality measurements and can be applied essentially to suspended solids and turbidity measurements and to the detection of strong phytoplankton

blooms. Quantitatively the results indicate a relationship between signal level in the MSS band 5 and the near surface suspended-solids concentration, whereas the ratio MSS 6/MSS 5 is useful for the detection of plankton blooms and surface vegetation. Due to the low signal-to-noise ratio, however, satisfactory results are obtained only for big suspended matter and phytoplankton concentrations. Deviations between ground measurements and "remotely sensed" suspended-solids concentrations are of the order of 13% in the range of concentrations between 20 – 80 mg/l. At lower concentrations the deviations become very high.

More promising results are obtained by using specially designed radiometers and multispectral scanners for the detection of upwelling spectral radiance from water. Neville and Gower (1977), Amann (1976) and Deschamps et al. (1977) have reported on the use of a dedicated nadir looking spectral radiometers

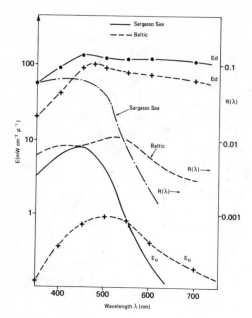

Fig. 31 *Up- and downwelling irradiance and internal reflection as a function of the wavelength*

wavelenth-scanning or narrow band filters at 466, 525, 550 and 600 nm with band width $\Delta\lambda = 10$ nm for the quantitative detection of chlorophyll. Such radiometers are designed to measure the ratio of upwelling to downwelling irradiance received at the level of the flying aircraft. This quantity, called albedo, is defined by:

$$A = \frac{E_u^{(a)}}{E_d^{(a)}} = \frac{\pi \cdot L_u^{(a)}}{E_d^{(a)}} \qquad (8.1)$$

where index (a) indicates that these values are valid above the water surface. $L_u^{(a)}$ is related to the upwelling radiance just below the water, L_u, by the relation

$$L_u^{(a)} = [\frac{t}{n^2} \cdot L_u + \rho L_s] \cdot T_a + L_p \qquad (8.2)$$

where n is the refractive index of water, t is the transmittance through the water surface from below (see Fig. 24b), L_s is the sky radiance (for a nadir-looking instrument it is the sky radiance from the zenith), ρ is the surface reflectance of the water, T_a is the transmittance through the atmosphere between water and the radiometer, L_p is the path radiance scattered by air molecules and dust particles into the aperture of the radiometer.

All quantities are more or less strongly depending on wavelength. Only the term with L_u contains the

desired information. By taking albedo differences $A(466) - A(525)$ and $A(550) - A(600)$, a great part of the undesired information can be eliminated as ρ, L_s, T_a and L_p depend only slowly on wave length*. Deschamps et al. conclude that the difference $A(550) - A(600)$ is sensitive to turbidity in general, whereas the difference $A(466) - A(525)$ is specifically sensitive to chlorophyll concentration. Fig. 32a shows their results of a theoretical evaluation of the difference $A(550) - A(600)$ as function of the particle scattering coefficient at $\lambda = 500$ nm, and Fig. 32b compares $A(466) - A(525)$ as function of chlorophyll concentration as obtained from the measurements and from theory. It is seen that at low chlorophyll concentrations the error due to presence of yellow substance can become considerable.

Recently NASA has developed a multispectral scanner, the so-called Colour Zone Coastal Scanner (CZCS), which will be launched end of next year (August, 1978) on board the NIMBUS-G satellite. NIMBUS-G will orbit in a sun-synchronous orbit at an altitude of 955 km and an inclination of 99.28° at an ascending note time of about 12 h mean solar time with a repetition cycle of 6 days, i.e. every 6 days a full coverage of the earth is provided. The CZCS has five bands in the visible spectrum, and one band in the thermal IR:

$$443 \pm 10 \quad nm$$
$$520 \pm 10 \quad nm$$
$$550 \pm 10 \quad nm$$
$$670 \pm 10 \quad nm$$
$$750 \pm 50 \quad nm$$
$$10.5 - 12.5 \, \mu$$

The spatial resolution is 825 m at nadir and the swath width 1566 km. The radiance signal (video signal) detected by the scanner will be digitized into 8 bits and recorded on a on-board tape recorder. The data will be transmitted to ground stations and processed into photographic products (calibrated black-and-white transparencies) and computer compatible tapes.

The optical bands of the CZCS have been chosen in order to provide information for the study of processes in coastal zones. The first four narrow channels are specifically designed for the quantitative detection of chlorophyll, suspended sediment and yellow substance. Their sensitivity is optimized over water. The fifth channel has nearly the same characteristics as the MSS 6 channel of the LANDSAT and will allow comparison with LANDSAT data. Channel 6 has been chosen to provide ocean surface temperature mapping.

---

* For L_p this is true only for low flying aircrafts (< 1000 m).

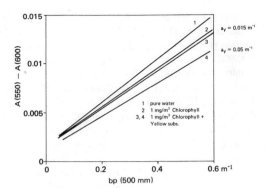

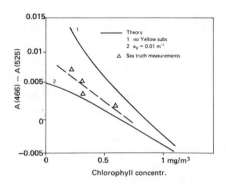

Fig. 32a Albedo difference as a function of the scattering parameter b_p (500 mm)

Fig. 32b Albedo difference as a function of chlorophyll concentration

During the preparation of this project NASA has experimented a prototype of the CZCS, called the Ocean Colour Scanner (OCS). In a series of experimental campaigns this scanner was flown on a U2 aircraft, while ground truth was provided by oceanographic research ships and fixed stations. The results from these campaigns have been used to develop interpretation algorithms which allow the quantitative determination of chlorophyll, suspended sediment and yellow substance.

As an example, Fig. 33 shows a comparison of measured and calculated chlorophyll concentrations from a campaign in the Gulf of Mexico (Hovis, Leung (1977)). The calculated curve is obtained from a very simple algorithm, which gives the chlorophyll concentration from a ratio of the radiance signals in two optical bands namely band 7 and 4, which correspond to 662 nm and 547 nm. The main problem of detecting quantitatively chlorophyll, sediment and yellow substance by radiometric measurements from space, consists in the evaluation of atmospheric scattered radiance which contributes to nearly 80% of the total signal in the blue range of the spectrum. In the equation (8.2) this scattered radiance is represented by the term L_p. Other difficulties arise from the evaluation of the atmospheric transmittance, T_a, which depends strongly on the aerosol content of the air and on the surface reflection term, $\rho L_s \cdot T_a$. The atmospheric scattering term, L_p, and the atmospheric transmittance, T_a, can be experimentally determined (Sturm (1975), Maracci, Sturm (1975)) from ground measurements at the moment of the satellite or airplane overflight. Gordon (1977) has proposed an algorithm which allows the correction of both additive terms of eq. (8.2): it can be assumed with very good approximation that for off-shore waters L_u becomes 0 for $\lambda = 750$ nm, due to the high

attenuation coefficient of water at this wave length.

In this case the scanner signal at the wave length 750 nm (band 5 of CZCS) gives the contribution of the undesired signal and also its variation with the scan angle. Gordon shows then how from these results the corrections can be made for the other wave length bands. The final result of the correction of the atmosphere's contribution to the signal is then the reflectance, R, defined in the preceding chapter.

Once $R(\lambda_i)$ is known for the remaining four narrow band channels of the CZCS λ_i, one has four equations of the form

$$R(\lambda_i) = C \cdot \frac{b_b(\lambda_i)}{a(\lambda_i) + b_b(\lambda_i)} \qquad i = 1, ..., 4 \qquad (8.3)$$

with

$$b_b(\lambda_i) = b_{b,w}(\lambda_i) + B_{b,cl}(\lambda_i) \cdot C_{cl} + B_{b,p}(\lambda_i) C_p$$

$$a(\lambda_i) = a_w(\lambda_i) + A_y(\lambda_i) \cdot C_y + A_{cl}(\lambda_i) C_{cl} + A_p(\lambda_i) C_p$$

$$(8.4)$$

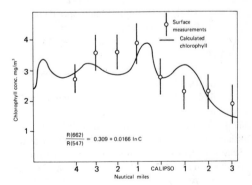

Fig. 33 Calculated rs measured chlorophyll (Gulf of Mexico)

where

$b_{b,w}(\lambda_i)$ = backward scattering coefficient of water at wave length λ_i,

$B_{b,cl}(\lambda_i)$ = specific backward scattering coefficient of chlorophyll at wave length λ_i $(m^{-1}/mg.m^{-3})$,

$B_{b,p}(\lambda_i)$ = specific backward scattering coefficient of non-chlorophyll particles at wave length λ_i $(m^{-1}/mg.l^{-1})$,

$a_w(\lambda_i)$ = absorption coefficient of water (m^{-1}),

$A_y(\lambda_i)$ = specific absorption coefficient of yellow substance,

$A_{cl}(\lambda_i)$ = specific absorption coefficient of chlorophyll $(m^{-1}/mg.m^{-3})$,

$A_p(\lambda_i)$ = specific absorption coefficient of non-chlorophyll particles,

C_{cl} = chlorophyll concentration (mg/m^3),

C_p = non-chlorophyll particle concentration (mg/m^3),

C_y = yellow substance concentration.

Assuming that the specific quantities of the various substances are known from four equations of type (8.3), three ratios $R(\lambda_i/R(\lambda_j)$ can be formed from which the three unknowns C_{cl}, C_p and C_y can be determined.

REFERENCES

Amann V. 1976, Flex Aircraft Mission, paper presented to the NIMBUS-G Experiment Team (NET) Meeting, Brussels, Sept. 29, 1976.

Austin R.W. 1974, in: Optical Aspects of Oceanography, Academic Press, New York, p. 317-343.

Benoit H., Stockmayer W.H., 1956, J. Phys. Radium, 17, p. 21.

Burt W.V., 1954, J. Meteorol., 11, p. 283-290.

Cabannes J., 1920, J. Phys., 6, p. 129-142.

Cabannes J. 1929, La Diffusion Moléculaire de la Lumière, Presses Universitaires de France.

Cox C., Munk W., 1955, Journ. Marine Res., 14, p. 63-78.

Cox C., Munk W., 1956, Bull. Scripps Inst. Oceanogr. Univ. Calif., 6, p. 401-488.

Cox C. 1974, in: Optical Aspects of Oceanography, Academic Press, New York, p. 51-75.

Collins J.R., 1925, Phys. Rev., 55, p. 470-472.

Deschamps Y., Lecomte P., Viollier M., 1977, Remote Sensing of Ocean Color and Detection of Chlorophyll Content, 11th Int. Symp. on Remote Sensing of Envir., Ann Arbor, Mich., April 25-29, 1977.

Einstein A., 1910, Ann. Physik, 33, p. 1275-1298.

Gordon D.C., 1970, Deep-Sea Res., 17, p. 175-185.

Gordon H.R., Brown O.B., Jacobs M.M., 1975, Appl. Optics, 14, p. 417.

Gordon H.R., 1977, A Possible Scheme for Correcting CZCS Imagery for Atmospheric Effects, paper submitted to the NIMBUS G-CZCS NET.

Hovis W.A., Leung K.C., 1977, Optical Eng., 16, p.158.

Jerlov N.G., 1955, Tellus, 7, 218-225.

Jerlov N.G., Fukuda M., 1960, Tellus, 12, p.348-355.

Jerlov N.G., 1968, Optical Oceanography; Elsevier Oceanographic Series, 5, Elsevier, Amsterdam, 1st ed. p. 197.

Jerlov N.G., 1974, in: Optical Aspects of Oceanography, Academic Press, New York, p. 78-94.

Jerlov N.G., 1976, Marine Optics, Elsevier, p. 231.

Kalle K., 1962, Kieler Meeresforschung, 18, p.128-131.

Kalle K., 1966, Mar. Biol. Annu. Rev., 4, p. 91-104.

Kattawar G.W., Plass G.N., 1969, Appl. Optics, 8, p. 455-466.

Kattawar G.W., Humphreys T.J., 1976, Appl. Optics, 15, p. 273.

Kullenberg G., 1968, Deep Sea Res., 15, p.423-432.

Kullenberg G., 1974, in: Optical Aspects of Oceanography, ed. N.G. Jerlov, Academic Press, p. 494.

Labs D., Nechel H., 1971, Solar Physics, 19, p. 3-15.

Lauscher F., 1955, in: Handbuch der Geophysik, Springer Berlin, 7, p. 723-763.

Le Grand Y., 1939, Ann. Inst. Oceanogr., 19, p.393-436.

Lenoble J., Saint-Guilly B., 1955, C.R., 240, p. 954-955.

Lisitsyn A.P., 1962, U.S.S.R. Acad. Sci. Com. of Sediment, Div. of Geol. and Geogr. Sci. Moscow, p. 175-231.

Lundgren B., 1976, University of Copenhagen, Internal Report: Radiance and Polarisation Measurements in the Meditarranean.

Maracci G., Sturm B., 1975, Measurements of Beam Transmittance and Path-Radiance for Correcting LANDSAT Data for Solar and Atmospheric Effects, Proceedings of the 5th Conf. on Space Optics, Marseille, Oct. 14-17, 1975.

Mie G., 1908, Ann. Phys., 25, p. 377.

Moon P., Spencer D.E., 1942, Illum. Eng., 37, p. 707-726.

Morel A., 1974, in: Optical Aspects of Oceanography, ed. N.G. Jerlov, Academic Press, p. 494.

Morel A., Prieur L., 1977, Analysis of variations in Ocean Color, Limnology and Oceanography, Vol. 22, No. 4, p. 709.

Neumann G., Hollman R., 1961, Union Geod. Geophys. Int. Monogr., 10, p. 72-83.

Neville R.A., Gower J.F.R., 1977, Passive Remote Sensing of Phytoplankton via Chlorophyll Fluorescence, to be published in J. of Geophys. Res.

Ochakowski Yu.E., 1966, U.S. Dep. Comm., Joint
Publ. Res. Ser. Rep., 36, (816), p. 16-24.

Palmer K.F., Williams D., 1974, J. Opt. Soc. Am.,
64, p. 1107-1110.

Polcyn F.C., Lyzenga D.R., 1975, Remote Bathymetry and Shoal Detection with ERTS, NASA
CR-142636.

Preissendörfer R.W., 1965, Radiative Transfer on Discrete Spaces, Pergamon, New York.

Prinz N., Prinz W., 1956, Physica, 22, p. 576-578.

Raschke E., Berechnung des durch Mehrfachstreuung
entstehenden Feldes solarer Strahlung in einem
System Ozean-Atmosphäre, BMBW - FBW71-20.

Rayleigh Lord, 1871, Philos. Mag., 41, p. 447-454.

Rogers H.R., 1975, Application of LANDSAT to the
Surveillance and Control of Lake Eutrophication
in the Great Lakes Basin, NASA CR-142852.

Rogers R.H., Reed L.E., Elliot Smith V., 1975,
Computer Mapping of Turbidity and Circulation
Patterns in Saginaw Bay, Michigan from LANDSAT
Data, NASA CR-142852.

Smith, Bahe, 1977, The Remote Sensing of Chlorophyll,NASA Contract No. UCSD 5-35406,
Scripps Inst. of Oceanography.

Smoluchowski M., 1908, Ann. Physik, 25, p. 205-226.

Sturm B., 1975, Determination of Beam Transmittance and Path Radiance in the Four Bands of the
ERTS-Satellite, Proceedings of DFVLR-Seminar
on Remote Sensing, April 7-11, Porz-Wahn, 1975.

Thekaekara M.P., 1972, Evaluating the Light from
the Sun, Optical Spectra.

Wezernak C.T., 1975, Water Quality Monitoring
Using ERTS-1 DATA, NASA CR-1422400,
ERNM 193300-55-F.

G. FRAYSSE
Commission of the European Communities
Joint Research Centre, Ispra Establishment, Ispra (Varese), Italy

European applications, requirements, perspectives

1 INTRODUCTION

In the title of this seminar, only the words "Agriculture and Hydrology" appear, but it includes lectures on Forestry, Land Use, Environmental Monitoring and Meteorology; the reasons are that all these disciplines (or applications) are closely interrelated and that remote sensing provides a new way of synoptic information on "land renewable resources".

During this seminar, the use of several remote sensing techniques was discussed: the most traditional photointerpretation of aero-space photography, automatic photointerpretation, together with digital processing of multispectral scanner and modern microwave sensors data (radar, passive microwave radiometer). We use the words "remote sensing" in their widest meaning, and we believe that this is particularly appropriate in Europe where complementary methods are often necessary.

At the end of this seminar, it seems useful to synthesize what has been said, to see how remote sensing can contribute to the improvement of European agriculture and to analyze the European situation in order to gain perspective and draw some guidelines for future efforts.

2 EUROPEAN AGRICULTURE

In the two first lectures, potential applications of remote sensing to EEC agriculture (management, planning, marketing, agronomic improvements) were presented. Although remote sensing methods applied to agriculture and forestry are nearly operational in the USA and in Canada, this is not yet the case in Europe, mainly due to the strong land and agricultural structures differences between the two continents.

While the average holding area is 158 hA in the USA, it is 18 hA in the EEC, i.e. 8.8 times bigger (ref. 1,2,3,4). Two supplementary differences must also be considered:

a) EEC holdings are seldom constituted of contiguous parcels, and, except in some regions, EEC farmers cultivate several different crops.
b) Modern agricultural practices do not require any more crop rotation in the various parcels; nevertheless large changes in sown areas (acreage and crop repartition) have still to be taken into account in most of the EEC regions (ref. 5).

Table 1 indicates the parcel dimension (ref. 6) in several European countries.

Table 1. Dimension of the parcels in several European countries

Country	Parcelling (Average parcels per holding)		
	Area (hA)	No. of parcels	Parcels (hA)
FRG	6.8	10.1	0.67 (1.5)
Denmark	15	1.3	
		(N.Schleswig: 5)	
Spain	8.75	15	1
France	14.2	60 (20-260)	0.42
			(0.19-1.23)
Ireland	4	2-40	
Italy	2.25	5.7	0.5
Netherlands	9.65	3.9	2.46
Sweden	9	2-3	
Switzerland	5.23	10	0.51

Only local studies are existing, concerning the average parcel area for each main crop. Such data would be essential in order to make a complete assessment of the exact potential of remote sensing applications to European agriculture and a precise evaluation of its economic advantages.

In order to justify the resolution improvement of LANDSAT-D, a study using statistical modelling was done at G.S.F.C. by Podwysocki (Fig. 2, ref. 7). The area which was investigated in Europe is situated in the Garonne area in the southwest of France. This study includes also the determination of cumulative frequency distribution of field length and field width vs. total number of fields. Table 2 summarizes these results for Kansas and France.

Table 2. Estimation of field size distribution

	Kansas	Garonne (France)	Ratio
Cumulative percentage of fields (50%) → field length	350 m	200 m	1.75
Cumulative percentage of fields (50%) → field width	220 m	150 m	1.47
Cumulative percentage of fields (50%) → field area	14 hA	6 hA	2.33

3   NEEDS AND REQUIREMENTS

In order to establish the requirements for future European operational remote sensing systems, the various objectives have to be classified according to the current state-of-the-art and to the possibilities offered by the new sensors and platforms, as it was explained in the previous lectures.

Table 3 summarizes the main objectives and tentatively indicates what type of sensor combination could offer a solution in European conditions. It must be noted that this table represents an average between very different cloud-cover conditions (northern and Southern Europe) and very different agricultural areas (structure, climate, soil, crops association, etc.).

It is obvious that the use of microwaves (radar) is more fundamental in those largely cloud covered countries and regions of Northern Europe than it is in Southern Italy and that spatial resolution constitutes a very stringent requirement for Europe. Other requirements (repetitivity, observation time, spectral channels, radiometric resolution, geometric accuracy, etc.) are not discussed here; they have been analyzed by the European Space Agency and summarized in ref. 5.

For yield forecasting, it has been shown in a recent study (ref. 8) that economic advantages could be expected, but a good forecasting accuracy has to be reached.

4   EUROPEAN RESEARCH

Most of the research on remote sensing applications in Europe was done with data collected by US or European sensors installed in European aircrafts or with data collected by US satellites. Basic laboratory research and ground-truth experiments was necessary in order to adapt US experiments to local conditions or to complement them.

Announcements of opportunity for participation in NASA experimental satellite missions (LANDSAT-1 and -2, HCMM, NIMBUS-G, SEASAT) were for many European scientists the starting point of their interest to remote sensing. But this very fast increase has lately pushed national and European organizations to increase their efforts to set up their own capability (sensors, data acquisition systems, satellites, image processing systems, software, etc.)

4.1   Aircrafts and airborne instrumentation

Several European organizations operate high-altitude (10 km) aircrafts, and medium- and low-altitude aircrafts equipped with US Commercial Multispectral Scanners or with European sensors (thermal infra-red M.S.S., electronic scanners, radiometers); European industry is now offering commercial products. But no aircraft like the US NASA U-2 flying at more than 20 km altitude is available in Europe for tests of equipment or fast data acquisition at the highest sub-satellite altitude. Some side-looking radars are operated also, but no synthetic aperture radar is yet available.

It seems that airborne remote sensing will remain an essential complement of earth observation satellites in Europe.

4.2   Meteorological satellites

Several organizations receive orbiting US meteorological satellites (NOAA-5), USSR satellites (METEOR-2), and geostationary US satellites (SMS). These satellites provide panchromatic and thermal infra-red images (resolution of about 1 km) or low resolution images (resolution of several km). Several research institutes use these data for environmental studies (air pollution) or for agricultural applications (thermal inertia of soils, rainfall determination).

Some of the most important parameters for yield forecasting models may be accessible through meteorological satellites; rainfall, soil moisture and daily-integrated sun energy. Real evapotranspiration is also a good descriptor of the yield and is only accessible through thermal-infra-red measurements. Sun energy can be measured with a greater accuracy than with a dense network of pyranometers (NOAA experiments in the Great Plains). Consequently, it is probable that the agricultural applications of meteorological satellites will increase in the next years.

METEOSAT is an ESA satellite launched in 1978.

Meteosat was built by a consortium of European Industries (Belgium, France, Germany, Italy, Spain, UK. Meteosat observes the earth and its cloud cover from a geostationary orbit both in visible light and in the infra-red.

Images are taken with a high resolution radiometer obtaining a definition at nadir of 2.5 km on the

Table 3. Agriculture and forestry objectives and requirements for a European Remote Sensing System

OBJECTIVES	LANDSAT-D TM (Vis. T.I.R.) High resolution	Very high resolution satellite (Vis. near I.R.)	Low resolution High repetitivity (Vis. T.I.R.) satellite	Subsampling with airborne observations	Medium resolution moisture radar satellite	High resolution multifrequency radar satellite for crop identification	Data Collecting System
Agriculture							
1. Crop census (inventory)		O		O		O	
2. Yield forecasting (cereals)	O	O	O	O	O	O	O
3. Reduction of losses by early identification of disease infestation				O			
4. Agriculture management (irrigation, fertilization, soils ...)	O	O	O	O	O		O
Forestry							
5. Forest census	O	O		O		O	
6. Timber volume estimate		O		O		O	
7. Determination of losses (stresses, burned areas, diseases)		O		O		O	O
8. Forestry management (cut areas, reafforestation)	O	O		O	O		

ground in daylight and 5 km in the infra-red. The spectral bands are $0.5 - 1\,\mu$, $10.5 - 12.5\,\mu$, $5.7 - 7.1\,\mu$ (water vapor absorption band). A complete picture achieved by scanning line by line (2500 lines for the infra-red, 5000 lines for the visible light channel) is taken every 30 minutes.

Meteosat provides high repetitivity images. Its usefulness for agriculture and hydrology is increased by the Data Collecting System which permits interrogation of earth Data Collecting Platforms (D.C.P.).

Meteosat data are received by the ESA receiving station (ESOC) at Darmstadt, Germany.

In collaboration with NASA and NOAA, CNES operates a Data Collecting System (ARGOS) using the satellite TIROS-N. This system is now operational.

4.3 US earth observation satellites

A large number of European scientists, as selected NASA investigators, or as buyers of data to US Geological Survey (EROS Data Center) has used large quantities of images or magnetic tapes collected by LANDSAT-1 (1972 - 1975), LANDSAT-2/3 (1975 - today), SKYLAB (1973). The results of these investigations have been presented in the main European specialized meetings (Frascati 1974,

Freiburg 1975, Oslo 1976, Bristol 1976, Graz 1977, Toulouse 1978, Paris 1979, etc., as well as in US or foreign symposia. Due to the poor resolution of LANDSAT, the results are not generally of immediate use.

This explains also why agricultural investigations using these satellites were only few, most of them are more adapted to airborne sensors possibilities. In this respect the European AGRESTE project using LANDSAT and aircraft data remains a good reference for the evaluation of the potentiality of the current LANDSAT generation because an extensive ground-truth has allowed precise quantitative determinations.

LANDSAT satellites data are received by the ESA receiving station of Fucino, Italy (built par Telespazio). Another station covering Northern Europe is operated at Kiruna (Sweden). These stations are part of the ESA receiving network (Earthnet) connected to the central archive at Frascati (Italy). The Norwegian authorities are also installing a national LANDSAT receiving station at Tromsö.

4.4 Basic scientific activities

Many laboratories support applied investigations by a basic scientific activity having short-, medium- or long-term aims. The following research fields may be indicated, but the list is not exhaustive:

- Reflectance properties of soils;
- Reflectance properties at different phenological stages of crops;
- Canopy models (wheat, rice, maize);
- Moisture evaluation in various types of soils;
- Evapotranspiration measurements;
- Scattering coefficients, radar response for different crops and soils.

In 1976, in a meeting at Lyngby (Denmark), a European Association of Remote Sensing Laboratories was created. This association groups now 120 laboratories. EARSeL aims are to promote remote sensing methods and to increase the collaboration between laboratories of different countries. It operates several working groups preparing reports and experiments. Two of these groups are concerned with agriculture; one is concerned with hydrology:

- Application of multi-spectral systems in agriculture and forestry;
- Microwave applications in agriculture and forestry;
- Utilization of NIMBUS-G radiometer data for snow;studies in the Alps, the Pyrénées and Northern Scandinavia.

EARSeL is a study group of the Parliamentary Assembly of the Council of Europe and is sponsored by the Commission of the European Communities and ESA.

4.5 US research satellites

The Joint Research Centre is carrying out, within the "TELLUS" project, an investigation with many national laboratories on the application of HCMM satellite data to soil moisture and heat budget evaluation in zones of agricultural and environmental interest (EEC countries). Two models were developed for bare and sparsely vegetated areas, using separately or together the day and night temperature measured by HCMM at the same geographical point, to determine thermal inertia, soil moisture and cumulative daily evaporation.

The utilization of HCMM data was prepared during preliminary joint aircraft flight experiments simulating HCMM operation. Unfortunately, HCMM power supply is giving some trouble, and data cannot be collected in the best conditions.

Such experiments are very useful, since soil moisture is a primary parameter for agriculture and hydrology.

Several individual scientists (ref. 9) in some European countries particpate also to HCMM investigations.

4.6 Industrial products

European industry offers now most of the equipment for data acquisition, ground-truth, data processing. A catalogue prepared by EUROSPACE for the Joint Research Centre is yearly updated and constitutes a useful tool for scientists and remote sensing users.

4.7 Documentation EARSeL

EARSeL has prepared a directory of its members, wherein facilities, staff and fields of research are described, yet a complete description of European research in remote sensing does not exist.

In 1976, the Department of Industry (UK) published an excellent catalog of all UK remote sensing activities. It contains UK information classified by disciplines, government departments, universities, industry and commerce, manufactures, libraries (44 universities, 26 national institutes, 25 government departments are engaged in remote sensing in the UK.) (ref. 10). The ESA Space Documentation Service at Frascati publishes regularly a bibliography "Earth Resources-Agriculture". This biliography covers both European and foreign literature (ref. 11).

The French CNRS also publishes a similar bibliography, more personalized (*Profils personnalisés*) which has the obvious advantage of a preliminary selection of papers in the enormous flow of US remote sensing literature.

Bibliographic services also exist in several other countries.

5 FUTURE EUROPEAN EARTH OBSERVATION SATELLITES

5.1 Spacelab R.S. Payload

Several European institutes are very active in the microwave field (passive or active). The microwave instrument to be flown on the first SPACELAB flight (1981) will be the first European microwave sensor to be placed in orbit. It will be operated in two modes:

- a main mode as a two frequency scattero meter (2FS) (useful for ocean wave spectra measurement);
- a secondary mode as a SAR for high resolution imaging (ground resolution 25 x 25 m) of the land surface; Swath 9 km; length of the imaged area 2500 km).

The SAR will operate during 10 minutes over Europe. It is only an experimental system but extremely useful to assess the value of space radar for land use applications. The microwave instrument will be provided by the German DFVLR to ESA.

It must be noted that a Zeiss RMK A30/23 aerial survey camera will also be flown on SPACELAB. It will be operated over preferred test areas in Europe and North America. The camera will be also provided in the framework of the German participation in the ESA-SPACELAB program (ref. 12).

Since the duration of the experiment will be short, one cannot expect important results in agriculture and hydrology which require essentially a high repetitivity; nevertheless some interesting correlations between visible imagery, ground-truth and radar return can be expected.

5.2 SPOT

SPOT (*Satellite Préopérationnel d'Observation de la Terre*) is a French project (CNES) of a high resolution earth observation satellite to be launched in 1984 by an ARIANE launcher; it should be the first remote sensing satellite totally European built and launched.

The mission requirements are the result of a working group. (ref. 13). Table 4 gives the main characteristics of SPOT and LANDSAT-D (1982). The first mission of SPOT will consider soil occupation, agriculture and forestry resources, water resources management and forecasting.

The satellite will be set on a sun synchronous circular orbit at an altitude of 832 km (at 45° latitude) with an inclination of 98.7°. The orbital cycle will be 26 days, the period, 101.5 minutes. The descending node is chosen at 10.30 a.m. local time.

Payload consists of 2 HRV (High Resolution Visible) sensors scanning 60 km each in the spectral bands 0.5 - 0.59 μ, 0.61 - 0.69 μ, 0.8 - 0.91 μ with a 20 m resolution and 0.5 - 0.9 μ with a 10 m resolution. The boresight angle of each sensor may be moved ± 27° around roll axis thus allowing stereoscopic observations and a high access rate (ref. 14).

Such as it is designed, the SPOT system is particularly well-suited for vegetation and water studies (choice of narrow spectral bands), cartographic applications (ground resolution) and geology (stereoscopy). One can expect from SPOT a decisive improvement in usefulness of remote sensing in European conditions. SPOT will carry on a Data Collecting package.

5.3 ESA's remote sensing preparatory program

In March 1978, ESA has presented the result of two industrial phase A studies on satellite systems (Land Application Satellite System (LASS) and Coastal Oceans Monitoring Satellite System (CoMMS)) (ref. 15).

The primary mission objective of LASS is to pro-

vide high resolution multispectral imagery for agriculture, land use and water management applications. LASS is a 1500 kg three-axis stabilized satellite, to be launched into a low-altitude, highly inclined, sun-synchronous orbit. It carries a payload composed of a set of optical imaging instruments covering the visible and near infra-red, the middle infra-red and the thermal infra-red regions, a synthetic aperture radar, an instrument data telemetry system performing multiplexing, formattings and transmission to the ground, of sensor data and a data collection package.

The main characteristics are:

OII: - 6 channels CCD pushbroom scanner:
Ground resolution 30 m
Swath width 200 km
2 optical systems having a 10° FOV

- Image space scanner:
2 channels in the near middle IR (1.65 - 2.2 μ)
Ground resolution 60 m
1 thermal IR channels
Ground resolution 120 m
2 visible channels with in-flight selectable filters
Ground resolution 60 m

SAR: - Single frequency C-band, single polarization (mainly for soil moisture detection)
Ground resolution: 100 m for 1 dB grey level resolution or 30 m for 2.5 dB grey level resolution (depending on ground processing mode)
Swath width 100 km
Boresight 20° off nadir

By its advanced characteristics (ground resolution, choice of spectral channels) and the integration of a SAR in the payload, LASS will offer an exceptional opportunity to European scientists working in agricultural and hydrological applications.

6 CONCLUSIONS

Agriculture, forestry, land use and hydrology are certainly the most important European remote sensing applications on land.

Airborne data remain essential as long as high resolution satellites are not available; they will constitute afterwards an important complement to LANDSAT-D, SPOT and LASS.

European requirements for an operational system are hard and complex.

Complex models integrating data from several sensors and platforms have to be built and tested, using airborne, ground experiments and simulation.

Europe is building its own capability in remote sensing, and rapid progress is expected in the coming years.

7 REFERENCES

1. Yearbook of Agricultural Statistics - EURO-STAT - 1976.
2. Situation of Agriculture in the EEC. Commission of the European Communities - January 1977.
3. Major Uses of Land in the USA 1969 - report no. 247, US Dept. of Agriculture, Economic Research Service.
4. Statistical Abstract of USA, 1976. US Dept. of Commerce, Bureau of the Census.
5. Mission Requirement Report, G. Fraysse, Agriculture and Forestry, European Space Agency, Remote Sensing Working Group, June 1977, RSP/OD/702/2.
6. Jacoby, Remembrement en Europe, 1960.
7. Melvin H. Podwysocki, An estimate of field size distributions for selected sites in the major grain producing countries, NASA, GSFC, X-923-76-93, April 1976.
8. Utilisation des données recueillies par télédétection pour les modèles de prévision de récoltes: avantages économiques. G. Fraysse, P. Malet, B. Susplugas. Proceedings of an Internation Conference on Earth Observation from Space and Management of Planetary Resources, held at Toulouse, 6 - 11 March 1978 (ESA-SP-134).
9. Mapping thermal inertia, soil moisture and evaporation from aircraft day-and-night thermal data. J. Dejace, J. Mégier, M. Kohl, G. Maracci, P. Reiniger, G. Tassone. International Symposium on Remote Sensing of Environment, Ann Arbor, May 1979.
10. Department of Industry, Remote Sensing of Earth Resources, London 1976.
11. ESA, Space Documentation Service, Standard Titles, No. 262, Earth Resources, Agriculture, August 1977.
12. The Earth observation programme of the European Space Agency. D. Lennertz, I. Pryke, ESA-SP-134, 1978.
13. Groupe Ad Hoc CNES Ressources Terrestres, Rapport Final Juin 1976, Partie "Groupe Occupation des Terres".
14. SPOT, French preoperational remote sensing satellite system, A. Scribot. Symposium on application of remote sensing to hydrology COSPAR, Bangalore, May - June 1979.
15. EARSeL News, Bulletin of the European Association of remote sensing laboratories, April 1979, Number 6.